Birkhäuser

Progress in Nonlinear Differential Equations and Their Applications
Volume 86

More information about this series at http://www.springer.com/series/4889

Alexandre N. Carvalho • Bernhard Ruf
Ederson Moreira dos Santos • Jean-Pierre Gossez
Sergio H.M. Soares • Thierry Cazenave
Editors

Contributions to Nonlinear Elliptic Equations and Systems

A Tribute to Djairo Guedes de Figueiredo on the Occasion of his 80th Birthday

Birkhäuser

Editors
Alexandre N. Carvalho
Instituto de Ciências Matemáticas e de Computação
Universidade de São Paulo
São Carlos, SP, Brazil

Ederson Moreira dos Santos
Instituto de Ciências Matemáticas e de Computação
Universidade de São Paulo
São Carlos, SP, Brazil

Sergio H.M. Soares
Instituto de Ciências Matemáticas e de Computação
Universidade de São Paulo
São Carlos, SP, Brazil

Bernhard Ruf
Dipartimento di Matematica
Universitá degli Studi di Milano
Milano, Italy

Jean-Pierre Gossez
Département de Mathématique
Université Libre de Bruxelles
Brussels, Belgium

Thierry Cazenave
Laboratoire Jacques-Louis Lions
Université Pierre et Marie Curie
Paris, France

ISSN 1421-1750 ISSN 2374-0280 (electronic)
Progress in Nonlinear Differential Equations and Their Applications
ISBN 978-3-319-19901-6 ISBN 978-3-319-19902-3 (eBook)
DOI 10.1007/978-3-319-19902-3

Library of Congress Control Number: 2015946754

Mathematics Subject Classification (2010): 35A15, 35J15, 35J30, 35J47, 35J60, and 35A15

Springer Cham Heidelberg New York Dordrecht London

Printed on acid-free paper

Springer International Publishing AG Switzerland is part of Springer Science+Business Media (www.springer.com)

Djairo G. de Figueiredo
Photo by Christina Figueiredo Prudencio

Preface

The ICMC-Summer Meeting on Differential Equations is a series of meetings that takes place at the Instituto de Ciências Matemáticas e de Computação (ICMC) of the Universidade de São Paulo (USP) in São Carlos. They happen every year, since 1996, in the southern hemisphere's summer holidays, a few days before Carnival.

After the 2013 Chapter of the meeting, the local organizers decided to dedicate the next meeting to the celebration of Djairo's 80th birthday. With Djairo's blessing, the conference was arranged from the 3rd to the 7th of February, 2014, and invitations were sent out. As anyone who has ever organized a meeting knows, the hardest part is to be able to gather a large number of well-known specialists in the area, and collaborators who will help with the organization. In the case of this meeting, we must say that it was all very easy. We rapidly had a large number of worldwide experts who happily agreed to participate and a large number of collaborators who agreed to help with the several different committees. Of course, that has to do with the fact that Djairo is so well cared for in the Differential Equations community.

The 2014 Chapter was the largest edition ever of the meeting with over 300 participants from 23 different countries, 3 different continents, and from all regions of Brazil. The celebration of Djairo's 80th birthday was a very high level meeting in a relaxed and friendly environment.

Just after the meeting, we decided to write this book to record this wonderful conference. We approached Birkhäuser as publishers and they were keen to accept. In compiling this volume, we have again seen the same enthusiastic behavior that we observed during the organization of the meeting: contributors were eager to pay tribute to Djairo.

Djairo Guedes de Figueiredo was born in the small town of "Limoeiro do Norte," state of Ceará, Northeast Brazil, in 1934. He graduated in Civil Engineering from the Universidade Federal do Rio de Janeiro, in 1956. In 1961, he obtained his Ph.D. in Mathematics from the Courant Institute at NYU under the guidance of Louis Nirenberg. His professional career started in 1961 at the Universidade de Brasília, in the newly built capital of Brazil. He went back to the US from 1965 to 1971, successively at the University of Wisconsin in Madison, the University of Chicago, the University of Maryland, and the University of Illinois at Chicago Circle. In 1971, he came back to Brazil at the Universidade de Brasília and finally moved in 1988 to UNICAMP, in Campinas, near São Paulo.

Djairo has a very active scientific life, publishing over a hundred research articles, which has made him the most cited Brazilian mathematician with almost 2500 citations in MathSciNet. His scientific contribution would be already enough to render him the dedication of a book like this one, but Djairo's contribution to Mathematics goes far beyond that. He published 29 monographs (many of them in Portuguese) that were very influential to many younger generations of Brazilian (and many other nationalities) mathematicians. His passion for mathematics has influenced entire generations of mathematicians in Brazil, South America, and throughout the world. His descendants, who include 26 Ph.D. students, are already in the fourth generation, and in four years should number 120 Ph.D.s.

It is safe to say that Djairo's influence in Brazilian mathematics made him one of the pillars of the subject in that country. He had a major influence in the development of the area of Analysis, especially in its applications to Nonlinear Elliptic Partial Differential Equations and Systems, in Brazil and throughout the world.

On a personal level, Djairo is a kind, friendly person with a great sense of humor. His natural leadership, kindness, uprightness of character, and family values are a very important part of his legacy to the younger generations.

We happily thank Djairo for the opportunity to gather together so many world-wide specialists, and for providing the bonds which brought us the family-like, high level research environment seen throughout the meeting. We are proud to be part of this small token of appreciation to Djairo's influence in our lives.

São Carlos, SP, Brazil — Alexandre N. Carvalho
Milano, Italy — Bernhard Ruf
São Carlos, SP, Brazil — Ederson Moreira dos Santos
Brussels, Belgium — Jean-Pierre Gossez
São Carlos, SP, Brazil — Sergio H.M. Soares
Paris, France — Thierry Cazenave
March 2015

Contents

Contributors

Francisco S.B. Albuquerque Centro de Ciências Exatas e Sociais Aplicadas, Universidade Estadual da Paraíba, Patos, PB, Brazil

Vieri Benci Dipartimento di Matematica, Università degli Studi di Pisa, Via Filippo Buonarroti 1/c, 56127 Pisa, Italy

Centro Linceo Interdisciplinare Beniamino Segre

Lucio Boccardo Dipartimento di Matematica, "Sapienza" Università di Roma, Roma, Italy

Denis Bonheure Département de Mathématique, Université Libre de Bruxelles, Bruxelles, Belgium

Marta Calanchi Dipartimento di Matematica, Università di Milano, Milano, Italy

Alfonso Castro Department of Mathematics, Harvey Mudd College, Claremont, CA, USA

Mónica Clapp Instituto de Matemáticas, Universidad Nacional Autónoma de México, México, D.F., Mexico

David G. Costa Department of Mathematical Sciences, University of Nevada Las Vegas, Las Vegas, NV, USA

J. M. do Ó Departamento de Matemática, Universidade Federal da Paraíba, João Pessoa, PB, Brazil

Grey Ercole Departamento de Matemática, Universidade Federal de Minas Gerais, Belo Horizonte, MG, Brazil

Jorge Faya Centro de Modelamiento Matemático, Universidad de Chile, Santiago, Chile

Giovany M. Figueiredo Universidade Federal do Pará, Faculdade de Matemática, Belém, PA, Brazil

J. Fleckinger Institut de Mathématique (UMR CNRS 5219), CEREMATH-UT1-Université de Toulouse, Toulouse, France

Donato Fortunato Dipartimento di Matematica, Università degli Studi di Bari Aldo Moro and NFN sezione di Bari, Bari, Italy

Marcelo F. Furtado Departamento de Matemática, Universidade de Brasília, Brasília, DF, Brazil

M. Garcia-Huidobro Facultad de Matematicas, Pontificia Universidad Catolica de Chile, Santiago, Chile

Filippo Gazzola Dipartimento di Matematica, Politecnico di Milano, Milano, Italy

Marco Ghimenti Dipartimento di Matematica, Università di Pisa, Pisa, Italy

J.-P. Gossez Département de Mathématique, C.P. 214, Université Libre de Bruxelles, Bruxelles, Belgium

Massimo Grossi Dipartimento di Matematica, Università di Roma La Sapienza, Roma, Italy

Alessandro Iacopetti Dipartimento di Matematica e Fisica, Universitá degli Studi di Roma Tre, Roma, Italy

S. Lorca Instituto de Alta Investigación, Universidad de Tarapacá, Arica, Chile

Fernando Codá Marques Princeton University, Princeton, NJ, USA

R. Manásevich DIM & CMM, FCFM, Universidad de Chile, Santiago, Chile

J. Mawhin Université Catholique de Louvain, IRMP, Louvain-la-Neuve, Belgium

Everaldo S. Medeiros Departamento de Matemática, Universidade Federal da Paraíba, João Pessoa, PB, Brazil

Jéssyca Lange Ferreria Melo Departamento de Matemática, Universidade Federal de Viçosa, Viçosa, MG, Brazil

Anna Maria Micheletti Dipartimento di Matematica, Università di Pisa, Pisa, Italy

Olímpio H. Miyagaki Departamento de Matemática, Universidade Federal de Juiz de Fora, Juiz de Fora, MG, Brazil

Ederson Moreira dos Santos Instituto de Ciências Matemáticas e de Computação, Universidade de São Paulo, São Carlos, SP, Brazil

Robson Nascimento Département de Mathématique, Université Libre de Bruxelles, Bruxelles, Belgium

Luigi Orsina Dipartimento di Matematica, "Sapienza" Università di Roma, Roma, Italy

Filomena Pacella Dipartimento di Matematica, Universitá di Roma "La Sapienza", Roma, Italy

Víctor Padrón Department of Mathematics, Normandale College, Bloomington, MN, USA

C.D. Pagani Dipartimento di Matematica, Politecnico di Milano, Milano, Italy

Rosa Pardo Departamento de Matemática Aplicada, Universidad Complutense de Madrid, Madrid, Spain

Marcone C. Pereira Instituto de Matemática e Estatística, Universidade de São Paulo, São Paulo, SP, Brazil

D. Pierotti Dipartimento di Matematica, Politecnico di Milano, Milano, Italy

Angela Pistoia Dipartimento di Metodi e Modelli Matematici, Universitá di Roma "La Sapienza", Roma, Italy

Peter Poláčik School of Mathematics, University of Minnesota, Minneapolis, MN, USA

J. Sánchez Departamento de Matemáticas, Universidad de La Serena, La Serena, Chile

Edcarlos D. Silva Instituto de Matemática e Estatística, Universidade Federal de Goiás, Goiânia, GO, Brazil

Ricardo P. Silva Instituto de Geociências e Ciências Exatas, Universidade Estadual Paulista, Rio Claro, SP, Brazil

Luis Silvestre Department of Mathematics, University of Chicago, Chicago, IL, USA

Marco Squassina Dipartimento di Informatica, Università degli Studi di Verona, Verona, Italy

Antonio Suárez Departamento de Ecuaciones Diferenciales y Análisis Numérico, Faculdade de Matemáticas, University de Sevilla, Sevilla, Spain

Eduardo V. Teixeira Universidade Federal do Ceará, Fortaleza, CE, Brazil

F.de Thélin Institut de Mathématique (UMR CNRS 5219), Université de Toulouse, Toulouse, France

Carlos Tomei Departamento de Matemática, PUC-Rio, Rio de Janeiro, RJ, Brazil

P. Ubilla Departamento de Matemática y C.C., Universidad de Santiago de Chile, Santiago, Chile

G. Verzini Dipartimento di Matematica, Politecnico di Milano, Milano, Italy

Yongda Wang Dipartimento di Matematica, Politecnico di Milano, Milano, Italy

Jianfu Yang Department of Mathematics, Jiangxi Normal University, Nanchang, Jiangxi, People's Republic of China

André Zaccur Departamento de Matemática, PUC-Rio, Rio de Janeiro, RJ, Brazil

A. Zilio Centre d'Analyse et de Mathématique Sociales, École des Hautes Études en Sciences Sociales, Paris, France

Nonexistence of positive classical solutions for the nonlinear Schrödinger equation with unbounded or decaying weights

Francisco S.B. Albuquerque and Everaldo S. Medeiros

Mathematics Subject Classification (2010): 35A01, 35A23, 35A25, 35B09, 35J20, 35J60

1 Introduction and main result

In this note we are concerned with *nonlinear Schrödinger equations* of the form

$$i\hbar\frac{\partial\psi}{\partial t} = -\hbar^2\Delta\psi + W(x)\psi - Q(x)\xi(|\psi|)\psi, \quad (x,t) \in \mathbb{R}^2 \times \mathbb{R} \tag{1}$$

or *nonlinear Klein–Gordon equations* of the form

$$i\hbar\frac{\partial^2\psi}{\partial t^2} = \hbar^2\Delta\psi + (W(x) - \mathfrak{m}^2)\psi - Q(x)\xi(|\psi|)\psi, \quad (x,t) \in \mathbb{R}^2 \times \mathbb{R}, \tag{2}$$

where $i = \sqrt{-1}$, $\hbar$ is the Planck's constant, $\mathfrak{m}$ is a positive number, $W(x), Q(x)$ are real-valued potentials, and $\xi : \mathbb{R}^+ \to \mathbb{R}$ is a suitable nonlinear term. Such

F.S.B. Albuquerque
Universidade Estadual da Paraíba, Centro de Ciências Exatas
e Sociais Aplicadas, 58700-070 Patos-PB, Brazil
e-mail: siberio_ba@uepb.edu.br

E.S. Medeiros (✉)
Universidade Federal da Paraíba, Departamento de Matemática,
58051-900 João Pessoa-PB, Brazil
e-mail: everaldo@mat.ufpb.br

A.N. Carvalho et al. (eds.), *Contributions to Nonlinear Elliptic Equations and Systems*, Progress in Nonlinear Differential Equations and Their Applications 86,
DOI 10.1007/978-3-319-19902-3_1

equations arise in various branches of mathematical-physics and mathematical-biology and they have been subject of extensive study in the past years, among others we refer to [5, 6, 11–15] and the references therein. Here our special interest is in the nonexistence of *standing wave solutions*, that is, solutions of type

$$\psi(x,t) = \exp(-i\mathscr{E}t/\hbar)u(x),$$

where $\mathscr{E} \in \mathbb{R}$ and $u > 0$ is a real function. It is known that ψ satisfies (1) or (2) if and only if the function u solves the following elliptic equation

$$-\Delta u + V(x)u = \lambda Q(x)f(u), \quad x \in \mathbb{R}^2, \tag{3}$$

where $V : \mathbb{R}^2 \to \mathbb{R}$ is the new potential, λ is a positive parameter, and $f : \mathbb{R} \to \mathbb{R}$ is a new nonlinearity. Precisely, the main purpose of the present work is to show that, by using an average argument, it is possible to find sufficient conditions for nonexistence of positive classical solutions for problem (3). We quote that this method has been used, for instance, in the papers [3, 4, 7, 17] and this approach proposed here is in the spirit of [3]. Throughout we will assume the following assumptions on V and Q:

(V) $V \in C(\mathbb{R}^2;\mathbb{R})$, $V > 0$ and there exists $a < -2$ such that

$$\lim_{|x|\to+\infty} \frac{V(x)}{|x|^a} < \infty;$$

(Q) $Q \in C(\mathbb{R}^2;\mathbb{R})$, $Q > 0$ and there exists $b > -2$ such that

$$\lim_{|x|\to+\infty} \frac{Q(x)}{|x|^b} > 0.$$

Remark 1.1. Important classes of weights satisfying the above hypotheses are $V(x) = |x|^\beta$ and $Q(x) = |x|^\gamma$, with $\beta < -2$ and $\gamma > -2$, which includes the Henon($\gamma > 0$) and singular weights type.

On the nonlinear term $f(s)$, we shall assume that $f : \mathbb{R} \to [0,+\infty)$ is a continuous function satisfying: there exists $\nu > 2$ and a positive constant C_0 such that for any $p \geq \nu - 1$ there holds

$$f(s) \geq C_0 s^p, \quad \text{for all } s \geq 0. \tag{4}$$

In the recent papers [2, 15], the authors studied the existence and multiplicity of solutions for problem (3) under similar conditions on the weights. In [2] for instance, the authors considered nonlinearity $f(s)$ with exponential critical growth as established in the paper by D. G. de Figueiredo et al, N. S. Trudinger, S. I. Pohozhaev, and J. Moser (see [8–10, 16]), that is, there exists a constant $\alpha_0 > 0$ such that

$$\lim_{s\to+\infty}\frac{f(s)}{e^{\alpha s^2}}=\begin{cases}0, & \forall\alpha>\alpha_0,\\ +\infty, & \forall\alpha<\alpha_0.\end{cases} \tag{5}$$

We point out that among other hypotheses on f and in order to use a variational approach to study problem (3), the authors also assumed the following conditions:

(f_1) there exists $\nu>2$ such that $\liminf_{s\to 0+}\frac{f(s)}{s^{\nu-1}}>0$;

(f_2) there exists $\theta>2$ such that

$$0<\theta F(s)\doteq\theta\int_0^s f(t)\mathrm{d}t\le sf(s),\quad \forall s>0.$$

It is worth mentioning that the main tool used in [2] by the authors to study problem (3) concerned the existence and multiplicity of solutions was a Trudinger–Moser type inequality in weighted Sobolev spaces, as well as an improvement of it, which has been obtained via classical and singular Trudinger–Moser inequality versions established in [9] and [1], respectively. For details, see [2, Theorem 1.1].

Remark 1.2. Using a straightforward computation, we can see that if $f(s)$ satisfies (5) with $\alpha<\alpha_0$ and the conditions (f_1) – (f_2) then (4) holds.

Our main result reads as follows:

Theorem 1.3. *Assume that* (V) – (Q) *hold. If f satisfies condition* (4), *then problem* (3) *has no positive C^2 solutions for λ large.*

Remark 1.4. We complement the multiplicity result obtained in [2, Theorem 1.5] in the sense that we just have established new ranges for the numbers a,b (see (V) – (Q)) for which we have given an additional information on the existence or nonexistence of solutions for problem (3). We also point out that nonexistence results involving exponential critical growth (situation covered by the condition (4) as pointed out in the Remark 1.2), only a partial result is known, that is, one due to D. G. de Figueiredo and B. Ruf [7], in which the nonexistence of a positive radial solution is proved for the following problem

$$\begin{cases}-\Delta u=h(u)e^{4\pi u^2} & \text{in } \Omega,\\ u=0 & \text{on } \partial\Omega,\end{cases}$$

where $\Omega=B_1(0)\subset\mathbb{R}^2$ and $h\in C^2(\mathbb{R})$ satisfies some suitable conditions. The proof of this result uses techniques of the theory of ordinary differential equations which we shall use in the proof of Theorem 1.3.

2 Proof of Theorem 1.3

In the proof we use an averaging process to reduce the problem to an ordinary differential inequality in order to get a contradiction via some elementary arguments. Before we need some technical lemmas. In the following, $B_r \subset \mathbb{R}^2$ denotes the open ball in centered at the origin with radius $r > 0$. We denote the spherical average $\bar{u}$ of a function $u \in C(\mathbb{R}^2)$ by

$$\bar{u}(r) \doteq \frac{1}{|\partial B_r|} \int_{\partial B_r} u(x) \mathrm{d}\sigma,$$

where $\mathrm{d}\sigma$ is the standard volume element on ∂B_r. It is standard to verify that if $u \in C^2(\mathbb{R}^2)$ then:

Lemma 2.1. *The following assertions are hold:*

(i) $\overline{u + v} = \bar{u} + \bar{v}$;
(ii) $\frac{d}{dr}(r\bar{u}'(r)) = r\Delta\bar{u}(r)$;
(iii) $\bar{u}''(r) + \frac{1}{r}\bar{u}'(r) = \Delta\bar{u}(r)$ *(Darboux's equation);*
(iv) $\overline{\Delta u}(r) = \Delta\bar{u}(r)$;
(v) $\bar{u}^p \leq \overline{u^p}$, *for all* $p > 1$ *(Jensen's inequality).*

Lemma 2.2. *Assume that* $(V) - (Q)$ *and* (4) *hold. Let* u *be a positive* C^2 *solution of problem* (3). *Setting* $w(t) = r^m\bar{u}(r)$ *with* $m = (b+2)/(p-1)$, $p \geq \nu - 1$ *and* $t = \log r$, *there exist real numbers* l_1 *and* l_2 *such that* w *satisfies*

$$w'' + l_1 w' + (l_2 - V(r)r^2)w + w^p \leq 0, \tag{6}$$

for t *sufficiently large, where* $V(r) = \max_{|x|=r} V(x)$.

Proof. Let $u \in C^2$ a nontrivial positive solution of problem (3). From hypothesis (Q) there exist $C_1, R_0 > 0$ such that

$$Q(x) \geq C_1|x|^b, \quad \text{for all } |x| \geq R_0.$$

From this and condition (4), it follows for any $p \geq \nu - 1$ that

$$\lambda Q(x)f(u) \geq \lambda C_0 C_1 C|x|^b u^p, \quad \text{for all } |x| \geq R_0.$$

Choosing λ sufficiently large such that $\lambda C_0 C_1 \geq 1$ we get

$$\lambda Q(x)f(u) \geq |x|^b u^p, \quad \text{for all } |x| \geq R_0.$$

Hence, u satisfies

$$\Delta u - V(x)u + |x|^b u^p \leq 0 \quad \text{in } \mathbb{R}^2 \setminus B_{R_0}. \tag{7}$$

By taking the spherical average in (7) and invoking Lemma 2.1, we obtain

$$\bar{u}''(r) + \frac{1}{r}\bar{u}'(r) - V(r)\bar{u}(r) + r^b \bar{u}^p(r) \leq 0, \quad \text{for } r > R_0, \tag{8}$$

where $V(r) = \max_{|x|=r} V(x)$. Setting $w(t) = r^m \bar{u}(r)$ with $m = (b+2)/(p-1)$ and $t = \log r$, we see that

$$\begin{aligned}
w'(t) &= mr^m \bar{u}(r) + r^{m+1}\bar{u}'(r), \\
w''(t) &= m^2 r^m \bar{u}(r) + mr^{m+1}\bar{u}'(r) + (m+1)r^{m+1}\bar{u}'(r) + r^{m+2}\bar{u}''(r), \\
l_1 w'(t) &= -2m^2 r^m \bar{u}(r) - 2mr^{m+1}\bar{u}'(r), \\
(l_2 - V(r)r^2)w(t) &= m^2 r^m \bar{u}(r) - V(r)r^{m+2}\bar{u}(r),
\end{aligned}$$

where $l_1 = -2m$ and $l_2 = m^2$. Thus, by using (8), we get (6) for t sufficiently large. The proof is completed. □

Proof (Proof of Theorem 1.3). We shall use similar arguments developed in [3]. Suppose by contradiction that u is a C^2 nontrivial positive solution of problem (3). We have three cases to consider:

Case 1. $w'(T) < 0$ for some T sufficiently large. We set $B(r) \doteq l_2 - V(r)r^2$. We claim that from the hypothesis (V), we have $B + w^{p-1} \geq 0$ at infinity. In fact, it just observes that

$$V(x)|x|^2 = \frac{V(x)}{|x|^a}|x|^{a+2} \to 0, \quad \text{as } |x| \to +\infty.$$

Thus,

$$V(x)|x|^2 \leq l_2, \quad \text{for all } |x| \gg 1,$$

which implies that $B(r) \geq 0$ at infinity. On the other hand, integrating (6) over $[T, t]$ for T large, we have

$$w'(t) \leq e^{-l_1(t-T)}w'(T) - e^{-l_1 t}\int_T^t (B + w^{p-1})we^{l_1 s}ds \leq e^{-l_1(t-T)}w'(T).$$

Since $l_1 < 0$, integrating the above inequality over $[T, t]$, we obtain

$$0 < w(t) \leq w(T) + \frac{1}{l_1}w'(T)\left(1 - e^{-l_1(t-T)}\right) \to -\infty, \quad \text{as } t \to +\infty,$$

which is a contradiction.

Case 2. w is non-decreasing and bounded at infinity. Then there exists $w_\infty > 0$ such that $w(t) \to w_\infty$ as $t \to \infty$. Thus, there exists a real sequence (t_n) with $\lim_{n\to\infty} t_n = \infty$ such that $w'(t_n), w''(t_n) \to 0$ as $n \to \infty$, which implies by taking the inferior limit in (6)

$$0 < w_\infty^p \le m^2 + \liminf_{n\to\infty} w(t_n)^p \le \liminf_{n\to\infty} \left[\left(B(e^{t_n}) + w(t_n)^{p-1} \right) w(t_n) \right] \le 0,$$

which is a contradiction.

Case 3. w is non-decreasing and unbounded at infinity. Setting $v(t) = e^{\frac{l_1}{2}t} w(t)$, we have

$$v''(t) + D(t)v \le 0, \tag{9}$$

where $D(t) \doteq B(e^t) - l_1^2/4 + w(t)^{p-1}$. Multiplying both sides of (9) by $\sin t$ and integrating by parts twice over $[2k\pi, (2k+1)\pi]$ with integer $k > 0$, we obtain

$$\int_{2k\pi}^{(2k+1)\pi} (D(t) - 1)v(t) \sin t \mathrm{d}t \le -v(2k\pi) - v((2k+1)\pi) \le 0. \tag{10}$$

Since $D(t) \to \infty$ as $t \to \infty$, we have in particular that $D > 1$ on $[2k\pi, (2k+1)\pi]$ for $k > 0$ sufficiently large, which contradicts inequality (10) and this completes the proof of Theorem 1.3. □

References

1. Adimurthi, Sandeep, K.: A singular Moser-Trudinger embedding and its applications. Nonlinear Differ. Equ. Appl. **13**, 585–603 (2007)
2. Albuquerque, F.S.B., Alves, C.O., Medeiros, E.S.: Nonlinear Schrödinger equation with unbounded or decaying radial potentials involving exponential critical growth in $\mathbb{R}^2$. J. Math. Anal. Appl. **409**, 1021–1031 (2014)
3. Bae, S., Byeon, J.: Standing waves of nonlinear Schrödinger equations with optimal conditions for potential and nonlinearity. Commun. Pure Appl. Anal. **12**, 831–850 (2013)
4. Benguria, R.D., Lorca, S., Yarur, C.S.: Nonexistence results for solutions of semilinear elliptic equations. Duke Math. J. **74**(3), 615–634 (1994)
5. Berestycki, H., Lions, P.-L.: Nonlinear scalar field equations. I. Arch. Ration. Mech. Anal. **82**, 313–345 (1983)
6. Berestycki, H., Lions, P.-L.: Nonlinear scalar field equations. II. Arch. Ration. Mech. Anal. **82**, 347–375 (1983)
7. de Figueiredo, D.G., Ruf, B.: On the existence and non-existence of solutions for elliptic equations with critical growth in $\mathbb{R}^2$. Commun. Pure Appl. Math. **48**, 639–655 (1995)
8. de Figueiredo, D.G., Miyagaki, O.H., Ruf, B.: Elliptic equations in $\mathbb{R}^2$ with nonlinearities in the critical growth range. Calc. Var. **3**, 139–153 (1995)
9. Moser, J.: A sharp form of an inequality by N. Trudinger. Indiana Univ. Math. J. **20**, 1077–1092 (1971)

10. Pohozhaev, S.I.: The Sobolev embedding in the case $pl = n$. In: Proceeding of Technical Science Conference on Advance Science Research, (1964–1965). Mathematics Section, pp. 158–170. Moscow Energy Institute, Moscow (1965)
11. Rabinowitz, P.H.: On a class of nonlinear Schrödinger equations. Z. Angew. Math. Phys. **43**, 270–291 (1992)
12. Strauss, W.A.: Mathematical Aspects of Classical Nonlinear Field Equations. Lecture Notes in Physics, vol. 98. Springer, Berlin (1979)
13. Stuart, C.A.: Self-trapping of an electromagnetic field and bifurcation from the essential spectrum. Arch. Ration. Mech. Anal. **113** 65–96 (1991)
14. Stuart, C.A., Zhou, H.S.: A variational problem related to self-trapping of an electromagnetic field. Math. Methods Appl. Sci. **113**, 1397–1407 (1996)
15. Su, J., Wang, Z.-Q., Willem, M.: Nonlinear Schrödinger equations with unbounded and decaying radial potentials. Commun. Contemp. Math. **9**, 571–583 (2007)
16. Trudinger, N.S.: On the embedding into Orlicz spaces and some applications. J. Math. Mech. **17**, 473–484 (1967)
17. Yarur, C.S.: Nonexistence of positive singular solutions for a class of semilinear elliptic systems. Electron. J. Differ. Equ. **8**, 1–22 (1996)

Hylomorphic solitons for the generalized KdV equation

Vieri Benci and Donato Fortunato

Dedicated to our friend Djairo De Figuereido on the occasion of his 80th birthday

Mathematics Subject Classification (2010): 74J35, 35C08, 35A15, 35Q74, 35B35

1 Introduction

The Korteweg–de Vries equation (KdV) was first introduced by Boussinesq (1877) and rediscovered by Diederik Korteweg and Gustav de Vries (1895). It is a model of waves on shallow water surfaces (see, e.g., [12]).

Many different variations of the KdV equation have been studied. The most common is the following one which is known as the generalized KdV equation (gKdV):

$$\frac{\partial u}{\partial t} + \frac{\partial^3 u}{\partial x^3} - \frac{\partial}{\partial x} W'(u) = 0 \tag{1}$$

where $u = u(t, x)$, and $W \in C^2(\mathbb{R})$. If $W(s) = -s^3$, then (1) reduces to the usual KdV equation

V. Benci (✉)
Dipartimento di Matematica, Università degli Studi di Pisa, Via Filippo Buonarroti 1/c, 56127 Pisa, Italy

Centro Linceo Interdisciplinare Beniamino Segre
e-mail: benci@dma.unipi.it

D. Fortunato
Dipartimento di Matematica, Università degli Studi di Bari Aldo Moro and INFN Sezione di Bari, Via Orabona 4, 70125 Bari, Italy
e-mail: fortunat@dm.uniba.it

A.N. Carvalho et al. (eds.), *Contributions to Nonlinear Elliptic Equations and Systems*, Progress in Nonlinear Differential Equations and Their Applications 86, DOI 10.1007/978-3-319-19902-3_2

$$\frac{\partial u}{\partial t} + \frac{\partial^3 u}{\partial x^3} + 6u\frac{\partial u}{\partial x} = 0 \tag{2}$$

If $W(s) = -\frac{s^{k+2}}{(k+2)(k+1)}$, then (1) reduces to the equation

$$\frac{\partial u}{\partial t} + \frac{\partial^3 u}{\partial x^3} + u^k\frac{\partial u}{\partial x} = 0 \tag{3}$$

known as the modified KdV equation (mKdV).

In this paper we are interested to the existence of solitary waves and solitons for the gKdV equation. Roughly speaking a *solitary wave* is a solution of a field equation whose energy travels as a localized packet and which preserves this localization in time. A *soliton* is a solitary wave which exhibits some form of stability so that it has a particle-like behavior.

Using the inverse scattering transform, it is possible to prove that KdV admits soliton solutions and to have an extremely powerful and precise information on them. However the inverse scattering techniques cannot be applied to the generalized KdV equation. In this paper, we shall use the method developed in [5] to prove that equation (1) admits solitons and we will show that they are *hylomorphic*. Following [2], a soliton is called *hylomorphic* if its stability is due to a particular interplay between the *energy* E and the *hylenic charge* $C := \int u^2 dx$ which is another integral of motion. More precisely, a soliton u_0 is hylomorphic if

$$E(u_0) = \min\left\{E(u) \mid \int u^2 dx = C(u_0)\right\}.$$

We will show that eq. (1) admits solitons provided that W satisfies suitable assumptions (see Theorem 4.6). In particular, if $W(s) = -\frac{s^{k+2}}{(k+2)(k+1)}$, hylomorphic solitons exist for $k = 1, 2, 3$. So, in this case, we get a different proof of well-known results (see [11]) and we show that the "usual" solitons of mKdV can be considered "hylomorphic".

In Th. 4.6, we obtain the existence of hylomorphic solitons under a very general set of assumptions on W; moreover, in contrast to other results on this topic, these assumptions are easy to verify.

2 Solitary waves and solitons

In this section we construct an abstract functional framework which allows to define solitary waves, solitons and hylomorphic solitons.

2.1 Solitary waves

Solitary waves and solitons are particular *states* of a dynamical system described by one or more partial differential equations. Thus, we assume that the states of this system are described by one or more *fields* which mathematically are represented by functions

$$\mathbf{u} : \mathbb{R}^N \to V$$

where V is a vector space with norm $|\cdot|_V$ and which is called the internal parameters space. We assume the system to be deterministic; this means that it can be described as a dynamical system (X, γ) where X is the set of the states and $\gamma : \mathbb{R} \times X \to X$ is the time evolution map. If $\mathbf{u}_0(x) \in X$, the evolution of the system will be described by the function

$$\mathbf{u}(t, x) := \gamma_t \mathbf{u}_0(x). \tag{4}$$

We assume that the states of X have "finite energy" so that they decay at ∞ sufficiently fast and that

$$X \subset L^1_{loc}\left(\mathbb{R}^N, V\right). \tag{5}$$

Thus we are led to give the following definition:

Definition 2.1. A dynamical system (X, γ) is called of FT type (field-theory-type) if X is a Hilbert space of functions of type (5).

The dynamical systems we shall consider will be tacitly assumed to be of FT type.

Let $\mathscr{T}$ be the group of translations in $\mathbb{R}^N$ and $U(V)$ the group of unitary transformations on V; set

$$G = \mathscr{T} \times U(V)$$

Given $(\tau, h) \in G$ we will consider the representation of G on $X \subset L^1_{loc}\left(\mathbb{R}^N, V\right)$ given by

$$\left[T_{(\tau,h)}\mathbf{u}\right](x) = h\mathbf{u}(x - \tau)$$

For example, take $X = L^2\left(\mathbb{R}^N, \mathbb{C}\right)$, $h = e^{i\theta} \in U(1)$, then

$$\left[T_{(\tau,h)}u\right](x) = e^{i\theta}u(x - \tau)$$

A solitary wave is a state of finite energy which evolves without changing its shape. This informal description can be formalized by the following definition:

Definition 2.2. A state $\mathbf{u}_0 \in X\backslash\{0\}$ is called solitary wave if there is a continuous trajectory

$$t \mapsto (\tau(t), h(t)) \in G$$

such that

$$\gamma_t \mathbf{u}_0(x) = h(t)\mathbf{u}_0(x - \tau(t))$$

For example, consider a solution of a field equation having the following form:

$$\mathbf{u}(t,x) = u_0(x - vt - x_0)e^{i(v\cdot x - \omega t)};\ u_0 \in L^2(\mathbb{R}^N); \tag{6}$$

$x_0, v \in \mathbb{R}^N, \omega \in \mathbb{R}$. Clearly $\mathbf{u}(t,x)$ is a solitary wave for every $t \in \mathbb{R}$. The evolution of a solitary wave is a translation plus a unitary change of the internal parameters (in this case the phase).

2.2 *Orbitally stable states and solitons*

The *solitons* are solitary waves characterized by some form of stability. To define them at this level of abstractness, we need to recall some well-known notions in the theory of dynamical systems.

Definition 2.3. A set $\Gamma \subset X$ is called *invariant* if $\forall \mathbf{u} \in \Gamma, \forall t \in \mathbb{R},\ \gamma_t \mathbf{u} \in \Gamma$.

Definition 2.4. Let (X, γ) be a dynamical system and let X be equipped with a metric d (it is not necessary to assume that $d(\mathbf{u}, \mathbf{v}) = \|\mathbf{u} - \mathbf{v}\|_X$). An invariant set $\Gamma \subset X$ is called stable (with respect to d), if $\forall \varepsilon > 0, \exists \delta > 0,\ \forall \mathbf{u} \in X$,

$$d(\mathbf{u}, \Gamma) \leq \delta,$$

implies that

$$\forall t \geq 0,\ d(\gamma_t \mathbf{u}, \Gamma) \leq \varepsilon.$$

Definition 2.5. Let (X, d) be a metric space and let $\mathcal{T}$ be the group of translations. A set $\Gamma \subset X$ is called $\mathcal{T}$-compact if for any sequence $\mathbf{u}_n(x) \in \Gamma$ there is a subsequence $\mathbf{u}_{n_k}$ and a sequence $\tau_k \in \mathcal{T}$ such that $\mathbf{u}_{n_k}(x - \tau_k)$ is convergent with respect to the metric d.

Now we give the definition of *orbitally stable state*:

Definition 2.6. Let (X, γ) be a dynamical system with X equipped with a metric d. A state $\mathbf{u} \in X$ is called *orbitally stable* (with respect to d) if $\mathbf{u} \in \Gamma \subset X$ where

(i) Γ is an invariant stable set with respect to d,
(ii) Γ is $\mathcal{T}$-compact (with respect to d).

This definition is usually present in the literature relative to the dynamics of PDE's (see, e.g., [6, 11], etc.).

Now we are able to give the definition of soliton:

Definition 2.7. Let (X, γ) be a dynamical system with X equipped with a metric d. A soliton is an orbitally stable solitary wave (with respect to d).

Remark 2.8. In our definition, since (X, γ) is a dynamical system, the map

$$t \mapsto \gamma_t \mathbf{u}$$

is continuous with respect to $\|\cdot\|_X$. In the above definitions, we have introduced a distance d, however we have not supposed that

$$d(\mathbf{u}, \mathbf{v}) = \|\mathbf{u} - \mathbf{v}\|_X$$

In fact, in some applications this is not true. As we will see in section 4, this is the case for equation (1) where we have

$$d(\mathbf{u}, \mathbf{v}) \leq M \|\mathbf{u} - \mathbf{v}\|_X$$

for a suitable constant $M > 0$.

2.3 Hylomorphic solitons

We now assume that the dynamical system (X, γ) has two constants of motion: the energy E and the hylenic charge C. At this level of abstraction, of course, the name energy and hylenic charge are conventional.

Definition 2.9. Let (X, γ) be a dynamical system where X is equipped with a metric d. A soliton $\mathbf{u}_0 \in X$ is called **hylomorphic** if the set Γ (given by Def. 2.6) has the following structure

$$\Gamma = \Gamma(e_0, c_0) = \{\mathbf{u} \in X \mid E(\mathbf{u}) = e_0, \ |C(\mathbf{u})| = c_0\} \tag{7}$$

where

$$e_0 = \min\{E(\mathbf{u}) \mid |C(\mathbf{u})| = c_0\}. \tag{8}$$

Notice that, by (8), we have that a hylomorphic soliton $\mathbf{u}_0$ minimizes the energy on

$$\mathfrak{M}_{c_0} = \{\mathbf{u} \in X \mid |C(\mathbf{u})| = c_0\}. \tag{9}$$

If $\mathfrak{M}_{c_0}$ is a manifold and E and C are differentiable, then $\mathbf{u}_0$ satisfies the following nonlinear eigenvalue problem:

$$E'(\mathbf{u}_0) = \lambda C'(\mathbf{u}_0).$$

3 Hylomorphic solitons for the nonlinear Schrödinger equation

The solitons for eq. (1), as we will see, are related to the solitons of the nonlinear Schrödinger equation. We recall that the orbital stability for the nonlinear Schrödinger equation has been proved in [6] (see also [1] for the general case and [9] with its references).

Here we shall use a method to prove the existence of hylomorphic solitons for (1) similar to the one presented in [5] (see also [3] and [4]). In this section we will resume this method.

The nonlinear Schrödinger equation is given by

$$i\frac{\partial\psi}{\partial t} = -\frac{1}{2}\frac{\partial^2\psi}{\partial x^2} + \frac{1}{2}W'(\psi) \tag{10}$$

where $\psi : \mathbb{R} \times \mathbb{R} \to \mathbb{C}$ and where $W : \mathbb{C} \to \mathbb{R}$ and

$$W'(\psi) = \frac{\partial W}{\partial \psi_1} + i\frac{\partial W}{\partial \psi_2}. \tag{11}$$

We assume that W depends only on $|\psi|$, namely

$$W(\psi) = F(|\psi|) \text{ and so } W'(\psi) = F'(|\psi|)\frac{\psi}{|\psi|}.$$

for some smooth function $F : [0,\infty) \to \mathbb{R}$. In the following we shall identify, with some abuse of notation, W with F.

The energy is given by

$$E = \int \left(\frac{1}{2}\left|\frac{\partial\psi}{\partial x}\right|^2 + W(\psi) \right) dx \tag{12}$$

Moreover the Schrödinger equation has an other important integral of motion

$$C = \int |\psi|^2 \, dx \tag{13}$$

to which we will refer as *charge*.

We make the following assumptions on the function W :

$$W(0) = W'(0) = 0 \tag{14}$$

$$W''(0) = 2E_0 > 0. \tag{15}$$

If we set

$$W(s) = E_0 s^2 + N(s), \tag{16}$$

then,

$$\exists s_0 \in \mathbb{R}^+ \text{ such that } N(s_0) < 0. \tag{17}$$

There exist q, r in $(2, +\infty)$, s. t.

$$|N'(s)| \leq c_1 s^{r-1} + c_2 s^{q-1} \tag{18}$$

$$N(s) \geq -cs^p,\ c \geq 0,\ 2 < p < 6 \text{ for } s \text{ large} \tag{19}$$

We can apply the abstract theory of section 2 setting:

- $X = H^1(\mathbb{R}, \mathbb{C})$, $\mathbf{u} = \psi$;
- $d(\psi, \varphi) = \|\psi - \varphi\|_{H^1}$.

Theorem 3.1. *Let W satisfy (14),...,(19). Then there exists $\delta_\infty > 0$ such that for every $\delta \in (0, \delta_\infty)$ there exist $c_\delta > 0$ and an orbitally stable state $\psi_\delta \in H^1(\mathbb{R}, \mathbb{C})$, such that ψ_δ minimizes the energy on the manifold*

$$\mathfrak{M}_{c_\delta} = \left\{ \mathbf{u} \in X \mid \int |\psi|^2 \, dx = c_\delta \right\}.$$

Moreover if $\delta_1 < \delta_2$ we have that $c_{\delta_1} > c_{\delta_2}$.

Proof. The proof is an immediate consequence of Th. 52 in [5] (see also [3]). □

By the above theorem we have that every ψ_δ is an orbitally stable state; the following theorem shows that it is a soliton.

Theorem 3.2. *Let $u_\delta \in H^1(\mathbb{R}, \mathbb{C})$ be a orbitally stable state as in Th 3.1. Then u_δ is a solution of the equation*

$$-\frac{1}{2}\frac{\partial^2 u}{\partial x^2} + \frac{1}{2}W'(u) = \omega u \tag{20}$$

and

$$\psi_{\delta}(t,x) := u_{\delta}(x)e^{-i\omega t} \tag{21}$$

solves (10). Namely u_{δ} is a (hylomorphic) soliton.

Proof. See Proposition 59 of [5] (see also [3]). □

4 Hylomorphic solitons for the generalized KdV equation

First of all let us show that the "good" solutions of equation (1) have two constants of motion.

Proposition 4.1. *Let W be a C^2 function and u be a smooth solution of equation (1) and assume that $u(t,.) \in H^1(\mathbb{R})$, $\frac{\partial u}{\partial t}(t,\cdot) \in L^2(\mathbb{R})$. Then u has two integrals of motion: the energy*

$$E = \int \left(\frac{1}{2}\left[\frac{\partial u}{\partial x}\right]^2 + W(u) \right) dx \tag{22}$$

and the charge

$$C = \frac{1}{2}\int u^2 dx \tag{23}$$

Proof. Since $\frac{\partial u}{\partial t}(t,\cdot) \in L^2(\mathbb{R})$ and

$$-\frac{\partial^3 u}{\partial x^3} + \frac{\partial}{\partial x}W'(u) = \frac{\partial u}{\partial t},$$

the integral

$$\int \left(-\frac{\partial^2 u}{\partial x^2} + W'(u) \right) \frac{\partial u}{\partial t} dx$$

is well defined and it equals the time derivative of $E(u(t))$:

$$\frac{d}{dt}E(u(t)) = \int \left(-\frac{\partial^2 u}{\partial x^2} + W'(u) \right) \frac{\partial u}{\partial t} dx \tag{24}$$

Moreover

$$\frac{\partial u}{\partial t} = -\frac{\partial^3 u}{\partial x^3} + \frac{\partial}{\partial x}(W'(u)) = -\frac{\partial}{\partial x}\left(\frac{\partial^2 u}{\partial x^2} - W'(u)\right) \tag{25}$$

Substituting (25) in (24), we get

$$\begin{aligned}\frac{d}{dt}E(u) &= \int \left(\frac{\partial^2 u}{\partial x^2} - W'(u)\right)\frac{\partial}{\partial x}\left(\frac{\partial^2 u}{\partial x^2} - W'(u)\right)dx \\ &= \frac{1}{2}\int \frac{\partial}{\partial x}\left[\left(\frac{\partial^2 u}{\partial x^2} - W'(u)\right)^2\right]dx = 0\end{aligned}$$

Then E is constant along the solution u.

Let us now show that also C is constant along u. By (25) we have

$$\frac{d}{dt}C(u) = \int u\frac{\partial u}{\partial t}dx = \int u\left(-\frac{\partial^3 u}{\partial x^3} + \frac{\partial}{\partial x}W'(u)\right)dx \tag{26}$$

Let us compute each piece separately:

$$\int u\left(-\frac{\partial^3 u}{\partial x^3}\right)dx = \int \frac{\partial u}{\partial x}\left(\frac{\partial^2 u}{\partial x^2}\right)dx = \frac{1}{2}\int \frac{\partial}{\partial x}\left(\frac{\partial u}{\partial x}\right)^2 = 0 \tag{27}$$

Moreover

$$\int u\frac{\partial}{\partial x}(W'(u))dx = -\int W'(u)\frac{\partial u}{\partial x}dx = -\int \frac{\partial}{\partial x}W(u)dx = 0 \tag{28}$$

Substituting (28) and (27) in (26) we get $\frac{d}{dt}C(u) = 0$. □

We will apply the abstract theory of section 2 setting:

- $X = H^2(\mathbb{R})$;
- $d(u, v) = \|\psi - \varphi\|_{H^1}$.

To this end, we need the following assumption which guarantees that also the weak solutions in $H^2(\mathbb{R})$ have the properties required by the theory.

Assumption 4.2 *We assume that the equation (1) defines a dynamical system on $X = H^2(\mathbb{R})$, namely, for any initial data $u_0 \in H^2(\mathbb{R})$ there is a unique (weak) solution in $C(\mathbb{R}, H^2(\mathbb{R}))$ of the Cauchy problem. Moreover we assume that the energy (22) and the charge (23) are conserved integrals.*

Remark 4.3. Clearly, assumption 4.2 depends on W. By the existence theory of Kato [7], the assumption

$$\limsup_{s\to\pm\infty} \frac{-W''(s)}{s^4} \leq 0 \tag{29}$$

implies the existence of a unique global solution of eq. (1) in $C(\mathbb{R}, H^2(\mathbb{R}))$ and the conservation of (22) and (23). For the well posedness of equation (1) see also [8, 10] and their references.

Theorem 4.4. *Let W satisfy the assumptions (14),...,(19). Then there exists $\delta_\infty > 0$ such that for every $\delta \in (0, \delta_\infty)$ there exist $c_\delta > 0$ and $u_\delta \in H^2(\mathbb{R})$ which minimizes the energy E on the manifold*

$$\mathfrak{M}_{c_\delta} = \left\{ u \in X \mid \int u^2 dx = c_\delta \right\}.$$

If $\delta_1 < \delta_2$ we have that $c_{\delta_1} > c_{\delta_2}$. Moreover, if also assumption 4.2 holds, then u_δ is an orbitally stable state.

Proof. The proof of this theorem is essentially the same as the proof of Th. 3.1 which can be found in [5], Th. 52 (see also [3]). The reason for this relies on the fact that the energy and the charge for eq. (10) given by (12) and (13) are formally the same as the energy and the charge of eq. (1) given by (22) and (23). The fact that in the first case ψ is complex while in the second case u is real valued does not affect the estimates.

Another difference concerns the space X which is $H^1(\mathbb{R}, \mathbb{C})$ for eq. (10) and $H^2(\mathbb{R})$ for eq. (1).

The proof of the theorem consists in minimizing a suitable functional K_δ on X. In the case of eq. (1), we minimize first the functional K_δ on $H^1(\mathbb{R})$ and then we can prove that the set Γ_δ of these minimizers is contained in $H^2(\mathbb{R})$. In fact the minimizers satisfy the following eigenvalue equation

$$-\frac{\partial^2 u}{\partial x^2} + W'(u) = \lambda_\delta u$$

and hence, since $W \in C^2$, by the standard elliptic regularization, we have that $\Gamma_\delta \subset H^2(\mathbb{R})$. □

Remark 4.5. Assumption (15) is not necessary. It is not restrictive to assume that

$$E_0 > 0$$

Proof. In fact consider the following equation

$$\frac{\partial u}{\partial t} + \frac{\partial^3 u}{\partial x^3} - \frac{\partial}{\partial x} W_0'(u) = 0 \tag{30}$$

where $W_0''(0) = -2E_0 < 0$. In this case it is convenient to consider the equation

$$\frac{\partial v}{\partial t} + \frac{\partial^3 v}{\partial x^3} - \frac{\partial}{\partial x} W'(v) = 0 \tag{31}$$

where $W(s) = W_0(s) + (E_0 + 1)\,s^2$. We have that

$$W''(0) = 2 > 0$$

and to every solution v of eq. (31) corresponds a solution

$$u(t,x) = v(t, x + ct) \;\; with \;\; c = 2\,(E_0 + 1)$$

of eq. (30). In fact

$$\begin{aligned}\frac{\partial u}{\partial t} + \frac{\partial^3 u}{\partial x^3} - \frac{\partial}{\partial x} W_0'(u) &= \frac{\partial v}{\partial t} + c\frac{\partial v}{\partial x} + \frac{\partial^3 v}{\partial x^3} - \frac{\partial}{\partial x} W_0'(v) \\ &= \frac{\partial v}{\partial t} + c\frac{\partial v}{\partial x} + \frac{\partial^3 v}{\partial x^3} - \frac{\partial}{\partial x}\left[W'(v) - 2\,(E_0 + 1)\,v\right] \\ &= \frac{\partial v}{\partial t} + \frac{\partial^3 v}{\partial x^3} - \frac{\partial}{\partial x} W'(v) + c\frac{\partial v}{\partial x} - 2\,(E_0 - 1)\,\frac{\partial v}{\partial x} \\ &= 0.\end{aligned}$$

□

We shall prove that the minimizer u_δ in Theorem 4.4 is a soliton (def. 2.7).

Theorem 4.6. *Under the assumptions and the notations of Th. 4.4, the minimizer u_δ is a (hylomorphic) soliton. Moreover it is a solution of the equation*

$$\frac{\partial^3 u_\delta}{\partial x^3} - \frac{\partial}{\partial x} W'(u_\delta) = c_\delta \frac{\partial u_\delta}{\partial x} \tag{32}$$

and

$$U_\delta(t,x) := u_\delta(x - c_\delta t)$$

solves (1).

Proof. By th. 4.4, the minimizer u_δ is an orbitally stable state. So, in order to show that it is a soliton (def. 2.7), we need to prove that u_δ is a solitary wave (def. 2.2).

Since u_δ is a minimizer of the energy E on the manifold $\mathfrak{M}_{c_\delta}$, there exists a Lagrange multiplier c_δ s.t.

$$E'(u_\delta) = -c_\delta C'(u_\delta).$$

The above equality can be written as follows:

$$-\frac{\partial^2 u_\delta}{\partial x^2} + W'(u_\delta) = -c_\delta u_\delta$$

So, if we take the derivative $\frac{\partial}{\partial x}$ on both sides, we get (32). Finally (32) implies that the travelling wave $u(t,x) = u_\delta(x - c_\delta t)$ solves (1) and consequently u_δ is a solitary wave. □

Corollary 4.7. *Equation (3) admits hylomorphic solitons for* $k = 1, 2, 3$.

Proof. Take

$$W(s) = E_0 s^2 - \frac{s^{k+2}}{(k+2)(k+1)}.\ E_0 > 0 \tag{33}$$

For $k = 1, 2, 3$ the function W satisfies (14,... 19) and (29). So, by Remark 4.3 also the assumption (4.2) is satisfied. Then, by Theorem 4.6, equation (1) with W as in (33), for $k = 1, 2, 3$, admits hylomorphic solitons. Then, by Remark 4.5, equation (3) for $k = 1, 2, 3$ admits hylomorphic solitons. □

Corollary 4.8. *If* $W(s) = -|s|^{k+2}$, *then equation (1) admits hylomorphic solitons for* $k \in (0, 4)$.

Proof. The proof is the same as for corollary 4.7. □

Results analogous to the above corollaries are contained in [11].

References

1. Bellazzini, J., Benci, V., Ghimenti, M., Micheletti, A.M.: On the existence of the fundamental eigenvalue of an elliptic problem in $\mathbb{R}^N$. Adv. Nonlinear Stud. **7**(3), 439–458 (2007)
2. Benci, V.: Hylomorphic solitons. Milan J. Math. **77**, 271–332 (2009)
3. Benci, V., Fortunato, D.: A minimization method and applications to the study of solitons. Nonlinear Anal. **75**(12), 4398–4421 (2012)
4. Benci, V., Fortunato, D.: Solitons in Schrödinger-Maxwell equations. J. Fixed Point Theory Appl. **15**(1), 101–132 (2014)
5. Benci, V., Donato Fortunato, D.: Variational methods in nonlinear field equations. Solitary waves, hylomorphic solitons and vortices. Springer Monographs in Mathematics. Springer, Cham (2014)
6. Cazenave, T., Lions, P.-L.: Orbital stability of standing waves for some nonlinear Schrödinger equations. Commun. Math. Phys. **85**(4), 549–561 (1982)
7. Kato, T.: On the Cauchy problem for the (generalized) Korteweg-de Vries equation. In: Studies in applied mathematics. Advance Mathematical Supplement Stud. vol. 8, pp. 93–128. Academic Press, New York (1983)
8. Kenig, C.E., Ponce, G., Vega, L.: Well-posedness and scattering results for the generalized Korteweg-de Vries equation via the contraction principle. Commun. Pure Appl. Math. **46**(4), 527–620 (1993)

9. Sulem, C., Sulem, P.-L.: The Nonlinear Schrödinger Equation. Self-focusing and Wave Collapse. Applied Mathematical Sciences, vol. 139. Springer, New York (1999)
10. Tsutsumi, M., Mukasa, T., Riichi Iino, R.: On the generalized Korteweg-de Vries equation. Proc. Jpn. Acad. **46**, 921–925 (1970)
11. Weinstein, M.I.: Lyapunov stability of ground states of nonlinear dispersive evolution equations. Commun. Pure Appl. Math. **39**(1), 51–67 (1986)
12. Whitham, G.B.: Linear and Nonlinear Waves. Pure and Applied Mathematics. Wiley, New York (1974)

A consequence of Djairo's Lectures on the Ekeland variational principle

Lucio Boccardo and Luigi Orsina

Dedicated to Djairo for his 40th birthday, twice

1 Introduction and statement of the result

In [1] we considered some properties of the minimizing sequences for integral functionals J. Thanks to the Ekeland Lemma, the subject of the lectures given by Djairo in Bangalore (see [3]), we proved the existence of a minimizing sequence compact in $L^s(\Omega)$ or in $C^{0,\alpha}$ for functionals which do not need to have a minimum, without using the integral representation of the relaxed functional J^*.

In this paper, we improve the study done in the paper [1], under the assumption that the functional J has a minimum belonging to $L^\infty(\Omega)$. Using again Ekeland's ε-variational principle, we prove that there exists a minimizing sequence u_n for J which uniformly converges to a minimum u.

Let us now make the precise assumptions on the functional J. Let Ω be an open, bounded subset of $\mathbb{R}^N$, $N \geq 2$, and let p be a real number, with $2 \leq p < N$. We will denote by p^* the Sobolev exponent of p, i.e., $p^* = \frac{Np}{N-p}$.

Let $j : \Omega \times \mathbb{R}^N \to \mathbb{R}$ be a Carathéodory function (i.e., measurable with respect to x for every $\xi \in \mathbb{R}^N$, and continuous with respect to ξ for almost every $x \in \Omega$) convex with respect to ξ, and such that

$$\alpha\, |\xi|^p \leq j(x, \xi) \leq \beta\, |\xi|^p , \tag{1}$$

for almost every $x \in \Omega$, for every $\xi \in \mathbb{R}^N$, where α, β are positive real numbers.

L. Boccardo (✉) • L. Orsina
Dipartimento di Matematica, "Sapienza" Università di Roma, P.le A. Moro 5, 00185 Roma, Italy
e-mail: boccardo@mat.uniroma1.it; orsina@mat.uniroma1.it

A.N. Carvalho et al. (eds.), *Contributions to Nonlinear Elliptic Equations and Systems*, Progress in Nonlinear Differential Equations and Their Applications 86, DOI 10.1007/978-3-319-19902-3_3

Let f in $L^m(\Omega)$, with $m \geq (p^*)'$, and let $J : W_0^{1,p}(\Omega) \to \mathbb{R}$ be defined by

$$J(v) = \int_\Omega j(x, \nabla v)\, dx - \int_\Omega f(x)\, v\, dx, \qquad v \in W_0^{1,p}(\Omega).$$

Under the assumptions on f and p, J is well defined on $W_0^{1,p}(\Omega)$.

We will further assume that there exists $a(x,\xi) = j_\xi(x,\xi)$ which satisfies the classical Leray–Lions assumptions (see [8]) and the standard strong monotonicity assumption

$$[a(x,\xi) - a(x,\eta)][\xi - \eta] \geq \alpha\, |\xi - \eta|^p \qquad \forall \xi\,, \eta \in \mathbb{R}^N. \tag{2}$$

Examples of functions j such that (2) holds true are $j(x,\xi) = a(x)\,|\xi|^p$, with a a measurable function such that $\alpha \leq a(x) \leq \beta$. Since the strong monotonicity condition is simpler to handle if $p \geq 2$ (the above assumption (2)), and is a little bit more involved if $1 < p < 2$, we confine ourselves to the former case.

Since J is both weakly lower semicontinuous and coercive on $W_0^{1,p}(\Omega)$, there exists a minimum u of J; we have the following results on the summability of such minima.

Theorem 1.1. *Let u be a minimum of J on $W_0^{1,p}(\Omega)$. Then*

(i) if $1 < m < \frac{N}{p}$, then u belongs to $L^\sigma(\Omega)$, $\sigma = \frac{(pm)^*}{p'}$ (see [2]);
(ii) If $m > \frac{N}{p}$, then u belongs to $L^\infty(\Omega)$ (see [7, 9]).

Let us now recall Ekeland's ε-variational principle (see [4–6]).

Lemma 1.2. *Let (V,d) be a complete metric space, and let $\mathscr{F} : V \to (-\infty, +\infty]$ be a lower semicontinuous function such that* $\inf_V \mathscr{F}$ *is finite. Let $\varepsilon > 0$ and $u \in V$ be such that*

$$\mathscr{F}(u) \leq \inf_{v \in V} \mathscr{F}(v) + \varepsilon\,.$$

Then there exists $v \in V$ such that

(i) $d(u,v) \leq \sqrt{\varepsilon}$;
(ii) $\mathscr{F}(v) \leq \mathscr{F}(u)$;
(iii) v minimizes the functional $\mathscr{G}(w) = \mathscr{F}(w) + \sqrt{\varepsilon}\, d(v,w)$.

Our main result is the following.

Theorem 1.3. *Let J be defined as above, with j satisfying* (1). *Let*

$$f \in L^m(\Omega) \quad m > \frac{N}{p}, \tag{3}$$

and let q be such that $q^* = m'$; we also suppose that J' satisfies (2). *Let u be a minimum of J on $W_0^{1,p}(\Omega)$, and let $\{\bar{u}_n\}$ be any minimizing sequence for J. Then the minimizing sequence $\{u_n\}$ built after $\{\bar{u}_n\}$ using the ε-variational principle satisfies*

$$\lim_{n\to+\infty} \|u_n - \bar{u}_n\|_{W_0^{1,q}(\Omega)} = 0\,, \tag{4}$$

$$\lim_{n\to\infty} \|u_n - u\|_{W_0^{1,p}(\Omega)} = 0\,, \tag{5}$$

and

$$\lim_{n\to\infty} \|u_n - u\|_{L^\infty(\Omega)} = 0\,. \tag{6}$$

The plan of the paper is as follows: we will prove Theorem 1.3 in Section 2, and in Section 3 we will show that adding a lower order term to J will allow us to prove the same result under the assumption that f belongs to $L^2(\Omega)$, and not to the possibly larger space $L^m(\Omega)$, $m > \frac{N}{p}$.

2 Proof of the main result

For $k > 0$ let us define

$$T_k(s) = \max(-k, \min(k, s))\,, \qquad G_k(s) = s - T_k(s)\,.$$

Before proving Theorem 1.3, let us note that since we know (see Theorem 1.1) that any minimum u belongs to $L^\infty(\Omega)$, there exists M such that $|u| \le M$. Since the sequence $\{u_n\}$, with $u_n = T_M(\bar{u}_n)$, satisfies

$$\int_\Omega j(x, \nabla T_M(\bar{u}_n))\,dx - \int_\Omega f(x)\,T_M(\bar{u}_n)\,dx \le \inf_{v\in W_0^{1,p}(\Omega)} J(v) + \varepsilon_n + \int_\Omega f(x)\,G_M(\bar{u}_n)\,dx\,,$$

and since

$$\lim_{n\to+\infty} \int_\Omega f(x)\,G_M(\bar{u}_n)\,dx = \int_\Omega f(x)\,G_M(u)\,dx = 0\,,$$

we have that

$$\int_\Omega j(x, \nabla u_n)\,dx - \int_\Omega f(x)\,u_n\,dx \le \inf_{v\in W_0^{1,p}(\Omega)} J(v) + \bar{\varepsilon}_n\,.$$

That is, the sequence $\{u_n\}$ is a minimizing sequence for J, and it is bounded in $L^\infty(\Omega)$.

Theorem 1.3 says more than that: thanks to the ε-variational principle, it is possible to build a minimizing sequence not only bounded in $L^\infty(\Omega)$ but also strongly convergent to u in the same space.

Proof (PROOF OF THEOREM 1.3). Note that if q is as in the statement, the assumption $m > \frac{N}{p}$ implies that

$$1 < q < \frac{N}{N-p+1} < p\,. \tag{7}$$

Let ε_n be a sequence of positive real numbers, converging to zero, and let $\bar{u}_n$ be such that, for every $n \in \mathbb{N}$,

$$J(\bar{u}_n) \leq \inf_{v \in W_0^{1,p}(\Omega)} J(v) + \varepsilon_n\,.$$

Let us now consider the complete metric space $W_0^{1,q}(\Omega)$, endowed with the distance

$$d_n(w, v) = \frac{1}{\sqrt{\varepsilon_n}} \left[\int_\Omega |\mathrm{d}w - \mathrm{d}v|^q \, dx \right]^{\frac{1}{q}}.$$

Thanks to Fatou Lemma, to the fact that $j(x, \xi) \geq 0$, and to the fact that f belongs to $W^{-1,q'}(\Omega)$ being $q^* = m'$, we have that J is strongly lower semicontinuous on $W_0^{1,q}(\Omega)$.

Thus, in view of Lemma 1.2, there exists a sequence $\{u_n\}$ in $W_0^{1,q}(\Omega)$ such that

$$\left[\int_\Omega |\mathrm{d}u_n - \mathrm{d}\bar{u}_n|^q \, dx \right]^{\frac{1}{q}} \leq \sqrt{\varepsilon_n}\,,$$

which proves (4), and

$$J(u_n) \leq J(\bar{u}_n) \leq \inf_{v \in W_0^{1,p}(\Omega)} J(v) + \varepsilon_n\,, \tag{8}$$

$$J(u_n) \leq J(w) + \sqrt{\varepsilon_n} \left[\int_\Omega |\mathrm{d}u_n - \mathrm{d}w|^q \, dx \right]^{\frac{1}{q}}, \qquad \forall w \in W_0^{1,q}(\Omega)\,. \tag{9}$$

Using the growth properties of J we have that u_n is bounded in $W_0^{1,p}(\Omega)$; indeed, by (1), we have

$$\alpha \int_\Omega |\nabla u_n|^p \, dx \leq \int_\Omega j(x, \nabla u_n) \, dx \leq \int_\Omega f(x)\, u_n \, dx + \inf_{v \in W_0^{1,p}(\Omega)} J(v) + \varepsilon_n\,,$$

which implies that the sequence $\{u_n\}$ is bounded in $W_0^{1,p}(\Omega)$ since f belongs to $W^{-1,p'}(\Omega)$. Thus, up to subsequences, still denoted by $\{u_n\}$, there exists a function u in $W_0^{1,p}(\Omega)$ such that

$$u_n \to u \qquad \text{weakly in } W_0^{1,p}(\Omega) \text{ and almost everywhere in } \Omega. \tag{10}$$

By the weak lower semicontinuity of J on $W_0^{1,p}(\Omega)$, and by (8), u is a minimum of J on this space.

Moreover, choosing $w = u_n - t\,\psi$ in (9), where t is a positive real number and ψ is a function in $W_0^{1,p}(\Omega)$, we obtain

$$J(u_n - t\,\psi) - J(u_n) + \sqrt{\varepsilon_n}\,t\left[\int_\Omega |\mathrm{d}\psi|^q\,dx\right]^{\frac{1}{q}} \geq 0\,.$$

Dividing by t, and letting t tend to zero, we get, since J is differentiable,

$$-\langle J'(u_n), \psi\rangle + \sqrt{\varepsilon_n}\left[\int_\Omega |\mathrm{d}\psi|^q\,dx\right]^{\frac{1}{q}} \geq 0\,,$$

so that

$$\langle J'(u_n), \psi\rangle \leq \sqrt{\varepsilon_n}\left[\int_\Omega |\mathrm{d}\psi|^q\,dx\right]^{\frac{1}{q}}. \tag{11}$$

Recalling that $J'(u) = 0$ since u is a minimum, we have

$$\langle J'(u_n) - J'(u), \psi\rangle \leq \sqrt{\varepsilon_n}\left[\int_\Omega |\mathrm{d}\psi|^q\,dx\right]^{\frac{1}{q}},$$

for every ψ in $W_0^{1,p}(\Omega)$. Observe that

$$\langle J'(u_n), \psi\rangle = \int_\Omega a(x, \mathrm{d}u_n)\mathrm{d}\psi\,dx - \int_\Omega f(x)\psi\,dx. \tag{12}$$

Choosing $\psi = u_n - u$, it is easy to prove (5) using (2). In order to prove (6), let $k > 0$, define $A_{k,n} = \{|u_n - u| \geq k\}$, and choose $\psi = G_k(u_n - u)$; we obtain, by (2), and by Hölder inequality,

$$\begin{aligned}
\alpha \int_\Omega |\mathrm{d}G_k(u_n - u)|^p\,dx &\leq \sqrt{\varepsilon_n}\left[\int_\Omega |\mathrm{d}G_k(u_n - u)|^q\,dx\right]^{\frac{1}{q}} \\
&\leq \sqrt{\varepsilon_n}\left[\left(\int_\Omega |\mathrm{d}G_k(u_n - u)|^p\,dx\right)^{\frac{q}{p}} \mathrm{meas}\,(A_{k,n})^{1-\frac{q}{p}}\right]^{\frac{1}{q}} \\
&= \sqrt{\varepsilon_n}\left[\int_\Omega |\mathrm{d}G_k(u_n - u)|^p\,dx\right]^{\frac{1}{p}} \mathrm{meas}\,(A_{k,n})^{\frac{1}{q}-\frac{1}{p}}\,,
\end{aligned}$$

which in turn yields

$$\alpha \left(\int_{\Omega} |\mathrm{d}G_k(u_n - u)|^p \, dx \right)^{\frac{1}{p'}} \leq \sqrt{\varepsilon_n} \, \mathrm{meas}\,(A_{k,n})^{\frac{1}{q} - \frac{1}{p}} .$$

Using Sobolev inequality, and choosing $h > k$ we arrive after straightforward passages, to

$$(h-k)^p \, \mathrm{meas}\,(A_{h,n})^{\frac{p}{p^*}} \leq C_1 \, \varepsilon_n^{\frac{p'}{2}} \, \mathrm{meas}\,(A_{k,n})^{(\frac{1}{q} - \frac{1}{p})p'} ,$$

which implies

$$\mathrm{meas}\,(A_{h,n}) \leq C_2 \, \frac{\varepsilon_n^{\frac{p^*}{p} \frac{p'}{2}}}{(h-k)^{p^*}} \mathrm{meas}\,(A_{k,n})^{\frac{p^*}{p}(\frac{1}{q} - \frac{1}{p})p'} .$$

Note that (7) implies that

$$\frac{p^*}{p} \left(\frac{1}{q} - \frac{1}{p} \right) p' > 1 ,$$

so that, by Lemma 4.1 of [9],

$$\|u_n - u\|_{L^\infty(\Omega)} \leq C_3 \, \varepsilon_n^A ,$$

for some positive constant A depending only on p and N. Recalling that ε_n converges to zero, we have the result. □

Remark 2.1. Assumption (3) was used only to ensure that the functional J is lower semicontinuous on $W_0^{1,q}(\Omega)$. Since the terms with f "cancel out" when calculating $J'(u_n) - J'(u)$, the summability of f is not necessary to prove that $u_n - u$ belongs to $L^\infty(\Omega)$.

Remark 2.2. We remark that from (11), choosing ψ and $-\psi$ it follows that u_n satisfies

$$-\sqrt{\varepsilon_n} \left[\int_{\Omega} |\mathrm{d}\psi|^q \, dx \right]^{\frac{1}{q}} \leq \langle J'(u_n), \psi \rangle \leq \sqrt{\varepsilon_n} \left[\int_{\Omega} |\mathrm{d}\psi|^q \, dx \right]^{\frac{1}{q}} . \tag{13}$$

Thus,

$$\langle J'(u_n) - J'(u_m), \psi \rangle \leq \sqrt{\varepsilon_n} \left[\int_{\Omega} |\mathrm{d}\psi|^q \, dx \right]^{\frac{1}{q}} + \sqrt{\varepsilon_m} \left[\int_{\Omega} |\mathrm{d}\psi|^q \, dx \right]^{\frac{1}{q}}$$

The choice of $\psi = G_k(u_n - u_m)$, and the same steps in the proof of Theorem 1.3, yield

$$\|u_n - u_m\|_{L^\infty(\Omega)} \le c\,(\varepsilon_n + \varepsilon_m)^A . \tag{14}$$

Note that we cannot say that $\{u_n\}$ is a Cauchy sequence in $L^\infty(\Omega)$, since the functions u_n may not belong to $L^\infty(\Omega)$, even if the difference of two of them is bounded. However, passing to the limit in (14) as m tends to infinity, the almost everywhere convergence (10) implies that

$$\|u_n - u\|_{L^\infty(\Omega)} \le c\,\varepsilon_n^A , \tag{15}$$

which implies that the functions u_n belong to $L^\infty(\Omega)$, since $u \in L^\infty(\Omega)$, and that the sequence $\{u_n\}$ uniformly converges to u. In other words, Theorem 1.3 can also be proved starting from (14).

3 The impact of a lower order term

Let the integral functional J be defined now by

$$J(v) = \int_\Omega j(x, \nabla v)\,dx + \frac{1}{2}\int_\Omega [f(x) - v]^2\,dx, \qquad v \in W_0^{1,p}(\Omega) \cap L^2(\Omega), \tag{16}$$

where

$$f \in L^2(\Omega). \tag{17}$$

Note that $W_0^{1,p}(\Omega) \cap L^2(\Omega) = W_0^{1,p}(\Omega)$ if $p \ge \frac{2N}{N+2}$. Since both $j(x, \nabla v)$ and $[f(x) - v]^2$ are positive, J is lower semicontinuous on $W_0^{1,q}(\Omega)$, for every $q \ge 1$.

Note that any minimum u of J does not belong to $L^\infty(\Omega)$, if $2 < \frac{N}{p}$, i.e., if $p < \frac{N}{2}$.

Then the minimizing sequence $\{u_n\}$ built after $\{\bar{u}_n\}$ using the ε-variational principle satisfies (11) with $q = 1$:

$$\langle J'(u_n), \psi\rangle \le \sqrt{\varepsilon_n}\int_\Omega |d\psi|\,dx, \qquad \forall\, w \in W_0^{1,1}(\Omega) .$$

Observe that now

$$\langle J'(u_n), \psi\rangle = \int_\Omega a(x, du_n)d\psi\,dx + \int_\Omega u_n(x)\psi\,dx - \int_\Omega f(x)\psi\,dx. \tag{18}$$

We can follow the same steps as in Remark 2.2 in order to prove inequalities (14) and (15), but now the assumption on f does not imply that $u \in L^\infty(\Omega)$. Therefore, in (15) the function u_n and its limit u may not belong to $L^\infty(\Omega)$; nevertheless, their difference belongs to $L^\infty(\Omega)$ and tends to zero in that space.

References

1. Boccardo, L., Ferone, V., Fusco, N., Orsina, L.: Regularity of minimizing sequences for functionals of the calculus of variations via the Ekeland principle. Differ. Integr. Equ. **12**(1), 119–135 (1999)
2. Boccardo, L., Giachetti, D.: A nonlinear interpolation result with application to the summability of minima of some integral functionals. Discrete Continuous Dyn. Syst. Ser. B **11**(1), 31–42 (2009)
3. de Figueiredo, D.G.: Lectures on the Ekeland variational principle with applications and detours. Tata Institute of Fundamental Research Lectures on Mathematics and Physics, vol. 81. Published for the Tata Institute of Fundamental Research, Bombay; by Springer, Berlin (1989)
4. Ekeland, I.: On the variational principle. J. Math. Anal. Appl. **47**, 324–353 (1974)
5. Ekeland, I.: Nonconvex minimization problems. Bull. Am. Math. Soc. (N.S.) **1**(3), 443–474 (1979)
6. Ekeland, I., Temam, R.: Analyse convexe et problèmes variationnels. Collection Études Mathématiques. Dunod; Gauthier-Villars, Paris-Brussels-Montreal, Que. (1974)
7. Ladyženskaja, O.A., Uralceva, N.N.: Équations aux dérivées partielles de type elliptique. Traduit par G. Roos. Monographies Universitaires de Mathématiques, vol. 31. Dunod, Paris (1968)
8. Leray, J., Lions, J.-L.: Quelques résulatats de Višik sur les problèmes elliptiques nonlinéaires par les méthodes de Minty-Browder. Bull. Soc. Math. Fr. **93**, 97–107 (1965)
9. Stampacchia, G.: Le problème de Dirichlet pour les équations elliptiques du second ordre à coefficients discontinus. Ann. Inst. Fourier (Grenoble) **15**(fasc. 1), 189–258 (1965)

Waveguide solutions for a nonlinear Schrödinger equation with mixed dispersion

Denis Bonheure and Robson Nascimento

Ao nosso amigo Djairo com admiração

Mathematics Subject Classification (2010): 35J60, 35B25, 35B40, 35J10

1 Introduction

The standard model for propagation of laser beams is the $2D$ Schrödinger equation with Kerr nonlinearity

$$i\partial_t\psi + \Delta\psi + |\psi|^2\psi = 0, \quad \psi(x,y,0) = \psi_0(x,y).$$

It is well known that this equation can become singular at finite time, see, for instance, [13] and the classical references therein. Karpman and Shagalov [16] studied the regularization and stabilization effect of a small fourth-order dispersion, namely they considered the equation

$$i\partial_t\psi + \Delta\psi + |\psi|^{2\sigma}\psi - \gamma\Delta^2\psi = 0, \tag{1}$$

for some $\gamma > 0$, the equation being now considered in $[0,\infty[\times\mathbb{R}^N$, $N \geq 1$. One of their results shows, by help of some stability analysis and numerical computations, that when $N\sigma \leq 2$, the waveguide solutions are stable for all γ and when $2 < N\sigma < 4$, they are stable for small values of γ. This result shows that when adding a small fourth-order dispersion term, a new critical exponent/dimension appears. In particular, the Kerr nonlinearity becomes subcritical in dimension 2 and 3 which is obviously an important feature of this extended model.

D. Bonheure (✉) • R. Nascimento
Département de Mathématique, Université Libre de Bruxelles, CP 214,
Boulevard du Triomphe, 1050 Bruxelles, Belgium
e-mail: denis.bonheure@ulb.ac.be; robson.nascimento@ulb.ac.be

A.N. Carvalho et al. (eds.), *Contributions to Nonlinear Elliptic Equations and Systems*, Progress in Nonlinear Differential Equations and Their Applications 86,
DOI 10.1007/978-3-319-19902-3_4

In [13], Fibich et al. have motivated the study of (1) by recalling that NLS (the nonlinear Schrödinger equation) arises from NLH (the nonlinear Helmholtz equation) as a paraxial approximation. But since NLS can become singular at a finite time, this suggests that some of the small terms, neglected in the paraxial approximation, plays in fact an important role to prevent the blow up. The natural question addressed by Fibich et al. is therefore whether nonparaxiality prevents the collapse. The small fourth-order dispersion coefficient γ is then shown to be part of the nonparaxial correction to NLS.

In [13], Fibich et al. showed the role of the new critical exponent $\sigma = 4/N$ in the global existence in time when applying the arguments of Weinstein [25]. The necessary Strichartz estimates follow from Ben-Artzi et al. [1]. A necessary condition for existence of waveguide solutions is given in [13, Lemma 4.1], see also the Derrick-Pohozahev identity in Section 6.

The purpose of this short note is to show that classical tools, available in the literature, allow to state the existence and some qualitative properties of least energy waveguide solutions. In particular, a small fourth-order dispersion coefficient does not affect the symmetry, uniqueness and nondegeneracy of the least energy waveguide solution at least for a Kerr nonlinearity in dimension $N \leq 3$.

From now on, we focus on standing wave solutions of (1), referred to as waveguide solutions in nonlinear optics, namely on solutions of (1) of the form

$$\psi(t,x) = \exp(i\alpha t)u(x).$$

This ansatz yields the semilinear elliptic equation

$$\gamma\Delta^2 u(x) - \Delta u(x) + \alpha u(x) = |u|^{2\sigma} u(x), \quad x \in \mathbb{R}^N. \tag{2}$$

By scaling the solutions as $v(x) = u(\frac{x}{\sqrt[4]{\gamma}})$, it is equivalent to consider the equation

$$\Delta^2 v(x) - \beta\Delta v(x) + \alpha v(x) = |v|^{2\sigma} v(x), \quad x \in \mathbb{R}^N. \tag{3}$$

where $\beta = \frac{1}{\sqrt{\gamma}}$.

It is standard that least energy solutions can be obtained by considering the minimization problem

$$m_{\mathbb{R}^N} := \inf_{u \in M_{\mathbb{R}^N}} J_{\mathbb{R}^N}(u) \tag{4}$$

where

$$J_{\mathbb{R}^N}(u) = \int_{\mathbb{R}^N} (|\Delta u|^2 + \beta|\nabla u|^2 + \alpha|u|^2)\, dx \tag{5}$$

and

$$M_{\mathbb{R}^N} := \{u \in H^2(\mathbb{R}^N) : \int_{\mathbb{R}^N} |u|^{2\sigma+2}\, dx = 1\}.$$

Indeed, if $u \in M_{\mathbb{R}^N}$ achieves the infimum $m = m_{\mathbb{R}^N}$, then u weakly solves

$$\Delta^2 u - \beta \Delta u + \alpha u = m|u|^{2\sigma} u. \tag{6}$$

Henceforth, if $m > 0$, then $v = (m)^{\frac{1}{2\sigma}} u$ solves (3). Moreover v is a least energy solution in the sense that it minimizes the action functional $E : H^2(\mathbb{R}^N) \to \mathbb{R}$ defined by setting

$$E(u) := \frac{1}{2} J_{\mathbb{R}^N}(u) - \frac{1}{2\sigma + 2} \int_{\mathbb{R}^N} |u|^{2\sigma+2}\, dx$$

among the set of (H^2 or smoother) solutions or equivalently within the Nehari manifold

$$\{u \in H^2(\mathbb{R}^N) : E'(u)(u) = 0\}.$$

We then prove the following results.

Theorem 1.1. *Assume $\alpha > 0$, $\beta > -2\sqrt{\alpha}$ and $2 < 2\sigma + 2 < \frac{2N}{N-4}$ if $N \geq 5$. Then problem* (3) *has a nontrivial least energy solution. If $\beta \geq 2\sqrt{\alpha}$, then any least energy solution does not change sign, is radially symmetric around some point and strictly radially decreasing.*

An existence statement (as well as the information on the sign of the minimizer) is also given in Section 3 and Section 4 when the equation is considered in a bounded domain with Navier boundary conditions. The symmetry properties of the solutions that match the symmetries of the domain are discussed in Section 4.

When β is large, the Laplacian is the driven term in the differential operator in (3) and we therefore expect to recover the uniqueness (up to translations) of the least energy solution. By scaling, we can discuss this issue by looking at least energy solutions of (2) for small γ. As a preliminary observation, we prove the strong convergence in H^1 to the unique least energy solution of NLS.

Theorem 1.2. *Assume $2 < 2\sigma + 2 < \frac{2N}{N-2}$ if $N \geq 3$. If $\gamma_k \to 0$ and u_k is a sequence of least energy solutions of* (2)*, then $(u_k)_k$ converges (after possible translations) in H^1 to u_0, where u_0 is the unique positive (radially symmetric) solution of the limit problem (2) with $\gamma = 0$.*

The positive solution of (2) with $\gamma = 0$ is unique up to translations. To ensure uniqueness, we have assumed that u_0 is the positive solution radially decreasing around 0. For the physical model (2) with $\sigma = 1$ in dimension $N \leq 3$, we can improve this convergence to strong convergence in H^2. The nondegeneracy of the least energy waveguide of NLS allows then to use the Implicit Function Theorem to prove uniqueness for small γ.

Theorem 1.3. *Assume $N \leq 3$ and $\sigma = 1$. Then there exists $\gamma_0 > 0$ such that if $0 < \gamma < \gamma_0$,* (2) *has a unique least energy solution (up to translations). Fixing its maximum at the origin, this solution is radially symmetric and strictly radially decreasing.*

An equivalent statement can be proved for the Navier boundary value problem in a ball (and a weaker statement holds for other bounded domains), see Section 6.

In the H^1 critical or supercritical regime, the least energy solution should disappear at the limit $\gamma \to 0$. In fact, if $\frac{2N}{N-2} \leq 2\sigma + 2 < \frac{2N}{N-4}$, $N \geq 5$, the least energy solutions are unbounded in H^2 when $\gamma \to 0$, see Section 6.

In contrast with Theorem 1.1, when β is small in (3), some of the usual properties of the least energy solution of NLS cannot hold. Namely, if one can prove that any least energy solution is radial in that case, then oscillations arise at infinity. These oscillations were suggested in [13]. We focus again on the model equation (2) with $\sigma = 1$ in dimension $N \leq 3$. We prove that least energy solutions among radial solutions do oscillate at infinity.

Theorem 1.4. *Suppose that $-2\sqrt{\alpha} < \beta < 2\sqrt{\alpha}$ and $N \leq 3$. Then every radial least energy solution of* (3) *with $\sigma = 1$ is sign-changing.*

This statement shows that when $\beta < 2\sqrt{\alpha}$, least energy solutions cannot be radial and monotone in contrast with the case $\beta \geq 2\sqrt{\alpha}$. We point out that on a bounded domain, we are not aware of an equivalent statement.

The paper is organized as follows. Section 2 deals with the functional framework and the formulation of the problem on a bounded domain. In Section 3, we prove the existence of a least energy solution in the whole space as well as in bounded domains. In Section 4, we consider the qualitative properties for large β. Section 5 is dedicated to the proof of Theorem 1.2 and Theorem 1.3 while Section 6 contains the proof of Theorem 1.4. In the last section, we give some concluding remarks.

Notes added in proofs: We thank Jean-Claude Saut for bringing to our attention the reference [5] which deals with an anisotropic mixed dispersion NLS also proposed in [13]. We believe that some arguments from [5] can be used to obtain the exponential decay of the ground state at least in some particular cases.

We also mention the very recent preprint [6] where the first theoretical proof of blow-up is obtained for the biharmonic NLS as well as a new *Fourrier rearrangement* is proposed in the Appendix. This rearrangement decreases the L^2-norm of $(-\Delta u)^s$ for every $s \geq 0$ and is therefore adequate to deal with polyharmonic as well as fractional equations. Applied to our problem, it completes Theorems 1.1 and 1.4 in the following way. Assuming $\beta \geq 0$ and $\sigma \in \mathbb{N}_0$ (including therefore the physical case $\sigma = 1$), there is a ground state solution of (C) which is radially symmetric. As a consequence of Theorem 1.4, assuming $\sigma \in \mathbb{N}_0$ and $0 \leq \beta < 2\sqrt{\alpha}$, this ground state is radially oscillatory at infinity. When σ is not an integer, the radial symmetry remains an open question in the range $\beta < 2\sqrt{\alpha}$ though the natural conjecture is that radial symmetry holds for every σ and every β in the range covered by Theorem 1.1.

2 Functional framework

In this section, we settle the functional setting. The natural space for (2) and (3) is $H^2(\mathbb{R}^N)$ or $H^2(\Omega) \cap H_0^1(\Omega)$ when we consider the boundary value problem in a bounded domain $\Omega \subset \mathbb{R}^N$ with Navier boundary conditions, namely

$$(P_\beta) \qquad \begin{cases} \Delta^2 u - \beta \Delta u + \alpha u = |u|^{2\sigma} u, & \text{in } \Omega, \\ u = \Delta u = 0, & \text{on } \partial\Omega. \end{cases}$$

We therefore set $H_\Omega := H^2(\Omega) \cap H_0^1(\Omega)$ and $H_{\mathbb{R}^N} := H^2(\mathbb{R}^N)$. We introduce the following conditions on α and β:

$(A1)$ $\alpha > 0$ and $\beta > -2\sqrt{\alpha}$;
$(A1')$ $\alpha > -\beta\lambda_1(\Omega) - \lambda_1^2(\Omega)$ and $-2\lambda_1(\Omega) < \beta$;

where $\lambda_1(\Omega)$ stands for the first eigenvalue of $-\Delta$ in $H_0^1(\Omega)$ when Ω is a bounded domain. Observe that when $|\Omega|$ is large, $\lambda_1(\Omega)$ is small. If β is negative, $(A1')$ is then more restrictive than $(A1)$. The following lemma follows from standard computations.

Lemma 2.1. *Assume Ω is a bounded smooth domain and $(A1)$ or $(A1')$ holds. Then H_Ω is a Hilbert space endowed with the inner product defined through*

$$\langle u, v \rangle = \int_\Omega (\Delta u \Delta v + \beta \nabla u \nabla v + \alpha u v)\, dx \quad \forall\, u, v \in H_\Omega.$$

Proof. From H^2 elliptic regularity [14, 18], we know that if $u \in H^2(\Omega) \cap H_0^1(\Omega)$, then

$$\|u\|_{H^2} \leq C \|\Delta u\|_{L^2}$$

for some $C > 0$ depending on Ω, so that H_Ω is a Hilbert space endowed with the inner product

$$\langle u, v \rangle_{H_\Omega} = \int_\Omega \Delta u \Delta v\, dx \quad \forall\, u, v \in H_\Omega.$$

It will be enough to show that there exists a constant $C > 0$ such that

$$\int_\Omega (|\Delta u|^2 + \beta |\nabla u|^2 + \alpha |u|^2)\, dx \geq C \|u\|_{H_\Omega} \qquad \forall\, u \in H_\Omega. \tag{7}$$

Obviously the inequality (7) holds true if we have $\alpha \geq 0$ and $\beta \geq 0$. For $u \in H_\Omega$, we can apply Young's inequality to obtain

$$
\begin{aligned}
\|u\|^2 &= \int_\Omega (|\Delta u|^2 + \beta|\nabla u|^2 + \alpha|u|^2)\,dx \\
&= \int_\Omega (|\Delta u|^2 - \beta u\Delta u + \alpha|u|^2)\,dx \\
&\geq \left(1 + \frac{\beta}{2\epsilon}\right)\int_\Omega |\Delta u|^2\,dx + \left(\alpha + \frac{\beta\epsilon}{2}\right)\int_\Omega |u|^2\,dx
\end{aligned}
\tag{8}
$$

for every $\epsilon > 0$. We have to distinguish two cases. If we can choose $\epsilon > 0$ such that both terms in the right-hand side of (8) are positive, then we are done. This ends the proof if $\beta > -2\sqrt{\alpha}$, namely if $(A1)$ holds. If

$$1 + \frac{\beta}{2\epsilon} > 0 \quad \text{and} \quad \alpha + \frac{\beta\epsilon}{2} < 0,$$

we write

$$\|u\|^2 \geq \left(1 + \frac{\beta}{2\epsilon}\right)\left[\int_\Omega |\Delta u|^2\,dx + g(\epsilon)\int_\Omega |u|^2\,dx\right],$$

where

$$g(\epsilon) = \frac{\alpha + \beta\epsilon/2}{1 + \beta/2\epsilon}.$$

Recalling Poincaré inequality

$$\int_\Omega |\Delta u|^2\,dx \geq \lambda_1^2(\Omega)\int_\Omega u^2\,dx \qquad \forall\, u \in H_\Omega,$$

we can complete the proof if

$$g(\epsilon) > -\lambda_1^2(\Omega)$$

for some $\epsilon > 0$. When $\beta > -2\lambda_1(\Omega)$, this condition can be fulfilled if

$$\alpha > -\beta\lambda_1(\Omega) - \lambda_1^2(\Omega)$$

while if $\beta \leq -2\lambda_1(\Omega)$, we recover the condition

$$-2\sqrt{\alpha} < \beta.$$

□

In the case $\Omega = \mathbb{R}^N$, the same arguments show that $(A1)$ implies

$$\langle u, v \rangle = \int_\Omega (\Delta u \Delta v + \beta \nabla u \nabla v + \alpha u v)\, dx$$

is a scalar product on $H_{\mathbb{R}^N}$. Elliptic regularity can be used here to ensure that

$$\left(1 + \frac{\beta}{2\epsilon}\right) \int_{\mathbb{R}^N} |\Delta u|^2\, dx + \left(\alpha + \frac{\beta\epsilon}{2}\right) \int_{\mathbb{R}^N} |u|^2\, dx$$

is a norm on $H^2(\mathbb{R}^N)$ as soon as $1 + \frac{\beta}{2\epsilon} > 0$ and $\alpha + \frac{\beta\epsilon}{2} > 0$. This yields the following lemma.

Lemma 2.2. *Assume that* (A1) *holds. Then the bilinear form*

$$\langle u, v \rangle = \int_\Omega (\Delta u \Delta v + \beta \nabla u \nabla v + \alpha u v)\, dx \quad \forall\, u, v \in H_{\mathbb{R}^N},$$

is an inner product on $H_{\mathbb{R}^N}$.

3 Existence of minimizers

In this section, we handle the minimization problem (4). We start with the simpler case of a bounded domain. In this case, the minimization problem writes

$$m_\Omega := \inf_{u \in M_\Omega} J_\Omega(u)$$

where

$$J_\Omega(u) = \int_\Omega (|\Delta u|^2 + \beta |\nabla u|^2 + \alpha |u|^2)\, dx$$

and

$$M_\Omega := \{u \in H_\Omega : \int_\Omega |u|^{2\sigma+2}\, dx = 1\}.$$

In the case of a bounded domain, it is standard to prove that m_Ω is achieved when $2\sigma + 2$ is a subcritical exponent because J_Ω is the square of a norm on H_Ω and we can rely on the compactness of the embedding of H_Ω into $L^{2\sigma+2}(\Omega)$. Moreover, since m_Ω is clearly positive, we deduce that $v = (m_\Omega)^{\frac{1}{2\sigma}} u$ solves (P_β). Moreover v is a least energy solution in the sense that it minimizes the action functional $E_\Omega : H_\Omega \to \mathbb{R}$ defined by

$$E_\Omega(u) := \frac{1}{2} J_\Omega(u) - \frac{1}{2\sigma+2} \int_\Omega |u|^{2\sigma+2}\, dx$$

among the set of (H^2 or smoother) solutions or equivalently within the Nehari manifold

$$\{u \in H_\Omega : E'_\Omega(u)(u) = 0\}.$$

Theorem 3.1. *Assume Ω is a bounded smooth domain and (A1) or (A1′) holds. Suppose moreover that $2 < 2\sigma + 2 < \frac{2N}{N-4}$ if $N \geq 5$. Then problem (P_β) has a nontrivial least energy solution.*

To handle the case of $\Omega = \mathbb{R}^N$, since we cannot use sign information, nor symmetry, we follow the celebrated method of concentration-compactness of P.L. Lions. We give a sketchy proof since classical arguments apply. All the details can easily be reconstructed from Kavian [17, Chapitre 8 - Exemple 8.5] with minor and obvious modifications with respect to the case treated therein.

Proof (Proof of the existence part in Theorem 1.1.). We introduce

$$M_\lambda = \{u \in H^2(\mathbb{R}^N) : \int_{\mathbb{R}^N} |u|^{2\sigma+2}\, dx = \lambda\}$$

where $\lambda > 0$ is fixed and we consider the minimization problem

$$m_\lambda := \inf_{u \in M_\lambda} J_{\mathbb{R}^N}(u)$$

where $J_{\mathbb{R}^N}(u)$ is defined as in (5).

Let $(u_k)_k \subset M_\lambda$ be such that $J_{\mathbb{R}^N}(u_k) \to m_\lambda$. Then, $(u_k)_k$ is bounded in $H^2(\mathbb{R}^N)$ and $\int_{\mathbb{R}^N} |u_k|^{2\sigma+2} = \lambda$. Thus, we can apply P.L. Lions' concentration-compactness lemma to the sequence $(\rho_k)_k = (\int_{\mathbb{R}^N} |u_k|^{2\sigma+2})_k$, see [21, Lemma I. 1]. Since $m_\lambda = \lambda^{\frac{1}{\sigma+1}} m_1$, we have $m_\lambda > 0$ for all $\lambda > 0$ and therefore, for all $R > 0$, the sequence

$$Q_k(R) := \sup_{y \in \mathbb{R}^N} \int_{B_R(y)} |u_k(x)|^{2\sigma+2}\, dx$$

does not converge to zero. Namely, vanishing is ruled out.

Since $2\sigma + 2 > 2$, we have, for $0 < \theta < \lambda$,

$$\lambda^{\frac{1}{\sigma+1}} < \theta^{\frac{1}{\sigma+1}} + (\lambda - \theta)^{\frac{1}{\sigma+1}},$$

which yields

$$m_\lambda < m_\theta + m_{\lambda-\theta}, \qquad \forall\, \theta \in]0,\lambda[. \tag{9}$$

Then dichotomy is ruled out using classical truncation arguments.

Therefore, the compactness holds for ρ_k, i.e., going to a subsequence of (u_k) if necessary, there exists a sequence $(y^k) \subset \mathbb{R}^N$ such that for every $\varepsilon > 0$ there exists $R > 0$ such that

$$\int_{B_R(y^k)} |u_k|^{2\sigma+2}\, dx > \lambda - \varepsilon.$$

Setting $w_k(x) := u_k(x + y^k)$, we have that (w_k) is also a minimizing sequence for m_λ. Then, up to a subsequence, w_k weakly converges in $H^2(\mathbb{R}^N)$ to $w \in M_\lambda$ and $J_{\mathbb{R}^N}(w) = m_\lambda$. This concludes the proof of the existence in Theorem 1.1. □

Remark 3.2. When $\beta \geq 2\sqrt{\alpha}$, we can avoid the use of the concentration-compactness lemma. Indeed, take a minimizing sequence $(u_k)_k \subset H^2(\mathbb{R}^N)$ for m. Then, let us set $f_k := -\Delta u_k + \beta u_k/2$ and define $v_k \in H^2(\mathbb{R}^N)$ to be the strong solution of $-\Delta v_k + \beta v_k/2 = |f_k|^*$ in $\mathbb{R}^N$, where $|f_k|^*$ denotes the Schwarz symmetrization of $|f_k|$. Thus for each $k \in \mathbb{N}$, we have $v_k \in H^2_{rad}(\mathbb{R}^N)$ which is the space of H^2 functions that are radially symmetric around the origin. Then a particular case of [3, Lemma 3.4] see also [4] implies

$$J\left(\frac{v_k}{|v_k|_{2\sigma+2}}\right) = \frac{\int_{\mathbb{R}^N} (-\Delta v_k + \beta v_k/2)^2\, dx - (\beta^2/4 - \alpha)\int_{\mathbb{R}^N} v_k^2\, dx}{|v_k|^2_{2\sigma+2}}$$
$$\leq \frac{\int_{\mathbb{R}^N} (-\Delta u_k + \beta u_k/2)^2\, dx - (\beta^2/4 - \alpha)\int_{\mathbb{R}^N} u_k^2\, dx}{|u_k|^2_{2\sigma+2}}.$$

Using the compact embedding of $H^2_{rad}(\mathbb{R}^N)$ into $L^{2\sigma+2}(\mathbb{R}^N)$, see, for instance, [20, Théorème II.1], it follows that $(v_k)_k$ weakly converges in H^2 to some $v \in M$ and the remaining arguments are standard.

4 Sign and symmetry

In order to investigate the symmetry properties of a fourth order equation with Navier boundary conditions or in the whole space, it is natural to ask if the equation may be rewritten as a cooperative system. If this is the case, then the moving plane procedure applies, see the work of Troy [23] in the case of a bounded domain or de Figueiredo-Yang [10] (if we assume exponential decay) and Busca-Sirakov [7] (without assuming exponential decay) when $\Omega = \mathbb{R}^N$. Observe that when $\alpha > 0$ and $|\beta| \geq 2\sqrt{\alpha}$, we can indeed write the equation as a cooperative system

$$-\Delta u + \frac{\beta}{2}u - v = 0, \quad -\Delta v + (\alpha - \frac{\beta^2}{4})u + \frac{\beta}{2}v = |u|^{2\sigma}u.$$

To prove that least energy solutions do not change sign, we use the minimality combined to the classical maximum principle for a single equation. The argument goes back to van der Vorst, see, for instance, [24]. We sketch it for completeness to emphasize the role of the assumption $|\beta| \geq 2\sqrt{\alpha}$.

Lemma 4.1. *Assume that $|\beta| \geq 2\sqrt{\alpha}$ and $-\lambda_1(\Omega) < \beta/2$ if Ω is bounded or $\beta > 0$ if $\Omega = \mathbb{R}^N$. If $u \in H_\Omega$ is a minimizer of* (4)*, then*

$$u > 0 \quad \text{and} \quad -\Delta u + \beta u/2 > 0 \quad \text{in } \Omega,$$

or else

$$u < 0 \quad \text{and} \quad -\Delta u + \beta u/2 < 0 \quad \text{in } \Omega.$$

Proof. Let $w \in H_\Omega$ be such that

$$\begin{cases} -\Delta w + \beta w/2 = |-\Delta u + \beta u/2|, & \text{in } \Omega, \\ w = 0, & \text{on } \partial\Omega. \end{cases}$$

Then

$$-\Delta(w \pm u) + \beta(w \pm u)/2 \geq 0.$$

Using the strong maximum principle we know that u has a fixed sign if $-\Delta u + \beta/2\, u$ does not change sign. We then argue by contradiction, suppose that $-\Delta u + \beta u/2$ changes sign. Then $|-\Delta u + \beta u/2| \neq 0$ and the strong maximum principle implies that $w > |u|$. For convenience denote by $|\cdot|_{2\sigma+2}$ the $L^{2\sigma+2}$ norm in Ω. Therefore

$$J_\Omega\left(\frac{w}{|w|_{2\sigma+2}}\right) = \frac{\int_\Omega (-\Delta w + \beta w/2)^2\,dx - (\beta^2/4 - \alpha)\int_\Omega w^2\,dx}{|w|_{2\sigma+2}^2}$$

$$< \frac{\int_\Omega (-\Delta u + \beta u/2)^2\,dx - (\beta^2/4 - \alpha)\int_\Omega u^2\,dx}{|u|_{2\sigma+2}^2}$$

which contradicts the minimality of u. Observe that the last inequality holds because the numerator is nonnegative. □

Remark 4.2. In the case of a bounded domain Ω and $0 < \alpha \leq \lambda_1(\Omega)^2$, we then know the sign of the least energy solutions of (P_β) for values of $\beta \in (-2\lambda_1(\Omega), -2\sqrt{\alpha}] \cup [2\sqrt{\alpha}, \infty)$. For Ω bounded, we do not know if the least energy solutions change sign for $\beta \in (-2\sqrt{\alpha}, 2\sqrt{\alpha})$. Section 6 deals with the case $\Omega = \mathbb{R}^N$ under the assumption that the minimizer is radial.

Proof (Proof of Theorem 1.1 continued). Existence was proved in Section 3 while we just proved in Lemma 4.1 that any least energy solution does not change sign.

Writing $f(u,v) = (\frac{\beta^2}{4} - \alpha)u - \frac{\beta}{2}v + |u|^{2\sigma}u$ and $g(u,v) = v - \frac{\beta}{2}u$, the equation is equivalent to the cooperative system

$$\Delta u + g(u,v) = 0, \quad \Delta v + f(u,v) = 0.$$

We are in the setting of Busca-Sirakov [7] and [7, Theorem 2] applies. Observe that clearly u and v must be symmetric with respect to the same point. □

In the case of a bounded domain, we have proved so far the following result for (P_β).

Theorem 4.3. *Assume Ω is a bounded smooth domain and $(A1)$ or $(A1')$ holds. Suppose moreover that $2 < 2\sigma + 2 < \frac{2N}{N-4}$ if $N \geq 5$. Then problem (P_β) has a nontrivial least energy solution. If in addition $|\beta| \geq 2\sqrt{\alpha}$ and $-\lambda_1(\Omega) < \beta/2$, then any least energy solution does not change sign. If Ω is a ball, then any least energy solution is radially symmetric and strictly radially decreasing.*

Proof. Existence has been achieved in Theorem 3.1 while the sign information follows from Lemma 4.1. If Ω is a ball, the symmetry of the minimizer follows from [23, Theorem 1]. □

We point out that the condition $|\beta| \geq 2\sqrt{\alpha}$ is crucial to rewrite the problem (P_β) as a cooperative system. In fact, we can deal more generally with smooth bounded or unbounded domain Ω with some symmetries. Then the symmetry properties of the solutions of constant sign can be deduced from the moving plane method adapted to cooperative systems in [23].

5 The effect of a small fourth order dissipation

In this section, we study the behaviour of minimizers of (4) when the coefficient of fourth order dissipation tends to zero. We assume throughout the section that $\alpha > 0$ and we choose the norm on $H^1(\mathbb{R}^N)$ defined through

$$\|u\|_{H^1}^2 = \int_\Omega (|\nabla u|^2 + \alpha |u|^2)\, dx.$$

We recall that the problem

$$\Delta^2 v(x) - \beta \Delta v(x) + \alpha v(x) = |v|^{2\sigma} v(x), \quad x \in \mathbb{R}^N$$

is equivalent to

$$\gamma \Delta^2 u(x) - \Delta u(x) + \alpha u(x) = |u|^{2\sigma} u(x), \quad x \in \mathbb{R}^N.$$

by scaling the solutions as $u(x) = v(\frac{x}{\sqrt{\beta}})$ where $\gamma = 1/\beta^2$. As before we consider the associated minimization problem

$$m_\gamma = \inf_{u \in M} J_\gamma(u)$$

where

$$M = \{u \in H_\Omega : \int_\Omega |u|^{2\sigma+2}\, dx = 1\}$$

and

$$J_\gamma(u) = \int_\Omega (\gamma |\Delta u|^2 + |\nabla u|^2 + \alpha |u|^2)\, dx.$$

When $\Omega = B_R$ or $\Omega = \mathbb{R}^N$, the results of the previous sections imply that when $\gamma \leq \frac{1}{4\alpha}$, any minimizer is radially symmetric and strictly radially decreasing (after a possible translation in the case $\Omega = \mathbb{R}^N$). In the case $\Omega = \mathbb{R}^N$, we assume from now on that the maximum of any minimizer has been translated to the origin.

For $\gamma = 0$, the associated minimization problem is

$$m_0 = \inf_{u \in M_0} J_0(u)$$

where

$$M_0 = \{u \in H_0^1(\Omega) : \int_\Omega |u|^{2\sigma+2}\, dx = 1\}$$

and

$$J_0(u) = \int_\Omega (|\nabla u|^2 + \alpha |u|^2)\, dx.$$

Assume $2 < 2\sigma + 2 < \frac{2N}{N-2}$ if $N \geq 3$, $\Omega = B_R$ or $\Omega = \mathbb{R}^N$ and let u_0 be the unique minimizer of J_0 in M_0. We refer to [9, 15, 19] for the uniqueness property (in the case $\Omega = \mathbb{R}^N$, we fix the maximum of the solution at the origin to achieve uniqueness). We first prove that if $\gamma_k \to 0$, then any sequence $(u_k)_k$ of minimizer of J_{γ_k} converge strongly in H^1 to u_0. A similar statement obviously holds for other bounded domains except that uniqueness of the minimizer does not hold in general so that in the conclusion, we can only state that we have convergence to one minimizer, see Theorem 5.3.

Proposition 5.1. *Assume* $2 < 2\sigma + 2 < \frac{2N}{N-2}$ *if* $N \geq 3$, $\Omega = B_R$ *or* $\Omega = \mathbb{R}^N$. *There exists* $C > 0$ *such that for every* $\gamma > 0$, *we have*

$$m_0 \leq m_\gamma \leq m_0 + C\gamma.$$

Moreover, if $\gamma_k \to 0$ *and* $(u_k)_k$ *is a sequence such that* $J_{\gamma_k}(u_k) = m_{\gamma_k}$, *then* $u_k \to u_0$ *strongly in* H^1.

Proof. The estimate of m_γ is clear since by elliptic regularity, we easily infer that $u_0 \in H^2(\Omega)$. Therefore, we have

$$m_\gamma \leq J_\gamma(u_0) = \gamma \int_\Omega |\Delta u_0|^2 \, dx + J_0(u_0) \leq C\gamma + m_0,$$

whereas taking any minimizer u_γ for m_γ, we get

$$m_\gamma = J_\gamma(u_\gamma) = \gamma \int_\Omega |\Delta u_\gamma|^2 \, dx + J_0(u_\gamma) \geq m_0.$$

Let $\gamma_k \to 0$ and $(u_k)_k$ be a sequence of minimizers for $m_k := m_{\gamma_k}$. Then

$$\int_\Omega (|\nabla u_k|^2 + \alpha |u_k|^2) \, dx \leq m_k \leq m_0 + C\gamma_k \to m_0.$$

Since we know that u_k is a radial function, it follows that u_k is bounded in $H^1_{rad}(\Omega)$ - the space of H^1 functions that are radially symmetric around the origin—so that up to a subsequence, u_k converges weakly in H^1 to some $u \in M$. The strong convergence in $L^{2\sigma+2}$ when $\Omega = \mathbb{R}^N$ follows from the compact embedding of $H^1_{rad}(\mathbb{R}^N)$ into $L^{2\sigma+2}(\mathbb{R}^N)$, see [20, 22].

Now, by weak lower semi-continuity, we have

$$\begin{aligned} m_0 \leq \int_\Omega (|\nabla u|^2 + \alpha |u|^2) \, dx &\leq \liminf_{k \to \infty} \int_\Omega (|\nabla u_k|^2 + \alpha |u_k|^2) \, dx \\ &\leq \limsup_{k \to \infty} \int_\Omega (|\nabla u_k|^2 + \alpha |u_k|^2) \, dx = m_0. \end{aligned}$$

Hence the convergence is strong in H^1 and u is a minimizer for m_0. By uniqueness, $u = u_0$ and the whole sequence converges. □

In the model case with a Kerr nonlinearity in dimension $N \leq 3$, we can improve this convergence.

Proposition 5.2. *Assume $\Omega = \mathbb{R}^N$, $\sigma = 1$ and $N \leq 3$. If $\gamma_k \to 0$ and $(u_k)_k$ is a sequence such that $J_{\gamma_k}(u_k) = m_{\gamma_k}$, then $u_k \to u_0$ strongly in H^2.*

Proof. To fix the ideas, we deal with the case $N = 3$, $N = 2$ being similar. The starting point is an a priori bound in H^1 and the strategy is to end up with an a priori H^4-bound. We already know from Proposition 5.1 that u_k converges to u_0 strongly in H^1. To improve the convergence, we use the Euler-Lagrange equation

$$\gamma_k \Delta^2 u_k - \Delta u_k + \alpha u_k = m_k u_k^3,$$

where $m_k = m_{\gamma_k}$. We can assume $\gamma_k \leq 1$ and $m_k \in [m_0, m_0 + C]$.

Bound in H^1. Since u_k is a minimizer, we can assume

$$\|u_k\|_{H^1} \leq m_0 + C.$$

This also provides an a priori bound in L^q for every $q \in [2, 6]$.

Bound in H^2. We denote $v_k = -\gamma_k \Delta u_k$. Then v_k solves

$$-\Delta v_k + \frac{1}{\gamma_k} v_k = w_k, \tag{10}$$

where $w_k := m_k u_k^3 - \alpha u_k$. Since $J_{\gamma_k}(u_k) \leq m_0 + C$, we infer that $v_k \to 0$ strongly in L^2. In particular, $(v_k)_k$ is bounded in L^2. Observe also that $(w_k)_k$ is a priori bounded in L^2. Now, by elliptic regularity, we infer that $v_k \in H^2(\mathbb{R}^3)$ with a bound that does not depend on k. Indeed, since $\frac{1}{\gamma_k} \geq 1$, we get this a priori bound as in Krylov [18, Chapter 1, Theorems 6.4 & 6.5]. Now, from this a priori H^2-bound on $(v_k)_k$ and the Euler equation

$$-\Delta u_k + \alpha u_k = m_k u_k^3 + \Delta v_k, \tag{11}$$

we deduce that $(u_k)_k$ is a priori bounded in $H^2(\mathbb{R}^3)$ as well.

Bound in H^4. It is straightforward to check that the H^2-bound on u_k implies that $w_k \in L^2(\mathbb{R}^3)$ and $\Delta w_k \in L^2(\mathbb{R}^3)$. Then, elliptic regularity implies w_k is bounded in H^2 as well. Using again (10), we now infer that $v_k \in H^4$ with a bound independent of k, arguing as in Krylov for H^{m+2} regularity [18, Chapter 1, Theorem 7.5 & Corollary 7.6]. Looking at (11) again, we have that the right-hand side is bounded in H^2, whence $u_k \in H^4$ with a bound independent of k.

Conclusion. Observe now that we can use the equation (11) to conclude. Since $-\Delta v_k = \gamma_k \Delta^2 u_k \to 0$ strongly in L^2, we conclude that

$$m_k u_k^3 + \Delta v_k \to m_0 u_0^3$$

strongly in L^2 and elliptic regularity applied to (11) implies that the convergence of u_k to u_0 is actually strong in H^2. □

Now that we have proved the strong convergence in H^2 to the unique minimizer for $\gamma = 0$, we can use its non degeneracy to apply the Implicit Function Theorem. This yields Theorem 1.3.

Proof (Proof of Theorem 1.3). We start by setting $X := H^2_{rad}(\mathbb{R}^3)$ and $Y := H^{-2}(\mathbb{R}^3)$. Let $F : \mathbb{R}^+ \times X \to Y$ be the operator defined (in the sense of distributions) by

$$F(\gamma, u) = \gamma \Delta^2 u - \Delta u + \alpha u - |u|^2 u.$$

Namely, for every $v \in H^2(\mathbb{R}^3)$, we have

$$F(\gamma, u)(v) = \int_{\mathbb{R}^3} (\gamma \Delta u \Delta v + \nabla u \nabla v + \alpha u v - |u|^2 u v)\, dx.$$

Obviously $F(0, \sqrt{m_0} u_0) = 0$. Also, F is continuously differentiable in a neighbourhood of $(0, \sqrt{m_0} u_0)$ with $D_u F(\gamma, u) \in \mathscr{L}(X, Y)$ defined by

$$D_u F(\gamma, u)v = \gamma \Delta^2 v - \Delta v + \alpha v - 3|u|uv, \qquad \forall\, v \in X,$$

i.e.

$$D_u F(\gamma, u)v[w] = \int_{\mathbb{R}^3} (\gamma \Delta v \Delta w + \nabla v \nabla w + \alpha v w - 3|u|uvw)\, dx, \qquad \forall\, v, w \in X.$$

We thus have in the distributional sense

$$L(v) := D_u F(0, \sqrt{m_0} u_0)v = -\Delta v + \alpha v - 3 m_0 u_0^2 v.$$

It is well known that the kernel of L is of dimension 3 when considered in $H^2(\mathbb{R}^3)$ and it is spanned by the partial derivatives of u_0. In particular, the kernel of L restricted to $H^2_{rad}(\mathbb{R}^3)$ is trivial and $L : X \to Y$ is one-to-one. We refer, for instance, to [8, 15, 19]. Moreover, it follows from the Open Mapping Theorem that $L^{-1} : Y \to X$ is continuous.

Since the linear map L is a homeomorphism, we can apply the Implicit Function Theorem. Namely, there exists $\gamma_0 > 0$ and an open set $U_0 \subset X$ that contains $\sqrt{m_0} u_0$ such that for every $\gamma \in [0, \gamma_0[$, the equation $F(\gamma, u) = 0$ has a unique solution $u_\gamma \in U_0$ and the curve

$$\Gamma : [0, \gamma_0[\to H^2(\mathbb{R}^3) : \gamma \mapsto u_\gamma$$

is of class C^1.

Now suppose that the uniqueness of least energy solutions fails in every interval $(0, \gamma)$. We can then construct two sequences in M of least energy solutions along a sequence γ_k converging to 0. We call them $(u_k)_k$ and $(v_k)_k$ whereas m_k is their common energy. By assumption, $u_k \neq v_k$. Since $\gamma_k \to 0$, we know that u_k and v_k are radially symmetric. Since these two sequences converge in H^2 to u_0 as $k \to \infty$, we have

$$\sqrt{m_k} u_k, \sqrt{m_k} v_k \to \sqrt{m_0} u_0,$$

where the convergence is strong in H^2. Then, for k large enough, there exist two solutions of the equation $F(\gamma_k, u) = 0$ in U_0 with $\gamma_k < \gamma_0$. This is a contradiction and ends the proof. □

We now state the counterpart of Theorem 1.3 for the boundary value problem in a bounded domain $\Omega \subset \mathbb{R}^N$ with Navier boundary conditions, namely

$$(P_\gamma) \qquad \begin{cases} \gamma \Delta^2 u - \Delta u + \alpha u = |u|^{2\sigma} u, & \text{in } \Omega, \\ u = \Delta u = 0, & \text{on } \partial\Omega. \end{cases}$$

We assume in the next statement that Ω is smooth. We have not searched to optimize the required regularity of the boundary. At some point, we need to take two partial derivatives into the equation. We assume enough regularity of the boundary so that the solution belongs at least to $H^6(\Omega)$. One could work with interior regularity which requires less regularity on the boundary but since our main motivation is to cover the case of a ball, working with global regularity is fine for our purpose as the ball has the regularity required.

Theorem 5.3. *Assume $\Omega \subset \mathbb{R}^N$ is a smooth bounded domain of class C^6 and $3 \leq 2\sigma + 2 < \frac{2N}{N-2}$ if $3 \leq N \leq 5$. If $\gamma_k \to 0$ and $(u_k)_k$ is a sequence of least energy solutions of (P_{γ_k}), then, up to a subsequence, u_k converges strongly in H^2 to some minimizer u_0 for m_0. If, in addition, Ω is a ball, then there exists $\gamma_0 > 0$ such that if $0 < \gamma < \gamma_0$, the problem (P_γ) has a unique least energy solution. This solution is radially symmetric and strictly radially decreasing.*

Proof. *Step 1. Global regularity.* Using elliptic regularity [14, Theorems 8.12 & 8.13], we easily infer that the solutions u_k are smooth, namely at least $H^6(\Omega)$. Indeed, one can write the equation as a double Dirichlet problem

$$-\Delta u_k = \phi_k, \quad u_k = 0 \text{ on } \partial\Omega,$$

$$-\gamma_k \Delta \phi_k + \phi_k = m_k |u_k|^{2\sigma} u_k - \alpha u_k, \quad \phi_k = 0 \text{ on } \partial\Omega.$$

Here γ_k stays fixed and we can start with the fact that $u_k \in H^2(\Omega)$, without caring about the dependence on k. Then the term $m_k |u_k|^{2\sigma} u_k - \alpha u_k \in L^2(\Omega)$ as it can be easily checked from the assumption on σ and the embedding of $H^2(\Omega)$ into $L^q(\Omega)$ for every $q \geq 1$ if $N \leq 4$ and $q \in [1, \frac{2N}{N-4}]$ if $N = 5$. We therefore infer from [14, Theorems 8.12] that $\phi_k \in H^2(\Omega)$ which in turn implies that $u_k \in H^4(\Omega)$ by [14, Theorems 8.13]. Now computing $\Delta(m_k|u_k|^{2\sigma}u_k - \alpha u_k)$, we realize that it is an L^2 function and therefore $m_k|u_k|^{2\sigma}u_k - \alpha u_k$ is an H^2 function. Indeed, the condition on σ ensures the required integrability of $|u_k|^{2\sigma-1}|\nabla u_k|^2$ and $|u_k|^{2\sigma}|\Delta u_k|$. We then conclude that ϕ_k belongs in fact at least to H^4 and therefore $u_k \in H^6(\Omega)$.

Step 2. Strong convergence in H^1. Arguing as in the proof of Proposition 5.1, we infer that there exists a minimizer $u_0 \in M_0$ and a subsequence that we still denote $(u_k)_k$ such that $u_k \to u_0$ strongly in H^1. If Ω is a ball, then u_0 is the unique minimizer and the whole sequence converge.

Step 3. Strong convergence in H^2. To improve the convergence, we argue as in the proof of Proposition 5.2. If $2\sigma + 1 \leq \frac{N}{N-2}$, then we can bootstrap using the H^{m+2} regularity theory. Due to the boundary condition, the argument of Krylov [18, Chapter 1] cannot be applied directly to get higher regularity in general, see [18, Chapter 8]. However, in our case, since we deal with Navier condition, we have that $u_k = \Delta u_k = 0$ on the boundary and therefore the equation (P_{γ_k}) tells that $\Delta^2 u_k = 0$ on the boundary as well. By Step 1, we can take the Laplacian inside the equation in (P_{γ_k}) and use the fact that Δu_k solves a boundary problem with Navier boundary conditions, namely

$$\begin{aligned} \gamma_k \Delta^2(\Delta u_k) - \Delta(\Delta u_k) + \alpha(\Delta u_k) = m_k f(u_k), \quad &\text{in } \Omega, \\ \Delta(\Delta u_k) = \Delta u_k = 0, \quad &\text{on } \partial\Omega, \end{aligned}$$

where

$$f(u_k) = (2\sigma + 1)\text{sign}(u_k)\left(2\sigma u_k^{2\sigma-1}|\nabla u_k|^2 + u_k^{2\sigma}\Delta u_k\right). \tag{12}$$

Then we can use the H^2 regularity for the Dirichlet problem associated with the systems

$$v_k = -\gamma_k \Delta u_k \qquad -\Delta v_k + \frac{1}{\gamma_k} v_k = w_k, \tag{13}$$

and

$$y_k = \Delta v_k = -\gamma_k \Delta^2 u_k \qquad -\Delta y_k + \frac{1}{\gamma_k} y_k = m_k f(u_k), \tag{14}$$

where $w_k = m_k|u_k|^{2\sigma}u_k - \alpha u_k$ and $f(u_k)$ is defined in (12). Applying [18, Chapter 8, Theorem 8.7] to the second equation of the first system (13), we get an H^2 a priori bound of v_k. Now turning to the Dirichlet problem

$$-\Delta u_k + \alpha u_k = m_k|u_k|^{2\sigma}u_k + \Delta v_k, \quad u_k = 0 \text{ on } \partial\Omega, \tag{15}$$

we deduce that u_k is a priori bounded in H^2 which leads to an L^2 bound for $f(u_k)$. Applying then [18, Chapter 8, Theorem 8.7] on the second equation of the system (14) gives an H^2 a priori bound of Δv_k. Whence v_k is a priori bounded in H^4. This allows to conclude that u_k is a priori bounded in H^4 because the right-hand side of Δ(15), namely

$$-\Delta(\Delta u_k) + \alpha\Delta u_k = m_k f(u_k) + \Delta^2 v_k,$$

is a priori bounded in L^2. The remaining steps are now as in the proof of Proposition 5.2.

If $\frac{N}{N-2} + 1 < 2\sigma + 2 < \frac{2N}{N-2}$, we can only start with a bound in $L^{\frac{2N}{(N-2)(2\sigma+1)}}$ on the right-hand side of

$$-\Delta v_k + \frac{1}{\gamma_k} v_k = w_k,$$

where we still use the notations $v_k = -\gamma_k \Delta u_k$ and $w_k = m_k|u_k|^{2\sigma}u_k - \alpha u_k$. We therefore need to improve this bound first. Arguing as above (still using [18, Chapter 8, Theorem 8.7]), we deduce an a priori bound in $W^{2,q}$ with $q = \frac{2N}{(N-2)(2\sigma+1)}$. Then Sobolev embeddings give a better integrability of w_k and we can bootstrap until we get an L^2 a priori bound on w_k. The strong convergence in H^2 is then achieved as in the proof of Proposition 5.2 taking into account the above remark concerning the way to obtain the higher order elliptic regularity. Observe that even if $\frac{N}{N-2} + 1 < 2\sigma + 2$, no additional bootstrap is necessary to derive the H^4 bound on u_k since once we get an a priori H^2 bound on u_k, the assumption on σ implies that $f(u_k)$ is a priori bounded in L^2.

Uniqueness in the case $\Omega = B_R$. When Ω is a ball, the arguments used in the proof of Theorem 1.3 are available. The nondegeneracy of u_0 allows to apply the Implicit Function Theorem to conclude the local uniqueness (in an H^2 neighbourhood of u_0) for γ small. The remaining arguments are then as in the proof of Theorem 1.3. □

We end up the analysis of the asymptotics for $\gamma \to 0$ by showing that the least energy solution blows up in H^2 when $2\sigma + 2$ is H^1 critical or supercritical. We focus on the case of $\Omega = \mathbb{R}^N$.

We first derive the Derrick-Pohozahev identity for minimizers. If u achieves m_γ in M, then, defining v_λ by $v_\lambda(x) = \lambda^{\frac{N}{2\sigma+2}} u(\lambda x)$, we infer that $f(\lambda) := J_\gamma(v_\lambda)$ achieves a local minimum at $\lambda = 1$. This yields a Derrick-Pohozahev identity

$$\gamma\,(2N - (2\sigma + 2)(N - 4)) \int_{\mathbb{R}^N} |\Delta u|^2\, dx + (2N - (2\sigma + 2)(N - 2)) \int_{\mathbb{R}^N} |\nabla u|^2\, dx$$
$$+\alpha(2N - (2\sigma + 2)N) \int_{\mathbb{R}^N} |u|^2\, dx = 0.$$

If $2\sigma + 2 \geq \frac{2N}{N-4}$, then u must be zero which is obviously a contradiction. This shows that m_γ is not achieved for $2\sigma + 2 \geq \frac{2N}{N-4}$.

For $\frac{2N}{N-2} \le 2\sigma + 2 < \frac{2N}{N-4}$, the first coefficient in the Derrick-Pohozahev identity is positive whereas the other two are nonpositive. We can then write

$$\gamma\,(2N - (2\sigma+2)(N-4)) \int_{\mathbb{R}^N} |\Delta u|^2\,dx \ge \alpha(2\sigma N) \int_{\mathbb{R}^N} |u|^2\,dx.$$

Now, from Gagliardo-Nirenberg inequality, we infer that for some $C > 0$,

$$1 = \left(\int_{\mathbb{R}^N} |u|^{2\sigma+2}\,dx\right)^{\frac{8}{4(2\sigma+2)-2\sigma N}} \le C\left(\int_{\mathbb{R}^N} |\Delta u|^2\,dx\right)^{\frac{2N}{4(2\sigma+2)-2\sigma N}} \int_{\mathbb{R}^N} |u|^2\,dx,$$

which implies

$$\gamma\,(2N - (2\sigma+2)(N-4)) \left(\int_{\mathbb{R}^N} |\Delta u|^2\,dx\right)^{1+\frac{2\sigma N}{4(2\sigma+2)-2\sigma N}} \ge \alpha(2\sigma N)C.$$

This shows that Δu blows up in $L^2(\mathbb{R}^3)$ when $\gamma \to 0$.

6 Sign-changing radial minimizer

In this section, we show that a radial least energy solution of (3) with $\sigma = 1$ is sign-changing when $-2\sqrt{\alpha} < \beta < 2\sqrt{\alpha}$. We assume $N = 3$ but the arguments apply in dimension $N = 2$ also.

We will require the decay of the radial derivatives. Arguing as in de Figueiredo et al [11, Theorem 2.2], one easily gets the following lemma.

Lemma 6.1. *Let $u \in H^m_{rad}(\mathbb{R}^3)$ and let $v :]0,\infty[\to \mathbb{R}$ be the function defined by $v(r) := u(x)$ with $r = |x|$. Then, $v \in H^m(]0,\infty[, r^2)$. Moreover, for a.e. $|x| \in]0,\infty[$ we have*

$$\left|D^j u(x)\right| \ge \left|v^{(j)}(|x|)\right|, \qquad \forall\, j = 0, 1, \dots, m.$$

In order to prove the Theorem 1.4 we adapt some arguments of Bonheure et al [2, Theorem 6].

Proof (Proof of Theorem 1.4). We suppose $N = 3$, the case $N = 2$ is similar.

Step 1. Classical regularity. We start by observing that by elliptic regularity, we have $u \in H^6(\mathbb{R}^3)$ which implies $u \in C^{4,1/2}(\mathbb{R}^3)$ and the solution can be understood in the classical sense. Indeed, we know that the solution is H^2, so that from the equation

$$-\Delta(-\Delta u) = |u|^2 u - \alpha u + \beta \Delta u,$$

we infer that $-\Delta(-\Delta u) \in L^2(\mathbb{R}^3)$. This implies that $-\Delta u, -\Delta(-\Delta u) \in L^2(\mathbb{R}^3)$ and henceforth $-\Delta u \in H^2(\mathbb{R}^3)$. Since $u \in H^2(\mathbb{R}^3)$, we conclude that $u \in H^4(\mathbb{R}^3)$. Looking again at the equation, we can now use the fact that the right-hand side is an H^2-function. Then $-\Delta u, -\Delta(-\Delta u) \in H^2(\mathbb{R}^3)$ and therefore $-\Delta u \in H^4(\mathbb{R}^3)$. At last, combining the fact that $u \in H^4(\mathbb{R}^3)$ and $-\Delta u \in H^4(\mathbb{R}^3)$, we deduce that $u \in H^6(\mathbb{R}^3)$. Here above, the required elliptic regularity theory can be found in [18, Chapter 1] and since we are in the whole space, this is just a consequence of simple Fourier analysis.

Step 2. Equation in radial coordinates and decay at infinity. Writing now the equation (3) in radial coordinates (the expression is especially simple in dimension $N = 3$), we compute that v, defined by $v(r) := u(x)$ for $r = |x|$, solves

$$v^{iv} + \frac{4}{r}v''' - \beta v'' - \frac{2\beta}{r}v' + \alpha v = |v|^2 v, \qquad r \in]0, \infty[. \tag{16}$$

The $H^5(\mathbb{R}^3)$ regularity yields

$$\lim_{|x|\to\infty} (u(x), \partial_{x_i} u(x), \partial^2_{x_i x_j} u(x), \partial^3_{x_i x_j x_k} u(x)) = (0, 0, 0, 0)$$

whatever $i, j, k \in \{1, 2, 3\}$. Then Lemma 6.1 implies that v satisfies

$$\lim_{r\to\infty} (v(r), v'(r), v''(r), v'''(r)) = (0, 0, 0, 0). \tag{17}$$

Step 3. Asymptotic analysis of the solution of the ordinary differential equation (16).

Claim 1 : Given $R > 0$ we can find $\overline{r} \geq R$ such that $v(\overline{r}) > 0$.

Let $R > 0$ be fixed. Consider the following Cauchy problem

$$(C1) \qquad \begin{cases} w^{iv}(r) - \beta w''(r) + \alpha w(r) = 0, & r > 0, \\ (w(r_0), w'(r_0), w''(r_0), w'''(r_0)) =: w_0, \end{cases}$$

where $r_0 > 0$ and $w_0 \in \mathbb{R}^4$. By using condition $(A1)$ we have that all the roots of the characteristic equation associated with $(C1)$ are complex, let us say $\pm a \pm ib$. We set $\Delta := 2\pi/b$. Then there exists $c > 0$ such that any solution of $(C1)$ satisfies

$$\sup_{[r_0, r_0+\Delta]} w, \quad \sup_{[r_0, r_0+\Delta]} (-w) \geq c|w_0|. \tag{18}$$

Moreover, there exists $M > 0$ such that any solution of $(C1)$ verifies

$$\|w\|_{C^3([r_0, r_0+\Delta])} \leq M|w_0|.$$

Again, we can also find $N > 0$ such that the solutions of

$$\begin{cases} \psi^{iv}(r) - \beta\psi''(r) + \alpha\psi(r) \;=\; h(r), \qquad r > 0, \\ (\psi(r_0), \psi'(r_0), \psi''(r_0), \psi'''(r_0)) =: 0, \end{cases}$$

satisfy

$$\|\psi\|_{C^3([r_0,r_0+\Delta])} \le N\|h\|_{L^\infty(r_0,r_0+\Delta)}.$$

Let us set $\delta > 0$ so that $c - \frac{MN\delta}{1-N\delta} > 0$. Denote by $v(r) = v(r; r_0, v_0)$ the solution of (16) with initial conditions

$$(v(r_0), v'(r_0), v''(r_0), v'''(r_0)) =: v_0, \qquad \text{where } r_0 > 0.$$

Now, let us fix $r_0 \ge R$ large enough so that $|v_0|$ is small enough to have

$$\sup_{r\in[r_0,r_0+\Delta]} |v(r)|^2, \quad \sup_{r\in[r_0,r_0+\Delta]} \frac{4}{r} \quad \text{and} \quad \sup_{r\in[r_0,r_0+\Delta]} \frac{2\beta}{r} < \delta.$$

We write

$$v = \psi + w,$$

where ψ solves

$$\begin{cases} \psi^{iv} - \beta\psi'' + \alpha\psi = |v|^2 v + \dfrac{2\beta}{r}v' - \dfrac{4}{r}v''', \; r > 0, \\ (\psi(r_0), \psi'(r_0), \psi''(r_0), \psi'''(r_0)) = 0, \end{cases}$$

and w is a solution of

$$\begin{cases} w^{iv}(r) - \beta w''(r) + \alpha w(r) = 0, \qquad r > 0, \\ (w(r_0), w'(r_0), w''(r_0), w'''(r_0)) = v_0. \end{cases}$$

Now, let us choose $\bar{r} \in [r_0, r_0 + \Delta]$ such that

$$w(\bar{r}) \ge c|v_0|.$$

Thus,

$$\|\psi\|_{C^3([r_0,r_0+\Delta])} \le N\delta\|v\|_{C^3([r_0,r_0+\Delta])},$$

which implies that

$$\|\psi\|_{C^3([r_0,r_0+\Delta])} \le \frac{N\delta}{1-N\delta}\|w\|_{C^3([r_0,r_0+\Delta])} \le \frac{MN\delta}{1-N\delta}|v_0|.$$

Then we obtain

$$v(\bar{r}) \ge c|v_0| - \|\psi\|_{L^\infty} \ge \left(c - \frac{MN\delta}{1-N\delta}\right)|v_0| > 0.$$

Claim 2 : Given $R > 0$ we can find $\underline{r} \ge R$ such that $v(\underline{r}) < 0$.

The proof of this claim is similar to that of Claim 1.

Conclusion. We have proved in the last step that u changes sign. In fact, we have even proved that u oscillate as $|x| \to +\infty$. □

7 Comments

This note provides some simple results for the model equation (3) with a Kerr nonlinearity and aims to partially complete the discussion on waveguide solutions in [13, Section 4.1]. The methods we used are standard. On the other hand, since radial solutions present oscillations for $-2\sqrt{\alpha} < \beta < 2\sqrt{\alpha}$, we expect that one needs new arguments to answer the question whether the least energy solutions are radial or not in this case. Also uniqueness is a challenging question if we are not in the asymptotic regime $\beta \to \infty$ (or equivalently $\gamma \to 0$).

We also mention that the important question about the decay at infinity of the least energy solutions will be addressed in a future work. We are only aware of [12] for a result in that direction. The analysis therein relies on the computation of the fundamental solution of the fourth-order operator in (3) with $\beta = 0$.

The analysis of the decay should also allow to extend the statement of Theorem 1.3 to the case $2 < 2\sigma + 2 < \frac{2N}{N-2}$ and $N \ge 3$. Indeed, the arguments we used are just fine for the Kerr nonlinearity whereas some technical adjustments are needed for a general subcritical power. In fact, one checks easily that our arguments apply in dimension $N \le 4$ if we assume $2 \le 2\sigma + 1 \le \frac{N}{N-2}$. The lower inequality on σ implies the required $C^{1,1}$ regularity of the function $s \mapsto |s|^{2\sigma}s$ whereas the upper inequality is used to start the bootstrap with an L^2-bound on $|u|^{2\sigma}u$ (here u is a solution).

The same remark holds for Theorem 1.4 which should be true with less restrictive assumptions. In dimension $N \le 8$, one can deal with $2 \le 2\sigma + 1 \le \frac{N}{N-4}$. The other cases will require more care and will be treated in a forthcoming work.

Acknowledgements D. Bonheure is supported by INRIA - Team MEPHISTO, MIS F.4508.14 (FNRS), PDR T.1110.14F (FNRS) & ARC AUWB-2012-12/17-ULB1- IAPAS.

References

1. Ben-Artzi, M., Koch, H., Saut, J.C.: Dispersion estimates for fourth order Schrödinger equations. C. R. Acad. Sci. Paris Sér. I Math. **330**, 87–92 (2000)
2. Bonheure, D., Habets, P., Sanchez, L.: Heteroclinics for fourth order symmetric bistable equations. Atti. del Seminario Matematico e Fisico dell'Universita di Modena **52**, 213–227 (2004)
3. Bonheure, D., Moreira dos Santos, E., Ramos, M.: Ground state and non-ground solutions of some strongly coupled elliptic systems. Trans. Am. Math. Soc. **364**, 447–491 (2012)
4. Bonheure, D., Moreira dos Santos, E., Tavares, H.: Hamiltonian elliptic systems: a guide to variational frameworks. Port. Math. (N.S.) **71**, Fasc. 3–4, 301–395 (2014)
5. Bouchel, O.: Remarks on NLS with higher order anisotropic dispersion. Adv. Differ. Equ. **13**, 1–2, 169–198 (2008)
6. Boulenger, T., Lenzmann, E.: Blowup for Biharmonic NLS. (2015) arXiv:1503.01741v1
7. Busca, J., Sirakov, B.: Results for semilinear elliptic systems in the whole space. J. Differ. Equ. **163**, 41–56 (2000)
8. Chang, S.M., Gustafson, S., Nakanishi, K., Tsai, T.P.: Spectra of linearized operators for NLS solitary waves. SIAM J. Math. Anal. **39**, 1070–1111 (2008)
9. Coffman, C.V.: Uniqueness of the ground state solution for $\Delta u - u + u^3 = 0$ and a variational characterization of other solutions. Arch. Ration. Mech. Anal. **46**, 81–95 (1972)
10. de Figueiredo, D.G., Yang, J.: Decay, symmetry and existence of solutions of semilinear elliptic systems. Nonlinear Anal. **33**, 211–234 (1998)
11. de Figueiredo, D.G., Moreira dos Santos, E., Miyagaki, O.H.: Sobolev spaces of symmetric functions and applications. J. Funct. Anal. **261**, 3735–3770 (2011)
12. Deng, Y.B., Li, Y.: Exponential decay of the solutions for nonlinear biharmonic equations. Commun. Contemp. Math. **9**, 753–768 (2007)
13. Fibich, G., Ilan, B., Papanicolaou, G.: Self-focusing with fourth order dispersion. SIAM J. Appl. Math. **62**, 1437–1462 (2002)
14. Gilbarg, D., Trudinger, N.S.: Elliptic Partial Differential Equations of Second Order. Springer, Heidelberg (1998)
15. Kabeya, Y., Tanaka, K.: Uniqueness of positive radial solutions of semilinear elliptic equations in $\mathbb{R}^N$ and Séré's non-degeneracy condition. Commun. Partial Differ. Equ. **24**, 563–598 (1999)
16. Karpman, I., Shagalov, A.G.: Stability of solitons described by nonlinear Schrödinger-type equations with higher-order dispersion. Phys. D **144**, 194–210 (2000)
17. Kavian, O.: Introduction à la théorie des Points Critiques. Springer, Heidelberg (1993)
18. Krylov, N.V.: Lectures on Elliptic and Parabolic Equations in Sobolev Spaces. Graduate Studies in Mathematics, vol. 96. American Mathematical Society, Providence (2008)
19. Kwong, M.K.: Uniqueness of positive solutions of $-\Delta u + u = u^p$ in $\mathbb{R}^N$. Arch Ration. Mech. Anal. **105**, 243–266 (1989)
20. Lions, P.L.: Symétrie et compacité dans les espaces de Sobolev. J. Funct. Anal. **49**, 315–334 (1982)
21. Lions, P.L.: The concentration-compactness principle in the calculus of variations. The locally compact case, part 1. Ann. Inst. Henri Poincaré, Analyse non-linéaire **1**, 109–145 (1984)
22. Strauss, W.A.: Existence of solitary waves in higher dimensions. Commun. Math. Phys. **57**, 149–162 (1977)
23. Troy, W.C.: Properties in systems of semilinear elliptic equations. J. Differ. Equ. **42**, 400–413 (1981)
24. van der Vorst, R.C.A.M.: Best constant for the embedding of the space $H^2 \cap H_0^1(\Omega)$ into $L^{2N/(N-4)}(\Omega)$. Differ. Integr. Equ. **6**, 259–276 (1993)
25. Weinstein, M.I.: Nonlinear Schrödinger equations and sharp interpolation estimates. Commun. Math. Phys. **87**, 567–576 (1983)

Fibers and global geometry of functions

Marta Calanchi, Carlos Tomei, and André Zaccur

Dedicated to Djairo, an example to follow in many directions

Mathematics Subject Classification (2010): 35B32, 35J91, 65N30

1 Introduction

When we teach the first courses in calculus and complex or real analysis, a great emphasis is given to geometric issues: we plot graphs, enumerate conformal mappings among special regions, identify homeomorphisms. Alas, this is far from being enough: mappings become too complicated soon. Still, the geometric approach, especially combined with numerical arguments, is very fruitful in some nonlinear contexts.

It is rather surprising that some infinite dimensional maps can be studied in a similar fashion—one may even think about their graphs! The examples which are amenable to such approach are very few, and they elicit the same sense of wonder that (the equally rare) completely integrable systems do: one is left with a feeling of deep understanding. This text is dedicated to some such examples.

The interested reader could hardly do better than going through the review papers by Church and Timourian [11, 12], which cover extremely well the material up to the mid nineties. Their approach is strongly influenced by the original Ambrosetti-Prodi view of the problem, which we describe in Section 2.2. In a nutshell, the global

M. Calanchi (✉)
Dipartimento di Matematica, Università di Milano, Via Saldini 50,
20133 Milano, Italy
e-mail: marta.calanchi@unimi.it

C. Tomei • A. Zaccur
Departamento de Matemática, PUC-Rio, R. Mq. de S. Vicente 225,
Rio de Janeiro 22453-900, Brazil
e-mail: carlos.tomei@gmail.com; zaccur.andre@gmail.com

A.N. Carvalho et al. (eds.), *Contributions to Nonlinear Elliptic Equations and Systems*, Progress in Nonlinear Differential Equations and Their Applications 86,
DOI 10.1007/978-3-319-19902-3_5

geometry of a proper function F is studied through certain properties of its critical set C together with its image $F(C)$, along with the stratification of C in terms of singularities.

This much less ambitious text is mainly an enumeration of techniques and of some recent developments, some of which have not been published. We mostly take the Berger-Podolak route [6] which has been extended by Podolak in [27] and, we believe, still allows for improvement. Instead of the critical set, we concentrate on the restriction of F to appropriate low dimensional manifolds (one dimensional, in the Ambrosetti-Prodi case), the so-called fibers.

Essentially, fibers are appropriate in the presence of *finite spectral interaction*, which roughly states that the function $F : X \to Y$ splits into a sum of linear and nonlinear terms, $F = L - N$ and N deforms L substantially only along a few eigenvectors spanning a subspace $V \subset X$. The domain splits into orthogonal subspaces, $X = H \oplus V$ and the hypotheses on the nonlinearities are naturally anisotropic. Different requests on H and V yield a global Lyapunov-Schmidt decomposition of F: on affine subspaces obtained by translating H, F is a homeomorphism and complications due to the nonlinear term manifest on fibers, which are graphs of functions from V to H.

Fibers are also convenient for the verification of properness of F. In particular, one may search for folds in nonlinear maps defined on functions with unbounded domains, which are natural in physical situations. Fibers also provide the conceptual starting point for algorithms that solve a class of partial differential equations, an idea originally suggested by Smiley [31, 32] and later implemented for finite spectral interaction of the Dirichlet Laplacian on rectangles in [7].

An abstract setup in the spirit of the characterization of folds as in [10, 11], or like the one we present in Section 4, provides a better understanding of the role of the hypotheses in the fundamental example of Ambrosetti and Prodi. Elliptic theory seems to be less relevant than one might think, it is just that it provides a context in which the required hypotheses are satisfied.

In Section 2, we present the seminal examples—the Dolph-Hammerstein homeomorphisms and the Ambrosetti-Prodi fold—in a manner appropriate for our arguments. Fibers and sheets are defined and constructed in Section 3. A global change of coordinates in Section 4 gives rise to adapted coordinates, in which the description of critical points is especially simple. A characterization of the critical points strictly in terms of spectral properties of the Jacobian DF is given. Also, the three natural steps to identify global folds become easy to identify. Further study of how to implement each step is the content of Sections 5, 6 and 7. The last section is dedicated to some examples.

The text is written as a guide: we try to convey the merits of a set of techniques, without providing details. Complete proofs will be presented elsewhere [8, 9].

Alas, we stop at folds. There are scattered results in which local or global cusps were identified: again, the excellent survey [12] covers the material up to the mid-nineties. So far, the description of cusps seems rather ad hoc. There are characterizations [12], but they are hard to verify and new ideas are needed. On the other hand, checking that maps are not global folds is rather simple, a matter of showing, for example, that some points in the image have more than two preimages. A numerical example is exhibited in Section 5.3.

2 The first examples in infinite dimension

Among the simplest continuous maps between Hilbert spaces are homeomorphisms, in particular linear isomorphisms. A second class of examples are folds.

2.1 *Homeomorphisms: Dolph and Hammerstein*

Dolph and Hammerstein [15, 16] obtained a simple condition under which nonlinear perturbation of linear isomorphisms are still homeomorphisms. A version of their results is the following.

Start with a real Hilbert space Y and a self-adjoint operator $L : X \subset Y \to Y$ for a dense subspace X of Y. Let $\sigma(L)$ be the spectrum of L.

Theorem 2.1. *Let $[-c, c] \cap \sigma(L) = \emptyset$ and suppose $N : Y \to Y$ is a Lipschitz map with Lipschitz constant $n < c$. Equip X with the graph topology, $\|x\|_X = \|x\|_Y + \|Lx\|_Y$. Then the map $F = L - N : X \to Y$ is a Lipschitz homeomorphism.*

Indeed, to solve $F(x) = y$, search for a fixed point of

$$C_y : Y \to Y, \quad C_y(z) = N(L^{-1}(z)) + y$$

which is a contraction because the operator $L^{-1} : Y \to Y$ has norm less than $1/c$ by standard spectral theory and then the map $N \circ L^{-1}$ is Lipschitz with constant less than $n/c < 1$. As usual, the fixed point varies continuously with y. Clearly, F is Lipschitz. To show the same for F^{-1}, keep track of the Banach iteration.

Notice that the statement allows for differential operators between Sobolev spaces. Very little is required from the spectrum of L. Clearly, for symmetric bounded operators one should take $X = Y$.

2.2 *Breaking the barrier: the Ambrosetti-Prodi theorem*

What about more complicated functions? Ambrosetti and Prodi [1] obtained an exquisite example. After refinements by Micheletti and Manes [24], Berger and Podolak [6] and Berger and Church [5], the result may be stated as follows. Let $\Omega \subset \mathbb{R}^n$ be a connected, open, bounded set with smooth boundary (for nonsmooth boundaries, see [34]). Let $H^2(\Omega)$ and $H^1_0(\Omega)$ be the usual Sobolev spaces and set

$X = H^2(\Omega) \cap H^1_0(\Omega)$ and $Y = H^0(\Omega) = L^2(\Omega)$. The eigenvalues of the Dirichlet Laplacian $-\Delta : X \subset Y \to Y$ are

$$\sigma(-\Delta) = \{0 < \lambda_1 < \lambda_2 \leq \ldots \to \infty\}.$$

Denote by ϕ_1 the (L^2-normalized, positive) eigenvector associated with λ_1 and split $X = H_X \oplus V_X$, $Y = H_Y \oplus V_Y$ in *horizontal* and *vertical* orthogonal subspaces, where $V_X = V_Y = \langle \phi_1 \rangle$, the one dimensional (real) vector space spanned by ϕ_1.

Theorem 2.2. *Let $F : X \to Y$ be $F = L - N$, where $L = -\Delta$, $N(u) = f(u)$, for a smooth, strictly convex function $f : \mathbb{R} \to \mathbb{R}$ satisfying*

$$\operatorname{Ran} f' = (a, b), \quad a < \lambda_1 < b < \lambda_2 .$$

Then there are global homeomorphisms $\zeta : X \to H_Y \oplus \mathbb{R}$ and $\xi : Y \to H_Y \oplus \mathbb{R}$ for which $\tilde{F}(z, t) = \xi \circ F \circ \zeta^{-1}(z, t) = (z, -t^2)$.

Said differently, the following diagram commutes.

$$\begin{array}{ccc} X & \xrightarrow{F} & Y \\ \zeta \downarrow & & \downarrow \xi \\ H_Y \oplus \mathbb{R} & \xrightarrow{(z, -t^2)} & H_Y \oplus \mathbb{R} \end{array}$$

Functions which admit such dramatic simplification are called *global folds*. The vertical arrows in the diagram above are (global) changes of variables and sometimes will be C^1 maps, but we will not emphasize this point.

The original approach by Ambrosetti and Prodi is very geometric [1]. In a nutshell, they show that F is a proper map whose critical set C (in the standard sense of differential geometry, the set of points $u \in X$ for which the derivative $DF(u)$ is not invertible) is topologically a hyperplane, together with its image $F(C)$. They then show that F is proper, its restriction to C is injective and $F^{-1}(F(C)) = C$. Finally, they prove that both connected components of $X - C$ are taken injectively to the same component of $Y - F(C)$. Their final result is a counting theorem: the number of preimages under F can only be 0, 1 or 2.

Berger and Podolak [6], on the other hand, construct a global Lyapunov-Schmidt decomposition for F. For $V_X = V_Y = \langle \phi_1 \rangle$, consider *affine horizontal (resp. vertical)* subspaces of X (resp. Y), i.e., sets of the form $H_X + t\phi_1$, for a fixed $t \in \mathbb{R}$ (resp. $y + V_Y$, for $y \in H_Y$). Let $P : Y \to H_Y$ be the orthogonal projection. The map $PF_t : H_X \to H_Y, PF_t(w) = PF(w + t\phi_1)$, is a bi-Lipschitz homeomorphism, as we shall see below. Thus, the inverse under F of vertical lines $y + V_Y$, for $y \in H_Y$ are curves $\alpha_y : \mathbb{R} \sim V_X \subset X \to H_X$, which we call *fibers*. Fibers stratify the domain X. Thus, to show that F is a global fold, it suffices to verify that each restriction $F : \alpha_y \to V_Y \sim \mathbb{R}$, essentially a map from $\mathbb{R}$ to $\mathbb{R}$, is a fold.

After such a remarkable example, one is tempted to push forward. This is not that simple: if the (generic) nonlinearity f is not convex, there are points in Y with four preimages [8], so the associated map $F : X \to Y$ cannot be a global fold (for a numerical example, see Section 5.3).

3 Fibers and height functions

Fibers come up in [6] and [32] for C^1 maps associated with second order differential operators and in [21] in the context of first order periodic ordinary differential equations. Due to the lack of self-adjointness, the construction in [21] is of a very different nature. We follow [27] and [34], which handle Lipschitz maps, allowing the use of piecewise linear functions in the Ambrosetti-Prodi scenario, namely f given by $f'(x) = a$ or b, depending if $x < 0$ or $x > 0$ [14, 19].

Let X and Y be Hilbert spaces, X densely included in Y. Let $L : X \subset Y \to Y$ be a self-adjoint operator with a simple, isolated, eigenvalue λ_p, with eigenvector $\phi_p \in X$ with $\|\phi_p\|_Y = 1$. Notice that λ_p may be located anywhere in the spectrum $\sigma(L)$ of L. As before, consider horizontal and vertical orthogonal subspaces,

$$X = H_X \oplus V_X\,,\ Y = H_Y \oplus V_Y\,,\ \text{ for } V_X = V_Y = \langle \phi_p \rangle$$

and the projection $P : Y \to H_Y$. Let $PF_t : H_X \to H_Y$ be the projection on H_Y of the restriction of F to the affine subspace $H_X + t\phi_p$, $PF_t(w) = PF(w + t\phi_p)$. In the same fashion, the nonlinearity $N : Y \to Y$ gives rise to maps $PN_t : H_Y \to H_Y$, which we require to be Lipschitz with constant n independent of $t \in \mathbb{R}$ so that

$$[-n, n] \cap \sigma(L) = \{\lambda_p\}\,. \tag{H}$$

The standard Ambrosetti-Prodi map fits these hypotheses. In this case, $X \subset Y$ are Sobolev spaces and the derivative $f' : \mathbb{R} \to \mathbb{R}$ is bounded by a and b. Set

$$\gamma = (a+b)/2, \quad L = -\Delta - \gamma, \quad N(u) = f(u) - \gamma u$$

and $\lambda_p = \lambda_1$, the smallest eigenvalue of $-\Delta$. Then the Lipschitz constant n of the maps PN_t satisfies $n < \gamma - a = b - \gamma < \lambda_2 - \gamma$, so that $\lambda_1 - \gamma \leq n$.

Theorem 3.1. *Let $F : X \to Y$ satisfy (H) above. Then for each $t \in \mathbb{R}$, the map PF_t is a bi-Lipschitz homeomorphism, and a C^k diffeomorphism if F is C^k. The Lipschitz constants for PF_t and $(PF_t)^{-1}$ are independent of t.*

Proof. The proof follows Theorem 2.1 once the potentially nasty eigenvalue λ_p is ruled out. Let c be the absolute value of the point in $\sigma(L) \setminus \{\lambda_p\}$ closest to 0, so that $0 \leq n < c$. The operator $L : X \to Y$ restricts to $L : H_X \to H_Y$, which is invertible self-adjoint, and again $L^{-1} : H_Y \to H_Y$ with $\|L^{-1}\| \leq 1/c$. The solutions $w \in H_X$ of

$PF_t(w) = g \in H_Y$ solve $PLu - PN(u) = Lw - PN_t(w) = g$ for $u = w + t\phi_p$. The solutions w correspond to the fixed points of $C_g : H_Y \to H_Y$, where

$$C_g(z) = PN_t(L^{-1}z) + g, \qquad \text{for } Lw = z \in H_Y.$$

The map C_g is a contraction with constant bounded by $n/c < 1$ (independent of t). Now follow the proof of Theorem 2.1. □

The attentive reader may have noticed that the effect of the nonlinearity N along the vertical direction is irrelevant for the construction of fibers.

The same construction applies when the interval $[-n, n]$ defined by the Lipschitz constant n of $PN_t : H_X \to H_Y$ interacts with an isolated subset I of $\sigma(L)$ — more precisely, $I = [-n, n] \cap \sigma(L)$ and there is an open neighborhood U of $I \subset \mathbb{R}$ for which $I = U \cap \sigma(L)$. In this case P is the orthogonal projection on I, which takes into account possible multiplicities. In the special situation when I consists of a finite number of eigenvalues (accounting multiplicity), we refer to *finite spectral interaction* between L and N.

We concentrate on the case when $I = \{\lambda_p\}$ consists of a simple eigenvalue. A more careful inspection of the constants in the Banach iteration in the proof above yields the following result [7, 34]. The image under F of horizontal affine subspaces of X are *sheets*. The inverse under F of vertical lines of Y are *fibers*.

Proposition 3.2. *If F is C^1, sheets are graphs of C^1 maps from H_Y to $\langle\phi_p\rangle$ and fibers are graphs of C^1 maps from $\langle\phi_p\rangle$ to H_X. Sheets are essentially flat, fibers are essentially steep.*

We define what we mean by essential flatness and steepness. Let $\nu(y)$ be the normal at a point $y \in Y$ of (the tangent space of) a sheet, and $\tau(u)$ be the tangent vector at $u \in X$ of a fiber. Then there is a constant $\epsilon \in (0, \pi/2)$ such that ϕ_p makes an angle less than ϵ (or greater than $\pi - \epsilon$, due to orientation) with both vectors.

4 Adapted coordinates and a plan

Suppose L and N interact at a simple eigenvalue λ_p. Write

$$F(u) = PF(u) + \langle F(u), \phi_p\rangle\phi_p = PF(u) + h(u)\phi_p$$

where the map $h : X \to \mathbb{R}$ is called the *height function*. In the diagram below, invertible maps are bi-Lipschitz [34] or C^k diffeomorphisms, depending if PF_t is Lipschitz or C^k. The smoothness of h and $h^a = h \circ \Phi$ follow accordingly.

$$\begin{array}{ccc} X = H_X \oplus V_X & \xrightarrow{F} & Y = H_Y \oplus V_Y \\ {\scriptstyle \Phi^{-1}=(PF_t, Id)} \searrow & & \nearrow {\scriptstyle F^a = F\circ\Phi = (Id, h^a)} \\ & Y & \end{array}$$

The map F has been put in *adapted coordinates* by the change of variables Φ:

$$F^a : Y \to Y , \quad (z, t) \mapsto (z, h^a(z, t)) .$$

Notice that fibers of F are taken to vertical lines in the domain of $F^a = F \circ \Phi$. Explicitly, the vertical lines $\{(z_0, t) : t \in \mathbb{R}\}$ parameterized by $z_0 \in H_Y$ correspond to fibers $u(z_0, t) = (PF_t)^{-1}(z_0) + t\phi_p = w(z_0, t) + t\phi_p$. Thus F^a is just a rank one nonlinear perturbation:

$$F^a(z, t) = (z, h^a(z, t)) \sim z + h^a(z + t\phi_p)\phi_p .$$

In a very strict sense, this is also true of F. In order to make F similar to an Ambrosetti-Prodi map, define $G = F^a \circ (-\Delta) : X \to Y$:

$$u \overset{-\Delta}{\longmapsto} z + t\phi_1 \overset{F^a}{\longmapsto} z + t\phi_1 + (h^a(z + t\phi_1) - t)\phi_1 = -\Delta u + \psi(u)\phi_1,$$

for some nonlinear functional ψ. We generalize slightly.

Proposition 4.1. *Let N be a C^1 map. Say L and N interact at a simple eigenvalue λ_p and L is invertible. Then, after a C^1 change of variables, the C^1 function $F = L - N : X \to Y$ becomes $G : X \to Y$, $G = L + \psi(u)\phi_p$, for some $\psi : X \to \mathbb{R}$.*

For Ambrosetti-Prodi operators $F(u) = -\Delta u - f(u)$, the nonlinear perturbation is given by a Nemitskii map $u \mapsto f(u)$. It is not surprising that once we enlarge the set of nonlinearities new global folds arise. For a map F given in adapted coordinates by $F^a(z, t) = (z, h^a(z, t))$, appropriate choices of the *adapted height function* h^a yields all sorts of behavior.

The critical set of $F : X \to Y$ is compatible with fibers as follows [6, 9].

Proposition 4.2. *Suppose the C^1 map $F : X \to Y$ admits fibers. Then u_0 is a critical point of F if and only if it is a critical point of the height function h along its fiber, or equivalently of the adapted height function h^a.*

Isolated local extrema have to alternate between maxima and minima. In particular, given the appropriate behavior at infinity at each fiber and the fact that all critical points are of the same type, we learn from a continuity argument that the full critical set C is connected, with a single point on each fiber [13].

The study of a function $F : X \to Y$ reduces to three steps:

1. Stratify X into fibers.
2. Verify the asymptotic behavior of F along fibers.
3. Classify the critical points of the restriction of F along fibers.

The following result is natural from this point of view [9]. Let $F : X \to Y$ satisfies (H) of Section 3, so that, by Proposition 3.2, X stratifies in one dimensional fibers $\{u(z, t) : t \in \mathbb{R}\}$, one for each $z \in H_Y$.

Proposition 4.3. *Suppose that, on each fiber,*

$$\lim_{t\to\pm\infty} \langle F(u(z,t)), \phi_p\rangle = \lim_{t\to\pm\infty} h(u(z,t)) = -\infty\,.$$

Suppose also that each critical point of h restricted to each fiber is an isolated local maximum. Then $F : X \to Y$ is a global fold, in the sense that there are homeomorphisms on domain and image that give rise to a diagram as in Theorem 2.2.

To verify that such limits exist, one might check hypotheses ($V\pm$) in Section 6.1, but there are alternatives. Similarly, there are ways of obtaining fibers which do not fit the construction presented in Section 3 (this is the case for perturbations of non-self-adjoint operators, Section 8.4). The upshot is that there is some loss in formulating the three-step recipe into a clear-cut theorem.

As trivial examples, $h^a(z,t) = -t^2$ is a global fold, whereas $h^a(z,t) = t^3 - t$ has a critical set consisting of two connected components having only (local) folds (from Section 7.1). More complicated singularities require the dependence on z: not every fiber of F (equivalently, vertical line in the domain of F^a) has the same number of critical points close to a cusp, for example. The reader is invited to check that $(z,t) \mapsto (z, t^3 - \langle z, \tilde{\phi}\rangle t)$ is a global cusp, for $\tilde{\phi}$ any fixed vector in H_Y. Higher order Morin singularities, considered in Section 7, are obtained in a similar fashion. From the Proposition 4.1, changes of variables on such maps yield nonlinear rank one perturbations of the Laplacian which are globally diffeomorphic to the standard normal forms of Morin singularities.

We consider the standard Ambrosetti-Prodi scenario in the light of this strategy. For the function $F(u) = -\Delta u - f(u)$ defined in Theorem 2.2, elliptic theory yields all sort of benefits — the smallest eigenvalue of the Jacobian $DF(u)$ is always simple, the ground state may be taken to be a positive function in X.

The hypotheses required for the construction of fibers in Theorem 3.1 do not imply the simplicity of the relevant eigenvalue: there are examples for which there is no naturally defined C^1 functional $\lambda_p : X \to \mathbb{R}$ because two eigenvalues collide. One might circumvent this difficulty by forcing the nonlinearity N to be smaller, but it turns out that this is not necessary. The hypotheses instead imply the simplicity of λ_p in an open neighborhood of the critical set C of F, and this is all we need, as we shall see in Section 7.1.

The positivity of the ground state and the convexity of the nonlinearity f are used in a combined fashion in the Ambrosetti-Prodi theorem to prove that along fibers the height function only has local maxima. Clearly, this is a property only of critical points. On the other hand, the nonlinearity $N(u) = f(u)$ is so rigid that the standard hypothesis of convexity of f is essentially necessary, as shown in [8]. More general nonlinearities require a better understanding of the singularities.

We now provide more technical details on each of the three steps.

5 Obtaining fibers in other contexts

For starters, what if L is not self-adjoint, or X is not Hilbert?

5.1 *Podolak's approach*

Suppose momentarily that X and Y are Banach spaces. Let $L : X \to Y$ be a Fredholm operator of index zero with kernel generated by a vector ϕ_X and let ϕ_Y be a vector not in Ran$L$. Podolak [27] considered the following scenario, for which she obtained a lower bound on the number of preimages for a region of Y of vectors with very negative component along ϕ_Y. Split $X = H_X \oplus V_X$ where $V_X = \langle \phi_X \rangle$ and H_X is any complement. Also, split $Y = H_Y \oplus V_Y$ where $H_Y =$ Ran$L$ and $V_Y = \langle \phi_Y \rangle$. In particular $L : H_X \to H_Y$ is an isomorphism. Also, define the associated projection $P : Y \to H_Y$. Write $u = w + t\phi_X$, $y = g + s\phi_Y$ for $w \in H_X$. The equation $F(u) = Lu - N(u) = y$ becomes

$$L(w + t\phi_X) - N(w + t\phi_X) = Lw - N(w + t\phi_X) = g + s\phi_Y,$$

and, as in Theorem 3.1, we are reduced to solving the map

$$C_g : H_Y \to H_Y, \quad C_g(z) = PN_t(L^{-1}z) + g\,, \qquad \text{for } Lw = z \in H_Y.$$

Her hypotheses imply that such maps are contractions.

5.2 *Transplanting fibers*

The estimates arising from spectral theorem in the Hilbert context are easy to obtain and possibly more effective. Podolak's hypotheses are harder to verify. There is a possibility: getting fibers in Hilbert spaces and transplanting them to Banach spaces. This happens, for example, when moving from the Ambrosetti-Prodi example as a map between Sobolev spaces [6] to a map between Hölder spaces [1]. The classification of singularities is simpler with additional smoothness (Section 7).

Proposition 5.1. *Let $F = L - N : X \to Y$ satisfy hypothesis (H) of Section 3. Consider the densely included Banach spaces $A \subset X$ and $B \subset Y$ allowing for the C^1 restriction $F : A \to B$ for which $V_X = V_Y \subset A$. Suppose that $DF(a) : A \to B$ is a Fredholm operator of index zero for each $a \in A$. Then fibers of $F : X \to Y$ either belong to A or do not intersect A.*

Said differently, if a point $u \in X$ belongs to A then the whole fiber does.

In the Ambrosetti-Prodi scenario, this proposition seems to be a consequence of elliptic regularity, which may be used to prove it. Regularity of eigenfunctions is irrelevant: fibers are the orbits of the vector field of their tangent vectors, which are inverses of the vertical vector under $DF(u)$, and necessarily lie in A [9]. Tangent vectors are indeed eigenfunctions $\phi_p(u)$ of $DF(u)$ at critical points u.

The fact that sheets and fibers are uniformly flat and steep (Proposition 3.2) allows one to modify vertical spaces ever slightly and still obtain space decompositions for which the Lyapunov-Schmidt decomposition, and hence the construction of fibers in Theorem 3.1, apply. In particular, transplants may be performed even when the eigenvector ϕ_p originally used to define the vertical spaces $V_X = V_Y$ do not have regularity, i.e., do not belong to $A \subset X$. We only have to require that A is dense in X, so that ϕ_p can be well approximated by a new vertical direction.

5.3 *Fibers and Numerics*

Finite spectral interaction is a very convenient context for numerics. Any question related to solving $F(u) = g$ for some fixed $g \in Y$ reduces to a finite dimensional problem in situations of finite spectral interaction, irrespective of additional hypotheses. If the interaction involves a simple eigenvalue λ_p, one simply has to look at the restriction of F to the (one dimensional) fiber associated with the affine vertical line through g.

Smiley and Chun realized the implications of this fact for numerical analysis [31, 32]. An implementation for functions $F(u) = -\Delta u - f(u)$ defined on rectangles $\Omega \subset \mathbb{R}^2$ was presented in [7]. In the forecoming sections, we will require more stringent hypotheses with the scope of obtaining very well-behaved functions F—we will mostly be interested in global folds. Such additional restrictions might improve on computations, but so far this has not seen to lead to substantial improvements on the available algorithms.

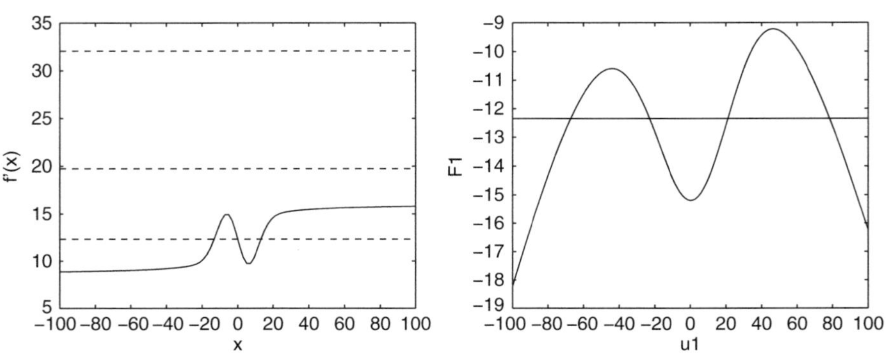

We present an example obtained from programs by José Cal Neto [7] and Otavio Kaminski. For $\Omega = [0,1] \times [0,2]$, $\lambda_1 \sim 12.337$ and $\lambda_2 \sim 19.739$. Consider

$$-u_{xx} - u_{yy} - f(u) = g\,, \quad (x,y) \in \Omega\,, \quad u = 0 \text{ in } \partial\Omega\,,$$

$$f'(x) = \frac{\lambda_2 - \lambda_1}{\pi}\left(\arctan(\frac{x}{10}) - \frac{2}{5}\,x\,e^{-(x/10)^2}\right) + \lambda_1\,, \quad f(0) \sim 47.12$$

$$g(x,y) = -100\big(x(x-1)y^2(y-2)\big) - 35\sin(\pi x)\sin(\frac{\pi y}{2})\,.$$

On the left, we show the graphs of f', which interacts only with λ_1. On the right, the height function h associated with the fiber obtained by inverting the vertical line through g. The height value -12.3 is reached by four preimages, displayed below. Notice the cameo appearance of the maximum principle: the four graphs sit one on top of the other as one goes up along the fiber (this is very specific of interactions with λ_1 of the Laplacian with Dirichlet conditions).

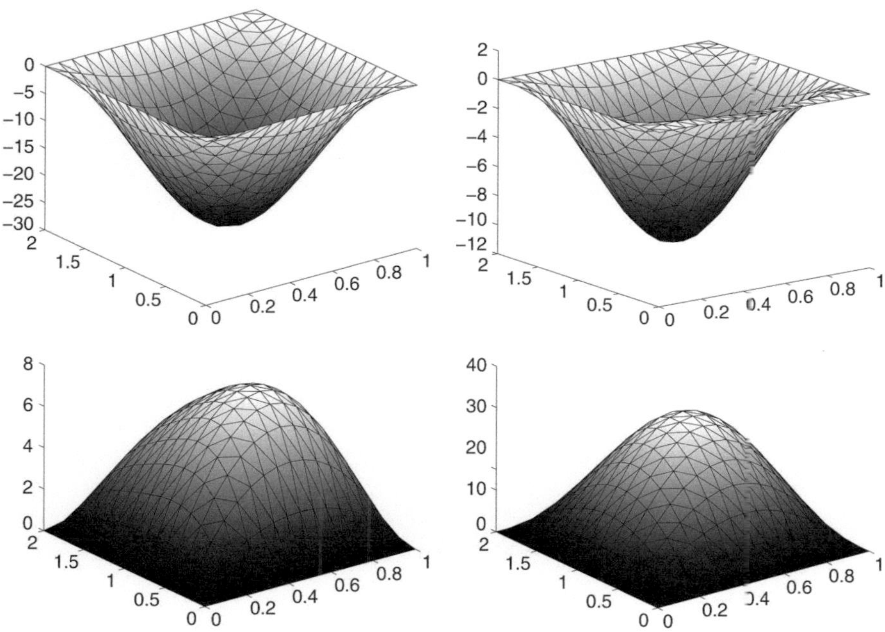

6 Asymptotics of F on fibers and vertical lines

We stick to one dimensional fibers and consider two issues.

1. How does F behave at infinity along fibers?
2. How do fibers look like at infinity ?

The first question, to say the very least, is tantamount to characterizing the image of F. The second is not relevant for the theoretical study of the global geometry of F, since a (global) coordinate system leading to a normal form (like $(z,t) \mapsto (z, -t^2)$) is insensitive to the shape of fibers. On the other hand, for numerical purposes, a uniform behavior at infinity of the fibers is informative.

6.1 F along fibers

The inverse of a vertical line $z_0 + V_Y, z_0 \in H_Y$ is the fiber $u(z_0,t) = w(z_0,t) + t\phi_p$:

$$F(u(z_0,t)) = z_0 + h^a(z_0,t)\,\phi_p \,. \qquad (*)$$

For a fixed $z_0 \in H_Y$, the C^1 map $t \mapsto h^a(z_0,t)$ is the *adapted height function* of the fiber associated with z_0. Clearly,

$$h^a(z_0,t) = \langle F(u(z_0,t)), \phi_p\rangle = \langle L(w(z_0,t) + t\phi_p) - N(u(z_0,t)), \phi_p\rangle$$

so that

$$h^a(z_0,t) = \lambda_p t - \langle N(u(z_0,t)), \phi_p\rangle.$$

In order to have

$$\lim_{t\to\pm\infty} \langle F(u(z_0,t)), \phi_p\rangle = \lim_{t\to\pm\infty} h^a(z_0,t) = -\infty$$

and some uniformity convenient to obtain properness as discussed in Section 6.4, we require an extra hypothesis:

For each $z_0 \in X$, there is a ball $U(z_0) \subset X$ and $\epsilon, T > 0, c_\pm$ such that, for $z \in U(z_0)$,

$$\langle N(u(z,t)), \phi_p\rangle > (\lambda_p + \epsilon)t + c_+ \,, \qquad \text{for } t > T \,, \qquad (V+)$$

$$\langle N(u(z,t)), \phi_p\rangle > (\lambda_p - \epsilon)t + c_- \,, \qquad \text{for } t < -T \,. \qquad (V-)$$

Notice that the asymptotic behavior on each fiber is the same.

6.2 Asymptotic geometry of fibers

Again, parameterize fibers as $u(z,t) = w(z,t) + t\phi_p$. Under mild hypotheses, the vectors $w(z,t)/t$ have a limit for $t \to \pm\infty$, which is independent of z. A version of this result was originally obtained by Podolak [27].

Proposition 6.1. *Suppose that $F : X \to Y$, $F = L - N$ satisfies hypothesis (H) of Section 3. Suppose also that, for every $u \in X$,*

$$\lim_{t\to+\infty} \frac{PN(tu)}{t} = N_\infty(u) \in Y.$$

Then there exist $w_+, w_- \in H_X$ such that, for every fiber $u(z,t) = w(z,t) + t\phi_p$,

$$\lim_{t\to+\infty} \|\frac{w(z,t)}{t} - w_+\|_X = 0\,, \quad \lim_{t\to-\infty} \|\frac{w(z,t)}{t} - w_-\|_X = 0$$

which are, respectively, the unique solutions of the equations

$$Lw - PN_\infty(w+\phi_p) = 0\,, \quad Lw + PN_\infty(-w-\phi_p) = 0.$$

It turns out that $N_\infty = PN_\infty$ satisfies the same Lipschitz bound that the functions PN_t in Theorem 3.1, which is why both equations are (uniquely) solvable.

Fibers are asymptotically vertical if and only if $\lim_{|t|\to\infty} w(z,t)/t = 0$, or equivalently, $PN_\infty(\pm\phi_p) = 0$. Indeed, in this case, $w = 0$ is the unique solution of both equations. This is what happens in the Ambrosetti-Prodi scenario, where $PN_\infty(u) = (b-\gamma)Pu^+ - (a-\gamma)Pu^-$ (recall $u = u^+ - u^-$), since $\phi_p = \phi_1 > 0$.

6.3 Comparing F on fibers and on vertical lines

One might wish to relate the heights of F along fibers and vertical lines, which are easier to handle. In [27] Podolak presented a scenario in which this is possible. We state a version of her result for the case $t \to +\infty$.

Theorem 6.2. *Let $X \subset Y$ be Hilbert spaces with X dense in Y. Let $L : X \to Y$ be a self-adjoint operator with $0 \in \sigma(L)$, a simple, isolated eigenvalue, associated with the normalized kernel vector ϕ_p. Set $H_Y = \langle\phi_p\rangle^\perp$. Take $N : Y \to Y$ and $F = L - N : X \to Y$ so that*

1. $\|N(u) - N(u_0)\|_Y \le \epsilon\|u - u_0\|_Y\,, \quad \lim_{t\to+\infty} N(tu)/t = N_\infty(u)$
2. $\langle N_\infty(\phi_p), \phi_p\rangle = -\lim_{t\to+\infty}\langle F(t\phi_p), \phi_p\rangle/t > 0$
3. $\epsilon\, \|\left(L|_{H_Y}\right)^{-1}\| < 1/2\,, \quad \epsilon^2\, \|\left(L|_{H_Y}\right)^{-1}\| < 1/2\, \langle N_\infty(\phi_p), \phi_p\rangle\,.$

Then, for each fiber (z_0, t) in adapted coordinates,

$$\left|\lim_{t\to+\infty} \frac{h^a(z_0,t)}{t} - \langle N_\infty(\phi_p), \phi_p\rangle\right| < \langle N_\infty(\phi_p), \phi_p\rangle.$$

The number $\langle N_\infty(\phi_p), \phi_p\rangle$ gives the asymptotic behavior of the height of F along the vertical line through the origin. The theorem implies that F along the upper part of each fiber converges to the same infinity that F along $\{t\phi_p,\ t \ge 0\}$.

A context in which these hypotheses apply is the Ambrosetti-Prodi operator with a piecewise nonlinearity $f(u) = (\lambda_p + c)u^+ - (\lambda_p - c)u^-$ for a sufficiently small number $c > 0$. However, for pairs $(\lambda_p - c_1, \lambda_p + c_2), p \neq 1$ in the Fučik spectrum of the (Dirichlet) negative second derivative, for which necessarily $c_1 \neq c_2$ (near λ_p), the condition involving ϵ^2 does not hold and indeed the thesis is not true.

6.4 Fibers and the properness of F

From a more theoretical point of view, fibers circumvent the fundamental issue of deciding if F is proper. For example [21], the map

$$F : C^1(\mathbb{S}^1) \to C^0(\mathbb{S}^1), \quad u \mapsto u' + \arctan(u)$$

is a diffeomorphism from the domain to the open region between two parallel planes,

$$\{\, y \in C^0(\mathbb{S}^1)\,,\ -\pi^2 < \int_0^{2\pi} y(\theta)d\theta < \pi^2 \,\}\,.$$

Indeed, fibers in this case are simply lines parallel to the vertical line of constant functions, and each is taken to such region.

Perhaps, it would be more appropriate to think of fibers as a tool to show properness [9]. As far as we know, for the Ambrosetti-Prodi map $F : X \to Y$ in unbounded domains, the properness has been proved only by making use of fibers (see Section 8).

Proposition 6.3. *The map $F : X \to Y$ satisfying hypotheses (H) of Section 3 and $(V\pm)$ above is proper if and only if the restriction of F to each fiber is proper.*

Points in the Fučik spectrum of the (Dirichlet) second derivative give rise to maps F which take the half-fiber $\{u(0,t), t \geq 0\}$ to a single point 0 [34], which shows that F is not proper, although the image of every vertical line has its vertical component taken to infinity.

A possible definition of a topological degree for F becomes innocuous — the relevant information is essentially the asymptotic behavior of F along each fiber.

7 Singularities

Generic singularities both of F and of each height function are very special — they are Morin singularities. Morin classified generic singularities of functions from $\mathbb{R}^n$ to $\mathbb{R}^n$ whose derivative at the singularity has one dimensional kernel [26]. This is sufficient for the study of critical points of height functions on one dimensional fibers, by Proposition 4.2. In order to do the same for the critical points of the whole function $F : X \to Y$, we need an equivalent classification for singularities of functions between infinite-dimensional spaces, which is very similar [13, 21, 29] — this is how we proceed next.

7.1 Morin theory in adapted coordinates

The first step in Morin's proof makes use of the implicit function theorem to write such a singularity at a point (z_0, t_0) in adapted coordinates, as in Section 4:

$$F^a : Y = H_Y \oplus V_Y \to Y = H_Y \oplus V_Y, \quad (z,t) \mapsto (z, h^a(z,t)).$$

Say F^a is C^{k+1}. The point (z_0, t_0) is a *Morin singularity of order k* if and only if

1. $D_t h^a(z_0,t_0) = \cdots = D_t^k h^a(z_0,t_0) = 0$, $D_t^{k+1} h^a(z_0,t_0) \neq 0$.
2. The Jacobian $D(h^a, D_t h^a, \ldots, D_t^{k-1} h^a)(z_0,t_0)$ has maximum rank.

Then, in a neighborhood of (z_0, t_0) there is an additional change of variables which converts F^a to the normal form

$$(\tilde{z}, x, t) \mapsto (\tilde{z}, x, t^{k+1} + x_1 t^{k-1} + \cdots + x_{k-1} t) \ .$$

Here the coordinates $(\tilde{z}, x)$ correspond to an appropriate splitting of $Y = \tilde{Y} \oplus \mathbb{R}^{k-1}$.

Morin singularities of order 1, 2, 3, and 4 are called, respectively, folds, cusps, swallowtails and butterflies.

Thus, the classification of critical points of F boils down to the study of a family of one dimensional maps, the height functions restricted on fibers. The first requirement is specific to each fiber (i.e., one checks it for every fixed z near z_0), whereas the second relates nearby fibers, i.e., one has to change z. Folds are structurally simpler than deeper singularities: the behavior along fibers near a fold point is always the same — essentially like $t \mapsto -t^2$, whereas this is not the case for cusps, where close to $t \mapsto t^3$ one finds $t \mapsto t^3 \pm \epsilon t$.

There is something unsatisfying in the fact that the relevant properties of the critical points of F requires knowledge of some version of the height function. This is circumvented by the next result [9].

Proposition 7.1. *Suppose $F : X \to Y$ is C^{k+1} and admits one dimensional fibers. Then there is an open neighborhood U of the critical set C with the properties below.*

1. *There is a unique C^k map $\lambda_p : U \to \mathbb{R}$ for which $\lambda_p = 0$ on C and is an eigenvalue of DF elsewhere.*
2. *There is a strictly positive C^k function $p : U \to \mathbb{R}^+$ such that*

$$\lambda_p(u(z,t)) = p(u(z,t))\, D_t h(u(z,t)) \ , \quad u(z,t) \in U \ .$$

A point $u_0 = u(z_0, t_0)$ is a Morin singularity of order k of F if and only if

1. $\lambda_p(u_0) = \cdots = D_t^{k-1}\lambda_p(u_0) = 0\,,\, D_t^k\lambda_p(u_0) \neq 0\,,$
2. *The image of $D(\lambda_p, \ldots, D_t^{k-2}\lambda_p)(u_0)$ together with $D_t\lambda_p(u_0)$ span $\mathbb{R}^n$.*

There is an analogous characterization in adapted coordinates.

7.2 Critical points of the height function

Consider a critical point $u_0 \in C \subset X$ and the fiber $u(z_0, t)$ through it, $u(z_0, t_0) = u_0$. From Proposition 7.1, u_0 is a (topological) fold of the height function h restricted to the fiber if and only if u_0 is a topologically simple root of $\lambda_p(u)$ along the fiber, i.e., λ_p is strictly negative on one side of u_0 and strictly positive on the other.

Once we reduce the issue to checking an eigenvalue along a fiber, *derivatives are irrelevant*: just study the quadratic form of the Jacobian. Clearly, this only handles topological equivalence between the function and a fold.

More explicitly, in standard Ambrosetti-Prodi contexts, $\lambda_1(u_0)$ is the minimum value of the quadratic form $\langle DF(u_0)v, v\rangle$. The derivative $D_t u(z_0, t_0)$ of the (C^1) fiber is the eigenfunction $\phi_1(u_0) > 0$, and it is easy to check that λ_1 increases with t by the convexity of the nonlinearity f. This should be compared with differentiability arguments, which require some estimate on $\phi_1(u_0)$ (say, boundedness).

The fact that all critical points are local maxima for height functions on fibers, as required in Proposition 4.3, suggests hypotheses to be checked only on the critical set of F. This is not the case in the original Ambrosetti-Prodi theorem: the statement of the theorem has the merit that it makes no reference to the critical set at all, an object which in principle is hard to identify. The convexity of the nonlinearity handles the difficulty and, rather surprisingly, is essentially necessary [8]. Further examples yielding local maximality are somewhat contrived.

8 Some examples

8.1 The non-autonomous case

The geometric formulation $F = L - N$ is not sufficient to accommodate situations of the form $F(u(x)) = -\Delta u(x) + f(x, u(x))$, the so-called non-autonomous case. Hammerstein [16] had already considered homeomorphisms of that form. A possibility is requiring that X and Y are function spaces defined on a domain Ω, so that the variable x makes sense. The formalism above carries over to this scenario without surprises.

More precisely, as usual X and Y are Hilbert spaces, X dense in Y. The linear operator $L : X \subset Y \to Y$ is self-adjoint with a simple eigenvalue λ_p associated with a normalized eigenvector ϕ_p. Let $P : Y \to H_Y = \langle \phi_p \rangle^\perp$ be the orthogonal projection.

From the nonlinear term $N : \Omega \times Y \to Y$, define as before $PN_t : H_Y \to H_Y, t \in \mathbb{R}$ and require a Lipschitz estimate,

$$\|PN_t(x, w_1) - PN_t(x, w_0)\|_Y \leq n\|w_1 - w_0\|_Y, \qquad \text{for } w_0, w_1 \in H_Y,$$

so that $[-n, n] \cap \sigma(L) = \{\lambda_p\}$, which is the same hypothesis (H) in Section 3. This obtains fibers for $F : X \to Y$ as in Theorem 3.1, which satisfy the same properties as those in the autonomous case, in particular, Proposition 3.2.

The hypothesis which obtain appropriate asymptotic behavior of F along fibers are the obvious counterparts of $(V+)$ and $(V-)$ in Section 6.1. For the classification of critical points, we simply do not distinguish between the autonomous and non-autonomous case: the subject has become a geometric issue.

8.2 Schrödinger operators on $\mathbb{R}^n$

As was surely known by Ambrosetti and Prodi (and [2] is an interesting example), the Laplacian with Dirichlet conditions might be replaced by more general self-adjoint operators. The approach in this text is flexible enough to handle nonlinear perturbations of Schrödinger operators on unbounded domains yielding global folds. In our knowledge there are no similar results in the literature. Tehrani [33] obtained counting results for Schrödinger operators in $\mathbb{R}^n$ in the spirit of those obtained by Podolak [27], indicated in Section 5.1 .

We state the by now natural hypotheses. Here $Y = L^2(\mathbb{R}^n)$.

1. The free operator $T = -\Delta + v(x) : X \subset Y \to Y$ is self-adjoint, with simple, isolated, smallest eigenvalue λ_1 and positive ground state ϕ_1.
2. $F : X \subset Y \to Y, F(u) = Tu - f(u)$ is a C^1 map.
3. The function $f \in C^2(\mathbb{R})$ satisfies $f(0) = 0, M \geq f'' > 0, f'(\mathbb{R}) = (a, b)$ and $a < \lambda_1 < b < \min\{\sigma(T) \setminus \{\lambda_1\}\}$.
4. The Jacobians $DF(u) : X \to Y$ are self-adjoint operators with eigenpair $(\lambda_1(u), \phi_1(u))$ sharing the properties of (λ_1, ϕ_1).

Theorem 8.1. *Under these hypotheses, the map $F : X \to Y$ is a global fold.*

Such hypotheses are satisfied for $v(x) = x^2/2$, the one dimensional quantum harmonic oscillator, as well as for the hydrogen atom in $\mathbb{R}^3$, for which $v(x) = -1/|x|$.

Hypotheses on the potential of a Schrödinger operator in order to obtain such properties are commonly studied in mathematical physics. The interested reader might consider [4, 18, 20, 28]. More about this in [9].

8.3 Perturbations of compact operators

We recall Mandhyan's second example of a global fold [22, 23], or better, a special case of the extension given by Church and Timourian [11].

For $\Omega \subset \mathbb{R}^n$ a compact subset, let $X = C^0(\Omega)$ and define the compact operator

$$K : X \to X, \quad K(u)(x) = \int_\Omega k(x,y)u(y)dy$$

where the kernel $k \in C^0(\Omega \times \Omega)$ is symmetric and positive. Let $\mu_1 > \mu_2$ be the largest eigenvalues of K. Now let $f : \mathbb{R} \to \mathbb{R}$ be a strictly convex C^2 function satisfying

$$0 < \lim_{x \to -\infty} f'(x) < 1/\mu_1 < \lim_{x \to \infty} f'(x) < 1/|\mu_2| .$$

Theorem 8.2. *Under these hypotheses for K and f, the map*

$$G : X \to X, \quad G(u)(x) = u(x) - Kf(u(y))$$

is a global fold.

This is the kind of nonlinear map obtained if one started from the Ambrosetti-Prodi original operator $F(u) = -\Delta u - f(u)$ and inverted the Laplacian. Actually, one could take another track: instead of inverting the linear part, one might consider the inversion of the nonlinear map $u \mapsto f(u)$, since f' is bounded away from zero. For maps $G(u) = Ku - f(u)$ obtained this way, we handle the case when K is a general compact symmetric operator K.

More precisely, let $\Omega \subset \mathbb{R}^n$, $B = C^0(\Omega)$ and $Y = L^2(\Omega)$. Let $K : B \to B$ and $K : Y \to Y$ be compact operators which preserve the cone of positive functions. Also, $K : Y \to Y$ has simple largest eigenvalue $\lambda_p = \|K\|$ and second largest eigenvalue λ_s . Let $f : \mathbb{R} \to \mathbb{R}$ be a strictly convex C^2 function, with $f(0) = 0$ if Ω is unbounded. Suppose

$$\lambda_s < a = \lim_{t \to -\infty} f'(t) < \lambda_p < b = \lim_{t \to \infty} f'(t) .$$

Theorem 8.3. *The map $F : B \to B$, $F(u) = Ku - f(u)$ is a global fold.*

The reader should notice that F is Lipschitz but not differentiable as a map from $L^2(\Omega)$ to itself. Still, the direct construction of fibers in $C^0(\Omega)$ is not a simple matter, because properness of F is not immediate. Transplanting fibers in this example is convenient, and was also used in Mandhyan's context.

8.4 *Folds as perturbations of non-self-adjoint operators*

McKean and Scovel [11, 25] studied the Riccati-like map on functions

$$u \in L^2([0,1]) \mapsto u + (D_2)^{-1} f(u) \in L^2([0,1]), \quad f(x) = x^2/2,$$

where $(D_2)^{-1}$ is the inverse of the second derivative acting on $W^{1,2}([0,1])$ and showed that the critical set consists of a countable union of (topological) hyperplanes. Church and Timourian [7] showed that the restriction of such map to a neighborhood of one specific critical component is (after global homeomorphic change of variables) a fold. The techniques employed are in the spirit of the original Ambrosetti-Prodi paper.

Fibers were relevant in [21], where perturbations of first order differential equations (clearly, non-self-adjoint operators) were shown to be global folds. An example is the map on periodic functions with (generic) convex nonlinearities f,

$$F : C^1(\mathbb{S}^1) \to C^0(\mathbb{S}^1), \quad u \mapsto u' + f(u) .$$

McKean and Scovel [25] and Kappeler and Topalov [17] considered the same map among Sobolev spaces, the celebrated *Miura map*, used as a change of variables between the Korteweg-deVries equation and its so called modified version.

More recently, a perturbation of a non-self-adjoint elliptic operator (as in [3], but with Lipschitz boundary) has been shown to yield a global fold [30].

Acknowledgements The first author thanks the Departamento de Matemática, PUC-Rio, for its warm hospitality. The second and third authors gratefully acknowledge support from CAPES, CNPq, and FAPERJ. We thank José Cal Neto and Otavio Kaminski for the numerical examples.

References

1. Ambrosetti, A., Prodi, G.: On the inversion of some differentiable mappings with singularities between Banach spaces. Ann. Mat. Pura Appl. **93**, 231–246 (1972)
2. Berestycki, H.: Le nombre des solutions de certain problèmes sémi-linéaires elliptiques. J. Funct. Anal. **40**, 1–29 (1981)
3. Berestycki, H., Nirenberg, L., Varadhan, S.R.S.: The principal eigenvalue and maximum principle for second-order elliptic operators in general domains. Commun. Pure Appl. Math. **47**, 47–92 (1994)
4. Berezin, F.A., Shubin, M.A.: The Schrödinger Equation. Kluwer, Dordrecht (1991)
5. Berger, M.S., Church, P.T.: Complete integrability and perturbation of a nonlinear Dirichlet problem. I. Indiana Univ. Math. J. **28**, 935–952 (1979)
6. Berger, M.S., Podolak, E.: On the solutions of a nonlinear Dirichlet problem. Indiana Univ. Math. J. **24**, 837–846 (1974)
7. Cal Neto, J.T., Tomei, C.: Numerical analysis of semilinear elliptic equations with finite spectral interaction. J. Math. Anal. Appl. **395** 63–77 (2012)

8. Calanchi, M, Tomei, C, Zaccur, A.: Abundance of cusps and a converse to the Ambrosetti-Prodi Theorem, preprint, (2015)
9. Calanchi, M., Tomei, C., Zaccur, A.: Fibers and global folds in infinite dimension, in preparation. (2015)
10. Church, P.T., Timourian, J.G.: Global fold maps in differential and integral equations. Nonlinear Anal. **18**, 743–758 (1992)
11. Church, P.T., Timourian, J.G.: Global structure for nonlinear operators in differential and integral equations I. Folds. In: Topological Nonlinear Analysis, II. Progress in Nonlinear Differential Equations and Their Applications, vol. 27, pp. 109–160. Birkhäuser, Boston (1997)
12. Church, P.T., Timourian, J.G.: Global structure for nonlinear operators in differential and integral equations II. Cusps. In: Topological Nonlinear Analysis, II. Progress in Nonlinear Differential Equations and Their Applications, vol. 27, pp. 161–246. Birkhäuser, Boston (1997)
13. Church, P.T., Dancer, E.N., Timourian, J.G.: The structure of a nonlinear elliptic operator. Trans. Am. Math. Soc. **338**, 1–42 (1993)
14. Costa, D.G., Figueiredo, D.G., Srikanth, P.N.: The exact number of solutions for a class of ordinary differential equations through Morse index computation. J. Differ. Equ. **96**, 185–199 (1992)
15. Dolph, C.L.: Nonlinear integral equations of the Hammerstein type. Trans. Am. Math. Sci. **66**, 289–307 (1949)
16. Hammerstein, A.: Nichtlineare Integralgleichungen nebst Anwendungen. Acta Math. **54**, 117–176 (1929)
17. Kappeler, T., Topalov, P.: Global fold structure of the Miura map on $L^2(\mathbb{S})$. Int. Math. Res. Not. **39**, 2039–2068 (2004)
18. Kato, T.: Perturbation Theory for Linear Operators. Classics in Mathematics. Springer, New York (1980)
19. Lazer, A.C., McKenna, P.J.: On the number of solutions of a nonlinear Dirichlet problem. J. Math Anal. Appl. **84**, 282–294 (1981)
20. Lieb, E.H., Loss, M.: Analysis. Graduate Studies in Mathematics, vol. 14. American Mathematical Society, Providence (2001)
21. Malta, I., Saldanha, N., Tomei, C.: Morin singularities and global geometry in a class of ordinary differential operators. Topol. Methods Nonlinear Anal. **10**, 137–169 (1997)
22. Mandhyan, I.: Examples of global normal forms for some simple nonlinear integral operators. Nonlinear Anal. **13**,1057–1066 (1989)
23. Mandhyan, I.: The diagonalization and computation of some nonlinear integral operators. Nonlinear Anal. **23**, 447–466 (1994)
24. Manes, A., Micheletti, A.M.: Un'estensione della teoria variazionale classica degli autovalori per operatori ellittici del secondo ordine. Boll. Unione Mat. Ital. **7**, 285–301 (1973)
25. McKean, H.P., Scovel, J.C.: Geometry of some simple nonlinear differential operators. Ann. Scuola Norm. Sup. Pisa Cl. Sci. **13**, 299–346 (1986)
26. Morin, B.: Formes canoniques des singularités d'une application différentiable. C. R. Acad. Sci. **260**, 5662–5665, 6503–6506 (1965)
27. Podolak, E.: On the range of operator equations with an asymptotically nonlinear term. Indiana Univ. Math. J. **25**(12), 1127–1137 (1976)
28. Reed, M., Simon, B.: Fourier Analysis, Self-Adjointness. Academic Press, New York (1975)
29. Ruf, B.: Singularity theory and bifurcation phenomena in differential equations. In: Topological Nonlinear Analysis, II. Progress in Nonlinear Differential Equations and Their Applications, vol. 27, pp. 315–395. Birkhäuser, Boston (1997)
30. Sirakov, B., Tomei, C., Zaccur, A.: Global folds given by perturbations of non-self-adjoint elliptic operators, in preparation (2015)
31. Smiley, M.W.: A finite element method for computing the bifurcation function for semilinear elliptic BVPs. J. Comput. Appl. Math. **70**, 311–327 (1996)

32. Smiley, M.W., Chun, C.: Approximation of the bifurcation equation function for elliptic boundary value problems. Numer. Methods Partial Differ. Equ. **16**, 194–213 (2000)
33. Tehrani, H.T.: A multiplicity result for the jumping nonlinearity problem. J. Differ. Equ. **188**, 272–305 (2003)
34. Tomei, C., Zaccur, A.: Geometric aspects of Ambrosetti-Prodi operators with Lipschitz nonlinearities. In: Analysis and Topology in Nonlinear Differential Equations. Progress in Nonlinear Differential Equations and Their Applications, vol. 85, pp. 445–456. Springer, Cham (2014)

Existence of infinitely many continua of radial singular polytropes with gain-loss function

Alfonso Castro and Víctor Padrón

Mathematics Subject Classification (2010): 85A15, 34B15, 35B33, 35J25, 35J60

1 Introduction

We consider the existence of *radial singular polytropes*, i.e. singular positive radial singular solutions to

$$\begin{cases} \Delta u + \lambda(u^p - u^q) = 0, & \|x\| < 1, \\ u(x) = 1, & \text{for } \|x\| = 1, \end{cases} \tag{1}$$

for $\lambda > 0$, $N/(N-2) < p < (N+2)/(N-2)$, $N > 2$, and $q < p$. A radial singular solution to (1) is a radially symmetric function $u : B := \{x \in \mathbb{R}^N; \|x\| \leq 1\} \to \mathbb{R}$ such that $u \in L^p(B) \cap H^{1,1}(B)$, $u(x) = 1$ for $\|x\| = 1$, $\lim_{x\to 0} u(x) = +\infty$, and

$$\int_B \nabla u(x) \cdot \nabla \varphi(x) - \lambda(u^p(x) - u^q(x))\varphi(x)dx = 0, \tag{2}$$

A. Castro (✉)
Department of Mathematics, Harvey Mudd College, Claremont, CA 91711, USA
e-mail: castro@g.hmc.edu

V. Padrón
Department of Mathematics, Normandale College, Bloomington, MN 55431, USA
e-mail: victor.padron@normandale.edu

A.N. Carvalho et al. (eds.), *Contributions to Nonlinear Elliptic Equations and Systems*, Progress in Nonlinear Differential Equations and Their Applications 86,
DOI 10.1007/978-3-319-19902-3_6

for any function $\varphi : B \to \mathbb{R}$ of class C^∞ and compact support in B. In particular, u is a solution to (1) in the sense of distributions.

The reader is referred to [2] for a derivation of (1) in the modeling of a static configuration in which the transfer of energy takes place entirely by thermal conduction. The heat diffusion coefficient, heat generation, and heat radiation are modeled, respectively, by

$$\kappa = \kappa_0 v^k\ , \quad \Gamma = \Gamma_0 v^m, \quad \Lambda = \Lambda_0 v^n\ , \tag{3}$$

where κ_0, k, Γ_0, m, Λ_0, and n are given constants, and v denotes the temperature.

The singular radial solutions to (1) are the functions $u : (0, 1] \to \mathbb{R}$ that satisfy (2), $u(1) = 1$, and $\lim_{r\to 0+} u(r) = +\infty$. By elliptic regularity, singular radial solutions $u(r)$ are regular solution of (1) in $0 < r \le 1$.

Motivated by the results in [1] we have the following extensions of the classification of regular solutions to (1) obtained in [2].

Theorem 1.1. *Let* $p \in (N/(N-2), (N+2)/(N-2))$ *and* $q \le -1$. *For each positive integer k there exists a two-parameter family of singular solutions* $u(\cdot, a, b, \lambda(a, b))$ *to* (1) *such that* $u(\cdot, a, b, \lambda(a, b)) - 1$ *has exactly k zeroes in* $(0, 1]$.

Theorem 1.2. *Let* $p \in (N/(N-2), (N+2)/(N-2))$ *and* $q > -1$. *For each positive integer k there exists a three-parameter family of singular solutions* $u(\cdot, u_0, u_0', r_0)$ *to* (1) *such that* $u(\cdot, u_0, u_0', r_0, \lambda(u_0, u_0', r_0)) - 1$ *has exactly k zeroes in* $(0, 1]$.

We note that while the singular solutions obtained in Theorem 1.1 are derived from solutions with large energy near zero for $\lambda = 1$, the solutions obtained in Theorem 1.2 come from solutions with negative energy near zero also for $\lambda = 1$. See Lemmas 2.5 and 2.7 and Figure 1 where numerical calculations have been used to infer the oscillations and singularity for the case $q > -1$.

The reader is referred to [3] for the role of (1) in radiative equilibrium of stars. See also [5]. For recent studies on related quasilinear equations, see [4].

2 Existence of a two parameter family of singular solutions

For $p \in (N/(N-2), (N+2)/(N-2))$, $N > 2$, and $q < p$ we consider the existence of solutions to

$$\begin{cases} u'' + \dfrac{N-1}{r}u' + u^p - u^q = 0, \quad 0 < r \le 1, \\ u(r_0) = u_0, \\ u'(r_0) = u_0', \end{cases} \tag{4}$$

for $0 < r_0 \le 1$, $u_0 > 0$, and $u_0' \in \mathbb{R}$.

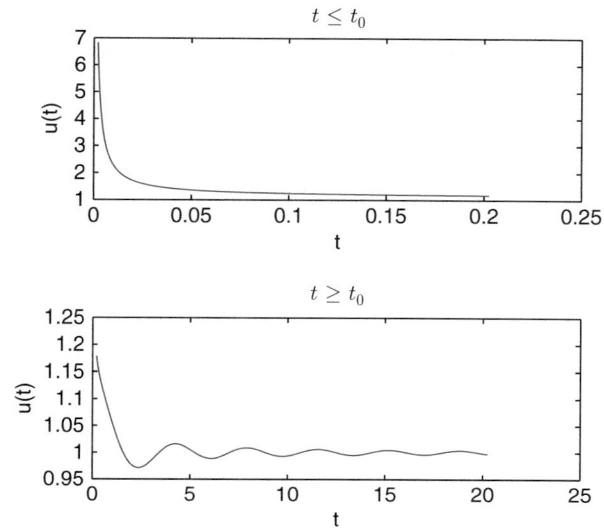

Fig. 1 Numerical simulation of a singular solution to (4) with $N = 3$, $p = 4$, $q = 1$, $t_0 = 0.2$, $u_0 = .1786$, and $u_0' = -0.3462$

From now on we write

$$f(t) = t^p - t^q, \text{ and } F(t) = \frac{t^{p+1}}{p+1} - \frac{t^{q+1}}{q+1}, \tag{5}$$

for $q + 1 \neq 0$. If $q + 1 = 0$, we replace the corresponding fraction by $\ln(t)$. For any solution $u(r)$ of (4) we define

$$E(r) = \frac{(u'(r))^2}{2} + F(u(r)), \tag{6}$$

a straightforward calculation shows that

$$E'(r) = \frac{-N+1}{r}(u'(r))^2. \tag{7}$$

Lemma 2.1. *For each* (u_0, u_0'), *there exists a unique function* $u : (0, 1] \to \mathbb{R}$ *satisfying* (4). *Such a solution depends continuously on* (u_0, u_0'), *and* $u(x) = u(\|x\|) \in L^p(B) \cap H^{1,1}(B)$ *and satisfies the partial differential equation in* (1) *in the sense of distribution.*

Proof. See Lemma 1 in [1]. □

We base our analysis of singular solutions to (1) on the existence and energy properties of singular solutions to (4), which are presented in the following lemmas.

Let $\theta = 2/(1-p)$, $A^{p-1} = -\theta(\theta + N - 2) = 2((N-2)(p-1)-2)/(p-1)^2$, $\tau_2 \in \left(\theta, \frac{2-N}{2}\right)$, and $\tau_1 \in (\theta, \tau_2)$. Since $p > N/(N-2)$, there exists $c > 0$ such that

$$\frac{(\theta A + \tau_2 c)^2}{2} + \frac{(A+c)^{p+1}}{p+1} + \frac{N-2}{2}(A+c)(\theta A + \tau_2 c) = 0. \tag{8}$$

Let $I = [a_1, a_2] \subset (0, \infty)$ be a compact interval, and $b_0 := a_2^{(\tau_2-\theta)/(\tau_1-\theta)}$. For $a \in I$ and $b \geq b_0$ we denote by $u(\cdot, a, b)$ the solution of (4) such that

$$\begin{aligned} u(\tilde{b}, a, b) &= (A+c)\tilde{b}^{\theta} + a\tilde{b}^{\tau_1} \\ u'(\tilde{b}, a, b) &= (\theta A + \tau_2 c)\tilde{b}^{\theta-1} + \tau_1 a \tilde{b}^{\tau_1 - 1}, \end{aligned} \tag{9}$$

with

$$\tilde{b} = b^{1/(\theta-\tau_2)}. \tag{10}$$

For future reference we note if $b \geq b_0$ and $a \in I$ then

$$ab^{\tau_1} \leq \tilde{b}^{\theta}. \tag{11}$$

For now on we denote

$$\begin{aligned} Q(r) &= r^{N-1}\left(rE(r) + \frac{N-2}{2}u(r)u'(r)\right), \\ \gamma_1 &= \left(\frac{N}{p+1} - \frac{N-2}{2}\right) > 0, \quad \gamma_2 = \left(\frac{N}{q+1} - \frac{N-2}{2}\right), \\ &\text{and } \Gamma(u) = \gamma_1 u^{p+1} - \gamma_2 u^{q+1} \end{aligned} \tag{12}$$

Let $0 < r_1 < r_2 \leq 1$. Multiplying (4) by $r^{N-1}u'$ and integrating on $[r_1, r_2]$, then multiplying (4) by $r^{N-1}u$ and integrating on $[r_1, r_2]$, and finally combining common terms one has the following *Pohozaev* identity

$$Q(r_2) = Q(r_1) + \int_{r_1}^{r_2} s^{N-1}\Gamma(u(s))\, ds. \tag{13}$$

Lemma 2.2. *There exists $b_1 \geq b_0$ such that, for $b \geq b_1$, $\lim_{r\to 0} u(r, b, a) = +\infty$ and $u - 1$ has no zero in $(0, \tilde{b})$.*

Proof. See Lemma 2 in [1]. Let $m \in (0, 1)$ such that

$$\theta m^{1-N}(m-1) < 1 \quad \text{and} \quad \tau_2 + 2^p(1-m)(A+c+1)^{p-1} < 0. \tag{14}$$

Suppose that there exists $r \in [m\tilde{b}, \tilde{b})$ such that $u'(r) > 0$. Since $u'(\tilde{b}) < 0$, there exists $r_1 \in (m\tilde{b}, \tilde{b})$ such that $u'(r_1) = 0$ and $u'(s) < 0$ for $s \in (r_1, \tilde{b}]$.

Let $b_1 \geq b_0$ such that $u^p - u^q > 0$. For $s \in [r_1, \tilde{b}]$,

$$u(s) \leq u(\tilde{b}) + u'(t)(s - \tilde{b}) \leq u(\tilde{b}) - u'(t)(1-m)\tilde{b} \tag{15}$$

Multiplying (4) by r^{N-1} and integrating one obtain

$$-u'(t) \leq -\left(\frac{\tilde{b}}{t}\right)^{N-1} u'(\tilde{b}) \leq -m^{1-N}u'(\tilde{b})$$

This, (14), (15), and the fact that $\tilde{b}u'(\tilde{b}) > \theta u(\tilde{b})$, implies

$$u(s) \leq 2u(\tilde{b}) \quad \text{for all} \quad s \in [r_1, \tilde{b}]. \tag{16}$$

Therefore

$$\begin{aligned}
0 = r_1^{N-1}u'(r_1) &\leq \tilde{b}^{N-1}u'(\tilde{b}) + \frac{2^p u^p(\tilde{b})}{N}(\tilde{b}^N - r_1^N) \\
&\leq \tilde{b}^{N-1}u'(\tilde{b}) + \frac{2^p u^p(\tilde{b})}{N}(1-m^N)\tilde{b}^N \\
&\leq \tilde{b}^{N-2}u(\tilde{b})\left[\frac{\tilde{b}u'(\tilde{b})}{u(\tilde{b})} + (1-m)2^p\tilde{b}^2u^{p-1}(\tilde{b})\right] \\
&\leq \tilde{b}^{N-2}u(\tilde{b})\left[\tau_2 + (1-m)2^p(A+c+1)^{p-1}\right] < 0.
\end{aligned}$$

This contradiction proves that $u' < 0$ on $[m\tilde{b}, \tilde{b}]$ and, hence, $u(t) \geq u(\tilde{b})$ for all $t \in [m\tilde{b}, \tilde{b}]$.

Also

$$Q(\tilde{b}) = \tilde{b}^{N-1}\left[\frac{\tilde{b}}{2}u'(\tilde{b})^2 + \tilde{b}\left(\frac{u(\tilde{b})^{p+1}}{p+1} - \frac{u(\tilde{b})^{q+1}}{q+1}\right) + \frac{N-2}{2}u(\tilde{b})u'(\tilde{b})\right].$$

It follows that

$$\begin{aligned}
\frac{\tilde{b}}{2}u'(\tilde{b})^2 &= \frac{(\theta A + \tau_2 c)^2}{2}\tilde{b}^{2\theta-1} + O(\tilde{b}^{\theta+\tau_1-2}), \\
\tilde{b}\left(\frac{u(\tilde{b})^{p+1}}{p+1} - \frac{u(\tilde{b})^{q+1}}{q+1}\right) &= \frac{(A+c)^{p+1}}{p+1}\tilde{b}^{2\theta-1} + O(\tilde{b}^{\theta+\tau_1-2}), \\
\frac{N-2}{2}u(\tilde{b})u'(\tilde{b}) &= \frac{N-2}{2}(A+c)(\theta A + \tau_2 c)\tilde{b}^{2\theta-1} + O(\tilde{b}^{\theta+\tau_1-2}).
\end{aligned}$$

This and (8) imply

$$Q(\tilde{b}) = O(\tilde{b}^{N+\theta+\tau_1-2}). \tag{17}$$

Therefore

$$\begin{aligned} Q(m\tilde{b}) &= Q(\tilde{b}) - \int_{m\tilde{b}}^{\tilde{b}} s^{N-1}\Gamma(u(s))\,ds \le Q(\tilde{b}) - Cu(\tilde{b})^{p+1}\tilde{b}^N \\ &= O(\tilde{b}^{N+\theta+\tau_1-2}) - O(\tilde{b}^{N+2\theta-2}) = -\tilde{b}^{N+2\theta-2}\left[1 + O(\tilde{b}^{\tau_1-\theta})\right] < 0. \end{aligned}$$

It follows from this and (13) that $Q(r) < 0$ for $r \in (0, m\tilde{b}]$. This implies that $u'(t) < 0$ for all $t \in (0, m\tilde{b}]$. Hence $\lim_{r\to 0} u(r, b, a) = +\infty$, which proves the lemma. □

Lemma 2.3. *Let K be a positive real number. Let ϕ satisfy*

$$\phi''(t) + \frac{N-1}{t}\phi'(t) + a(t)\phi(t) = 0, \ t > t_0 \ge 0, \ \phi(t_0) > 0. \tag{18}$$

If $\phi'(t_0) \le 0$ and $a(t) \ge K$ while $\phi(t) \ge 0$, then $\phi(t) = 0$ for some $t \in [t_0, t_0 + \frac{\mu_1}{\sqrt{K}}]$.

Proof. See Lemma 1 in [2]. □

Lemma 2.4. *There exists $r_1 > \tilde{b}$ such that*

$$u(r_1, a, b) = 1,$$

and $u(r, a, b) > 1$ for $0 < r < r_1$.

Proof. Let $d > 1$ be such that $F(d) = E(\tilde{b})$. It follows from (7) that $u(r, a, b) \le d$ for $r \ge \tilde{b}$. Let $I_d = [1, d]$.

Since $f'(1) = p - q > 0$, there exists $\epsilon_1 > 0$ and $\delta_1 > 0$ such that

$$f'(x) > \delta_1 \tag{19}$$

for $x \in [1, 1 + \epsilon_1]$. Hence, if $x \in (1, 1 + \epsilon_1] - \{1\}$, then by the mean value theorem we have

$$\frac{f(x)}{x-1} = f'(\xi) > \delta_1, \tag{20}$$

where $\xi \in (1, 1 + \epsilon_1)$.

On the other hand, by compactness, $\frac{f(x)}{x-1}$ is bounded away from zero on $I_d - (1, 1 + \epsilon_1)$. This and (20) imply that there exists $\delta > 0$ such that

$$\frac{f(x)}{x-1} \ge \delta \ \text{ for all } \ x \in I_d, \tag{21}$$

where the function $f(x)/(x-1)$ is extended to 1 as $f'(1) = p - q$.

Letting $\phi = u - 1$ it follows that $\phi(r)$ satisfies the following equation

$$\phi''(r) + \frac{N-1}{r}\phi'(r) + \frac{f(u(r))}{\phi}\phi(r) = 0, \ r > \tilde{b},$$

and $\phi'(\tilde{b}) = u'(\tilde{b}) < 0$, and for $r > \tilde{b}$ with $\phi \geq 1$,

$$\frac{f(u)}{\phi} > \delta.$$

It follows from Lemma 2.3 that $\phi(r_1) = 0$ for some $r_1 \in [\tilde{b}, \tilde{b} + \frac{\mu_1}{\sqrt{\delta}}]$. This proves the lemma. □

Lemma 2.5. *If $q \leq -1$, then there exist positive numbers*

$$s_1(a,b) < s_2(a,b) < \cdots < s_k(a,b) < \cdots \to \infty$$

such that $u(s_i(a,b), a, b) = 1$, for $i = 1, 2, \ldots$, and $u(\cdot, a, b) \neq 1$ on (s_i, s_{i+1}), for $i = 0, 1, \ldots$ where $s_0 = 0$.

Proof. Let $\bar{r} > 0$ such that $u(\bar{r}, a, b) > 1$, and $d > 1$ such that $F(d) = E(\bar{r})$. From the definition of F it follows that there exists $d' < 1$ such that $F(d') = F(d)$, and $F(t) \leq F(d)$ if and only if $t \in I_d := [d', d]$. By (7), for $r > \bar{r}$, we have

$$F(u(r)) \leq E(\bar{r}) - \frac{[u'(r)]^2}{2} \leq F(d).$$

Thus

$$u(r) \in I_d \ \text{ for any } \ r > \bar{r}. \tag{22}$$

Since $f'(1) = p - q > 0$, there exists $\epsilon_1 > 0$ and $\delta_1 > 0$ such that

$$f'(x) > \delta_1 \tag{23}$$

for $x \in [1 - \epsilon_1, 1 + \epsilon_1]$. Hence, if $x \in [1 - \epsilon_1, 1 + \epsilon_1] - \{1\}$ then by the mean value theorem we have

$$\frac{f(x)}{x-1} = f'(\xi) > \delta_1, \tag{24}$$

where $\xi \in (1 - \epsilon_1, 1 + \epsilon_1)$.

On the other hand, by compactness, $\frac{f(x)}{x-1}$ is bounded away from zero on $I_d - (1-\epsilon_1, 1+\epsilon_1)$. This and (24) imply that there exists $\delta > 0$ such that

$$\frac{f(x)}{x-1} \geq \delta \text{ for all } \quad x \in I_d, \tag{25}$$

where the function $f(x)/(x-1)$ is extended to 1 as $f'(1) = p - q$.

From (22) and (25) letting $w(r) = u(r) - 1$ we see that

$$\frac{f(u(r))}{w(r)} > \delta, \quad \text{for all} \quad r > \bar{r}. \tag{26}$$

Let $\mu_1 < \mu_2 < \cdots < \mu_k \cdots \to \infty$ be the radial eigenvalues of the negative Laplacian operator with zero Dirichlet boundary data on the unit ball in $\mathbb{R}^N$. Thus, $h(r) := J(\sqrt{\delta}r)$ vanishes at

$$\frac{\mu_1}{\sqrt{\delta}}, \frac{\mu_2}{\sqrt{\delta}}, \ldots$$

Therefore, by (26) and the Sturm Comparison Theorem, $w(r)$ has a zero in $\left[\frac{\mu_k}{\sqrt{\delta}}, \frac{\mu_{k+1}}{\sqrt{\delta}}\right]$. This establishes the existence of infinitely many zeroes for $u(r) - 1$. Since $d \neq 1$ and $f(1) = 0$ by uniqueness of solutions to initial value problems if $u(r) = 1$ then $u'(r) \neq 0$. Thus the zeroes of $u(r) - 1$ form a discrete set $\{s_1 < s_2 < \ldots\}$. Since $u > 1$ on $(0, s_1)$ then $u'(s_1) \leq 0$. Since $f(1) = 0$ by uniqueness of solutions to initial value problems $u'(s_1) \neq 0$. Thus $u'(s_1) < 0$. Hence there exists $a > 0$ such that $u < 0$ on the interval $(s_1, s_1 + a)$. Hence $u < 0$ on (s_1, s_2). This in turn implies that $u'(s_2) \geq 0$. Using again uniqueness of solutions for initial value problems we see that $u'(s_2) > 0$. Inductively it follows that $u < 1$ on (s_i, s_{i+1}) for i odd and $u > 1$ on (s_i, s_{i+1}) for i even. □

Lemma 2.6. *If $q > -1$ there exists $b_1 = b_1(a)$ such that for $b \geq b_1$, $u(\cdot, a, b) - 1$ has only one zero.*

Proof. Suppose by contradiction that, for large b, $u(\cdot, a, b) - 1$ has more than one zero. We can assume, without loss of generality, that there exists $r_1 > \tilde{b}$ such that $0 < u(r_1) < 1$, $u'(r_1) = 0$, and $u'(r) < 0$ for $0 < r < r_1$.

By the Pohozaev's identity

$$r_1^N F(u(r_1)) = Q(\tilde{b}) + \int_{\tilde{b}}^{r_1} s^{N-1} \Gamma(u(s))\, ds.$$

Since both $F(u)$ and $\Gamma(u)$ are bounded below, this implies that there exists a constant C such that

$$Q(\tilde{b}) \leq C.$$

This is a contradiction, since by (17), $\lim_{b\to\infty} Q(\tilde{b}) = \infty$. □

Lemma 2.6 shows that there is no oscillatory singular solutions to (4) of the type $u(\cdot, a, b)$, these solutions decrease towards zero with unbounded energy as $b \to \infty$. In the next lemma we will show the existence of oscillatory singular solutions to (4) with initial negative energy.

If $q > -1$, there exists $\bar{u} \in (1, +\infty)$ such that $F(\bar{u}) = 0$, $F(u) < 0$ for $u \in (0, \bar{u})$, and $F(u) > 0$ for $u > \bar{u}$. In addition, Γ is bounded from below. Thus we let $\Gamma_{min} = \min\{\Gamma(u); u \in (0, +\infty)\}$.

Lemma 2.7. *Let $q > -1$. If $u_0 \in (1, \bar{u})$, $u_0' < 0$, and $r_0 > 0$ are such that*

$$\frac{u_0'^2}{2} + F(u_0) < 0, \quad \frac{N-2}{2} u_0 u_0' - \frac{\Gamma_{min}}{N} r_0 < 0, \quad \text{and} \quad u_0' + f(\bar{u}) r_0 < 0, \tag{27}$$

and u is the solution to (4) *such that $u(r_0) = u_0$, $u'(r_0) = u_0'$, then there exist positive numbers*

$$s_1(u_0, u_0', r_0) < s_2(u_0, u_0', r_0) < \cdots < s_k(u_0, u_0', r_0) < \cdots \to \infty$$

such that $\lim_{r \to 0} u(r) = \infty$, $u(s_i) = 1$, *for* $i = 1, 2, \ldots,$ *and* $u(\cdot) \neq 1$ *on* (s_i, s_{i+1}), *for* $i = 0, 1, \ldots$ *where* $s_0 = 0$.

Proof. By (27), and (13) we have

$$\begin{aligned} Q(\bar{u}_0) &= Q(r_0) - \int_{\bar{r}_0}^{r_0} s^{N-1} \Gamma(u(s))\, ds \\ &\leq r_0^N E(r_0) + r_0^{N-1} \frac{N-2}{2} u_0 u_0' - \frac{\Gamma_{min}}{N} \left(r_0^N - \bar{u}_0^N \right) \\ &\leq r_0^{N-1} \frac{N-2}{2} u_0 u_0' - \frac{\Gamma_{min}}{N} r_0^N < 0. \end{aligned} \tag{28}$$

By (27), for $r_1 \in (0, r_0)$ with $u(r) \in (u_0, \bar{u})$ for all $r \in [r_1, r_0]$ we have

$$r_1^{N-1} u'(r_1) = r_0^{N-1} u_0' + \int_{r_1}^{r_0} s^{N-1} f(u(s)) ds \leq r_0^{N-1} (u_0' + r_0 f(\bar{u})) < 0. \tag{29}$$

From (29) we see that u is strictly decreasing on $[u^{-1}(\bar{u}), r_0]$, and from (28) we conclude that u cannot have critical points on $(0, u^{-1}(\bar{u})]$. Thus u is strictly decreasing on $(0, r_0)$. Also from (28) we infer that $\limsup_{r \to 0+} Q(r) < 0$. Hence $\lim_{r \to 0+} u(r) = +\infty$. That is, u is a singular solution to (4).

Arguing as the proof of Lemma 2.5, the existence of the sequence $\{s_i\}$ follows from the fact that $E(r) \leq E(r_0) < 0$ for all $r \in [r_0, +\infty)$. □

3 Proof of Theorems 1.1 and 1.2

Proof of Theorem 1.1

Proof. Let $a \in I$, and $b \geq b_1$ be as in Lemma 2.2. Taking $\lambda = (s_k(a, b))^2$ as in Lemma 2.5 and $v_\lambda(r) = u(\lambda^{1/2}r, a, b)$, it is readily verified that v_λ is a singular solution to (1) with exactly k nodal sets in $\bar{B}$. □

Proof of Theorem 1.2

Proof. Let u_0, u_0', r_0 be as in (27). Let $s_k(u_0, u_0', r_0)$ be as in Lemma 2.7. Taking $\lambda = (s_k(a, b))^2$ as in Lemma 2.5 and $v_\lambda(r) = u(\lambda^{1/2}r, a, b)$, it is readily verified that v_λ is a singular solution to (1) with exactly k nodal sets in $\bar{B}$. □

Acknowledgements This work was partially supported by a grant from the Simons Foundations # 245966 to Alfonso Castro.

References

1. Ardila, V., Caicedo, J., Castro, A.: Existence of non-degenerate continua of singular radial solutions for several classes of semilinear elliptic problems. Milan J. Math. **82**(2), 313–330 (2014)
2. Castro, A., Padrón, V.: Classification of radial solutions arising in the study of thermal structures with thermal equilibrium or no flux at the boundary. Mem. Am. Math. Soc. **208**(976), vi+72 (2010)
3. Chandrasekhar, S.: An Introduction to the Study of Stellar Structure. Dover Publications, New York, NY (1957)
4. Franca, M.: Ground states and singular ground states for quasilinear elliptic equation in the subcritical case. Funkcial. Ekvac. **48**(3), 331–349 (2005)
5. Ibañez, S.M.H., Rosenzweig, P.: Analytical criteria for nonlinear instability of thermal structures. Phys. Plasmas **2**(11), 4127–4133 (1995)

Branches of positive solutions for subcritical elliptic equations

Alfonso Castro and Rosa Pardo

1 Introduction

In this paper we prove the existence of positive solutions to the boundary-value problem:

$$\begin{cases} -\Delta u = \lambda u + g(u), & \text{in } \Omega, \\ \quad u = 0, & \text{on } \partial\Omega, \end{cases} \tag{1}$$

where $\Omega \subset \mathbb{R}^N$, $N \geq 2$, is a bounded C^2 domain, g is a subcritical nonlinearity, and λ is a real parameter. For simplicity we assume $N > 2$, but our techniques fit well to the case $N = 2$.

Let λ_1, ϕ_1 stand for the first eigenvalue, first eigenfunction of the eigenvalue problem $-\Delta\phi_1 = \lambda_1\phi_1$ in Ω, $\phi_1 = 0$ on $\partial\Omega$. In Theorem 2.1, we provide sufficient conditions guarantying that either for any $\lambda < \lambda_1$ there exists at least a positive solution to (1), or for any continuum (λ, u_λ) of positive solution to (1), there exists a $\lambda^* < 0$ such that $\lambda^* < \lambda < \lambda_1$ and

$$\|\nabla u_\lambda\|_{L^2(\Omega)} \to \infty, \qquad \text{as} \quad \lambda \to \lambda^*.$$

A. Castro
Department of Mathematics, Harvey Mudd College, Claremont, CA 91711, USA
e-mail: castro@math.hmc.edu

R. Pardo (✉)
Departamento de Matemática Aplicada, Universidad Complutense de Madrid, 28040 Madrid, Spain
e-mail: rpardo@mat.ucm.es

A.N. Carvalho et al. (eds.), *Contributions to Nonlinear Elliptic Equations and Systems*, Progress in Nonlinear Differential Equations and Their Applications 86,
DOI 10.1007/978-3-319-19902-3_7

In case Ω is convex, we provide sufficient conditions guarantying that for any $\lambda < \lambda_1$ there exists at least a positive solution to (1), see [2, Theorem 2.4].

In [3, Theorem 1.1], we provide a-priori $L^\infty(\Omega)$ bounds for a classical positive solutions to the boundary-value problem:

$$\begin{cases} -\Delta u = f(u), & \text{in } \Omega, \\ u = 0, & \text{on } \partial\Omega, \end{cases} \tag{2}$$

where Ω is a bounded C^2 domain, and f is a subcritical nonlinearity. Our main result in [3] is:

Theorem 1.1. *Assume that $\Omega \subset \mathbb{R}^N$ is a bounded domain with C^2 boundary. Assume that the nonlinearity f is locally Lipschitzian and satisfies the following conditions*

(H1) $\dfrac{f(s)}{s^{2^\star-1}}$ *is nonincreasing for any $s \geq 0$, where $2^\star = \frac{2N}{N-2}$,*

(H2) *There exists a constant $C_1 > 0$ such that* $\limsup_{s\to\infty} \dfrac{\max_{[0,s]} f}{f(s)} \leq C_1$

(H3) *There exists a constant $C_2 > 0$ two constant $C_2, C_3 > 0$ and a non-increasing function $H : \mathbb{R}^+ \to \mathbb{R}^+$ such that*

(H3.1) $$\liminf_{s\to+\infty} \frac{2NF(s) - (N-2)sf(s)}{sf(s)H(s)} \geq C_2 > 0,$$

where $F(s) = \int_0^s f(t)\,dt$, and

(H3.2) $$\lim_{s\to+\infty} \frac{\frac{f(s)}{s^{2^\star-1}}}{\left[H(s)\right]^{\frac{2}{N-2}}} = 0,$$

(H4) $\liminf_{s\to\infty} \dfrac{f(s)}{s} > \lambda_1$, *where λ_1 is the first eigenvalue of $-\Delta$ acting on $H_0^1(\Omega)$.*

Then, there exists a uniform constant C, depending only on Ω and f, such that for every $u > 0$, classical solution to (2),

$$\|u\|_{L^\infty(\Omega)} \leq C.$$

Theorem 1.1 widens the known ranges of subcritical nonlinearities for which positive solutions to (2) are a priori bounded, see [5] and [6]. We prove that functions such as $f_1(s) = s^{2^\star-1}/\ln(s+2)^\alpha$ satisfy our hypotheses for $\alpha > 2/(N-2)$ with $H(s) = 1/\ln(s+2)$, see [3, Corollary 2.2], but not those of [5] neither of [6]. Moreover, $g_1(s) = s^{2^\star-1}/\ln(s+2)^\alpha$ can be considered as a subcritical nonlinearity in (1), see Corollary 2.3.

2 On the a-priori bounds for any regular domain

In this section we state our main result on the existence of positive solutions for the nonlinear eigenvalue problem (1). We prove that there exists a branch of positive solutions bifurcating from the trivial solution.

Theorem 2.1. *Assume that $\Omega \subset \mathbb{R}^N$ is a bounded domain with C^2 boundary. Assume that the nonlinearity g is locally Lipschitzian and satisfies the conditions* (H1)–(H3). *Assume also that $g : \mathbb{R}^+ \to \mathbb{R}^+$ satisfies the following conditions*

(H5) $\displaystyle\lim_{s\to 0} \frac{g(s)}{s} = 0.$

(H6) $\displaystyle\lim_{s\to\infty} \frac{s}{g(s)H(s)} = 0.$

(H7) $\displaystyle\lim_{s\to\infty} \frac{1}{s^2 H(s)} = 0.$

Then, the following holds:

(i) If there exists one positive solution to (1), *then $\lambda < \lambda_1$.*
(ii) If $0 \le \lambda < \lambda_1$, then there exists at least a positive solution to (1).
(iii) Either there exists at least a positive solution to (1) *for any $\lambda < \lambda_1$, or for any continuum (λ, u_λ) of positive solution to* (1), *there exists a $\lambda^* < 0$ such that $\lambda^* < \lambda < \lambda_1$ and $\|\nabla u_\lambda\|_{L^2(\Omega)} \to \infty$, as $\lambda \to \lambda^*$.*

Remark 2.2. From hypothesis, $H : \mathbb{R}^+ \to \mathbb{R}^+$ is a non-increasing function, then $0 \le \lim_{s\to\infty} H(s) < \infty$ or equivalently $\infty \ge \lim_{s\to\infty} \frac{1}{H(s)} > 0$.

Proof (Proof of Theorem 2.1).

(i) Assume there exists a positive solution (λ, u) to (1). Multiplying (1) by $\phi_1 > 0$, integrating by parts on Ω, and due to $g : \mathbb{R}^+ \to \mathbb{R}^+$ we deduce

$$(\lambda_1 - \lambda) \int_\Omega u\phi_1 = \int_\Omega g(u)\phi_1 > 0, \tag{3}$$

and consequently, $\lambda < \lambda_1$, which concludes this part of the proof.

(ii) From the Crandall-Rabinowitz's Theorem, and thanks to hypothesis (H5), $(\lambda_1, 0)$ is a bifurcation point of positive solutions, and there exists a continuum of positive solutions $\mathscr{C} = \{(\lambda, u_\lambda) \in \mathbb{R} \times C^{1,\alpha}(\overline{\Omega})\}$ emanating from $(\lambda_1, 0)$ and solving (1), see [4]. From the Maximum Principle and the Hopf Maximum Principle, if $u \ge 0$, $u \ne 0$, is a solution to (1), and thanks to $\lambda < \lambda_1$, then $u > 0$ in Ω, and $\frac{\partial u}{\partial n} < 0$ on in $\partial\Omega$. Consequently, $\mathscr{C} \subset \{(\lambda, u_\lambda) \in \mathbb{R} \times C^{1,\alpha}(\overline{\Omega}) : u > 0 \text{ in } \Omega,\ \frac{\partial u}{\partial n} < 0 \text{ on } \partial\Omega\}$. From Rabinowitz's global bifurcation theorem, the continuum of solutions emanating from $(\lambda_1, 0)$ is either unbounded or goes to another bifurcation point from the trivial solution set, see [7]. Due to $\lambda < \lambda_1$, the continuum cannot meet another bifurcation point from the trivial solution and therefore $\mathscr{C}$ is unbounded.

In order to apply Theorem 1.1 to (1) we verify that the nonlinearity $f(s) = \lambda s + g(s)$ satisfy hypotheses (H1)–(H4) for $\lambda \geq 0$, assuming that g satisfy (H1)–(H3) and (H5)–(H7).

(H1) Obviously, hypothesis (H1) holds for $f(s) = \lambda s + g(s)$ with $\lambda \geq 0$.

(H2) From Remark 2.2, and by (H6) we conclude that $\lim_{s\to\infty} \frac{s}{g(s)} = 0$. Now, from definition of f, dividing it by $g(s)$ up and down, due to $\lim_{s\to\infty} \frac{s}{g(s)}$ and thanks to hypothesis (H2) on $g(s)$ we can write

$$\limsup_{s\to+\infty} \frac{\max_{[0,s]} f}{f(s)} \leq \limsup_{s\to+\infty} \frac{|\lambda| s}{\lambda s + g(s)} + \limsup_{s\to+\infty} \frac{\max_{[0,s]} g}{\lambda s + g(s)}$$
$$= \limsup_{s\to+\infty} \frac{|\lambda| \frac{s}{g(s)}}{\lambda \frac{s}{g(s)} + 1} + \limsup_{s\to+\infty} \frac{\max_{[0,s]} g}{\lambda \frac{s}{g(s)} + 1} \leq C_1 > 0.$$

(H3.1) Let $G(s) = \int_0^s g(t)\,dt$, then from definition, dividing by $sg(s)H(s)$ up and down, thanks to hypothesis (H6) and due to $\lim_{s\to\infty} \frac{s}{g(s)} = 0$, we can write

$$\liminf_{s\to+\infty} \frac{2NF(s) - (N-2)sf(s)}{sf(s)H(s)} = \liminf_{s\to+\infty} \frac{2NG(s) - (N-2)sg(s) + 2\lambda s^2}{\lambda s^2 H(s) + sg(s)H(s)}$$
$$= \liminf_{s\to+\infty} \frac{\dfrac{2NG(s)-(N-2)sg(s)}{sg(s)H(s)} + \dfrac{2\lambda s}{g(s)H(s)}}{\dfrac{\lambda s}{g(s)} + 1}$$
$$= \liminf_{s\to+\infty} \frac{2NG(s)-(N-2)sg(s)}{sg(s)H(s)} \geq C_2 > 0.$$

From definition of f, by hypothesis (H7) and (H3.2)

$$\lim_{s\to+\infty} \frac{\dfrac{f(s)}{s^{2^\star-1}}}{\left[H(s)\right]^{\frac{2}{N-2}}} = \lim_{s\to+\infty} \left[\frac{\lambda}{\left(s^2\left[H(s)\right]^{1+\delta_2(N-2)/2}\right)^{\frac{2}{N-2}}} + \frac{\dfrac{g(s)}{s^{2^\star-1}}}{\left[H(s)\right]^{\frac{2}{N-2}}} \right] = 0.$$

(H4) By (H6) $\lim_{s\to+\infty} \frac{g(s)H(s)}{s} = +\infty$. From Remark 2.2 and (H6) we conclude that $\lim_{s\to\infty} \frac{g(s)}{s} = \infty$, and consequently

$$\liminf_{s\to\infty} \frac{f(s)}{s} = \liminf_{s\to\infty} \frac{\lambda s + g(s)}{s} > \lambda_1.$$

Therefore, hypotheses (H1)–(H4) hold for $f(s) = \lambda s + g(s)$ with $\lambda \geq 0$.

From Theorem 1.1, whenever $\lambda \in [0, \lambda_1)$, the positive solutions are a-priori bounded. Consequently, the projection of $\mathscr{C}$ in the parameter space must contain the whole interval $[0, \lambda_1)$, which proves part (ii).

(iii) From the proof of part (ii), we can assert that the continuum of solutions $\mathscr{C} \subset \mathbb{R} \times C^{1,\alpha}(\overline{\Omega})$ is an unbounded set. We conclude that either for any $\lambda < \lambda_1$ there exists a positive solution to (1), or the projection of $\mathscr{C}$ in the parameter space is bounded, and therefore, $\mathscr{C}$ is unbounded in $C^{1,\alpha}(\overline{\Omega})$. From elliptic regularity, in particular $\|u_\lambda\|_{L^\infty(\Omega)} \to \infty$, as $\lambda \to \lambda^*$.

We reason by contradiction, assuming that $\|\nabla u_\lambda\|_{L^2(\Omega)} \leq C$, as $\lambda \to \lambda^* < 0$. From Poincaré inequality, $\|u_\lambda\|_{L^2(\Omega)} \leq C$, as $\lambda \to \lambda^*$. Multiplying (1) by u and integrating by parts on Ω we can write

$$\int_\Omega |\nabla u|^2 + |\lambda| \int_\Omega u^2 = \int_\Omega u g(u) \leq C, \tag{4}$$

for some constant C independent of u.

Moreover, from hypothesis (H3.3), and Remark 2.2, for any $\varepsilon > 0$ there exists a constant C_ε such that

$$g(s)^{\frac{1}{2^\star-1}} \leq \varepsilon s + C_\varepsilon, \qquad \text{for any} \quad s > 0,$$

therefore

$$\int_\Omega g(u)^{1+\frac{1}{2^\star-1}}\, dx \leq \varepsilon \int_\Omega u g(u)\, dx + C_\varepsilon \leq C\varepsilon + C_\varepsilon, \tag{5}$$

and consequently,

$$\begin{aligned}\int_\Omega \left|g\big(u(x)\big)\right|^q dx &\leq \int_\Omega \left|g\big(u(x)\big)\right|^{1+\frac{1}{2^\star-1}} \left|g\big(u(x)\big)\right|^{q-1-\frac{1}{2^\star-1}} dx \\ &\leq (C\varepsilon + C_\varepsilon)\ \left\|g\big(u(\cdot)\big)\right\|_\infty^{q-1-\frac{1}{2^\star-1}},\end{aligned} \tag{6}$$

for some $q > N/2$. Assuming for a while that g is non-decreasing, then

$$\int_\Omega \left|g\big(u(x)\big)\right|^q dx \leq (C\varepsilon + C_\varepsilon) \left[g\big(\|u\|_\infty\big)\right]^{q-1-\frac{1}{2^\star-1}}. \tag{7}$$

Therefore, from elliptic regularity,

$$\|u\|_{W^{2,q}(\Omega)} \leq C\|\Delta u\|_{L^q(\Omega)} \leq (C\varepsilon + C_\varepsilon) \left[g\big(\|u\|_\infty\big)\right]^{1-\frac{1}{q}-\frac{1}{(2^\star-1)q}}. \tag{8}$$

Let us restrict $q \in (N/2, N)$. From Sobolev embeddings

$$\|u\|_{W^{1,q^*}(\Omega)} \leq C\|u\|_{W^{2,q}(\Omega)} \leq (C\varepsilon + C_\varepsilon) \left[g\big(\|u\|_\infty\big)\right]^{1-\frac{1}{q}-\frac{1}{(2^\star-1)q}}. \tag{9}$$

where $1/q^* = 1/q - 1/N$ and $q^* > N$.

Moreover, from Morrey's Theorem, see [1, Theorem 9.12]

$$|u(x_1) - u(x_2)| \le C|x_1 - x_2|^{1-N/q^*} \|\nabla u\|_{L^{q^*}(\Omega)}, \qquad \text{a.e.} \quad x_1, x_2 \in \Omega.$$

Therefore, for any x_1, x_2 such that $|x_1 - x_2| \le R$,

$$|u(x_1) - u(x_2)| \le (C\varepsilon + C_\varepsilon)\, R^{2-\frac{N}{q}} \left[g\big(\|u\|_\infty\big)\right]^{1-\frac{1}{q}-\frac{1}{(2^\star-1)q}}. \tag{10}$$

From now on, we shall argue by contradiction. Let $\{u_k\}_k$ be a sequence of classical positive solutions to (2) and assume that

$$\lim_{k\to\infty} \|u_k\| = +\infty, \qquad \text{where} \quad \|u_k\| := \|u_k\|_\infty. \tag{11}$$

Let $C, \delta > 0$ be as in [3, Theorem 2.10]. Let $x_k \in \overline{\Omega_\delta}$ be such that

$$u_k(x_k) = \max_{\Omega_\delta} u_k = \max_{\Omega} u_k.$$

By taking a subsequence if needed, we may assume that there exists $x_0 \in \overline{\Omega_\delta}$ such that

$$\lim_{k\to\infty} x_k = x_0 \in \overline{\Omega_\delta}, \qquad \text{and} \quad d_0 := dist(x_0, \partial\Omega) \ge \delta > 0. \tag{12}$$

Then, for any $x \in \Omega$ such that $|x - x_k| \le R_k$,

$$|u_k(x) - u_k(x_k)| \le (C\varepsilon + C_\varepsilon)\, R_k^{2-\frac{N}{q}} \left[g\big(\|u_k\|\big)\right]^{1-\frac{1}{q}-\frac{1}{(2^\star-1)q}}. \tag{13}$$

Let us choose R_k such that $B_k = B(x_k, R_k) \subset \Omega$, and

$$u_k(x) \ge \frac{1}{2}\, \|u_k\| \qquad \text{for any} \quad x \in B(x_k, R_k).$$

and there exists $y_k \in \partial B(x_k, R_k)$ such that

$$u_k(y_k) = \frac{1}{2}\, \|u_k\|. \tag{14}$$

Particularizing $x = y_k$ in (14)

$$(C\varepsilon + C_\varepsilon)\, R_k^{2-\frac{N}{q}} \left[g\big(\|u_k\|\big)\right]^{1-\frac{1}{q}-\frac{1}{(2^\star-1)q}} \ge \frac{1}{2} u_k(x_k) = \frac{1}{2}\|u_k\|, \tag{15}$$

or equivalently

$$R_k \geq \left(\frac{1}{(C\varepsilon + C_\varepsilon)} \frac{\|u_k\|}{\left[g\left(\|u_k\|\right)\right]^{1-\frac{1}{q}-\frac{1}{(2^\star-1)q}}} \right)^{1/\left(2-\frac{N}{q}\right)}. \tag{16}$$

We define

$$R_k = \left(\frac{\|u_k\|}{\left[g\left(\|u_k\|\right)\right]^{1-\frac{1}{q}-\frac{1}{(2^\star-1)q}}} \right)^{1/\left(2-\frac{N}{q}\right)}. \tag{17}$$

Let us choose a constant C such that $B(x_k, CR_k) \subset B(x_0, d_0) \subset \Omega$, see (12). Let us denote it again by $B(x_k, R_k)$ by redefining R_k if necessary. Therefore,

$$u_k(x) \geq \frac{1}{2}\, \|u_k\| \qquad \text{for any} \quad x \in B(x_k, R_k). \tag{18}$$

Consequently, and taking into account that g is non-decreasing

$$\int_\Omega u_k g(u_k)\, dx \geq \frac{1}{2}\, \|u_k\| \int_{B(x_k,R_k)} g(u_k)\, dx \geq \frac{1}{2}\, \|u_k\| g\left(\frac{1}{2}\, \|u_k\|\right) R_k^N.$$

At this moment, let us observe that, from hypothesis (H1),

$$\frac{g\left(\frac{1}{2}s\right)}{g(s)} = \left(\frac{1}{2}\right)^{2^\star-1} \frac{\dfrac{g\left(\frac{1}{2}s\right)}{\left(\frac{1}{2}s\right)^{2^\star-1}}}{\dfrac{g(s)}{s^{2^\star-1}}} \geq \left(\frac{1}{2}\right)^{2^\star-1}, \qquad \text{for all} \quad s \geq s_\varepsilon, \tag{19}$$

consequently, for $q > N/2$,

$$\frac{g\left(\frac{1}{2}\|u_k\|\right)^{\frac{2}{N}-\frac{1}{q}}}{g\left(\|u_k\|\right)^{\frac{2}{N}-\frac{1}{q}}} \geq C, \qquad \text{for all } k \quad \text{big enough.} \tag{20}$$

Substituting R_k^N, rearranging terms twice, using (20), and finally using hypothesis (H3.2) and Remark 2.2 to observe that $g(s)/s^{2^\star-1} \to 0$ as $s \to \infty$, we obtain

$$\int_{\Omega} u_k g(u_k)\, dx \geq \|u_k\| g\left(\frac{1}{2}\|u_k\|\right) \left(\frac{\|u_k\|}{g(\|u_k\|)^{1-\frac{1}{q}-\frac{1}{(2^\star-1)q}}}\right)^{\frac{N}{2-\frac{N}{q}}}$$

$$= \left(\left[\|u_k\| g\left(\frac{\|u_k\|}{2}\right)\right]^{\frac{2}{N}-\frac{1}{q}} \frac{\|u_k\|}{g(\|u_k\|)^{1-\frac{1}{q}-\frac{1}{(2^\star-1)q}}}\right)^{\frac{N}{2-\frac{N}{q}}}$$

$$= \left[\frac{\|u_k\|^{1+\frac{2}{N}-\frac{1}{q}} g\left(\frac{\|u_k\|}{2}\right)^{\frac{2}{N}-\frac{1}{q}}}{g(\|u_k\|)^{1-\frac{1}{q}-\frac{1}{(2^\star-1)q}}}\right]^{\frac{N}{2-\frac{N}{q}}} \geq C\left[\frac{\|u_k\|^{1+\frac{2}{N}-\frac{1}{q}}}{g(\|u_k\|)^{1-\frac{2}{N}-\frac{1}{(2^\star-1)q}}}\right]^{\frac{N}{2-\frac{N}{q}}}$$

$$= C\left[\frac{\|u_k\|^{2^\star-1}}{g(\|u_k\|)}\right]^{\left(1+\frac{2}{N}-\frac{1}{q}\right)\frac{1}{2^\star-1}\left(\frac{N}{2-\frac{N}{q}}\right)} \to \infty \quad \text{as } k \to \infty$$

which contradicts (4), ending this part of the proof, achieved assuming that g is non-decreasing.

Next, we shall discuss the general case. Let us denote by $\psi(s) := g(s)/s^{2^\star-1}$, from hypothesis (H1), $\psi(s)$ is non-increasing. Therefore, whenever $\frac{1}{2}\ \|u_k\| \leq u_k(x) \leq \|u_k\|$, we obtain

$$g(u_k(x)) \leq \psi(u_k(x))\|u_k\|^{2^\star-1} \leq \psi\left(\frac{\|u_k\|}{2}\right)\|u_k\|^{2^\star-1} = 2^{2^\star-1} g\left(\frac{\|u_k\|}{2}\right), \tag{21}$$

and

$$g(u_k(x)) \geq \psi(u_k(x))\left(\frac{\|u_k\|}{2}\right)^{2^\star-1} \geq \psi(\|u_k\|)\left(\frac{\|u_k\|}{2}\right)^{2^\star-1} = \left(\frac{1}{2}\right)^{2^\star-1} g(\|u_k\|). \tag{22}$$

Choosing R_k such that $B(x_k, R_k) \subset \Omega$, and (18) is satisfied, from (6) we obtain

$$\int_{B(x_k,R_k)} |g(u_k(x))|^q\, dx \leq (C\varepsilon + C_\varepsilon)\left[g\left(\frac{1}{2}\|u_k\|\right)\right]^{q-1-\frac{1}{2^\star-1}}, \tag{23}$$

and from elliptic regularity

$$\|u_k\|_{W^{2,q}\Omega} \leq (C\varepsilon + C_\varepsilon)\left[g\left(\frac{1}{2}\|u_k\|\right)\right]^{1-\frac{1}{q}-\frac{1}{(2^\star-1)q}}. \tag{24}$$

Defining

$$R_k = \left(\frac{\|u_k\|}{g\left(\frac{1}{2}\|u_k\|\right)^{1-\frac{1}{q}-\frac{1}{(2^\star-1)q}}} \right)^{\frac{1}{2-\frac{N}{q}}}$$

Redefining R_k if necessary, we can assert that $B(x_k, R_k) \subset B(x_0, d_0) \subset \Omega$. Consequently, taking into account (22),

$$\int_\Omega u_k g(u_k)\, dx \geq C\, \|u_k\| g\big(\|u_k\|\big) R_k^N$$

Substituting R_k^N, and rearranging terms we obtain

$$\begin{aligned}
\int_\Omega u_k g(u_k)\, dx &\geq C\, \|u_k\| g\big(\|u_k\|\big) \left(\frac{\|u_k\|}{g\left(\frac{1}{2}\|u_k\|\right)^{1-\frac{1}{q}-\frac{1}{(2^\star-1)q}}} \right)^{\frac{N}{2-\frac{N}{q}}} \\
&= C \left(\left[\|u_k\| g\big(\|u_k\|\big) \right]^{\frac{2}{N}-\frac{1}{q}} \frac{\|u_k\|}{g\left(\frac{1}{2}\|u_k\|\right)^{1-\frac{1}{q}-\frac{1}{(2^\star-1)q}}} \right)^{\frac{N}{2-\frac{N}{q}}} \\
&= C \left(\frac{\|u_k\|^{1+\frac{2}{N}-\frac{1}{q}} \left[g\big(\|u_k\|\big) \right]^{\frac{2}{N}-\frac{1}{q}}}{g\left(\frac{1}{2}\|u_k\|\right)^{1-\frac{1}{q}-\frac{1}{(2^\star-1)q}}} \right)^{\frac{N}{2-\frac{N}{q}}}
\end{aligned}$$

From hypothesis (H2),

$$\frac{g(s)}{g\left(\frac{1}{2}s\right)} \geq C - \varepsilon, \qquad \text{for all} s \geq s_\varepsilon, \tag{25}$$

consequently, for $q > N/2$,

$$\frac{g\big(\|u_k\|\big)^{\frac{2}{N}-\frac{1}{q}}}{g\left(\frac{1}{2}\|u_k\|\right)^{\frac{2}{N}-\frac{1}{q}}} \geq C, \qquad \text{for all } k \text{ big enough.} \tag{26}$$

Hence, by (H2), (H3.2), and Remark 2.2, and argüing as before, we can assert that

$$\int_{\Omega} u_k g(u_k)\,dx \geq C \left(\frac{\|u_k\|^{1+\frac{2}{N}-\frac{1}{q}}}{g\left(\frac{1}{2}\|u_k\|\right)^{1-\frac{2}{N}-\frac{1}{(2^\star-1)q}}} \right)^{\frac{N}{2-\frac{N}{q}}}$$

$$\geq C \left(\frac{\|u_k\|^{1+\frac{2}{N}-\frac{1}{q}}}{g\left(\|u_k\|\right)^{1-\frac{2}{N}-\frac{1}{(2^\star-1)q}}} \right)^{\frac{N}{2-\frac{N}{q}}} \to \infty$$

as $k \to \infty$, which contradicts (4), ending the proof. □

Corollary 2.3. *Assume that $\Omega \subset \mathbb{R}^N$ is a bounded domain with C^2 boundary. Let us consider any $u > 0$, classical solution to*

$$\begin{cases} -\Delta u = \lambda u + \dfrac{u^{2^\star-1}}{\ln(2+u)^\alpha}, & \text{in } \Omega, \\ u = 0, & \text{on } \partial\Omega, \end{cases} \tag{27}$$

with $\alpha > 2/(N-2)$. Then, the conclusions of Theorem 2.1 hold: if there exists one positive solution to (27)*, then $\lambda < \lambda_1$. If $0 \leq \lambda < \lambda_1$, then there exists at least a positive solution to* (27)*. Moreover, either* (27) *has a positive solution for any $\lambda < \lambda_1$, or for any continuum (λ, u_λ) of positive solution to* (27)*, there exists a $\lambda^* < 0$ such that $\lambda^* < \lambda < \lambda_1$ and $\|\nabla u_\lambda\|_{L^2(\Omega)} \to \infty$, as $\lambda \to \lambda^*$.*

Proof. We prove that $g_1(s) = s^{2^\star-1}/\ln(s+2)^\alpha$ with $\alpha > 2/(N-2)$ satisfy our hypotheses for $H(s) = 1/\ln(s+2)$. Hypotheses (H1)–(H2) and (H4)–(H8) hold trivially. To prove (H3), see [2, proof of Corollary 2]. □

Acknowledgements This work was partially supported by a grant from the Simons Foundation # 245966 to Alfonso Castro. The second author is supported by Spanish MINISTERIO DE ECONOMIA Y COMPETITIVIDAD (MEC) under Project MTM2012-31298.

References

1. Brezis, H.: Functional Analysis, Sobolev Spaces and Partial Differential Equations. Universitext. Springer, New York (2011)
2. Castro, A., Pardo, R.: Branches of positive solutions of subcritical elliptic equations in convex domains. In: AIMS Proceedings (to appear)
3. Castro, A., Pardo, R.: A priori bounds for positive solutions of subcritical elliptic equations. Rev. Math. Complut. **28**(3), 715–731 (2015)
4. Crandall, M.G., Rabinowitz, P.H.: Bifurcation from simple eigenvalues. J. Funct. Anal. **8**, 321–340 (1971)

5. de Figueiredo, D.G., Lions, P.-L., Nussbaum, R.D.: A priori estimates and existence of positive solutions of semilinear elliptic equations. J. Math. Pures Appl. (9) **61**(1), 41–63 (1982)
6. Gidas, B., Spruck, J.: A priori bounds for positive solutions of nonlinear elliptic equations. Commun. Partial Differ. Equat. **6**(8), 883–901 (1981)
7. Rabinowitz, P.H.: Some global results for nonlinear eigenvalue problems. J. Funct. Anal. **7**, 487–513 (1971)

Multiple solutions to anisotropic critical and supercritical problems in symmetric domains

Mónica Clapp and Jorge Faya

Para Djairo, em seu aniversário, com grande afeto e admiração.

1 Introduction and statement of results

Consider the anisotropic critical or supercritical problem

$$(\wp_p) \qquad \begin{cases} -\mathrm{div}(a(x)\nabla u) + b(x)u = c(x)|u|^{p-2}u & \text{in } \Omega, \\ u = 0 & \text{on } \partial\Omega, \end{cases}$$

where Ω is a bounded smooth domain in $\mathbb{R}^N$, $N \geq 3$, $a \in \mathscr{C}^{1,\alpha}(\overline{\Omega})$, $b, c \in \mathscr{C}^{0,\alpha}(\overline{\Omega})$, a and c are strictly positive functions on $\overline{\Omega}$ and $p \geq 2^*$, with $2^* := \frac{2N}{N-2}$ the critical Sobolev exponent.

The remarkable results of Brezis and Nirenberg [5] and Bahri and Coron [2] in the 1980s triggered numerous investigations on the critical problem, and many results are available nowadays for $(\wp_{2^*})$ with constant coefficients $a = c = 1$, $b \in \mathbb{R}$. Yet, still little is known on the anisotropic case.

Substantially less is known for the supercritical problem $p > 2^*$. A fruitful approach which has been applied in recent years to treat this problem consists in reducing it to a critical or subcritical problem of the form $(\wp_p)$, either by considering rotational symmetries, or by means of maps which preserve the Laplace operator, or by a combination of both. This approach has allowed to obtain existence and multiplicity results in some domains for supercritical problems with constant coefficients. We refer the reader to the survey [9], and the references therein.

M. Clapp (✉)
Instituto de Matemáticas, Universidad Nacional Autónoma de México,
Circuito Exterior, C.U., 04510 México D.F., Mexico
e-mail: monica.clapp@im.unam.mx

J. Faya
Centro de Modelamiento Matemático, Universidad de Chile,
Blanco Encalada 2120, Santiago, Chile
e-mail: jfaya@dim.uchile.cl

A.N. Carvalho et al. (eds.), *Contributions to Nonlinear Elliptic Equations and Systems*, Progress in Nonlinear Differential Equations and Their Applications 86,
DOI 10.1007/978-3-319-19902-3_8

Bahri and Coron [2] established the existence of at least one positive solution to the critical problem $(\wp_{2^*})$ with $a = c = 1$, $b = 0$, in every domain Ω having nontrivial reduced homology with $\mathbb{Z}/2$-coefficients. Moreover, if Ω is invariant under the action of a group G of linear isometries of $\mathbb{R}^N$ and every G-orbit in Ω is infinite, the critical problem is known to have infinitely many solutions [6]. Neither of these conditions is enough to guarantee existence in the supercritical case. Passaseo exhibited an example of a domain Ω having the homotopy type of a k-dimensional sphere and infinite $O(k+1)$-orbits, in which $(\wp_p)$ has no solution for $p \geq 2^*_{N,k} := \frac{2(N-k)}{N-k-2}$ [17, 18]. Examples with richer cohomology were exhibited in [10]. The exponent $2^*_{N,k}$ is called the $(k+1)$-st critical exponent in dimension N. Note that $2^*_{N,0} = 2^*$.

Here we shall obtain some multiplicity results for the anisotropic problem $(\wp_p)$ with $p = 2^*_{N,k}$ under some symmetry assumptions.

As usual, we write $O(N)$ for the group of linear isometries of $\mathbb{R}^N$. We denote by $Gx := \{gx : g \in G\}$ the G-orbit of x and by $\#Gx$ its cardinality. If G is a closed subgroup of $O(N)$, a subset X of $\mathbb{R}^N$ is said to be G-invariant if $Gx \subset X$ for every $x \in X$ and a function $u : X \to \mathbb{R}$ is called G-invariant if it is constant on every Gx with $x \in X$.

If Ω, a and b are G-invariant, we set

$$\mu^G_{a,b}(\Omega) := \inf_{\substack{u \in H^1_0(\Omega)^G \\ u \neq 0}} \frac{\int_\Omega \left(a(x)\,|\nabla u|^2 + b(x)u^2\right)}{\int_\Omega |\nabla u|^2}, \tag{1}$$

where $H^1_0(\Omega)^G := \{u \in H^1_0(\Omega) : u \text{ is } G\text{-invariant}\}$.

1.1 The critical problem

Anisotropic critical problems of the form $(\wp_p)$ have been studied by Egnell [12] and, more recently, by Hadiji et al. [13, 14]. They obtained existence and multiplicity results under some assumptions which involve flatness of the coefficient functions at some local maximum or minimum point in the interior of Ω. Our results involve only some symmetry assumptions. We shall prove the following results.

Theorem 1.1. *If Ω, a, b and c are G-invariant and if $\#Gx = \infty$ for every $x \in \Omega$, then the critical problem $(\wp_{2^*})$ has infinitely many G-invariant solutions.*

Next, we fix a closed subgroup Γ of $O(N)$ and a nonempty Γ-invariant bounded smooth domain D in $\mathbb{R}^N$ such that $\#\Gamma x = \infty$ for every $x \in D$. We assume that the functions a, b and c are Γ-invariant and that $\mu^\Gamma_{a,b}(D) > 0$. We prove the following multiplicity result.

Theorem 1.2. *There exists an increasing sequence (ℓ_m) of positive real numbers, depending only on Γ, D, a, b and c, with the following property: If Ω is a bounded* *smooth domain which contains D and if it is invariant under the action of a closed subgroup G of Γ for which $\mu^G_{a,b}(\Omega) > 0$ and the inequality*

$$\min_{x\in\overline{\Omega}} \frac{a(x)^{\frac{N}{2}} \#Gx}{c(x)^{\frac{N-2}{2}}} > \ell_m$$

holds true, then the critical problem $(\wp_{2^*})$ has at least m pairs of G-invariant solutions $\pm u_1, \ldots, \pm u_m$ such that u_1 is positive, $u_2, \ldots, u_m$ change sign, and

$$\int_\Omega c(x)\,|u_k|^{2^*} \leq \ell_k S^{N/2} \qquad \textit{for every } k = 1, \ldots, m,$$

where S is the best Sobolev constant for the embedding $D^{1,2}(\mathbb{R}^N) \hookrightarrow L^{2^*}(\mathbb{R}^N)$.

For example, we may fix a bounded smooth domain D_0 whose closure is contained in the half-space $(0,\infty) \times \mathbb{R}^{N-2}$ and set

$$D := \{(y,z) \in \mathbb{C} \times \mathbb{R}^{N-2} : (|y|, z) \in D_0\}.$$

Then D is invariant under the action of the group $\Gamma := \mathbb{S}^1$ of unit complex numbers on $\mathbb{C} \times \mathbb{R}^{N-2}$ given by $e^{i\theta}(y,z) := (e^{i\theta}y, z)$. If $G_n := \{e^{2\pi ik/n} : k = 0, \ldots, n-1\}$ is the cyclic subgroup of Γ of order n, then $\#G_n x = n$ for every $x \in (\mathbb{C} \smallsetminus \{0\}) \times \mathbb{R}^{N-2}$. Therefore, for every G_n-invariant bounded smooth domain Ω in $\mathbb{C} \times \mathbb{R}^{N-2}$ with

$$D \subset \Omega \subset (\mathbb{C} \smallsetminus \{0\}) \times \mathbb{R}^{N-2} \quad \text{and} \quad \left(\min_{x\in\overline{\Omega}} \frac{a(x)^{\frac{N}{2}}}{c(x)^{\frac{N-2}{2}}}\right) n > \ell_m,$$

Theorem 1.2 yields at least m pairs of solutions to problem $(\wp_{2^*})$ in Ω.

Theorem 1.2 extends a similar result obtained in [7] for $a = c = 1$ and $b = 0$. As in that paper, the proof relies on the fact that symmetries provide an energy threshold below which the symmetric Palais-Smale condition holds true. We shall prove a representation theorem for symmetric Palais-Smale sequences of the functional associated with problem $(\wp_{2^*})$, which extends Struwe's global compactness result [20] and relates the symmetries of the concentration points to those of the solution to the limit problem concentrating at those points.

1.2 The supercritical case

Next, we consider problem $(\wp_p)$ with $p = 2^*_{N,k}$ in a domain Ω of the following form: we write $k = k_1 + \cdots + k_d$ with $k_1, \ldots, k_d \in \mathbb{N}$ and assume that $N \geq k + d + 2$. We consider domains

$$\Omega := \{(x^1, \ldots, x^d, x') \in \mathbb{R}^{k_1+1} \times \cdots \times \mathbb{R}^{k_d+1} \times \mathbb{R}^{N-k-d} : (|x^1|, \ldots, |x^d|, x') \in \Theta\},$$

where Θ is a bounded smooth domain in $\mathbb{R}^{N-k}$ whose closure is contained in $(0,\infty)^d \times \mathbb{R}^{N-k-d}$. We assume that a, b and c are radial in x^i, i.e. they can be written as

$$a(x^1,\ldots,x^d,x') = \alpha\left(\left|x^1\right|,\ldots,\left|x^d\right|,x'\right),$$
$$b(x^1,\ldots,x^d,x') = \beta\left(\left|x^1\right|,\ldots,\left|x^d\right|,x'\right),$$
$$c(x^1,\ldots,x^d,x') = \gamma\left(\left|x^1\right|,\ldots,\left|x^d\right|,x'\right).$$

Passaseo showed that if $d = 1$, Θ is a ball centered on $(0,\infty)\times\{0\}$, $b = 0$ and $a = c = 1$, problem $(\wp_{2^*_{N,k}})$ does not have a nontrivial solution. Examples of domains Θ where $(\wp_{2^*_{N,k}})$ has multiple or even infinitely many solutions in Ω have been exhibited by Wei and Yan [22] and Kim and Pistoia [15].

We consider $O(N-k-d)$ as a subgroup of $O(N-k)$ by making it act on the second factor of $\mathbb{R}^d \times \mathbb{R}^{N-k-d} \equiv \mathbb{R}^{N-k}$, i.e. $g(x'',x') := (x'',gx')$ for all $(x'',x') \in \mathbb{R}^d \times \mathbb{R}^{N-k-d}$, $g \in O(N-k-d)$. We shall prove the following results.

Theorem 1.3. *Let G be a closed subgroup of $O(N-k-d)$. If Θ, α, β and γ are G-invariant and $\#Gx = \infty$ for every $x \in \Theta$, then the supercritical problem $(\wp_{2^*_{N,k}})$ has infinitely many solutions in Ω of the form $u(x^1,\ldots,x^d,x') = v\left(\left|x^1\right|,\ldots,\left|x^d\right|,x'\right)$, where v is G-invariant.*

Next, we fix a closed subgroup Γ of $O(N-k-d)$ and a nonempty Γ-invariant bounded smooth domain D contained in $(0,\infty)^d \times \mathbb{R}^{N-k-d}$ such that $\#\Gamma y = \infty$ for every $y \in D$. We assume that the functions α, β and γ are Γ-invariant and that $\mu^{\Gamma}_{\alpha,\beta}(\Theta) > 0$. Let $\varrho : (0,\infty)^d \times \mathbb{R}^{N-k-d} \to \mathbb{R}$ be the function given by

$$\varrho(y_1,\ldots,y_d,y') := y_1^{k_1}\cdots y_d^{k_d}, \qquad y_i \in (0,\infty),\ \ y' \in \mathbb{R}^{N-k-d}. \tag{2}$$

Under these assumptions, we obtain the following result.

Theorem 1.4. *There exists an increasing sequence (ℓ_m) of positive real numbers, depending only on Γ, D, a, b and c, with the following property: If Θ is a bounded smooth domain in $\mathbb{R}^{N-k}$ such that $D \subset \overline{\Theta} \subset (0,\infty)^d \times \mathbb{R}^{N-k-d}$, and if Θ is invariant under the action of a closed subgroup G of Γ for which $\mu^{G}_{\alpha,\beta}(\Theta) > 0$ and*

$$\min_{y\in\overline{\Theta}} \frac{\varrho(y)\alpha(y)^{\frac{N-k}{2}}\#Gy}{\gamma(y)^{\frac{N-k-2}{2}}} > \ell_m,$$

then the supercritical problem $(\wp_{2^*_{N,k}})$ has at least m pairs of solutions $\pm u_1,\ldots,\pm u_m$ in Ω of the form

$$u_j(x^1,\ldots,x^d,x') = v_j\left(\left|x^1\right|,\ldots,\left|x^d\right|,x'\right),$$

where u_1 is positive, $u_2, \dots, u_m$ change sign, and v_j is G-invariant (in the variable x') and satisfies

$$\int_{\Theta} \varrho(y)\gamma(y) \left|v_j\right|^{2^*} \leq \ell_j S^{N/2} \qquad \text{for every } j = 1, \dots, m.$$

Theorem 1.4 is similar to a result obtained in [10] for $a = 1$ and $b = 0$ in domains of a different kind, arising from the Hopf fibrations.

The last two theorems will follow immediately from the former two. We shall prove these four theorems in the next section. The last section is devoted to the proof a symmetric global compactness result.

2 Proofs of the main results

2.1 *The proof of Theorems 1.1 and 1.2*

By the principle of symmetric criticality [16] the G-invariant solutions to problem $(\wp_{2^*})$ are the critical points of the functional

$$J_{a,b,c}(u) := \frac{1}{2} \int_{\Omega} \left[a(x) \left|\nabla u\right|^2 + b(x)u^2\right] - \frac{1}{2^*} \int_{\Omega} c(x) \left|u\right|^{2^*}$$

on the subspace $H_0^1(\Omega)^G$ of G-invariant functions in $H_0^1(\Omega)$.

Definition 2.1. A G-invariant Palais-Smale sequence for $J_{a,b,c}$ at the level τ is a sequence (u_n) such that

$$u_n \in H_0^1(\Omega)^G, \qquad J_{a,b,c}(u_n) \to \tau, \qquad J'_{a,b,c}(u_n) \to 0 \text{ in } H^{-1}(\Omega).$$

We shall say that $J_{a,b,c}$ satisfies condition $(PS)_\tau^G$ in $H_0^1(\Omega)$ if every G-invariant Palais-Smale sequence for $J_{a,b,c}$ at the level τ contains a subsequence which converges in $H_0^1(\Omega)$.

In the following section we shall prove that $J_{a,b,c}$ satisfies condition $(PS)_\tau^G$ below a certain level, see Corollary 3.2.

Proof (Proof of Theorem 1.1). Corollary 3.2 asserts that $J_{a,b,c}$ satisfies condition $(PS)_\tau^G$ for every $\tau \in \mathbb{R}$ whenever $\#Gx = \infty$ for all $x \in \Omega$. Arguing as in [19, Theorem 9.38], it is easy to check that the functional $J_{a,b,c}$ satisfies all hypotheses of the symmetric mountain pass theorem [19, Theorem 9.12]. Consequently, it has an unbounded sequence of critical values. □

To prove Theorem 1.2 we need the following result. Note that, if $\mu_{a,b}^G(\Omega) > 0$, then

$$\|u\|_{a,b}^2 := \int_\Omega \left[a(x)\,|\nabla u|^2 + b(x)u^2\right]$$

is a norm in $H_0^1(\Omega)^G$ which is equivalent to the standard one.

Theorem 2.2. *Let W be a finite dimensional subspace of $H_0^1(\Omega)^G$. If $\mu_{a,b}^G(\Omega) > 0$ and $J_{a,b,c}$ satisfies condition $(PS)_\tau^G$ in $H_0^1(\Omega)$ for every $\tau \leq \sup_W J_{a,b,c}$, then $J_{a,b,c}$ has at least* $\dim W - 1$ *pairs of sign changing critical points $u \in H_0^1(\Omega)^G$ such that $J_{a,b,c}(u) \leq \sup_W J_{a,b,c}$.*

Proof. The proof, up to minor modifications, may be found in [8, section 3]. □

Now the same argument used to prove the main result in [7] yields Theorem 1.2. We include the proof for the reader's convenience.

Proof (Proof of Theorem 1.2). We divide the proof into four steps.

STEP 1: *We define ℓ_m and show that the sequence (ℓ_m) is strictly increasing.* Let $\mathscr{P}_1(D)$ be the collection of all nonempty Γ-invariant bounded smooth domains contained in D, and define

$$\mathscr{P}_m(D) := \{(D_1, \ldots, D_m) : D_i \in \mathscr{P}_1(D),\ D_i \cap D_j = \emptyset \text{ if } i \neq j\}.$$

Note that $\mathscr{P}_m(D) \neq \emptyset$ for each $m \in \mathbb{N}$. Since $\#\Gamma x = \infty$ for all $x \in D_i$, Corollary 3.2 asserts that $J_{a,b,c}$ satisfies condition $(PS)_\tau^\Gamma$ in $H_0^1(D_i)$ for every $\tau \in \mathbb{R}$. Hence, the mountain pass theorem [1] yields a nontrivial least energy Γ-invariant solution ω_{D_i} to problem $(\wp_{2^*})$ in D_i which satisfies.

$$J_{a,b,c}(\omega_{D_i}) = \max_{t \geq 0} J_{a,b,c}(t\,\omega_{D_i}). \tag{3}$$

Extending ω_{D_i} by zero outside D_i we have that $\omega_{D_i} \in H_0^1(\Omega)^G$. Set $\tau_\infty := \frac{1}{N} S^{N/2}$ and define

$$\tau_m := \inf\left\{\sum_{i=1}^m J_{a,b,c}(\omega_{D_i}) : (D_1, \ldots, D_m) \in \mathscr{P}_m(D)\right\} \quad \text{and} \quad \ell_m := \tau_\infty^{-1}\tau_m.$$

Note that $J_{a,b,c}(\omega_{D_i}) \geq J_{a,b,c}(\omega_D) = \tau_1 > 0$. Therefore, for any $(D_1, \ldots, D_m) \in \mathscr{P}_m(D)$ with $m \geq 2$ we have that

$$\sum_{i=1}^m J_{a,b,c}(\omega_{D_i}) = \sum_{i=1}^{m-1} J_{a,b,c}(\omega_{D_i}) + J_{a,b,c}(\omega_{D_m}) \geq \tau_{m-1} + \tau_1 > \tau_{m-1}.$$

It follows that $\tau_m > \tau_{m-1} > 0$ for all $m \geq 2$. Hence, (ℓ_m) is strictly increasing. The following steps are devoted to showing that (ℓ_m) has the desired property. So let us fix $m \in \mathbb{N}$, a closed subgroup G of Γ and a bounded smooth domain Ω, such that $D \subset \Omega$, $\mu_{a,b}^G(\Omega) > 0$ and

$$\kappa := \min_{x \in \overline{\Omega}} \frac{a(x)^{N/2}\#Gx}{c(x)^{(N-2)/2}} > \ell_m. \tag{4}$$

STEP 2: *We show that, for any given $\varepsilon > 0$, $J_{a,b,c}$ has m pairs of critical points $\pm v_1, \ldots, \pm v_m$ in $H_0^1(\Omega)^G$ such that v_1 is positive, $v_2, \ldots, v_m$ change sign,*

$$J_{a,b,c}(v_k) < \tau_m \quad \textit{if} \;\; k = 1, \ldots, m-1, \qquad \textit{and} \qquad J_{a,b,c}(v_m) \leq \tau_m + \varepsilon.$$

Inequality (4) implies that $\tau_m < \kappa\, \tau_\infty$. Hence, we may assume that $\varepsilon \in (0, \tau_1)$ and $\tau_m + \varepsilon < \kappa\, \tau_\infty$. We choose $(D_1, \ldots, D_m) \in \mathscr{P}_m(D)$ such that

$$\tau_m \leq \sum_{i=1}^{m} J_{a,b,c}(\omega_{D_i}) < \tau_m + \varepsilon$$

and, for each $k = 1, \ldots, m$, we consider the subspace W_k of $H_0^1(\Omega)^G$ generated by $\{\omega_{D_1}, \ldots, \omega_{D_k}\}$. Since $D_i \cap D_j = \emptyset$ for $i \neq j$, the functions ω_{D_i} and ω_{D_j} are orthogonal in $H_0^1(\Omega)^G$. Therefore, $\dim W_k = k$, and identity (3) implies that

$$\sigma_k := \sup_{W_k} J_{a,b,c} \leq \sum_{i=1}^{k} J_{a,b,c}(\omega_{D_i}) < \kappa\, \tau_\infty.$$

By Corollary 3.2, $J_{a,b,c}$ satisfies $(PS)_\tau^G$ in $H_0^1(\Omega)$ for all $\tau \leq \sigma_k$. Hence, the mountain pass theorem [1] yields a positive critical point $v_1 \in H_0^1(\Omega)^G$ of $J_{a,b,c}$ such that $J_{a,b,c}(v_1) \leq \sigma_1$ and, applying Theorem 2.2 to each W_k with $k \geq 2$, we obtain $k-1$ pairs of sign changing critical points $\pm v_{k,2}, \ldots, \pm v_{k,k} \in H_0^1(\Omega)^G$ such that

$$J_{a,b,c}(v_{k,i}) \leq \sigma_k \qquad \text{for every } \; i = 2, \ldots, m.$$

Now, for each $k \geq 2$, we inductively choose $v_k \in \{v_{k,2}, \ldots, v_{k,k}\}$ such that $v_k \neq v_j$ for all $1 \leq j < k$. Note that

$$\sigma_k + (m-k)\tau_1 \leq \sum_{i=1}^{k} J_{a,b,c}(\omega_{D_i}) + \sum_{i=k+1}^{m} J_{a,b,c}(\omega_{D_i}) < \tau_m + \varepsilon.$$

Since $\varepsilon \in (0, \tau_1)$, this implies that

$$J_{a,b,c}(v_k) \leq \sigma_k < \tau_m \;\text{ if }\; k < m, \qquad \text{and} \qquad J_{a,b,c}(v_m) \leq \sigma_m < \tau_m + \varepsilon,$$

as claimed.

STEP 3: *We show that $J_{a,b,c}$ has m pairs of critical points $\pm v_1, \ldots, \pm v_m$ in $H_0^1(\Omega)^G$ such that v_1 is positive, $v_2, \ldots, v_m$ change sign, and*

$$J_{a,b,c}(v_k) \leq \tau_m \quad \textit{for all} \;\; k = 1, \ldots, m.$$

Let $\varepsilon_n > 0$ be such that $\varepsilon_n \to 0$. By Step 2, for each $n \in \mathbb{N}$, there are m pairs of critical points $\pm w_{n,1}, \ldots, \pm w_{n,m}$ of $J_{a,b,c}$ in $H_0^1(\Omega)^G$ such that $w_{n,1}$ is positive, $w_{n,2}, \ldots, w_{n,m}$ are sign changing, $J_{a,b,c}(w_{n,k}) < \tau_m$ *if* $k = 1, \ldots, m-1$, and $J_{a,b,c}(w_{n,m}) < \tau_m + \varepsilon_n$. Now, if for some n_0 we have that $J_{a,b,c}(w_{n_0,m}) \leq \tau_m$, then the functions $v_k := w_{n_0,k}$ verify the statement. If, on the other hand, $J_{a,b,c}(w_{n,m}) > \tau_m$ for all n, then $J_{a,b,c}(w_{n,m}) \to \tau_m$. Since $J'_{a,b,c}(w_{n,m}) = 0$ and $J_{a,b,c}$ satisfies $(PS)^G_{\tau_m}$, there exists a $v_m \in H^1_0(\Omega)^G$ such that, after passing to a subsequence, $w_{n,m} \to v_m$ as $n \to \infty$. It follows that v_m is a critical point of $J_{a,b,c}$ with $J_{a,b,c}(v_m) = \tau_m$. Note that v_m is positive if $m = 1$ and it is sign changing if $m \geq 2$. Moreover, since $J_{a,b,c}(w_{1,k}) < \tau_m$ *for all* $k = 1, \ldots, m-1$, setting $v_k := w_{1,k}$ for $k = 1, \ldots, m-1$, we have that $v_m \neq \pm v_k$ for every $k = 1, \ldots, m-1$ and the statement is proved.

STEP 4: *We show that* $J_{a,b,c}$ *has* m *pairs of critical points* $\pm u_1, \ldots, \pm u_m$ *in* $H^1_0(\Omega)^G$ *such that* u_1 *is positive,* $u_2, \ldots, u_m$ *change sign, and*

$$J_{a,b,c}(u_k) \leq \tau_k \quad \textit{for each } k = 1, \ldots, m.$$

Since $\kappa > \ell_m$ we have that $\kappa > \ell_k$ for $k = 1, \ldots, m$. Applying Step 3 to each $k = 1, \ldots, m$, we obtain k pairs of critical points $\pm v_{k,1}, \ldots, \pm v_{k,k}$ of $J_{a,b,c}$ in $H^1_0(\Omega)^G$ such that $v_{k,1}$ is positive, $v_{k,2}, \ldots, v_{k,k}$ change sign, and $J_{a,b,c}(v_{k,i}) \leq \tau_k$ for all $i = 1, \ldots, k$. Set $u_1 := v_{1,1}$ and, for $2 \leq k \leq m$, choose $u_k \in \{v_{k,2}, \ldots, v_{k,k}\}$ inductively, such that $u_k \neq u_i$ for every $i = 2, \ldots, k-1$. These u_k's have the desired properties.

Finally, note that

$$\int_\Omega c(x)\,|u_k|^{2^*} = NJ_{a,b,c}(u_k) \leq N\tau_k = N\ell_k\tau_\infty = \ell_k S^{N/2} \quad \text{for all } k = 1, \ldots, m.$$

This concludes the proof of Theorem 1.2. □

2.2 The proof of Theorems 1.3 and 1.4

If $u(x^1, \ldots, x^d, x') = v\left(\left|x^1\right|, \ldots, \left|x^d\right|, x'\right)$ with $v \in \mathscr{C}^2(\Theta)$, a straightforward computation shows that

$$\operatorname{div}(a(x)\nabla u) = \frac{1}{\varrho(y)}\operatorname{div}(\varrho(y)a(y)\nabla v)$$

where $x = (x^1, \ldots, x^d, x')$, $y = \left(\left|x^1\right|, \ldots, \left|x^d\right|, x'\right)$ and ϱ is the function defined in (2). Consequently, u satisfies

$$-\operatorname{div}(a(x)\nabla u) + b(x)u = c(x)|u|^{p-2}u \quad \text{in } \Omega$$

if and only if v satisfies

$$-\operatorname{div}(\varrho(y)\alpha(y)\nabla v) + \varrho(y)\beta(y)v = \varrho(y)\gamma(y)|v|^{p-2}v \quad \text{in } \Theta. \tag{5}$$

Note that $p = 2^*_{N,k} := \frac{2(N-k)}{N-k-2}$ is the critical exponent in dimension $N-k = \dim \Theta$. With these remarks, Theorems 1.3 and 1.4 follow immediately from Theorem 1.1 and 1.2, respectively.

Note that, in the special case when a, b, c are constant, the functions $\varrho\alpha, \varrho\beta, \varrho\gamma$ do not have local maxima nor minima in Ω : their extrema are attained on the boundary of Ω. Hence, Egnell's results [12] do not yield solutions to problem $(\wp_{2^*_{N,k}})$ in this case.

3 Representation of *G*-invariant Palais-Smale sequences

Throughout this section we assume that Ω is a G-invariant bounded smooth domain in $\mathbb{R}^N$, $N \geq 3$, and that $a, b, c \in \mathscr{C}^0(\overline{\Omega})$ are G-invariant functions, such that a and c are positive on $\overline{\Omega}$. We consider the functional

$$J_{a,b,c}(u) := \frac{1}{2}\int_\Omega \left[a(x)\,|\nabla u|^2 + b(x)u^2\right] - \frac{1}{2^*}\int_\Omega c(x)\,|u|^{2^*}$$

defined in $H_0^1(\Omega)^G$.

When G is the trivial group and $a = b = c \equiv 1$, Struwe [20] gave a complete description of the Palais-Smale sequences for $J_{a,b,c}$ in terms of the solutions to the limit problem

$$(\wp_\infty) \qquad -\Delta u = |u|^{2^*-2}u, \qquad u \in D^{1,2}(\mathbb{R}^N).$$

Roughly speaking, he showed that the Palais-Smale sequences which do not converge approach a sum of a (possibly trivial) solution to problem $(\wp_{2^*})$ plus nontrivial solutions to the limit problem $(\wp_\infty)$ which concentrate at certain points of the domain. In the symmetric case the concentration will occur at G-orbits of Ω. The aim of this section is to give a precise description of the relation between the symmetries of the concentration points and those of the corresponding solution to the limit problem.

Recall that the G-isotropy subgroup of a point $x \in \mathbb{R}^N$ is defined as

$$G_x := \{g \in G : gx = x\}.$$

The G-orbit Gx of x is G-homeomorphic to the homogeneous space G/G_x, see [3, 11]. In particular, if we denote by $|G/G_x|$ the index of G_x in G, then $|G/G_x| = \#Gx$.

We write $J_\infty : D^{1,2}(\mathbb{R}^N) \to \mathbb{R}$ for the energy functional associated with problem $(\wp_\infty)$, given by

$$J_\infty(u) := \frac{1}{2}\|u\|^2 - \frac{1}{2^*}|u|_{2^*}^{2^*},$$

where

$$\|u\|^2 := \int_{\mathbb{R}^N} |\nabla u|^2 \qquad \text{and} \qquad |u|_{2^*}^{2^*} := \int_{\mathbb{R}^N} |u|^{2^*}.$$

We shall prove the following result.

Theorem 3.1. *Let (u_n) be a G-invariant Palais-Smale sequence for $J_{a,b,c}$ at the level τ. Then, after passing to a subsequence, there exist a (possibly trivial) G-invariant solution u to problem $(\wp_{2^*})$, an integer $m \geq 0$, m closed subgroups $G_1, \dots, G_m$ of finite index in G, m sequences $(y_{1,n}), \dots, (y_{m,n})$ in Ω, m sequences $(\varepsilon_{1,n}), \dots, (\varepsilon_{m,n})$ in $(0, \infty)$, and m nontrivial solutions $\hat{u}_1, \dots, \hat{u}_m$ to problem $(\wp_\infty)$, with the following properties:*

(i) $G_{y_{i,n}} = G_i$ *for all* $n \in \mathbb{N}$, *and* $y_{i,n} \to y_i$ *in* $\overline{\Omega}$ *as* $n \to \infty$ *for each* $i = 1, \dots, m$.
(ii) $\varepsilon_{i,n}^{-1}\operatorname{dist}(y_{i,n}, \partial\Omega) \to \infty$ *and* $\varepsilon_{i,n}^{-1}|gy_{i,n} - g'y_{i,n}| \to \infty$ *as* $n \to \infty$ *for all* $[g'] \neq [g]$ *in* G/G_i *and* $i = 1, \dots, m$.
(iii) $\hat{u}_i$ *is* G_i*-invariant for each* $i = 1 \dots m$.
(iv) $\lim\limits_{n\to\infty} \left\| u_n - u - \sum\limits_{i=1}^m \sum\limits_{[g]\in G/G_i} \left(\frac{a(y_i)}{c(y_i)}\right)^{\frac{N-2}{4}} \varepsilon_{i,n}^{\frac{2-N}{2}} \hat{u}_i \left(g^{-1}\left(\frac{\cdot - gy_{i,n}}{\varepsilon_{i,n}}\right)\right) \right\| = 0.$
(v) $J_{a,b,c}(u) + \sum\limits_{i=1}^m |G/G_i| \left(\frac{a(y_i)^{N/2}}{c(y_i)^{(N-2)/2}}\right) J_\infty(\hat{u}_i) = \tau.$

Since $J_{a,b,c}(u) \geq 0$ for every solution u to problem $(\wp_{2^*})$ and $J_\infty(\hat{u}) \geq \frac{1}{N}S^{\frac{N}{2}}$ for every nontrivial solution $\hat{u}$ to problem $(\wp_\infty)$, the next statement is an immediate consequence of Theorem 3.1.

Corollary 3.2. *The functional $J_{a,b,c}$ satisfies condition $(PS)_\tau^G$ for every*

$$\tau < \left(\min_{x\in\overline{\Omega}} \frac{a(x)^{\frac{N}{2}}\#Gx}{c(x)^{\frac{N-2}{2}}}\right)\frac{1}{N}S^{\frac{N}{2}}, \tag{6}$$

In particular, if $\#Gx = \infty$ for all $x \in \overline{\Omega}$, then $J_{a,b,c}$ satisfies condition $(PS)_\tau^G$ for every $\tau \in \mathbb{R}$.

Theorem 1 in [6] is a special case of Theorem 3.1, but there is a gap in the proof of Proposition 4 in [6]. The argument given here fills that gap. The key to do this is given by Lemma 3.3 below.

We recall some basic facts about G-actions, which may be found, for instance, in [3, 11], and introduce some notation. Isotropy groups satisfy $G_{gx} = gG_xg^{-1}$. Thus, every subgroup K of G which is conjugate to the isotropy group of a point $x \in \mathbb{R}^N$ satisfies that $K = G_{gx}$ for some $g \in G$. The conjugacy class (G_x) of an isotropy group G_x is called a G-isotropy class of $\mathbb{R}^N$. The set of G-isotropy classes of $\mathbb{R}^N$ is finite. Conjugacy classes of subgroups of G are partially ordered as follows:

$$(K_1) \leq (K_2) \iff \text{there exists } g \in G \text{ such that } gK_1g^{-1} \subset K_2. \tag{7}$$

We write

$$(\mathbb{R}^N)^K := \{x \in \mathbb{R}^N : gx = x \ \text{ for all } \ g \in K\}$$

for the K-fixed point set of $\mathbb{R}^N$.

Lemma 3.3. *Given sequences (ε_n) in $(0,\infty)$ and (ξ_n) in $\mathbb{R}^N$, there exist a sequence (y_n) in $\mathbb{R}^N$ and a closed subgroup K of G such that, after passing to a subsequence, the following statements hold true:*

(s1) The sequence $(\varepsilon_n^{-1}\operatorname{dist}(G\xi_n, y_n))$ is bounded.
(s2) $G_{y_n} = K$ for all $n \in \mathbb{N}$.
(s3) If $|G/K| < \infty$, then $\varepsilon_n^{-1}|gy_n - g'y_n| \to \infty$ for any $g, g' \in G$ with $g'g^{-1} \notin K$.
(s4) If $|G/K| = \infty$, then there is a closed subgroup K' of G such that $K \subset K'$, $|G/K'| = \infty$ and $\varepsilon_n^{-1}|gy_n - g'y_n| \to \infty$ for any $g, g' \in G$ with $g'g^{-1} \notin K'$.

Proof. Set

$$V := \{x \in \mathbb{R}^N : \#Gx < \infty\}.$$

Note that V is a G-invariant linear subspace of $\mathbb{R}^N$. There are two cases:

CASE 1: *The sequence $(\varepsilon_n^{-1}\operatorname{dist}(\xi_n, V))$ is unbounded.*

Passing to a subsequence we may assume that $\varepsilon_n^{-1}\operatorname{dist}(\xi_n, V) \to \infty$. Since the set of G-isotropy classes of $\mathbb{R}^N$ is finite, we may also assume that there exists a subgroup K of G and an element $y_n \in G\xi_n$ such that $G_{y_n} = K$ for every $n \in \mathbb{N}$. So, clearly, (s1) and (s2) are satisfied. Since $\xi_n \notin V$, we have that $|G/K| = \#G\xi_n = \infty$. Thus, it remains to prove (s4).

To this end, let $V^\perp$ the orthogonal complement of V in $\mathbb{R}^N$ and $y_n^\perp$ be the orthogonal projection of y_n onto $V^\perp$. Since $y_n \notin V$ we have that $y_n^\perp \neq 0$ and, after passing to a subsequence,

$$\varrho_n := \frac{y_n^\perp}{|y_n^\perp|} \to \varrho \in V^\perp.$$

Define $K' := G_\varrho$. Since V and $V^\perp$ are G-invariant, we have that $K \subset K'$ and, since $\varrho \notin V$, we have that $|G/K'| = \#G\varrho = \infty$. Now, let $g, g' \in G$ be such that $[g] \neq [g']$ in G/K'. Then $|g\varrho - g'\varrho| > 0$, and may choose $n_0 \in \mathbb{N}$ such that

$$|\varrho_n - \varrho| < \frac{1}{4}\left|g\varrho - g'\varrho\right| \qquad \forall n \geq n_0.$$

It follows that

$$\begin{aligned}\left|g\varrho - g'\varrho\right| &\leq |g\varrho - g\varrho_n| + \left|g\varrho_n - g'\varrho_n\right| + \left|g'\varrho_n - g'\varrho\right| \\ &= \left|g\varrho_n - g'\varrho_n\right| + 2\left|\varrho_n - \varrho\right| \leq \left|g\varrho_n - g'\varrho_n\right| + \frac{1}{2}|g\varrho - g'\varrho| \qquad \forall n \geq n_0.\end{aligned}$$

Therefore

$$\frac{1}{2}\left|g\varrho - g'\varrho\right|\left|y_n^{\perp}\right| \leq \left|gy_n^{\perp} - g'y_n^{\perp}\right| \leq \left|gy_n - g'y_n\right| \qquad \forall n \geq n_0.$$

Multiplying this inequality by ε_n^{-1} we obtain

$$\frac{1}{2}\left|g\varrho - g'\varrho\right|\varepsilon_n^{-1}\mathrm{dist}(y_n, V) \leq \varepsilon_n^{-1}\left|gy_n - g'y_n\right| \qquad \forall n \geq n_0$$

and, since $\varepsilon_n^{-1}\mathrm{dist}(y_n, V) = \varepsilon_n^{-1}\mathrm{dist}(\xi_n, V) \to \infty$, statement (s4) follows.

CASE 2: *The sequence* $(\varepsilon_n^{-1}\mathrm{dist}(\xi_n, V))$ *is bounded.*

Let $\mathfrak{F}$ be the set of G-isotropy classes (G_x) with $x \in V$ such that the sequence $(\varepsilon_n^{-1}\mathrm{dist}(G\xi_n, (\mathbb{R}^N)^{G_x}))$ contains a bounded subsequence.

We claim that $\mathfrak{F} \neq \emptyset$. To prove this, let $x_n \in V$ be the orthogonal projection of ξ_n onto V. Since the set of G-isotropy classes in V is finite and every G-orbit in V is finite, after passing to a subsequence, we may assume that there exists a closed subgroup L of G such that $G_{x_n} = L$ for every $n \in \mathbb{N}$. Observe that $(\mathbb{R}^N)^L \subset V$, because $L \subset G_x$ for every $x \in (\mathbb{R}^N)^L$ and $\#Gx = |G/G_x| \leq |G/L| = \#G_{x_n} < \infty$. It follows that

$$\varepsilon_n^{-1}\mathrm{dist}(\xi_n, (\mathbb{R}^N)^L) \leq \varepsilon_n^{-1}|\xi_n - x_n| = \varepsilon_n^{-1}\mathrm{dist}(\xi_n, V)$$

and, since the right-hand side is bounded, we conclude that $(L) \in \mathfrak{F}$.

Since $\mathfrak{F}$ is finite and nonempty, we may choose an element $(K) \in \mathfrak{F}$ which is maximal with respect to the partial order defined in (7). After passing to a subsequence, there exist $\zeta_n \in G\xi_n$ and $c > 0$ such that

$$\varepsilon_n^{-1}\mathrm{dist}(\zeta_n, (\mathbb{R}^N)^K) < c \qquad \forall n \in \mathbb{N}.$$

We define y_n to be the orthogonal projection of ζ_n onto $(\mathbb{R}^N)^K$. Then (s1) is trivially satisfied, because

$$\varepsilon_n^{-1}\mathrm{dist}(G\xi_n, y_n) \leq \varepsilon_n^{-1}|\zeta_n - y_n| = \varepsilon_n^{-1}\mathrm{dist}(\zeta_n, (\mathbb{R}^N)^K) < c \qquad \forall n \in \mathbb{N}. \tag{8}$$

To prove (s2) note that $y_n \in (\mathbb{R}^N)^K \subset V$. Hence, $K \subset G_{y_n}$. The previous inequality implies that $(G_{y_n}) \in \mathfrak{F}$ and, since (K) is maximal, we conclude that $G_{y_n} = K$.

Since $|G/K| < \infty$, we are left with proving (s3). Arguing by contradiction, assume there exist $g, g' \in K$ such that $[g] \neq [g']$ in G/K and $(\varepsilon_n^{-1}|gy_n - g'y_n|)$ is bounded. Let $\hat{g} := g^{-1}g'$, L be the subgroup of G generated by $K \cup \{\hat{g}\}$, $W_1 := (\mathbb{R}^N)^L$ and W_2 be the orthogonal complement of W_1 in $(\mathbb{R}^N)^K$. Write

$$y_n = y_n^1 + y_n^2 \qquad \text{with } y_n^i \in W_i, \;\; i = 1, 2.$$

Since $\hat{g} \notin K = G_{y_n}$, we have that $\hat{g}y_n \neq y_n$ for all $n \in \mathbb{N}$. Therefore $y_n \notin W_1$ and, hence, $y_n^2 \neq 0$ for all $n \in \mathbb{N}$. After passing to a subsequence we have that

$$\frac{y_n^2}{|y_n^2|} \to y \in W_2 \smallsetminus W_1.$$

If the sequence $(\varepsilon_n^{-1} y_n^2)$ were unbounded, since we are assuming that $(\varepsilon_n^{-1}|gy_n - g'y_n|)$ is bounded, we would have—after passing to a subsequence—that

$$\left|\frac{\hat{g}y_n^2}{|y_n^2|} - \frac{y_n^2}{|y_n^2|}\right| = \frac{\varepsilon_n^{-1}|\hat{g}y_n - y_n|}{\varepsilon_n^{-1}|y_n^2|} = \frac{\varepsilon_n^{-1}|gy_n - g'y_n|}{\varepsilon_n^{-1}|y_n^2|} \to 0$$

and, hence, that $\hat{g}y = y$. This means that $y \in (\mathbb{R}^N)^L = W_1$, which is not true. Therefore, $(\varepsilon_n^{-1} y_n^2)$ must be bounded. After passing to a subsequence, we may assume that there exists a closed subgroup L_1 of G such that $G_{y_n^1} = L_1$ for all $n \in \mathbb{N}$. Since $y_n^1 \in W_1$ we have that $K \subset L \subset L_1$. Moreover, inequality (8) yields that

$$\varepsilon_n^{-1} \mathrm{dist}(G\xi_n, (\mathbb{R}^N)^{L_1}) \leq \varepsilon_n^{-1} \left|\zeta_n - y_n^1\right| \leq \varepsilon_n^{-1} \left|\zeta_n - y_n\right| + \varepsilon_n^{-1} \left|y_n^2\right| < c_1,$$

for some constant $c_1 > 0$. This shows that $(L_1) \in \mathfrak{F}$. Using again that (K) is a maximal element in $\mathfrak{F}$, we conclude that $K = L = L_1$, which contradicts the fact that $\hat{g} \notin K$. This proves (s3) and concludes the proof of the lemma. □

The proof of the following result is similar to that of the Brezis-Lieb lemma [4].

Lemma 3.4. *Let $p \in [1, \infty)$, (c_n) be a bounded sequence in $L^\infty(\mathbb{R}^N)$ and (u_n) be a bounded sequence in $L^p(\mathbb{R}^N)$, such that $c_n(x) \to \overline{c}(x)$ and $u_n(x) \to u(x)$ a.e. in $\mathbb{R}^N$. Then $u \in L^p(\mathbb{R}^N)$ and*

$$\lim_{n\to\infty} \left(\int_{\mathbb{R}^N} c_n \left|u_n\right|^p - c_n \left|u_n - u\right|^p \right) = \int_{\mathbb{R}^N} \overline{c} \left|u\right|^p.$$

Proof. By Fatou's lemma, $u \in L^p(\mathbb{R}^N)$. Let $\varepsilon > 0$. Then there exists $C > 0$ such that

$$\left||u_n(x)|^p - |u_n(x) - u(x)|^p\right| - \varepsilon \left|u_n(x) - u(x)\right|^p \leq C|u(x)|^p \qquad \forall x \in \mathbb{R}^N.$$

Set $M := \sup\{\|c_n\|_\infty, \|\overline{c}\|_\infty : n \in \mathbb{N}\}$. It follows that

$$\begin{aligned} v_n(x) :&= \left|c_n(x)\,|u_n(x)|^p - c_n(x)\,|u_n(x) - u(x)|^p - \overline{c}(x)\,|u(x)|^p\right| - \varepsilon\,|c_n(x)|\,|u_n(x) - u(x)|^p \\ &\leq |c_n(x)|\,\left(\left||u_n(x)|^p - |u_n(x) - u(x)|^p\right| - \varepsilon\,|u_n(x) - u(x)|^p\right) + |\overline{c}(x)|\,|u(x)|^p \\ &\leq M(C+1)|u(x)|^p \qquad \text{a.e. in } \mathbb{R}^N. \end{aligned}$$

Hence, $v_n^+ := \max\{v_n, 0\}$ satisfies $\left|v_n^+(x)\right| \leq M(C+1)|u(x)|^p$ a.e. in $\mathbb{R}^N$ and Lebesque's dominated convergence theorem yields

$$\lim_{n\to\infty} \int_{\mathbb{R}^N} v_n^+ = 0.$$

Since (u_n) is bounded in $L^p(\mathbb{R}^N)$, we have that

$$\int_{\mathbb{R}^N} |c_n |u_n|^p - c_n |u_n - u|^p - \bar{c}\,|u|^p| \leqq M\varepsilon \int_{\mathbb{R}^N} |u_n - u|^p + \int_{\mathbb{R}^N} v_n^+ \leqq C'\varepsilon + \int_{\mathbb{R}^N} v_n^+$$

for some positive constant C'. Letting $n \to \infty$ we conclude that

$$\lim_{n\to\infty} \int_{\mathbb{R}^N} |c_n |u_n|^p - c_n |u_n - u|^p - \bar{c}\,|u|^p| \leqq C'\varepsilon \qquad \text{for every } \varepsilon > 0.$$

Hence,

$$\lim_{n\to\infty} \left| \int_{\mathbb{R}^N} c_n |u_n|^p - c_n |u_n - u|^p - \bar{c}\,|u|^p \right| = 0,$$

as claimed. □

Lemma 3.5. *Let (a_n) be a bounded sequence in $L^\infty(\mathbb{R}^N)$ and $\bar{a} \in L^\infty(\mathbb{R}^N)$ be such that $a_n \to \bar{a}$ in $L^\infty_{loc}(\mathbb{R}^N)$, and let (u_n) be a sequence in $D^{1,2}(\mathbb{R}^N)$ such that $u_n \rightharpoonup u$ weakly in $D^{1,2}(\mathbb{R}^N)$. Then*

$$\lim_{n\to\infty} \left(\int_{\mathbb{R}^N} a_n |\nabla u_n|^2 - a_n |\nabla (u_n - u)|^2 \right) = \int_{\mathbb{R}^N} \bar{a}\,|\nabla u|^2.$$

Proof. We write

$$\begin{aligned} &a_n |\nabla u_n|^2 - a_n |\nabla (u_n - u)|^2 - \bar{a}\,|\nabla u|^2 = a_n \nabla (2u_n - u) \nabla u - \bar{a}\,|\nabla u|^2 \\ &= (a_n - \bar{a})\nabla (2u_n - u) \nabla u + 2\bar{a}\nabla (u_n - u) \nabla u. \end{aligned} \tag{9}$$

Fix $R > 0$. Then, there is a positive constant C such that

$$\begin{aligned} \left| \int_{\mathbb{R}^N} (a_n - \bar{a})\nabla (2u_n - u) \nabla u \right| &\leq \left| \int_{B_R(0)} (a_n - \bar{a})\nabla (2u_n - u) \nabla u \right| \\ &\quad + \left| \int_{\mathbb{R}^N \smallsetminus B_R(0)} (a_n - \bar{a})\nabla (2u_n - u) \nabla u \right| \\ &\leq C\,|a_n - \bar{a}|_{L^\infty(B_R(0))} + C \int_{\mathbb{R}^N \smallsetminus B_R(0)} |\nabla u|^2. \end{aligned}$$

It follows that

$$\limsup_{n\to\infty}\left|\int_{\mathbb{R}^N}(a_n-\overline{a})\nabla(2u_n-u)\nabla u\right|\leq C\int_{\mathbb{R}^N\smallsetminus B_R(0)}|\nabla u|^2$$

and, letting $R\to\infty$, we conclude that

$$\lim_{n\to\infty}\int_{\mathbb{R}^N}(a_n-\overline{a})\nabla(2u_n-u)\nabla u=0. \tag{10}$$

On the other hand, since $u_n-u\rightharpoonup 0$ weakly in $D^{1,2}(\mathbb{R}^N)$, we have that

$$\lim_{n\to\infty}\int_{\mathbb{R}^N}\overline{a}\nabla(u_n-u)\nabla u=0. \tag{11}$$

Equations (9), (10) and (11) yield

$$\begin{aligned}&\lim_{n\to\infty}\int_{\mathbb{R}^N}\left(a_n|\nabla u_n|^2-a_n|\nabla(u_n-u)|^2-\overline{a}|\nabla u|^2\right)\\&=\lim_{n\to\infty}\int_{\mathbb{R}^N}(a_n-\overline{a})\nabla(2u_n-u)\nabla u+\lim_{n\to\infty}\int_{\mathbb{R}^N}\overline{a}\nabla(u_n-u)\nabla u=0,\end{aligned}$$

as claimed. □

Lemma 3.6. *Let (c_n) be a bounded sequence in $L^\infty(\mathbb{R}^N)$ and $\overline{c}\in L^\infty(\mathbb{R}^N)$ be such that $c_n\to\overline{c}$ in $L^\infty_{loc}(\mathbb{R}^N)$. Let (u_n) be a sequence in $D^{1,2}(\mathbb{R}^N)$ such that $u_n\rightharpoonup u$ weakly in $D^{1,2}(\mathbb{R}^N)$ with $u\in L^\infty_{loc}(\mathbb{R}^N)$. Then*

$$c_n|u_n|^{2^*-2}u_n-c_n|u_n-u|^{2^*-2}(u_n-u)\longrightarrow\overline{c}|u|^{2^*-2}u\quad in\ \ (D^{1,2}(\mathbb{R}^N))'.$$

Proof. The proof is similar to that of Lemma 8.9 in [23]. □

If $b\equiv 0$, to simplify notation we write

$$J_0:=J_{a,0,c}.$$

We use the previous lemmas to prove the following result.

Proposition 3.7. *Let (u_n) be a G-invariant Palais-Smale sequence for J_0 at the level $\tau>0$ such that $u_n\rightharpoonup 0$ weakly in $H^1_0(\Omega)^G$. Then, after passing to a subsequence, there exist a closed subgroup K of finite index in G, a sequence (y_n) in Ω, a sequence (ε_n) in $(0,\infty)$, a nontrivial solution $\hat{u}$ to the limit problem $(\wp_\infty)$ and a G-invariant Palais-Smale sequence (v_n) for J_0 with the following properties:*

(i) $G_{y_n}=K$ *for all* $n\in\mathbb{N}$, *and* $y_n\to y_0$ *in* $\overline{\Omega}$ *as* $n\to\infty$.
(ii) $\varepsilon_n^{-1}\mathrm{dist}(y_n,\partial\Omega)\to\infty$ *and* $\varepsilon_n^{-1}|gy_n-g'y_n|\to\infty$ *as* $n\to\infty$ *if* $[g']\neq[g]$ *in* G/K.

(iii) $\hat{u}$ *is* K*-invariant.*

(iv) $\lim_{n\to\infty}\left\| u_n - v_n - \sum_{[g]\in G/K}\left(\frac{a(y_0)}{c(y_0)}\right)^{\frac{N-2}{4}} \varepsilon_n^{\frac{2-N}{2}}\hat{u}\left(g^{-1}\left(\frac{\cdot - gy_n}{\varepsilon_n}\right)\right)\right\| = 0.$

(v) $\lim_{n\to\infty} J_0(u_n) = \lim_{n\to\infty} J_0(v_n) + |G/K|\left(\frac{a(y_0)^{N/2}}{c(y_0)^{(N-2)/2}}\right) J_\infty(\hat{u}).$

Proof. Let (u_n) be a G-invariant Palais-Smale sequence for J_0 at the level $\tau > 0$. As usual, we consider $H_0^1(\Omega)$ as a subspace of $D^{1,2}(\mathbb{R}^N)$ by defining $u \in H_0^1(\Omega)$ as cero outside Ω. We subdivide the proof into three steps.

STEP 1: *We define sequences* (ε_n) *in* $(0,\infty)$ *and* (y_n) *in* $\mathbb{R}^N$*, and a subgroup* K *of finite index in* G *such that* $G_{y_n} = K$.

Since (u_n) is a Palais-Smale sequence, (u_n) is bounded in $H_0^1(\Omega)$ and

$$\int_\Omega c\,|u_n|^{2^*} = N\left(J_0(u_n) - \frac{1}{2}J_0'(u_n)u_n\right) \to N\tau > 0.$$

Set

$$\delta := \min\left\{\frac{N\tau}{2}, \left(\frac{S\min_{x\in\overline{\Omega}} a(x)}{2\left(\max_{x\in\overline{\Omega}} c(x)\right)^{\frac{N-2}{N}}}\right)^{\frac{N}{2}}\right\}. \tag{12}$$

Since $\delta \le \frac{N\tau}{2}$, there are bounded sequences (ε_n) in $(0,\infty)$ and (ξ_n) in $\mathbb{R}^N$ such that, after passing to a subsequence,

$$\sup_{x\in\mathbb{R}^N}\int_{B_{\varepsilon_n}(x)} c|u_n|^{2^*} = \int_{B_{\varepsilon_n}(\xi_n)} c|u_n|^{2^*} = \delta.$$

For these sequences we choose K and (y_n) as in Lemma 3.3. Then, $G_{y_n} = K$ and there exists a positive constant C_1 such that

$$\varepsilon_n^{-1}\mathrm{dist}(G\xi_n, y_n) < C_1$$

for all $n \in \mathbb{N}$. Therefore, (y_n) is bounded and there exists $g_n \in G$ such that $B_{\varepsilon_n}(g_n\xi_n) \subset B_{C\varepsilon_n}(y_n)$ with $C := C_1 + 1$. Since c and u_n are G-invariant, we conclude that

$$\delta = \int_{B_{\varepsilon_n}(\xi_n)} c|u_n|^{2^*} = \int_{B_{\varepsilon_n}(g_n\xi_n)} c|u_n|^{2^*} \le \int_{B_{C\varepsilon_n}(y_n)} c|u_n|^{2^*}. \tag{13}$$

Next, we will show that $|G/K| < \infty$. Arguing by contradiction, if we assume that $|G/K| = \infty$, then property (s4) of Lemma 3.3 asserts that there exists a closed subgroup K' of G such that $K \subset K'$, $|G/K'| = \infty$ and $\varepsilon_n^{-1}|gy_n - g'y_n| \to \infty$ for any $[g],[g'] \in G/K'$ with $[g] \ne [g']$. Hence, for each $m \in \mathbb{N}$, we may choose $g_1,\ldots,g_m \in G$ such that $[g_i] \ne [g_j]$ in G/K' and

$$B_{C\varepsilon_n}(g_iy_n) \cap B_{C\varepsilon_n}(g_jy_n) = \emptyset \quad \text{for } i \ne j \text{ and } n \text{ sufficiently large.}$$

Using again that the functions c and u_n are G-invariant, from inequality (13) we obtain that

$$m\delta \leq \sum_{i=1}^{m} \int_{B_{C\varepsilon_n}(g_i y_n)} c|u_n|^{2^*} \leq \int_{\Omega} c|u_n|^{2^*} = N\tau + o(1),$$

for every $m \in \mathbb{N}$. This is a contradiction. We conclude that $|G/K| < \infty$.

STEP 2: *We show that $\varepsilon_n^{-1}\mathrm{dist}(y_n, \partial\Omega) \to \infty$ and that $y_n \in \Omega$, and we define a nontrivial K-invariant solution $\hat{u} \in D^{1,2}(\mathbb{R}^N)$ to problem $(\wp_\infty)$.*

For $z \in \Omega_n := \{z \in \mathbb{R}^N : \varepsilon_n z + y_n \in \Omega\}$ set

$$\tilde{u}_n(z) := \varepsilon_n^{\frac{N-2}{2}} u_n(\varepsilon_n z + y_n), \qquad a_n(z) := a(\varepsilon_n z + y_n), \qquad c_n(z) := c(\varepsilon_n z + y_n).$$

Since $G_{y_n} = K$ and u_n, a and c are G-invariant, we have that $\tilde{u}_n, a_n$ and c_n are K-invariant. Note also that

$$\int_{\Omega} a|\nabla u_n|^2 = \int_{\Omega_n} a_n|\nabla \tilde{u}_n|^2 \qquad \text{and} \qquad \int_{\Omega} c|u_n|^{2^*} = \int_{\Omega_n} c_n|\tilde{u}_n|^{2^*}.$$

Hence, $(\tilde{u}_n)$ is bounded in $D^{1,2}(\mathbb{R}^N)$ and, therefore, a subsequence satisfies that $\tilde{u}_n \rightharpoonup \tilde{u}$ weakly in $D^{1,2}(\mathbb{R}^N)$, $\tilde{u}_n \to \tilde{u}$ strongly in $L^2_{loc}(\mathbb{R}^N)$, and $\tilde{u}_n \to \tilde{u}$ a.e. in $\mathbb{R}^N$. It follows that $\tilde{u}$ is K-invariant. A standard argument, using definition (12) and inequality (13), shows that $\tilde{u} \neq 0$, see, e.g., [6, 20, 21, 23].

Since (y_n) and (ε_n) are bounded, passing to a subsequence, we have that $y_n \to y_0$ in $\mathbb{R}^N$ and $\varepsilon_n \to \varepsilon$ in $[0, \infty)$. If we suppose that $\varepsilon \neq 0$ then, since $u_n \rightharpoonup 0$ weakly in $H_0^1(\Omega)$, we would have that $\tilde{u} = 0$, which is a contradiction. Therefore, $\varepsilon = 0$. It follows that

$$a_n \to a_0 := a(y_0) \qquad \text{and} \qquad c_n \to c_0 := c(y_0) \qquad \text{in } L^\infty_{loc}(\mathbb{R}^N).$$

Let $\varphi \in \mathscr{C}_c^\infty(\mathbb{R}^N)$ be such that $\mathrm{supp}(\varphi) \subset \Omega_n$ for all n sufficiently large. Then, setting $\varphi_n(x) := \varphi(\varepsilon_n^{-1}(x - y_n))$, we have that

$$\int_{\Omega_n} a_n \nabla\tilde{u}_n \cdot \nabla\varphi - \int_{\Omega_n} c_n|\tilde{u}_n|^{2^*-2}\tilde{u}_n\varphi = J_0'(u_n)\varphi_n = o(1),$$

because (φ_n) is bounded in $H_0^1(\Omega)$. Since

$$\begin{aligned}
&\left|\int_{\mathbb{R}^N} a_n\nabla\tilde{u}_n \cdot \nabla\varphi - \int_{\mathbb{R}^N} a_0\nabla\tilde{u} \cdot \nabla\varphi\right| \\
&\leq \left|\int_{\mathbb{R}^N} (a_n - a_0)\nabla\tilde{u}_n \cdot \nabla\varphi\right| + \left|\int_{\mathbb{R}^N} a_0(\nabla\tilde{u}_n \cdot \nabla\varphi - \nabla\tilde{u} \cdot \nabla\varphi)\right| \\
&\leq C\,|a_n - a_0|_{L^\infty(\mathrm{supp}(\varphi))} + a_0\left|\int_{\mathbb{R}^N} (\nabla\tilde{u}_n \cdot \nabla\varphi - \nabla\tilde{u} \cdot \nabla\varphi)\right| = o(1)
\end{aligned}$$

and, similarly,

$$\left| \int_{\mathbb{R}^N} c_n |\tilde{u}_n|^{2^*-2} \tilde{u}_n \varphi - \int_{\mathbb{R}^N} c_0 |\tilde{u}|^{2^*-2} \tilde{u} \varphi \right| = o(1),$$

we conclude that

$$\int_{\mathbb{R}^N} a_0 \nabla \tilde{u} \cdot \nabla \varphi - \int_{\mathbb{R}^N} c_0 |\tilde{u}|^{2^*-2} \tilde{u} \varphi = 0$$

for every $\varphi \in \mathscr{C}_c^\infty(\mathbb{R}^N)$ such that $\operatorname{supp}(\varphi) \subset \Omega_n$ for n sufficiently large. Arguing as in [20, 21, 23] one shows that, if the sequence $(\varepsilon_n^{-1} \operatorname{dist}(y_n, \Omega))$ were bounded, then $\tilde{u}$ would be a solution to problem

$$-\Delta u = \frac{c_0}{a_0} |u|^{2^*-2} u, \qquad u \in D_0^{1,2}(\mathbb{H}),$$

in some half-space $\mathbb{H}$ contained in $\mathbb{R}^N$, contradicting the fact that this problem does not have a nontrivial solution in a strictly starshaped proper subdomain of $\mathbb{R}^N$, see [21, Theorem 1.3]. Hence $\varepsilon_n^{-1} \operatorname{dist}(y_n, \partial\Omega) \to \infty$. This implies that $y_n \in \Omega$, for otherwise $B_{C\varepsilon_n}(y_n) \subset \mathbb{R}^N \smallsetminus \Omega$, contradicting (13). It also implies that $\tilde{u}$ is a solution to problem

$$-\Delta u = \frac{c_0}{a_0} |u|^{2^*-2} u, \qquad u \in D^{1,2}(\mathbb{R}^N). \tag{14}$$

Therefore,

$$\hat{u} := \left(\frac{c_0}{a_0} \right)^{\frac{N-2}{4}} \tilde{u}$$

is a nontrivial, K-invariant solution to problem $(\wp_\infty)$.

STEP 3: *We define a sequence* (v_n) *which satisfies* (iv) *and* (v) *and is a* G-*invariant* Palais-Smale sequence for J_0.

Let $G/K := \{[g_1], \ldots, [g_m]\}$. Set

$$r_n := \frac{1}{4} \min\{\operatorname{dist}(y_n, \partial\Omega), \ \left| g_i(y_n) - g_j(y_n) \right| : i, j = 1, \ldots, m, \ i \neq j\}.$$

Choose a radially symmetric function $\chi \in \mathscr{C}_c^\infty(\mathbb{R}^N)$ such that $0 \leq \chi \leq 1$, $\chi(x) = 1$ if $|x| \leq 1$ and $\chi(x) = 0$ if $|x| \geq 2$ and define

$$v_n(x) := u_n(x) - \sum_{i=1}^m \varepsilon_n^{\frac{2-N}{2}} \tilde{u} \left(g_i^{-1} \left(\frac{x - g_i y_n}{\varepsilon_n} \right) \right) \chi(r_n^{-1}(x - g_i y_n)).$$

Since $\tilde{u}$ is K-invariant and $G_{y_n} = K$ for all $n \in \mathbb{N}$, we have that v_n is G-invariant and $v_n \in H_0^1(\Omega)$. Similarly, the functions

$$w_n^j(x) := u_n(x) - \sum_{i=j}^{m} \varepsilon_n^{\frac{2-N}{2}} \tilde{u}\left(g_i^{-1}\left(\frac{x - g_i y_n}{\varepsilon_n}\right)\right), \qquad j = 1, \ldots, m,$$

are G-invariant functions in $D^{1,2}(\mathbb{R}^N)$. Note that $r_n \varepsilon_n^{-1} \to \infty$. Using this fact, an easy computation shows that

$$\left\| u_n - v_n - \sum_{i=1}^{m} \varepsilon_n^{\frac{2-N}{2}} \tilde{u}\left(g_i^{-1}\left(\frac{\cdot - g_i y_n}{\varepsilon_n}\right)\right) \right\| = \left\| w_n^1 - v_n \right\| \to 0, \tag{15}$$

i.e. (v_n) satisfies (iv).

Next, we rescale w_n^j and use the G-invariance of u_n to obtain

$$\begin{aligned}
\tilde{w}_n^j(z) :&= \varepsilon_n^{\frac{N-2}{2}} w_n^j(\varepsilon_n z + g_j y_n) \\
&= \varepsilon_n^{\frac{N-2}{2}} u_n(\varepsilon_n z + g_j y_n) - \sum_{i=j+1}^{m} \tilde{u}\left(g_i^{-1}\left(z + \frac{g_j y_n - g_i y_n}{\varepsilon_n}\right)\right) - \tilde{u}(g_j^{-1} z) \\
&= \tilde{u}_n(g_j^{-1} z) - \sum_{i=j+1}^{m} \tilde{u}\left(g_i^{-1}\left(z + \frac{g_j y_n - g_i y_n}{\varepsilon_n}\right)\right) - \tilde{u}(g_j^{-1} z).
\end{aligned}$$

Since $\tilde{u}_n \rightharpoonup \tilde{u}$ weakly in $D^{1,2}(\mathbb{R}^N)$ and $\varepsilon_n^{-1} \left| g_j y_n - g_i y_n \right| \to \infty$ for every $i \neq j$, we have that

$$\tilde{u}_n \circ g_j^{-1} - \sum_{i=j+1}^{m} \tilde{u}\left(g_i^{-1}\left(z + \frac{g_j y_n - g_i y_n}{\varepsilon_n}\right)\right) \rightharpoonup \tilde{u} \circ g_j^{-1} \quad \text{weakly in } D^{1,2}(\mathbb{R}^N).$$

Without loss of generality we may assume that $a, c \in \mathscr{C}^0(\mathbb{R}^N) \cap L^\infty(\mathbb{R}^N)$. Using Lemma 3.5 we obtain that

$$\begin{aligned}
&\int_{\mathbb{R}^N} a|\nabla w_n^j|^2 = \int_{\mathbb{R}^N} a_n |\nabla \tilde{w}_n^j|^2 \\
&= \int_{\mathbb{R}^N} a_n \left| \nabla \left(\tilde{u}_n \circ g_j^{-1} - \sum_{i=j+1}^{m} \tilde{u}\left(g_i^{-1}\left(\cdot + \frac{g_j y_n - g_i y_n}{\varepsilon_n}\right)\right)\right)\right|^2 \\
&\quad - \int_{\mathbb{R}^N} a_0 \left|\nabla(\tilde{u} \circ g_j^{-1})\right|^2 + o(1)
\end{aligned}$$

$$= \int_{\mathbb{R}^N} a \left| \nabla \left(u_n - \sum_{i=j+1}^{m} \varepsilon_n^{\frac{2-N}{2}} \tilde{u} \left(g_i^{-1} \left(\frac{\cdot - g_i y_n}{\varepsilon_n} \right) \right) \right) \right|^2 - \int_{\mathbb{R}^N} a_0 \, |\nabla \tilde{u}|^2 + o(1)$$

$$= \int_{\mathbb{R}^N} a |\nabla w_n^{j+1}|^2 - \int_{\mathbb{R}^N} a_0 \, |\nabla \tilde{u}|^2 + o(1).$$

These identities for $j = 1, \ldots, m$, together with (15), yield

$$\int_{\Omega} a|\nabla v_n|^2 = \int_{\mathbb{R}^N} a|\nabla w_n^1|^2 = \int_{\Omega} a|\nabla u_n|^2 - m \int_{\mathbb{R}^N} a_0 \, |\nabla \tilde{u}|^2 + o(1).$$

Similarly, using Lemma 3.4, we conclude that

$$\int_{\Omega} c|v_n|^{2^*} = \int_{\Omega} c|u_n|^{2^*} - m \int_{\mathbb{R}^N} c_0 \, |\tilde{u}|^{2^*} + o(1).$$

These last two identities yield

$$J_0(u_n) = J_0(v_n) + m \left(\frac{1}{2} \int_{\mathbb{R}^N} a_0 \, |\nabla \tilde{u}|^2 - \frac{1}{2^*} \int_{\mathbb{R}^N} c_0 \, |\tilde{u}|^{2^*} \right) + o(1)$$

$$= J_0(v_n) + |G/K| \, \frac{a_0^{N/2}}{c_0^{(N-2)/2}} J_\infty(\hat{u}) + o(1).$$

This proves (v).

Since $J'_\infty(\hat{u}) = 0$, a similar argument using Lemma 3.6 shows that

$$o(1) = J'_0(u_n) = J'_0(v_n) + o(1) \quad \text{in} \quad (D^{1,2}(\mathbb{R}^N))'.$$

This proves that (v_n) is a G-invariant Palais-Smale sequence for J_0 and concludes the proof of Proposition 3.7. □

Proof (Proof of Theorem 3.1). Let (u_n) be a G-invariant Palais-Smale sequence for $J_{a,b,c}$ at the level τ. Since a and c are continuous and positive on $\overline{\Omega}$,

$$\|u\|_a^2 := \int_{\Omega} a(x) \, |\nabla u|^2 \qquad \text{and} \qquad |u|_{c,2^*}^{2^*} := \int_{\Omega} c(x) \, |u|^{2^*}$$

are norms in $H_0^1(\Omega)$ and $L^{2^*}(\Omega)$, respectively, which are equivalent to the standard ones. So, for some positive constants C_i,

$$C_1 + o(1) \, \|u_n\|_a \geq 2J_{a,b,c}(u_n) - J'_{a,b,c}(u_n)u_n = (1 - \frac{2}{2^*}) \, |u_n|_{c,2^*}^{2^*} \geq C_2(|u_n|_2^2)^{2^*/2}$$

and, hence,

$$\|u_n\|_a^2 = 2J_{a,b,c}(u_n) - \int_{\Omega} b(x)|u_n|^2 + \frac{2}{2^*} \, |u|_{c,2^*}^{2^*} \leq C_3 + o(1) \, \|u_n\|_a \, .$$

This proves that the sequence (u_n) is bounded in $H_0^1(\Omega)^G$. After passing to a subsequence, we have that $u_n \rightharpoonup u$ weakly in $H_0^1(\Omega)^G$, $u_n \to u$ strongly in $L^2(\Omega)$, and $u_n \to u$ a.e. in Ω. Set $u_n^1 := u_n - u$. A standard argument shows that u is a solution to problem $(\wp_{2^*})$ and that

$$J_{a,b,c}(u_n) = J_{a,b,c}(u) + J_0(u_n^1) + o(1) \qquad \text{and}$$
$$J'_{a,b,c}(u_n) = J'_{a,b,c}(u) + J'_0(u_n^1) + o(1) = J'_0(u_n^1) + o(1).$$

Therefore, (u_n^1) is a G-invariant Palais-Smale sequence for J_0 at the level $\tau - J_{a,b,c}(u)$ and $u_n^1 \rightharpoonup u$ weakly in $H_0^1(\Omega)^G$. So, we may apply Proposition 3.7 to (u_n^1). As in [20], the conclusion of Theorem 3.1 follows after applying that proposition a finite number of times, because 0 is the smallest level at which there is a G-invariant Palais-Smale sequence. □

Acknowledgements M. Clapp was partially supported by CONACYT grant 237661 and PAPIIT-DGAPA-UNAM grant IN104315 (Mexico). J. Faya was partially supported by FONDECYT postdoctoral grant 3150172 (Chile).

References

1. Ambrosetti, A., Rabinowitz, P.H.: Dual variational methods in critical point theory and applications. J. Funct. Anal. **14**, 349–381 (1973)
2. Bahri, A., Coron, J.M.: On a nonlinear elliptic equation involving the critical Sobolev exponent: the effect of the topology of the domain. Commun. Pure Appl. Math. **41**, 253–294 (1988)
3. Bredon, G.E.: Introduction to Compact Transformation Groups. Pure and Applied Mathematics, vol. 46. Academic, New York/London (1972)
4. Brezis, H., Lieb, E.: A relation between pointwise convergence of functions and convergence of functionals. Proc. Am. Math. Soc. **88**, 486–490 (1983)
5. Brézis, H., Nirenberg, L.: Positive solutions of nonlinear elliptic equations involving critical Sobolev exponents. Commun. Pure Appl. Math. **36**, 437–477 (1983)
6. Clapp, M.: A global compactness result for elliptic problems with critical nonlinearity on symmetric domains. In: Lupo, D., Pagani, C.D., Ruf, B. (eds.) Nonlinear Equations: Methods, Models and Applications. Progress in Nonlinear Diferential Equations and Their Applications, vol. 54, pp. 117–126. Birkhauser, Boston (2003)
7. Clapp, M., Faya, J.: Multiple solutions to the Bahri-Coron problem in some domains with nontrivial topology. Proc. Am. Math. Soc. **141**, 4339–4344 (2013)
8. Clapp, M., Pacella, F.: Multiple solutions to the pure critical exponent problem in domains with a hole of arbitrary size. Math. Z. **259**, 575–589 (2008)
9. Clapp, M., Pistoia, A.: Symmetries, Hopf fibrations and supercritical elliptic problems. Contemp. Math. (to appear)
10. Clapp, M., Faya, J., Pistoia, A.: Nonexistence and multiplicity of solutions to elliptic problems with supercritical exponents. Calc. Var. Partial Differ. Equat. **48**, 611–623 (2013)
11. tom Dieck, T.: Transformation Groups. De Gruyter Studies in Mathematics, vol. 8. Walter de Gruyter, Berlin/New York (1987)
12. Egnell, H.: Semilinear elliptic equations involving critical Sobolev exponents. Arch. Ration. Mech. Anal. **104**, 27–56 (1988)

13. Hadiji, R., Yazidi, H.: Problem with critical Sobolev exponent and with weight. Chin. Ann. Math. Ser. B **28**, 327–352 (2007)
14. Hadiji, R., Molle, R., Passaseo, D., Yazidi, H.: Localization of solutions for nonlinear elliptic problems with critical growth. C. R. Math. Acad. Sci. Paris **343**, 725–730 (2006)
15. Kim, S., Pistoia, A.: Supercritical problems in domains with thin toroidal holes Discrete Contin. Dyn. Syst. **34**, 4671–4688 (2014)
16. Palais, R.: The principle of symmetric criticality. Commun. Math. Phys. **69**, 19–30 (1979)
17. Passaseo, D.: Nonexistence results for elliptic problems with supercritical nonlinearity in nontrivial domains. J. Funct. Anal. **114**, 97–105 (1993)
18. Passaseo, D.: New nonexistence results for elliptic equations with supercritical nonlinearity. Differ. Integr. Equat. **8**, 577–586 (1995)
19. Rabinowitz, P.H.: Minimax Methods in Critical Point Theory with Applications to Differential Equations. CBMS Regional Conference Series in Mathematics, vol 65. American Mathematical Society, Providence, RI (1986)
20. Struwe, M.: A global compactness result for elliptic boundary value problems involving limiting nonlinearities. Math. Z. **187**, 511–517 (1984)
21. Struwe, M.: Variational Methods. Springer, Berlin/Heidelberg (1996)
22. Wei, J., Yan, S.: Infinitely many positive solutions for an elliptic problem with critical or supercritical growth. J. Math. Pures Appl. **96**, 307–333 (2011)
23. Willem, M.: Minimax Theorems. Progress in Nonlinear Diferential Equations and Their Applications, vol. 24. Birkhäuser, Boston (1996)

Compactness and non-compactness for Yamabe-type problems

Fernando Codá Marques

Dedicated to the 80th birthday of Prof. Djairo de Figueiredo

1 Introduction

Let (M^n, g) be a smooth compact Riemannian manifold of dimension $n \geq 3$. The classical Yamabe Problem consists in finding metrics of constant scalar curvature in the conformal class of g. The corresponding question in dimension two is known as the Uniformization Theorem. The existence of a solution is one of the greatest achievements of geometric analysis, and follows from the works of Yamabe [54], Trudinger [50], Aubin [5], and Schoen [45]. We refer the reader to [31] for an account of these results.

The Yamabe problem can be also formulated for other curvature quantities leading to interesting fully nonlinear or higher-order elliptic equations. This is a very active field. We refer the reader to the survey articles of A. Chang [11] and of Chang and Yang [12] for a discussion of the partial differential equations of conformal geometry. There are Yamabe-type problems also for manifolds with boundary and for coupled systems of equations. We will discuss compactness results in these different settings after reviewing what is known in the classical Yamabe problem.

The *conformal class* of g is defined to be

$$[g] = \{\tilde{g} = u^{\frac{4}{n-2}} g : u \in C^{\infty}(M), u > 0\}.$$

If $\tilde{g} = u^{\frac{4}{n-2}} g$, $u > 0$, we can compute the scalar curvature as

$$R_{\tilde{g}} = -\frac{4(n-1)}{n-2} u^{-\frac{n+2}{n-2}} \Big(\Delta_g u - \frac{n-2}{4(n-1)} R_g u \Big).$$

F.C. Marques (✉)
Princeton University, Princeton, NJ 08544-1000, USA
e-mail: coda@math.princeton.edu

A.N. Carvalho et al. (eds.), *Contributions to Nonlinear Elliptic Equations and Systems*, Progress in Nonlinear Differential Equations and Their Applications 86,
DOI 10.1007/978-3-319-19902-3_9

Here R_g and $R_{\tilde{g}}$ denote the scalar curvatures of g and $\tilde{g}$, respectively, and Δ_g is the Laplace-Beltrami operator associated with g. The linear operator $L_g = \Delta_g - \frac{n-2}{4(n-1)} R_g$ is usually called the *conformal Laplacian* of g.

It follows that in order to find a constant scalar curvature metric in $[g]$ one needs to prove the existence of a positive solution u to the partial differential equation

$$L_g(u) + c(n)Ku^{\frac{n+2}{n-2}} = 0 \tag{1}$$

for some constant K, where $c(n) = \frac{n-2}{4(n-1)}$.

A nice feature of the problem is that it is variational. The constant scalar curvature metrics $\tilde{g} \in [g]$ are the critical points of the functional

$$Q(\tilde{g}) = \frac{\int_M R_{\tilde{g}}\, dv_{\tilde{g}}}{\left(\int_M dv_{\tilde{g}}\right)^{\frac{n-2}{n}}},$$

called the *normalized total scalar curvature functional*, when restricted to the conformal class $[g]$. Equivalently, the conformal factor u is a critical point of the energy

$$Q(u^{\frac{4}{n-2}} g) = c(n)^{-1} \frac{\int_M \left(|\nabla_g u|^2 + c(n) R_g u^2\right) dv_g}{\left(\int_M u^{\frac{2n}{n-2}} dv_g\right)^{\frac{n-2}{n}}}.$$

The number

$$Q(M, g) = \inf_{\tilde{g} \in [g]} Q(\tilde{g}).$$

is called the *Yamabe quotient* of M and it is always achieved as the energy of a minimizing metric. If the Yamabe quotient is negative, the Maximum Principle applied to equation (1) implies that the solution (of negative constant scalar curvature) is unique, while if it is zero, the solution (of zero scalar curvature) is unique up to a constant factor. For manifolds of positive Yamabe quotient the uniqueness fails in general, and a significant part of the research work after the existence was settled is dedicated to understand the structure of the set of solutions. When the Yamabe quotient is positive, the constant K in (1) must be positive also and we normalize it to be $n(n-1)$.

2 Compactness and noncompactness

In his 1988 topics course at Stanford (see also [46] and [47]), R. Schoen proposed the Compactness Conjecture:

The set

$$\mathcal{M}_g = \{\tilde{g} \in [g] : R_{\tilde{g}} = n(n-1)\}$$

of solutions to the Yamabe Problem, in the positive Yamabe quotient case, is compact (in any C^k topology) unless the manifold is conformally equivalent to the standard sphere.

By the Harnack inequality and standard elliptic estimates this is equivalent to establishing a priori estimates for solutions of the equation (1). Note that for the standard sphere the set of solutions must be noncompact, since the group of conformal transformations of S^n is noncompact itself.

The program Schoen outlined to prove this conjecture made essential use of the Positive Mass Theorem of General Relativity. The Positive Mass Theorem is known to hold in dimensions $n \leq 7$ (Schoen and Yau [48]), and in any dimension for spin manifolds (Witten [53]). Despite recent attempts, the Positive Mass Conjecture remains an open problem for nonspin high-dimensional manifolds.

Schoen proved the conjecture in the locally conformally flat case [46] and in the three-dimensional case (the argument can be found in [49]). In dimensions 4 and 5, the conjecture was proved by O. Druet (see [15]).

Basic blow-up analysis implies that a sequence of solutions that blows-up must concentrate at some points of the manifold. These are the blow-up points. In appropriate coordinates around these points the solution (up to a dimensional factor) can be well approximated by a bubble:

$$u_\varepsilon(x) = \left(\frac{\varepsilon}{\varepsilon^2 + |x|^2}\right)^{\frac{n-2}{2}},$$

for some small $\varepsilon > 0$. The term bubble comes from the fact that these functions appear as conformal factors when we write the spherical metric (after a dilation) as a conformal deformation of the Euclidean metric.

Schoen formulated the *Weyl Vanishing Conjecture*, that predicts the location of possible blow-up points:

If $\bar{x} \in M$ is a blow-up point of a sequence of solutions $\tilde{g}_k = u_k^{\frac{4}{n-2}} g$ to the Yamabe Problem, then the Weyl tensor of the metric g should satisfy

$$\nabla^i W_g(\bar{x}) = 0$$

for all $0 \leq i \leq [\frac{n-6}{2}]$.

There have been many contributions to these problems (see [36, 37, 39]). For arbitrary n, it follows from the works of the author [39] and Y. Y. Li and L. Zhang [36] that compactness holds under the assumption that the Weyl tensor vanishes nowhere to second order.

For general manifolds the story turned out to be very different. In a paper of the author with Khuri and Schoen [30], the Compactness and the Weyl Vanishing Conjectures are proved for spin manifolds of dimension $n \leq 24$ (the spin assumption is here only to make sure that the Positive Mass Theorem is valid). The dimension 24 comes from the behavior of a quadratic form discovered in [30]. It is sharp in light of the counterexamples constructed in higher dimensions: Brendle [7] for

$n \geq 52$, Brendle and the author [9] for $25 \leq n \leq 52$. The underlying manifold in these examples is the sphere S^n, that is spin in any dimension. Non-smooth blow-up examples had been obtained before by A. Ambrosetti and A. Malchiodi in [4], and by M. Berti and Malchiodi in [6]. In [16], O. Druet and E. Hebey have also obtained blow-up examples for Yamabe-type equations. We refer the reader to [10] for an account of these results.

For any $p \in [1, \frac{n+2}{n-2}]$ we define

$$\Phi_p = \{u > 0, u \in C^\infty(M) : L_g u + K u^p = 0 \text{ on } M\}.$$

Although the geometric problem corresponds to the critical exponent $p = \frac{n+2}{n-2}$, the consideration of the subcritical equations is useful for other purposes like computing the total Leray-Schauder degree of the problem.

In [30], the following theorem is proved (again the spin assumption can be dropped if the positive mass conjecture is valid in full generality):

Theorem 2.1. *Suppose* $3 \leq n \leq 24$. *If* (M^n, g) *is spin and is not conformally diffeomorphic to* (S^n, g_0), *then for any* $\varepsilon > 0$ *there exists a constant* $C > 0$ *depending only on* g *and* ε *such that*

$$C^{-1} \leq u \leq C \quad \textit{and} \quad \| u \|_{C^{2,\alpha}} \leq C,$$

for all $u \in \bigcup_{1+\varepsilon \leq p \leq \frac{n+2}{n-2}} \Phi_p$, *where* $0 < \alpha < 1$.

The basic tool in trying to rule out blow-up is a general Pohozaev identity that can be written in geometric form as:

Proposition 2.2 (Pohozaev Identity). *Let* (Ω^n, g) *be a Riemannian domain,* $n \geq 3$. *If* X *is a vector field on* Ω, *then*

$$\frac{n-2}{2n} \int_\Omega X(R_g)\, dv_g + \int_\Omega \langle \mathscr{D}_g X, T_g \rangle\, dv_g = \int_{\partial\Omega} T_g(X, \eta_g)\, d\sigma_g.$$

Here $T_g = Ric_g - \frac{R_g}{n} g$ *is the traceless Ricci tensor,* $(\mathscr{D}_g X)_{ij} = X_{i;j} + X_{j;i} - \frac{2}{n} div_g X\, g_{ij}$ *is the conformal Killing operator, and* η_g *is the outward unit normal to* $\operatorname{div}\Omega$.

The idea is to apply this identity with the radial vector field $X = r\frac{\partial}{\partial r}$ to the constant scalar curvature metrics in a neighborhood of the blow-up points. The left-hand side gives information about the local geometry, in particular about the Weyl tensor, while the right-hand side is linked to the mass of a certain asymptotically flat manifold (obtained by multiplying the original metric by a Green's function of the conformal Laplacian). The proof is by contradiction and uses that this mass is positive, but it only works when the leading order term coming from the left-hand side has a definite sign.

It turns out that this is encoded in a canonical quadratic form $\mathscr{P}_n$ defined in some vector space $\mathscr{V}_n$ (a subspace of the space of symmetric matrices with polynomial entries), depending only on the dimension. This was discovered in [30], where it is proved that the quadratic form is positive definite if and only if $n \leq 24$ thereby explaining the dimensional phenomenon.

The main theorems of [7] and [9] put together give:

Theorem 2.3. *Suppose $n \geq 25$. There exists a smooth Riemannian metric g on S^n and a sequence of positive functions $u_k \in C^\infty(S^n)$ $(k \in \mathbb{N})$ with the following properties:*

(i) *g is not conformally flat,*
(ii) *u_k is a solution of the Yamabe equation (1) for all $k \in \mathbb{N}$,*
(iii) $Q(u_k^{\frac{4}{n-2}} g) \nearrow Q(S^n, g_0)$ *as $k \to \infty$,*
(iv) $\sup_{S^n} u_k \to \infty$ *as $k \to \infty$.*

Here g_0 denotes the standard metric on S^n. The metric g can be chosen arbitrarily close to g_0.

The proof uses Lyapunov-Schmidt reduction to reduce the construction to solving a finite dimensional variational problem, and a glueing procedure to construct the metric g. This can only work because the quadratic form $\mathscr{P}_n$ fails to have a sign when $n \geq 25$.

3 Manifolds with boundary

Compactness questions can be studied in a variety of settings, for instance for manifolds with boundary. In that setting one looks for conformal metrics with constant scalar curvature and constant mean curvature at the boundary. The case of zero scalar curvature is particularly interesting because it leads to a linear equation in the interior with a critical Neumann-type nonlinear boundary condition:

$$\begin{cases} \Delta_g u - c(n) R_g u = 0 \text{ in } M, \\ \frac{\partial u}{\partial \eta} - d_n H_g u + K u^{\frac{n}{n-2}} = 0 \text{ on } \partial M, \end{cases} \tag{2}$$

where η is the inward pointing unit normal to ∂M, H_g is the boundary mean curvature with respect to g, $d_n = \frac{n-2}{2}$ and K is a constant.

The study of this and of the analogous problem of constant scalar curvature and zero mean curvature were initiated by Escobar in 1992 [21, 22]. He solved almost all cases while the remaining ones were the subject of subsequent papers [1, 8, 14, 38, 40]. Similarly one defines a Yamabe quotient:

$$Q(M, \partial M) = \inf_{u \neq 0 \text{ on } \partial M} \frac{\int_M |\nabla_g u|^2 + \frac{n-2}{4(n-1)} R_g u^2 dv_g + \frac{n-2}{2} \int_{\partial M} H_g u^2 d\sigma_g}{\left(\int_{\partial M} |u|^{\frac{2(n-1)}{n-2}} d\sigma_g \right)^{\frac{n-2}{n-1}}}$$

The problem of compactness was studied by many people, including V. Felli and M. Ould Ahmedou in [24, 25] and Han and Li in [27]. Under a generic condition it was proved by Almaraz [3]. Almaraz showed that the trace-free part of the second fundamental form must vanish at a boundary blow-up point if $n \geq 7$. Hence he is able to prove compactness for the set of metrics such that this quantity is nonzero everywhere:

Theorem 3.1 ([3]). *Let (M^n, g) be a compact Riemannian manifold with boundary, $n \geq 7$. Suppose $Q(M, \partial M) > 0$ and that the trace-free 2nd fundamental form of ∂M is nonzero everywhere. Then, given a small $\gamma_0 > 0$, there exists a constant $C > 0$ such that for any solution of*

$$\begin{cases} \Delta_g u - c(n) R_g u = 0 \text{ in } M, \\ \frac{\partial u}{\partial \eta} - d_n H_g u + K u^p = 0 \text{ on } \partial M, \end{cases} \tag{3}$$

with $p \in [1 + \gamma_0, \frac{n}{n-2}]$ satisfies

$$C^{-1} \leq u \leq C, \quad ||u||_{C^{2,\alpha}(M)} \leq C.$$

The proof of this result is based on a local argument with a Pohozaev-type identity and avoids the use of any positive mass assumption. The trace-free second fundamental form plays the role of the Weyl tensor as the first nontrivial obstruction to blow-up in high dimensions.

Inspired by the case of manifolds without boundary, we expect that there should be a critical dimension n_0 such that compactness holds for problem (2) if $n < n_0$ (at least under the spin assumption) and fails when $n \geq n_0$.

In [2], Almaraz constructed blow-up examples when $n \geq 25$ (hence the critical dimension must satisfy $n_0 \leq 25$):

Theorem 3.2 ([2]). *Let $n \geq 25$. There exists a Riemannian metric on the unit ball $M = B^n$ and a sequence of positive solutions u_k of equations (2), with $K > 0$, and with the following properties:*

- *g is not conformally flat,*
- *∂B^n is umbilic with respect to g,*
- *$\sup_{B^n} u_k \to \infty$ as $k \to \infty$.*

It should be an interesting problem to try to prove compactness for $n \leq 24$. Another problem is to understand the compactness/noncompactness phenomenon for the equivariant Yamabe problem of Hebey-Vaugon [28]. It should share some similarities with the boundary case.

4 Stability question and systems

There has been considerable interest in recent years in proving compactness results for Yamabe-type systems [17, 18, 20]. For instance, a priori estimates for electrostatic Klein-Gordon-Maxwell systems in three dimensions were obtained in [18]. Recently, stability for the Einstein-Lichnerowicz constraints system with respect to the physics data was proved on the three-sphere [19]. A general reference is Hebey [29].

We discuss a result of Druet and Hebey [17], about systems of the form

$$\Delta_g u_i - \sum_{j=1}^{p} A_{ij}(x) u_i + |U|^{\frac{4}{n-2}} u_i = 0 \tag{4}$$

on a compact Riemannian manifold (M^n, g). Here $U = (u_1, \ldots, u_p)$ and $|U|^2 = \sum_{i=1}^{p} u_i^2$. More generally they consider the *stability problem* where one is allowed to vary the matrix potential A. The system is called *analytically stable* if for every sequence of C^1 symmetric matrix functions $A_\alpha : M \to Sym(\mathbb{R}^p, \mathbb{R}^p)$ converging in C^1 to a map $A : M \to Sym(\mathbb{R}^p, \mathbb{R}^p)$, and for any sequence of associated nonnegative nontrivial solutions U_α bounded in $W^{1,2}$, there is a subsequence U_β that converges in C^2 to a solution U of the system (4). If C^2 convergence is replaced by weak convergence in $W^{1,2}$, then the system is called *weakly stable*.

The interesting feature that relates to the geometric Yamabe problem is that the behavior is dictated by the difference $A_n = A - \frac{(n-2)}{4(n-1)} R_g \mathrm{Id}$, where Id denotes the identity matrix. This kind of perturbation goes back at least to the work of Aubin [5]. The theorem of Druet and Hebey applies to the case where the matrix A_n is either positive definite or negative definite everywhere. More generally, they make the following assumption (H'): for any $x \in M$ and any $k \in \{1, \ldots, p\}$, there does not exist an orthonormal frame $(e_1, \ldots, e_k)$ of $A_n(x)$-isotropic vectors ($\langle A_n(x) \cdot e_i, e_i \rangle = 0$) in $\mathrm{Vect}_+(\mathbb{R}^p)$ spanning a subspace V with $A_n(x) \cdot V \subset V$. Here $\mathrm{Vect}_+(\mathbb{R}^p)$ denotes the space of vectors in $\mathbb{R}^p$ with nonnegative coordinates.

Druet and Hebey prove the following result:

Theorem 4.1 ([17]). *Assume $n \geq 4$, the assumption (H') and*

$$\mathrm{Ker}(\Delta_g - A) \cap L^2(M, \mathrm{Vect}_+(\mathbb{R}^p)) = \{0\}.$$

The system (4) is analytically stable if $n \neq 6$ and weakly stable if $n = 6$. If $n = 6$, there are examples that are analytically unstable.

The situation when one does not assume an energy bound for the sequence of solutions and also the three-dimensional case were studied in [20] under the assumption that A_n is negative definite. The consideration of such strongly coupled elliptic systems provides a natural background for the interplay between geometry and asymptotic analysis.

Even in the scalar case the issue of understanding the effect on compactness of perturbing the geometric Yamabe equation is nontrivial [23, 41]. In [23], Esposito, Pistoia, and Vétois prove that the a priori estimates fail for perturbations of the linear potential $-c(n)R_g$ if $n \geq 4$. Sign-changing solutions that blow-up are constructed in [43] and [44], while blow-up constructions for the prescribed scalar curvature problem are done in [32, 33]. The constructions of blow-up examples are again based on finite-dimensional Lyapunov-Schmidt reduction.

5 Higher order Yamabe problems

There are other linear operators that satisfy conformal covariance besides the conformal Laplacian. The literature is vast, and we refer the reader to the surveys of A. Chang [11] and of Chang and Yang [12] for a discussion of conformally covariant partial differential equations.

An example is the fourth-order Paneitz operator discovered in 1983 [42]:

$$P_g = \Delta_g^2 - \mathrm{div}_g(a_n R_g g + b_n \mathrm{Ric}_g)d + \frac{n-4}{2}Q_g,$$

where $a_n = \frac{(n-2)^2+4}{2(n-1)(n-2)}$, $b_n = -\frac{4}{n-2}$ and

$$Q_g = -\frac{1}{2(n-1)}\Delta_g R_g + \frac{n^3-4n^2+16n-16}{8(n-1)^2(n-2)^2}R_g^2 - \frac{2}{(n-2)^2}|\mathrm{Ric}_g|^2$$

is the Q-curvature.

The problem of finding conformal metrics with constant Q-curvature is equivalent to finding positive solutions of the fourth-order equation:

$$P_g u = c u^{\frac{n+4}{n-4}}. \tag{5}$$

In [52], Wei and Zhao proved that compactness fails for this problem in dimensions $n \geq 25$:

Theorem 5.1. *Suppose $n \geq 25$. There exists a smooth Riemannian metric g on S^n and a sequence of positive functions $u_k \in C^\infty(S^n)$ $(k \in \mathbb{N})$ with the following properties:*

(i) g is not conformally flat,
(ii) u_k is a solution of the equation (5) with $c = \frac{n-4}{2}$ for all $k \in \mathbb{N}$,
(iii) $\sup_{S^n} u_k \to \infty$ as $k \to \infty$.

Remarkably, the dimension is the same as in the classical Yamabe problem.

There are several important equations that lead to interesting Yamabe-type problems. The equations are typically either fully nonlinear or of higher order, and their study is analytically challenging. Several compactness results for fully nonlinear Yamabe problems have been obtained (for instance, [13, 26, 34, 35, 51]). There have been many important advances but the general picture of compactness/noncompactness is yet to be understood.

References

1. Almaraz, S.: An existence theorem of conformal scalar-flat metrics on manifolds with boundary. Pac. J. Math. **248**(1), 1–22 (2010)
2. Almaraz, S.: Blow-up phenomena for scalar-flat metrics on manifolds with boundary. J. Differ. Equ. **251**(7), 1813–1840 (2011)
3. Almaraz, S.: A compactness theorem for scalar-flat metrics on manifolds with boundary. Calc. Var. Partial Differ. Equ. **41**(3–4), 341–386 (2011)
4. Ambrosetti, A., Malchiodi, A.: A multiplicity result for the Yamabe problem on S^n. J. Funct. Anal. **168**, 529–561 (1999)
5. Aubin, T.: Équations différentielles non linéaires et probléme de Yamabe concernant la courbure scalaire. J. Math. Pures Appl. **55**, 269–296 (1976)
6. Berti, M., Malchiodi, A.: Non-compactness and multiplicity results for the Yamabe problem on S^n. J. Funct. Anal. **180**, 210–241 (2001)
7. Brendle, S.: Blow-up phenomena for the Yamabe equation. J. Am. Math. Soc. **21**, 951–979 (2008)
8. Brendle, S., Chen, S.: An existence theorem for the Yamabe problem on manifolds with boundary. J. Eur. Math. Soc. **16**(5), 991–1016 (2014)
9. Brendle, S., Marques, F.C.: Blow-up phenomena for the Yamabe equation II. J. Differ. Geom. **81**, 225–250 (2009)
10. Brendle, S., Marques, F.C.: Recent progress on the Yamabe problem. In: Surveys in Geometric Analysis and Relativity. Advanced Lectures in Mathematics, vol. 20, pp. 29–47. International Press, Somerville (2011)
11. Chang, S.-Y.A.: Non-linear elliptic equations in conformal geometry. In: Zurich Lectures in Advanced Mathematics. European Mathematical Society, Zürich (2004)
12. Chang, S.-Y.A., Yang, P.C.: Non-linear partial differential equations in conformal geometry. In: Proceedings of the International Congress of Mathematicians, vol. I (Beijing, 2002), pp. 189–207. Higher Ed. Press, Beijing (2002)
13. Chang, S.-Y.A., Gursky, M., Yang, P.: An a priori estimate for a fully nonlinear equation on four-manifolds. J. Anal. Math. **87**, 151–186 (2002)
14. Chen, S.: Conformal Deformation to Scalar Flat Metrics with Constant Mean Curvature on the Boundary in Higher Dimensions (2009). arXiv:0912.1302 [math.DG]
15. Druet, O.: Compactness for Yamabe metrics in low dimensions. Int. Math. Res. Not. **23**, 1143–1191 (2004)
16. Druet, O., Hebey, E.: Blow-up examples for second order elliptic PDEs of critical Sobolev growth. Trans. Am. Math. Soc. **357**, 1915–1929 (2004)
17. Druet, O., Hebey, E.: Stability for strongly coupled critical elliptic systems in a fully inhomogeneous medium. Anal. Partial Differ. Equ. **2**(3), 305–359 (2009)
18. Druet, O., Hebey, E.: Existence and a priori bounds for electrostatic Klein-Gordon-Maxwell systems in fully inhomogeneous spaces. Commun. Contemp. Math. **12**(5), 831–869 (2010)
19. Druet, O., Premoselli, B.: Stability of the Einstein-Lichnerowicz Constraints System (2013). arXiv:1312.6574 [math.AP]
20. Druet, O., Hebey, E., Vétois, J.: Bounded stability for strongly coupled critical elliptic systems below the geometric threshold of the conformal Laplacian. J. Funct. Anal. **258**(3), 999–1059 (2010)

21. Escobar, J.F.: Conformal deformation of a Riemannian metric to a scalar flat metric with constant mean curvature on the boundary. Ann. Math. (2) **136**(1), 1–50 (1992)
22. Escobar, J.F.: The Yamabe problem on manifolds with boundary. J. Differ. Geom. **35**, 21–84 (1992)
23. Esposito, P., Pistoia, A., Vétois, J.: The effect of linear perturbations on the Yamabe problem. Math. Ann. **358**(1–2), 511–560 (2014)
24. Felli, V., Ould Ahmedou, M.: Compactness results in conformal deformations of Riemannian metrics on manifolds with boundaries. Math. Z. **244**, 175–210 (2003)
25. Felli, V., Ould Ahmedou, M.: A geometric equation with critical nonlinearity on the boundary. Pac. J. Math. **218**(1), 75–99 (2005)
26. Guan, P., Viaclovsky, J., Wang, G.: Some properties of the Schouten tensor and applications to conformal geometry. Trans. Am. Math. Soc. **355**(3), 925–933 (2003)
27. Han, Z-C., Li, Y.Y.: The Yamabe problem on manifolds with boundary: existence and compactness results. Duke Math. J. **99**(3), 489–542 (1999)
28. Hebey, E., Vaugon, M.: Le problème de Yamabe équivariant. Bull. Sci. Math. **117**, 241–286 (1993)
29. Hebey, E.: Compactness and stability for nonlinear elliptic equations. In: Zurich Lectures in Advanced Mathematics. European Mathematical Society (EMS), Zürich (2014)
30. Khuri, M.A., Marques, F.C., Schoen, R.M.: A compactness theorem for the Yamabe problem. J. Differ. Geom. **81**, 143–196 (2009)
31. Lee, J., Parker, T.: The Yamabe problem. Bull. Am. Math. Soc. **17**, 37–91 (1987)
32. Leung, M.C.: Construction of blow-up sequences for the prescribed scalar curvature equation on Sn. I. Uniform cancellation. Commun. Contemp. Math. **14**(2), 1250008, 31 pp. (2012)
33. Leung, M.C.: Construction of blow-up sequences for the prescribed scalar curvature equation on Sn. II. Annular domains. Calc. Var. Partial Differ. Equ. **46**(1–2), 1–29 (2013)
34. Li, A., Li, Y.Y.: On some conformally invariant fully nonlinear equations. Commun. Pure Appl. Math. **56**(10), 1416–1464 (2003)
35. Li, Y.Y., Nguyen, L.: A compactness theorem for a fully nonlinear Yamabe problem under a lower Ricci curvature bound. J. Funct. Anal. **266**(6), 3741–3771 (2014)
36. Li, Y., Zhang, L.: Compactness of solutions to the Yamabe problem II. Calc. Var. Partial Differ. Equ. **25**, 185–237 (2005)
37. Li, Y., Zhang, L.: Compactness of solutions to the Yamabe problem III. J. Funct. Anal. **245**(2), 438–474 (2006)
38. Marques, F.: Existence results for the Yamabe problem on manifolds with boundary. Indiana Univ. Math. J. **54**, 1599–1620 (2005)
39. Marques, F.C.: A priori estimates for the Yamabe problem in the non-locally conformally flat case. J. Differ. Geom. **71**, 315–346 (2005)
40. Marques, F.C.: Conformal deformations to scalar-flat metrics with constant mean curvature on the boundary. Commun. Anal. Geom. **15**(2), 381–405 (2007)
41. Micheletti, A.M., Pistoia, A., Vétois, J.: Blow-up solutions for asymptotically critical elliptic equations on Riemannian manifolds. Indiana Univ. Math. J. **58**(4), 1719–1746 (2009)
42. Paneitz, S.M.: A quartic conformally covariant differential operator for arbitrary pseudo-Riemannian manifolds. SIGMA Symmetry Integrability Geom. Methods Appl. **4**, Paper 036, 3 pp. (2008)
43. Pistoia, A., Vétois, J.: Sign-changing bubble towers for asymptotically critical elliptic equations on Riemannian manifolds. J. Differ. Equ. **254**(11), 4245–4278 (2013)
44. Robert, F., Vétois, J.: Sign-changing blow-up for scalar curvature type equations. Commun. Partial Differ. Equ. **38**(8), 1437–1465 (2013)
45. Schoen, R.: Conformal deformation of a Riemannian metric to constant scalar curvature. J. Differ. Geom. **20**, 479–495 (1984)
46. Schoen, R.: On the number. Please check and confirm. of constant. In: Lawson, H.B., Tenenblat, K. (eds.) Differential Geometry: A Symposium in Honor of Manfredo Do Carmo, pp. 311–320. Wiley, New York (1991)

47. Schoen, R.: A report on some recent progress on nonlinear problems in geometry. Surv. Differ. Geom. **1**, 201–241 (1991)
48. Schoen, R., Yau, S.-T.: On the proof of the positive mass conjecture in General Relativity. Commun. Math. Phys. **65**, 45–76 (1979)
49. Schoen, R., Zhang, D.: Prescribed scalar curvature on the n-sphere. Calc. Var. Partial Differ Equ. **4**, 1–25 (1996)
50. Trudinger, N.: Remarks concerning the conformal deformation of Riemannian structures on compact manifolds. Ann. Scuola Norm. Sup. Pisa Cl. Sci. **22**(3), 165–274 (1968)
51. Trudinger, N., Wang, X.-J.: The intermediate case of the Yamabe problem for higher order curvatures. Int. Math. Res. Not. **2010**(13), 2437–2458 (2010)
52. Wei, J., Zhao, C.: Non-compactness of the prescribed Q-curvature problem in large dimensions. Calc. Var. Partial Differ. Equ. **46**(1–2), 123–164 (2013)
53. Witten, E.: A new proof of the positive energy theorem. Commun. Math. Phys. **80**, 381–402 (1981)
54. Yamabe, H.: On a deformation of Riemannian structures on compact manifolds. Osaka Math. J. **12**, 21–37 (1960)

Asymptotics of ground states for fractional Hénon systems

David G. Costa, Olímpio H. Miyagaki, Marco Squassina, and Jianfu Yang

Dedicated to Professor Djairo G. de Figueiredo on the occasion of his 80th birthday

Mathematics Subject Classification (2010): 35B06, 35B09, 35J15

1 Introduction and main results

Let $s \in (0, 1)$, $N > 2s$ and $B = \{x \in \mathbb{R}^N : |x| < 1\}$. Consider the fractional system of Hénon type

D.G. Costa (✉)
Department of Mathematical Sciences, University of Nevada Las Vegas,
Box 454020, Las Vegas, NV, USA
e-mail: costa@unlv.nevada.edu

O.H. Miyagaki
Departamento de Matemática, Universidade Federal de Juiz de Fora,
CEP 36036-330, Juiz de Fora, Brazil
e-mail: olimpio@ufv.br

M. Squassina
Dipartimento di Informatica, Università degli Studi di Verona,
Strada Le Grazie 15, 37134 Verona, Italy
e-mail: marco.squassina@univr.it

J. Yang
Department of Mathematics, Jiangxi Normal University,
Nanchang, Jiangxi 330022, China
e-mail: jfyang_2000@yahoo.com

A.N. Carvalho et al. (eds.), *Contributions to Nonlinear Elliptic Equations and Systems*, Progress in Nonlinear Differential Equations and Their Applications 86,
DOI 10.1007/978-3-319-19902-3_10

$$\begin{cases} (-\Delta)^s u = \dfrac{2p}{p+q}|x|^\alpha u^{p-1} v^q & \text{in } B, \\ (-\Delta)^s v = \dfrac{2q}{p+q}|x|^\alpha u^p v^{q-1} & \text{in } B, \\ u > 0, \ v > 0 & \text{in } B, \\ u = v = 0 & \text{in } \partial B, \end{cases} \tag{1}$$

where $(-\Delta)^s$ stands for the fractional Laplacian. Recently, a great attention has been focused on the study of nonlinear problems involving the fractional Laplacian, in view of concrete real-world applications. For instance, this type of operators arises in the thin obstacle problem, optimization, finance, phase transitions, stratified materials, crystal dislocation, soft thin films, semipermeable membranes, flame propagation, conservation laws, ultra-relativistic limits of quantum mechanics, quasi-geostrophic flows, multiple scattering, minimal surfaces, materials science and water waves, see, e.g., [1, 3, 11, 26, 28, 30]. See also [20] and the references therein. In a smooth bounded domain $B \subset \mathbb{R}^N$, the operator $(-\Delta)^s$ can be defined by using the eigenvalues $\{\lambda_k\}$ and corresponding eigenfunctions $\{\varphi_k\}$ of the Laplace operator $-\Delta$ in B with zero Dirichlet boundary values, normalized by $\|\varphi_k\|_{L^2(B)} = 1$, for all $k \in \mathbb{N}$, that is,

$$-\Delta\varphi_k = \lambda_k\varphi_k \text{ in } B, \qquad \varphi_k = 0 \text{ on } \partial B.$$

We define the space $H_0^s(B)$ by

$$H_0^s(B) := \Big\{u = \sum_{k=1}^{\infty} u_k\varphi_k \text{ in } L^2(B) : \sum_{k=1}^{\infty} u_k^2\lambda_k^s < \infty\Big\},$$

equipped with the norm

$$\|u\|_{H_0^s(B)} := \Big(\sum_{k=1}^{\infty} u_k^2\lambda_k^s\Big)^{1/2}.$$

Thus, for all $u \in H_0^s(B)$, the fractional Laplacian $(-\Delta)^s$ can be defined as

$$(-\Delta)^s u(x) := \sum_{k=1}^{\infty} u_k\lambda_k^s\varphi_k(x), \quad x \in B.$$

We wish to point out that a different notion of fractional Laplacian, available in the literature, is given by $(-\Delta)^s u = \mathcal{F}^{-1}(|\xi|^{2s}\mathcal{F}(u)(\xi))$, where $\mathcal{F}$ denotes the Fourier transform. This is also called the integral fractional Laplacian. This definition, in bounded domains, is really different from the spectral one. In the case of the integral notion, due to the strong nonlocal character of the operator, the Dirichlet datum is given in $\mathbb{R}^N \setminus B$ and not simply on ∂B. Recently, Caffarelli and Silvestre [12] developed a local interpretation of the fractional Laplacian given in $\mathbb{R}^N$ by considering a Dirichlet to Neumann type operator in the domain $\{(x,t) \in \mathbb{R}^{N+1} : t > 0\}$. A similar extension, in a bounded domain with zero Dirichlet boundary condition, was established, for instance, by Cabrè and Tan in [10], Tan [32], Capella, Dàvila, Dupaigne, and Sire [14], and by Brändle, Colorado, de Pablo, and Sànchez [7]. For any $u \in H_0^s(B)$, the solution $w \in H^1_{0,L}(C_B)$ of

$$\begin{cases} -\mathrm{div}(y^{1-2s}\nabla w) = 0 \text{ in } C_B := B \times (0,\infty), \\ w = 0 \text{ on } \partial_L C_B := \partial B \times (0,\infty), \\ w = u \text{ on } B \times \{0\}, \end{cases} \tag{2}$$

is called the s-harmonic extension $w = E_s(u)$ of u, and it belongs to the space

$$H^1_{0,L}(C_B) = \left\{ w \in L^2(C_B) : w = 0 \text{ on } \partial_L C_B : \int_{C_B} y^{1-2s}|\nabla w|^2 dxdy < \infty \right\}.$$

It is proved (see [7, Section 4.1–4.2]) that

$$-k_s \lim_{y \to 0+} y^{1-2s} \frac{\partial w}{\partial y} = (-\Delta)^s u,$$

where $k_s = 2^{1-2s}\Gamma(1-s)/\Gamma(s)$. Here $H^1_{0,L}(C_B)$ is a Hilbert space endowed with the norm

$$\|u\|_{H^1_{0,L}(C_B)} = \left(k_s \int_{C_B} y^{1-2s}|\nabla w|^2 dxdy \right)^{1/2}.$$

In the local case, the so-called Hénon problem

$$(HP) \qquad \begin{cases} -\Delta u = |x|^\alpha u^{p-1} & \text{in } B, \\ u > 0 & \text{in } B, \\ u = 0 & \text{on } \partial B, \end{cases}$$

was first studied in [29] after being introduced by Hénon in [24] in connection with the research of rotating stellar structures. This problem has been studied by several authors, e.g. [4, 13, 31] and the references therein. For this class of problems, moving plane methods [22] cannot be applied, and numerical calculations [15] suggest that the existence of non-radial solutions is in fact possible. In [13] the authors have shown that the maximum point x_p of a ground state solution for the Hénon equation (HP) approaches a point $x_0 \in \partial B$ as $p \to 2^*$, where $2^* = 2N/(N-2)$. This result was extended to local Hénon type variational systems in [33], as well as for scalar nonlocal Hénon type equations in [18]. The main goal of this paper is to get a similar result for the nonlocal Hénon system (1).

We reformulate the nonlocal systems (1) into a local system, by using the local reduction, that is, we set

$$(LS)\qquad \begin{cases} -\mathrm{div}(y^{1-2s}\nabla w_1) = 0 \text{ in } C_B = B\times(0,\infty),\\ -\mathrm{div}(y^{1-2s}\nabla w_2) = 0 \text{ in } C_B = B\times(0,\infty),\\ w_1 = w_2 = 0 \text{ on } \partial_L C_B = \partial B\times(0,\infty),\\ w_1 = u \geq 0 \text{ on } B\times\{0\},\\ w_2 = v \geq 0 \text{ on } B\times\{0\},\\ k_s y^{1-2s}\frac{\partial w_1}{\partial \nu} = \frac{2p}{p+q}|x|^\alpha u^{p-1}v^q \text{ on } B\times\{0\},\\ k_s y^{1-2s}\frac{\partial w_2}{\partial \nu} = \frac{2q}{p+q}|x|^\alpha u^{p}v^{q-1} \text{ on } B\times\{0\}.\end{cases}$$

Here $u(x) = w_1(x,0)$, $v(x) = w_2(x,0)$, and the outward normal derivative should be understood as

$$y^{1-2s}\frac{\partial w}{\partial \nu} = -\lim_{y\to 0^+} y^{1-2s}\frac{\partial w}{\partial y}.$$

Let us define the space $H := H^1_{0,L}(C_B)\times H^1_{0,L}(C_B)$ and the functional $I : H \to \mathbb{R}$,

$$I(w_1,w_2) = \frac{k_s}{2}\int_{C_B} y^{1-2s}(|\nabla w_1|^2+|\nabla w_2|^2)dxdy - \frac{2}{p+q}\int_B |x|^\alpha w_1(x,0)^p w_2(x,0)^q dx.$$

A weak solution to system (LS) is a vector $(w_1,w_2)\in H$ verifying $I'(w_1,w_2)(h,k) = 0$ for all $(h,k)\in H$,

$$I'(w_1,w_2)(h,k) = k_s\int_{C_B} y^{1-2s}(\nabla w_1\cdot\nabla h + \nabla w_2\cdot\nabla k)\,dxdy$$

$$-\frac{2p}{p+q}\int_B |x|^\alpha w_1(x,0)^{p-1}w_2(x,0)^q h dx - \frac{2q}{p+q}\int_B |x|^\alpha w_1(x,0)^p w_2(x,0)^{q-1} k dx.$$

For the nonlocal scalar problem

$$\begin{cases} (-\Delta)^s u = |x|^\alpha u^p & \text{in } B,\\ u > 0 & \text{in } B,\\ u = 0 & \text{in } \partial B,\end{cases} \tag{3}$$

we have

$$(LE)\qquad \begin{cases} -\mathrm{div}(y^{1-2s}\nabla w) = 0 \text{ in } C_B = B\times(0,\infty),\\ w = 0 \text{ on } \partial_L C_B = \partial B\times(0,\infty),\\ k_s y^{1-2s}\frac{\partial w}{\partial \nu} = |x|^\alpha u^{p-1} \text{ on } B\times\{0\}.\end{cases}$$

For this problem consider the associated minimization problem

$$\mathscr{S}^{\alpha}_{s,p}(C_B) = \inf_{w\in H^1_{0,L}(C_B)} \frac{k_s \displaystyle\int_{C_B} y^{1-2s}|\nabla w|^2 dxdy}{\left(\displaystyle\int_B |x|^{\alpha}|w(x,0)|^p dx\right)^{2/p}}.$$

Then $\mathscr{S}^0_{s,2^*_s}(C_B)$, where $2^*_s := 2N/(N-2s)$, is never achieved [7] and $\mathscr{S}^0_{s,2^*_s}(\mathbb{R}^{N+1}_+)$ is attained by the w which are the s-harmonic extensions of

$$u_\varepsilon(x) = \frac{\varepsilon^{\frac{N-2s}{4}}}{(\varepsilon+|x|^2)^{\frac{N-2s}{2}}}, \quad \varepsilon > 0,\ x\in\mathbb{R}^N.$$

Let $U(x) = (1+|x|^2)^{\frac{2s-N}{2}}$ and let W be the extension of U, namely

$$\mathscr{W}(x,y) = E_s(U) = cy^{2s}\int_{\mathbb{R}^N} \frac{U(z)}{(|x-z|^2+y^2)^{\frac{N+2s}{2}}}dz.$$

For the system (*LS*) consider the following minimization problem

$$\mathscr{S}^{\alpha}_{s,p,q}(C_B) = \inf_{w_1,w_2\in H^1_{0,L}(C_B)} \frac{k_s \displaystyle\int_{C_B} y^{1-2s}(|\nabla w_1|^2+|\nabla w_2|^2)dxdy}{\left(\displaystyle\int_B |x|^{\alpha}|w_1(x,0)|^p|w_2(x,0)|^q dx\right)^{\frac{2}{p+q}}} \tag{4}$$

Theorem 1.1. *For any* $\alpha > 0$, $\mathscr{S}^{\alpha}_{s,p,q}(C_B)$ *is achieved if* $2 < p+q < 2^*_s$.

Proof. Since B is bounded and $\alpha > 0$ we have $|x|^{\alpha}|u|^r \le C|u|^r$. The trace operator from $H^1_{0,L}(C_B)$ to $L^r(B)$ is continuous if $1/r \ge 1/2 - s/N$, and compact if strict inequality holds, see [7, Theorem 4.4] see also [5, 10]. Then the trace operator t_r : $H^1_{0,L}(C_B) \to L^r(|x|^{\alpha}, B)$ is compact for $r < 2N/(N-2s)$. Taking a minimizing sequence $(w_{1,n}, w_{2,n})$, there is $(w_1,w_2)\in H$ with $w_{i,n} \rightharpoonup w_i$, as $n\to\infty$. Then

$$w_{1,n} \to w_1 \text{ in } L^{p+q}(|x|^{\alpha}, B), \quad p+q < 2^*_s,$$
$$w_{2,n} \to w_2 \text{ in } L^{p+q}(|x|^{\alpha}, B), \quad p+q < 2^*_s.$$

By Young inequality we conclude that

$$\int_B |x|^{\alpha}|w_{1,n}(x,0)|^p|w_{2,n}(x,0)|^q dx \to \int_B |x|^{\alpha}|w_1(x,0)|^p|w_2(x,0)|^q dx, \text{ as } n\to\infty.$$

This implies that $\mathscr{S}^{\alpha}_{s,p,q}(C_B)$ is achieved if $2 < p+q < 2^*_s$. □

Remark 1.2. If $(w_{1,n}, w_{2,n})$ is a minimizing sequence to $S^{\alpha}_{s,p,q}(C_B)$, then it is readily seen that the sequence $(|w_{1,n}|, |w_{2,n}|)$ is minimizing too. Thus, we can assume that the minimizer (w_1, w_2) is nonnegative, that is, $w_{1,n}, w_{2,n} \geq 0$. By maximum principle we have $w_{1,n}, w_{2,n} > 0$. Finally, invoking the regularity theory we infer that $w_{1,n}, w_{2,n} \in C^{\gamma}(C_B)$, for some $\gamma \in (0,1)$. Notice that (w_1, w_2) is a weak solution for (LS). Indeed, by Lagrange multiplier theorem, considering the constraint

$$\mathscr{M} := \left\{(w_1, w_2) \in H : \int_B |x|^{\alpha} w_1(x,0)^p w_2(x,0)^q dx = 1\right\},$$

there exists $\lambda \in \mathbb{R}$ such

$$F'(w_1, w_2)(h,k) = \lambda G'(w_1, w_2)(h,k), \quad \forall (h,k) \in H,$$

where

$$F(w_1, w_2) = \frac{k_s}{2} \int_{C_B} y^{1-2s}(|\nabla w_1|^2 + |\nabla w_2|^2) dxdy,$$

$$G(w_1, w_2) = \int_B |x|^{\alpha} w_1(x,0)^p w_2(x,0)^q dx - 1.$$

Then, for all $(h,k) \in H$, we have

$$\begin{aligned} k_s \int_{C_B} & y^{1-2s}(\nabla w_1 \cdot \nabla h + \nabla w_2 \cdot \nabla k) dxdy \\ &= \lambda p \int_B |x|^{\alpha} w_1(x,0)^{p-1} w_2(x,0)^q h dx + \lambda q \int_B |x|^{\alpha} w_1(x,0)^p w_2(x,0)^{q-1} k dx. \end{aligned}$$

By choosing $(h,k) = (w_1, w_2)$, we get

$$\begin{aligned} k_s \int_{C_B} y^{1-2s}(|\nabla w_1|^2 + |\nabla w_2|^2) dxdy &= \lambda(p+q) \int_B |x|^{\alpha} w_1(x,0)^p w_2(x,0)^q h dx \\ &= \lambda(p+q). \end{aligned}$$

Therefore $\lambda > 0$ and $(\hat{w}_1, \hat{w}_2) = (\beta w_1, \beta w_2)$ with $\beta = \left(\frac{\lambda(p+q)}{2}\right)^{\frac{1}{p+q-2}}$ is a weak solution of (LS).

Now, we state the asymptotic behavior of ground states when $p + q \to 2^*_s$.

Theorem 1.3. *Let $\alpha > 0$, p_ε, $q > 1$ with $p_\varepsilon + q < 2^*_s$, $p_\varepsilon \to p$ as $\varepsilon \to 0$ and $p + q = 2^*_s$. Let $(w_{1,\varepsilon}, w_{2,\varepsilon}) \in H$ be a solution to the minimization problem* (4). *Then there exists $x_0 \in \partial B$ such that*

i) $k_s y^{1-2s}(|\nabla w_{1,\varepsilon}|^2 + |\nabla w_{2,\varepsilon}|^2) \rightharpoonup \mu \delta_{(x_0,0)}$ *in the sense of measure, ii)* $|u_{1,\varepsilon}|^{p_\varepsilon}|u_{2,\varepsilon}|^q \rightharpoonup \gamma\delta_{x_0}$ *in the sense of measure,*

where $\mu > 0, \gamma > 0$ *satisfy* $\mu \geq S\gamma^{2/2_s^*}$ *and* δ_{x_0} *is the Dirac mass at* x_0.

Let $(w_{1,\varepsilon}, w_{2,\varepsilon})$ be a minimizer of $\mathscr{S}^{\alpha}_{s,p_\varepsilon,q}(C_B)$ which exists because $2 < p_\varepsilon + q < 2_s^*$. By regularity results (see, e.g., [7, 9, 14]), $(w_{1,\varepsilon}, w_{2,\varepsilon})$ is Hölder continuous. We will show that there exists $x_\varepsilon,\ y_\varepsilon \in \overline{B}$ with

$$M_{i,\varepsilon} = w_{i,\varepsilon}(x_\varepsilon, 0) = \max_{(x,y)\in\overline{B}\times(0,\infty)} w_{i,\varepsilon}(x,y).$$

Let $\lambda_\varepsilon > 0$ and $\bar{\lambda}_\varepsilon > 0$ be such that $\lambda_\varepsilon^{\frac{N-2s}{2}} M_{1,\varepsilon} = 1$ and $\bar{\lambda}_\varepsilon^{\frac{N-2s}{2}} M_{2,\varepsilon} = 1$, where

$$\lambda_\varepsilon, \bar{\lambda}_\varepsilon \to 0, \quad \text{as } p_\varepsilon + q \to 2_s^*.$$

We state another description of the phenomenon exhibited in Theorem 1.3.

Theorem 1.4. *There hold*

i) $M_{1,\varepsilon} = \mathscr{O}_\varepsilon(1) M_{2,\varepsilon}$ *as* $\varepsilon \to 0$, *hence,* $\lambda_\varepsilon = \mathscr{O}_\varepsilon(1)\bar{\lambda}_\varepsilon$ *as* $\varepsilon \to 0$.

ii) $\operatorname{dist}(x_\varepsilon, \partial B) \to 0$ *and* $\dfrac{\operatorname{dist}(x_\varepsilon, \partial B)}{\lambda_\varepsilon} \to \infty$ *as* $p_\varepsilon + q \to 2_s^*$;

iii) $\displaystyle\lim_{p_\varepsilon+q\to 2_s^*} k_s \int_{C_B} y^{1-2s}(|\nabla \mathscr{T}_{1,\varepsilon}|^2 + |\nabla \mathscr{T}_{2,\varepsilon}|^2)dxdy = 0,$

where we have set $\mathscr{T}_{i,\varepsilon}(x,y) := w_{1,\varepsilon}(x,y) - \lambda_\varepsilon^{\frac{2s-N}{2}} \mathscr{W}\left(\frac{x-x_\varepsilon}{\lambda_\varepsilon}, \frac{y}{\lambda_\varepsilon}\right)$, for $i = 1, 2$.

2 Preliminaries

For any $u_i \in H_0^s(B)$, there is a unique extension $w_i = E_s(u_i) \in H^1_{0,L}(C_B)$ of u_i. The extension operator is an isometry between $H_0^s(B)$ and $H^1_{0,L}(C_B)$, that is (see [5, 7, 18])

$$\|E_s(u_i)\|_{H^1_{0,L}(C_B)} = \|u_i\|_{H_0^s(B)}, \quad i = 1, 2.$$

Let us set

$$x_0 := \left(1 - \frac{1}{|\ln\varepsilon|}, 0, \ldots, 0\right) \in \mathbb{R}^N, \qquad z_0 := (x_0, 0) \in \mathbb{R}^{N+1}.$$

Let us denote $B_\rho := \{x \in \mathbb{R}^N : |x - x_0| < \rho\}$ and

$$\mathbb{A}_\rho := \{(x,y) \in \mathbb{R}^{N+1} : |(x,y) - z_0| < \rho\}, \quad \mathbb{B}_\rho := \{(x,y) \in \mathbb{R}^{N+1} : |(x,y)| < \rho\}.$$

Let $\varphi \in C_0^\infty(C_B)$ be a cut-off function satisfying

$$\varphi(x,y) := \begin{cases} 1 \text{ if } (x,y) \in \mathbb{A}_{\frac{1}{2|\ln\varepsilon|}} \\ 0 \text{ if } (x,y) \notin \mathbb{A}_{\frac{1}{|\ln\varepsilon|}}, \end{cases}$$

with $0 \leq \varphi(x,y) \leq 1$ and $|\nabla\varphi(x,y)| \leq C|\ln\varepsilon|$, for $(x,y) \in C_B$. If $\mathscr{W}$ is the extension of the function U previously introduced, we have (see [5]) $|\nabla\mathscr{W}(x,y)| \leq Cy^{-1}\mathscr{W}(x,y)$, for $(x,y) \in \mathbb{R}_+^{N+1}$. The extension of $U_\varepsilon(x) = (\varepsilon + |x|^2)^{(2s-N)/2}$ has the form

$$\mathscr{W}_\varepsilon(x,y) = \varepsilon^{\frac{2s-N}{2}}\mathscr{W}\Big(\frac{x-x_0}{\sqrt{\varepsilon}}, \frac{y}{\sqrt{\varepsilon}}\Big), \qquad \varepsilon > 0.$$

Notice that $\varphi\mathscr{W}_\varepsilon \in H^1_{0,L}(C_B)$ for ε small enough. The following lemma is proved in [18, Lemma 3.1]

Lemma 2.1. *There holds*

$$\frac{\displaystyle\int_{C_B} k_s y^{1-2s}|\nabla(\varphi\mathscr{W}_\varepsilon)|^2 dxdy}{\Big(\displaystyle\int_B |x|^\alpha|\varphi(x,0)\mathscr{W}_\varepsilon(x,0)|^p dx\Big)^{2/p}} = \mathscr{S}^0_{s,2_s^*}(C_B) + o_\varepsilon(1),$$

as $p \to 2_s^*$, *and* $\varepsilon \to 0$.

A minimizer of $\mathscr{S}^\alpha_{s,p,q}(C_B)$ exists as $2 < p+q < 2_s^*$ and arguing as in [2, Theorem 5] we have

$$\mathscr{S}^\alpha_{s,p,q}(C_B) = C_{p,q}\,\mathscr{S}^\alpha_{s,p+q}(C_B), \qquad C_{p,q} := \Big[\Big(\frac{p}{q}\Big)^{\frac{q}{p+q}} + \Big(\frac{p}{q}\Big)^{-\frac{p}{p+q}}\Big], \tag{5}$$

where we have set

$$\mathscr{S}^\alpha_{s,p+q}(C_B) := \inf_{w\in H^1_{0,L}(C_B)} \frac{\displaystyle\int_{C_B} k_s y^{1-2s}|\nabla w|^2 dxdy}{\Big(\displaystyle\int_B |x|^\alpha|w(x,0)|^{p+q}dx\Big)^{2/(p+q)}}.$$

In particular

$$\mathscr{S}_{s,p,q}(C_B) := \mathscr{S}^0_{s,p,q}(C_B) = C_{p,q}\mathscr{S}^0_{s,p+q}(C_B) = C_{p,q}\mathscr{S}_{p+q}(C_B).$$

Furthermore, if w_0 realizes $\mathscr{S}^\alpha_{s,p+q}(C_B)$, then $(u_0, v_0) = (Bw_0, Cw_0)$ realizes $\mathscr{S}^\alpha_{s,p,q}(C_B)$, for

$$B, C > 0, \quad B = \sqrt{p/q}\,C.$$

Setting $\hat{u}_\varepsilon = \sqrt{p_\varepsilon}\varphi\mathscr{W}_\varepsilon$ and $\hat{v}_\varepsilon = \sqrt{q}\varphi\mathscr{W}_\varepsilon$ and applying identity (5), we have

$$\frac{\displaystyle\int_{C_B} k_s y^{1-2s}(|\nabla \hat{u}_\varepsilon|^2 + |\nabla \hat{v}_\varepsilon|^2)dxdy}{\left(\displaystyle\int_B |x|^\alpha |\hat{u}_\varepsilon(x,0)|^{p_\varepsilon}|\hat{v}_\varepsilon(x,0)|^q dx\right)^{2/(p_\varepsilon+q)}}$$

$$= C_{p_\varepsilon,q}\frac{\displaystyle\int_{C_B} k_s y^{1-2s}|\nabla(\varphi\mathscr{W}_\varepsilon)|^2 dxdy}{\left(\displaystyle\int_B |x|^\alpha |\varphi(x,0)\mathscr{W}_\varepsilon(x,0)|^{p_\varepsilon+q}dx\right)^{2/(p_\varepsilon+q)}} = C_{p_\varepsilon,q}\mathscr{S}^0_{s,2^*_s}(C_B) + o_\varepsilon(1),$$

as $p_\varepsilon + q \to 2^*_s$ for $\varepsilon \to 0$. Following [18, Lemma 3.2], we have

Lemma 2.2. *Let $(u_\varepsilon, v_\varepsilon)$ be a minimizer of $\mathscr{S}^\alpha_{s,p_\varepsilon,q}(C_B)$ and $p_\varepsilon + q \to 2^*_s$ for $\varepsilon \to 0$. Then we have*

$$\lim_{\varepsilon\to 0}\frac{\displaystyle\int_{C_B} k_s y^{1-2s}(|\nabla u_\varepsilon|^2 + |\nabla v_\varepsilon|^2)dxdy}{\left(\displaystyle\int_B |x|^\alpha |u_\varepsilon(x,0)|^{p_\varepsilon}|v_\varepsilon(x,0)|^q dx\right)^{2/(p_\varepsilon+q)}} = C_{p,q}\mathscr{S}^0_{s,2^*_s}(C_B) = C_{p,q}\mathscr{S}^C_{s,2^*_s}(\mathbb{R}^{N+1}_+),$$

$$\lim_{\varepsilon\to 0}\frac{\displaystyle\int_{C_B} k_s y^{1-2s}(|\nabla u_\varepsilon|^2 + |\nabla v_\varepsilon|^2)dxdy}{\left(\displaystyle\int_B |u_\varepsilon(x,0)|^{p_\varepsilon}|v_\varepsilon(x,0)|^q dx\right)^{2/(p_\varepsilon+q)}} = C_{p,q}\mathscr{S}^0_{s,2^*_s}(C_B) = C_{p,q}\mathscr{S}^C_{s,2^*_s}(\mathbb{R}^{N+1}_+).$$

Proof. We already know that $\mathscr{S}^0_{s,2^*_s}(C_B) = \mathscr{S}^0_{s,2^*_s}(\mathbb{R}^{N+1}_+)$. Notice that, by (5), we get by Lemma 2.1

$$\frac{\displaystyle\int_{C_B} k_s y^{1-2s}(|\nabla u_\varepsilon|^2 + |\nabla v_\varepsilon|^2)dxdy}{\left(\displaystyle\int_B |u_\varepsilon(x,0)|^{p_\varepsilon}|v_\varepsilon(x,0)|^q dx\right)^{2/(p_\varepsilon+q)}} \le \frac{\displaystyle\int_{C_B} k_s y^{1-2s}(|\nabla u_\varepsilon|^2 + |\nabla v_\varepsilon|^2)dxdy}{\left(\displaystyle\int_B |x|^\alpha |u_\varepsilon(x,0)|^{p_\varepsilon}|v_\varepsilon(x,0)|^q dx\right)^{2/(p_\varepsilon+q)}}$$

$$\le C_{p_\varepsilon,q}\frac{\displaystyle\int_{C_B} k_s y^{1-2s}|\nabla(\varphi W_\varepsilon)|^2 dxdy}{\left(\displaystyle\int_B |x|^\alpha |\varphi(x,0)W_\varepsilon(x,0)|^{p_\varepsilon+q}dx\right)^{2/(p_\varepsilon+q)}} = C_{p_\varepsilon,q}\mathscr{S}^0_{s,2^*_s}(C_B) + o(\varepsilon),$$

as $p_\varepsilon + q \to 2_s^*$, for $\varepsilon \to 0$. On the other hand, we infer that

$$\frac{\int_{C_B} k_s y^{1-2s}(|\nabla u_\varepsilon|^2 + |\nabla v_\varepsilon|^2)dxdy}{\left(\int_B |u_\varepsilon(x,0)|^{p_\varepsilon}|v_\varepsilon(x,0)|^q dx\right)^{2/(p_\varepsilon+q)}} \geq \mathcal{S}^0_{s,p_\varepsilon,q}(C_B)$$

$$= C_{p_\varepsilon,q}\mathcal{S}^0_{s,p_\varepsilon+q}(C_B) \geq C_{p_\varepsilon,q}\mathcal{S}^0_{s,2_s^*}(C_B).$$

The last inequality is due to Hölder inequality. This concludes the proof. □

Corollary 2.3. *Let $p + q = 2_s^*$. Then the infimum $\mathcal{S}^\alpha_{s,p,q}(C_B)$ cannot be achieved.*

Proof. Observe that, for all $\alpha \geq 0$, there holds $\mathcal{S}^\alpha_{s,p,q}(C_B) = C_{p,q}\,\mathcal{S}^\alpha_{s,2_s^*}(C_B)$. Suppose, by contradiction, that $\mathcal{S}^\alpha_{s,p,q}(C_B)$ is achieved by a function $(w_1, w_2) \in H$. Without loss of generality, we may assume that $w_1 \geq 0$ and $w_2 \geq 0$. By Lemma 2.2, we get

$$C_{p,q}\mathcal{S}^0_{s,2_s^*}(C_B) = \mathcal{S}^\alpha_{s,p,q}(C_B) = \frac{\int_{C_B} k_s y^{1-2s}(|\nabla w_1|^2 + |\nabla w_2|^2)dxdy}{\left(\int_B |x|^\alpha w_1(x,0)^p w_2(x,0)^q dx\right)^{2/(p+q)}}$$

$$\geq \frac{\int_{C_B} k_s y^{1-2s}(|\nabla w_1|^2 + |\nabla w_2|^2)dxdy}{\left(\int_B w_1(x,0)^p w_2(x,0)^q dx\right)^{2/(p+q)}} \geq \mathcal{S}^0_{s,p,q}(C_B) = C_{p,q}\mathcal{S}^0_{s,2_s^*}(C_B),$$

so that $\mathcal{S}^0_{s,2_s^*}(C_B)$ is achieved at $(w_1, w_2) \in H$, being

$$C_{p,q}\mathcal{S}^0_{s,2_s^*}(C_B) = \frac{\int_{C_B} k_s y^{1-2s}(|\nabla w_1|^2 + |\nabla w_2|^2)dxdy}{\left(\int_B w_1(x,0)^p w_2(x,0)^q dx\right)^{2/(p+q)}}.$$

By setting $\tilde{w}_i(x,t) := w_i(x,t)$ for $(x,t) \in B \times (0,\infty)$ and $\tilde{w}_i(x,t) := 0$ for $(x,t) \in \mathbb{R}^N \setminus B \times (0,\infty)$ we get the minimizer $(\tilde{w}_1, \tilde{w}_2) \in \mathcal{S}^0_{2_s^*}(\mathbb{R}^{N+1}_+)$. A contradiction, since $w_i > 0$, by the maximum principle. □

Definition 2.4. A sequence $(w_{1,n}, w_{2,n}) \subset H$ is said to be tight if, for all $\eta > 0$, there is $\rho_0 > 0$ with

$$\sup_{n\in\mathbb{N}} \int_{\{y>\rho_0\}} \int_B k_s y^{1-2s}(|\nabla w_{1,n}|^2 + |\nabla w_{2,n}|^2)dxdy \leq \eta.$$

The following concentration compactness principle [27] can be adapted from [5, Theorem 5.1]

Proposition 2.5. *Let $(w_{1,n}, w_{2,n}) \subset H$ be tight and weakly convergent to (w_1, w_2) in H. Let us denote $u_{i,n} = \mathrm{Tr}(w_{i,n})$ and $u_i = \mathrm{Tr}(w_i)$, $p + q = 2^*_s$. Let μ, ν be two nonnegative measures such that*

i) $k_s y^{1-2s}(|\nabla w_{1,n}|^2 + |\nabla w_{2,n}|^2) \rightharpoonup \mu$ *in the sense of measure,*
ii) $|u_{1,n}|^p |u_{2,n}|^q \rightharpoonup \nu$ *in the sense of measure.*

Then there exist an at most countable set I and points $\{x_i\}_{i \in I} \subset B$ such that

$$\mu \geq k_s y^{1-2s}(|\nabla w_1|^2 + |\nabla w_2|^2) + \sum_{k \in I} \mu_k \delta_{(x_k,0)}, \qquad \nu = |u_1|^p |u_2|^q + \sum_{k \in I} \nu_k \delta_{x_k}, \tag{6}$$

with $\mu_k > 0$, $\nu_k > 0$ and $\mu_k \geq C_{p,q} \mathscr{S}^0_{s,2^*_s} \nu_k^{2/2^*_s}$.

Finally, we give an explicit form to the solutions of the problem

$$\begin{cases} (-\Delta)^s u = \dfrac{2p}{p+q} u^{p-1} v^q & \text{in } \mathbb{R}^N, \\ (-\Delta)^s v = \dfrac{2q}{p+q} u^p v^{q-1} & \text{in } \mathbb{R}^N, \\ u > 0, \ v > 0 & \text{in } \mathbb{R}^N, \end{cases} \tag{7}$$

where $p + q = 2^*_s$. Let $u, v \in L^{2^*_s}(\mathbb{R}^N)$ be solutions of the following problem

$$\begin{cases} u = \dfrac{2p}{p+q} \displaystyle\int_{\mathbb{R}^N} \frac{u^{p-1}(y) v^q(y)}{|x-y|^{N-2s}} \, dy, \\ v = \dfrac{2q}{p+q} \displaystyle\int_{\mathbb{R}^N} \frac{u^p(y) v^{q-1}(y)}{|x-y|^{N-2s}} \, dy. \\ u > 0, \ v > 0 \quad \text{in } \mathbb{R}^N. \end{cases} \tag{8}$$

Denote by

$$\tilde{u}(x) := \frac{1}{|x|^{N-2s}} u\left(\frac{x}{|x|^2}\right), \qquad \tilde{v}(x) := \frac{1}{|x|^{N-2s}} v\left(\frac{x}{|x|^2}\right),$$

the Kelvin transform of u and v, respectively. Hence, $(\tilde{u}, \tilde{v})$ is also a solution of (8). We may prove as in [16, Theorem 4.5] that problems (7) and (8) are equivalent, that is if (u, v) with $u, v \in H^s(\mathbb{R}^N)$ is a weak solution of (7), then (u, v) is a solution of (8), while if (u, v) with $u, v \in L^{\frac{2N}{N-2s}}(\mathbb{R}^N)$ solves (8), then (u, v) is a solution of (7).

Now we show that $L^{\frac{2N}{N-2s}}(\mathbb{R}^N)$ solution (u, v) of the following problem is radially symmetric.

$$\begin{cases} u(x) = \displaystyle\int_{\mathbb{R}^N} \frac{u^{p-1}(y)v^q(y)}{|x-y|^{N-2s}}\,dy, \\ v(x) = \displaystyle\int_{\mathbb{R}^N} \frac{u^p(y)v^{q-1}(y)}{|x-y|^{N-2s}}\,dy. \\ \quad u > 0,\ v > 0 \quad \text{in } \mathbb{R}^N. \end{cases} \tag{9}$$

Let $\Sigma_\lambda = \{x = (x_1, \cdots, x_N) : x_1 > \lambda\}$, $x^\lambda = (2\lambda - x_1, x_2, \cdots, x_N)$ and $u_\lambda(x) = u(x^\lambda)$.

Lemma 2.6. *Let* (u, v) *be a solution of* (9). *Then* (u, v) *is radially symmetric with respect to some point.*

Proof. The result is proved by the moving plane methods developed for integral equations, see [17]. The argument is now standard, we sketch the proof. For details, we refer to similar arguments in [35]. We have

$$u_\lambda(x) - u(x) = \int_{\Sigma_\lambda} \Big(\frac{1}{|x-y|^{N-2s}} - \frac{1}{|x^\lambda - y|^{N-2s}}\Big)\Big(u_\lambda^{p-1}(y)v_\lambda^q(y) - u^{p-1}(y)v^q(y)\Big)\,dy$$

and

$$v_\lambda(x) - v(x) = \int_{\Sigma_\lambda} \Big(\frac{1}{|x-y|^{N-2s}} - \frac{1}{|x^\lambda - y|^{N-2s}}\Big)\Big(u_\lambda^{p}(y)v_\lambda^{q-1}(y) - u^{p}(y)v^{q-1}(y)\Big)\,dy.$$

Next, we claim that there exist $K \geq 0$, such that if $\lambda < -K$, there holds

$$u(x) \geq u_\lambda(x) \quad \text{and} \quad v(x) \geq v_\lambda(x).$$

Indeed, define

$$\Sigma_\lambda^u = \{x \in \Sigma_\lambda : u(x) \leq u_\lambda(x)\}, \quad \Sigma_\lambda^v = \{x \in \Sigma_\lambda : v(x) \leq v_\lambda(x)\}$$

and $\Sigma_\lambda^- = \Sigma_\lambda \setminus \big(\Sigma_\lambda^u \cup \Sigma_\lambda^v\big)$, we can deduce as [35] that

$$\begin{aligned} u_\lambda(x) - u(x) \leq &\int_{\Sigma_\lambda^v} \frac{1}{|x-y|^{N-2s}} u_\lambda^{p-1}(y)\Big(v_\lambda^q(y) - v^q(y)\Big)\,dy \\ &+ \int_{\Sigma_\lambda^u} \frac{1}{|x-y|^{N-2s}} v_\lambda^{q}(y)\Big(u_\lambda^{p-1}(y) - u^{p-1}(y)\Big)\,dy. \end{aligned}$$

By the Hardy-Littlewood-Sobolev inequality,

$$\|u_\lambda(x) - u(x)\|_{L^{2_s^*}(\Sigma_\lambda^u)} \leq C\|u_\lambda^{p-1} v_\lambda^{q-1}(v_\lambda - v)\|_{L^{\frac{2_s^* N}{N+2s2_s^*}}(\Sigma_\lambda^v)} + C\|u_\lambda^{p-2} v_\lambda^{q}(u_\lambda - u)\|_{L^{\frac{2_s^* N}{N+2s2_s^*}}(\Sigma_\lambda^u)}.$$

By Hölder's inequality,

$$\|u_\lambda(x) - u(x)\|_{L^{2_s^*}(\Sigma_\lambda^u)} \leq C\|u_\lambda\|^{p-1}_{L^{2_s^*}(\Sigma_\lambda^v)} \|v_\lambda\|^{q-1}_{L^{2_s^*}(\Sigma_\lambda^v)} \|(v_\lambda - v)\|_{L^{2_s^*}(\Sigma_\lambda^v)} + C\|u_\lambda\|^{p-2}_{L^{2_s^*}(\Sigma_\lambda^u)} \|v_\lambda\|^{q}_{L^{2_s^*}(\Sigma_\lambda^u)} \|(u_\lambda - u)\|_{L^{2_s^*}(\Sigma_\lambda^u)}.$$

Choose $K > 0$ large and for $\lambda < -K$, we have

$$\|u_\lambda(x) - u(x)\|_{L^{2_s^*}(\Sigma_\lambda^u)} \leq \frac{1}{4}\|u_\lambda(x) - u(x)\|_{L^{2_s^*}(\Sigma_\lambda^u)} + \frac{1}{4}\|v_\lambda(x) - v(x)\|_{L^{2_s^*}(\Sigma_\lambda^v)}.$$

Similarly,

$$\|v_\lambda(x) - v(x)\|_{L^{2_s^*}(\Sigma_\lambda^u)} \leq \frac{1}{4}\|u_\lambda(x) - u(x)\|_{L^{2_s^*}(\Sigma_\lambda^u)} + \frac{1}{4}\|v_\lambda(x) - v(x)\|_{L^{2_s^*}(\Sigma_\lambda^v)}.$$

The claim follows easily. Now, we may proceed as the proof of [35, Theorem 1.1]. □

It is known [16] that a positive solution $U \in L^{2_s^*}(\mathbb{R}^N)$ of the problem

$$(-\Delta)^s u = u^{\frac{N+2s}{N-2s}} \quad \text{in } \mathbb{R}^N, \tag{10}$$

is given by

$$U(x) = C\Big(\frac{t}{t^2 + |x - x_0|^2}\Big)^{\frac{N-2s}{2}},$$

for some constant $C = C(N, s) > 0$, some $t > 0$ and $x_0 \in \mathbb{R}^N$.

Lemma 2.7. *Let* (u, v) *be a nontrivial weak solution of problem* (7). *There exist* $A, B > 0$ *such that* $u = AU$ *and* $v = BU$.

Proof. We known that the solutions (u, v) of (7) are solutions of (8). By Lemma 2.6, any solution (u, v) of (8) is radially symmetric and monotone decreasing about some point. Let $(\tilde{u}, \tilde{v})$ be the Kelvin transform of (u, v) with the pole $p \neq 0$

$$\tilde{u}(x) = \frac{1}{|x-p|^{N-2s}} u\Big(\frac{x-p}{|x-p|^2} + p\Big), \qquad \tilde{v}(x) = \frac{1}{|x-p|^{N-2s}} v\Big(\frac{x-p}{|x-p|^2} + p\Big).$$

We remark that $(\tilde{u}, \tilde{v})$ is a solution of (8) too, and then $(\tilde{u}, \tilde{v})$ is radially symmetric with respect to some point q. Following the argument on page 280 in [23], we can see that if $p = q$, then (u, v) is constant, which is not true in our case. Hence, $p \neq q$. Now, using the Kelvin transform

$$K(f)(x) = \frac{1}{|x|^{N-2s}} f\Big(\frac{x}{|x|^2}\Big),$$

we deduce as in [6, proof of Lemma 7] that $u = AU$ and $v = BU$. □

3 Proof of Theorem 1.3

Choose p_k such that $p_k + q \to 2_s^*$, as $k \to \infty$. Let $(w_{1,k}, w_{2,k}) \in H$ be a nonnegative solution to

$$\mathscr{S}^{\alpha}_{s,p_k,q}(C_B) = \frac{k_s \displaystyle\int_{C_B} y^{1-2s}(|\nabla w_{1,k}|^2 + |\nabla w_{2,k}|^2) dxdy}{\Big(\displaystyle\int_B |x|^{\alpha} |w_{1,k}(x,0)|^{p_k} |w_{2,k}(x,0)|^q dx\Big)^{\frac{2}{p_k+q}}}. \tag{11}$$

Up to the factor $((p_k + q)\lambda_k/2)^{1/(p_k+q-2)}$ depending upon the Lagrange multiplier λ_k, $(w_{1,k}, w_{2,k})$ solves

$$\begin{cases} -\text{div}(y^{1-2s}\nabla w_{1,k}) = 0, \quad -\text{div}(y^{1-2s}\nabla w_{2,k}) = 0, & \text{in } C_B, \\ k_s y^{1-2s} \frac{\partial w_{1,k}}{\partial \nu} = \frac{2p_k}{p_k+q} |x|^{\alpha} w_{1,k}(x,0)^{p_k-1} w_{2,k}(x,0)^q, & \text{on } B \times \{0\}, \\ k_s y^{1-2s} \frac{\partial w_{2,k}}{\partial \nu} = \frac{2q}{p_\varepsilon+q} |x|^{\alpha} w_{1,k}(x,0)^{p_k} w_{2,k}(x,0)^{q-1}, & \text{on } B \times \{0\}, \\ w_{1,k} = w_{2,k} = 0, & \text{on } \partial_L C_B. \end{cases} \tag{12}$$

In particular, we get

$$\int_{C_B} k_s y^{1-2s}(|\nabla w_{1,k}|^2 + |\nabla w_{2,k}|^2) dxdy = 2 \int_B |x|^{\alpha} w_{1,k}(x,0)^{p_k} w_{2,k}(x,0)^q dx. \tag{13}$$

One may now set, for every $x \in B$ and $y > 0$,

$$\tilde{w}_{i,k}(x,y) := C_k w_{i,k}(x,y), \qquad C_k = \Big(\int_B w_{1,k}(x,0)^{p_k} w_{2,k}(x,0)^q dx\Big)^{-\frac{1}{p_k+q}}, \quad i = 1, 2. \tag{14}$$

We have

$$\int_B \tilde{w}_{1,k}(x,0)^{p_k}\tilde{w}_{2,k}(x,0)^q dx = 1, \qquad \text{for all } k \in \mathbb{N},$$

and by (11) and Lemma 2.2, we have

$$\int_{C_B} k_s y^{1-2s}(|\nabla \tilde{w}_{1,k}|^2 + |\nabla \tilde{w}_{2,k}|^2)dxdy = C_{p,q}\mathscr{S}^0_{s,2_s^*} + o_k(1), \quad \text{as } k \to \infty.$$

The sequence C_k converges to some $C > 0$, whenever $k \to \infty$. This can be proved by comparison with the term $\int_B |x|^\alpha w_{1,k}(x,0)^{p_k} w_{2,k}(x,0)^q dx$, which converges to a constant in view of formulas (11), (13) and Lemma 2.2. In fact, taking into account the Sobolev trace inequality, we have

$$0 < \sigma \le \int_B |x|^\alpha w_{1,k}(x,0)^{p_k} w_{2,k}(x,0)^q dx \le C_k^{-p_k-q} \le C\|w_{1,k}\|^{p_k}_{L^{2_s^*}(B)}\|w_{2,k}\|^{q}_{L^{2_s^*}(B)} \le C.$$

The sequence $(\tilde{w}_{1,k}, \tilde{w}_{2,k})$ is bounded in H. Furthermore, it is *tight*. This fact can be proved by arguing as in [5, Lemma 3.6]. By Proposition 2.5, there exist nonnegative measures μ, ν, a pair of functions $(w_1, w_2) \in H$, an at most countable set J and points with $\{x_i\}_{i\in J} \subset B$ such that

i) $\tilde{w}_{i,k} \rightharpoonup w_i, \quad i = 1, 2,$
ii) $k_s y^{1-2s}(|\nabla \tilde{w}_{1,k}|^2 + |\nabla \tilde{w}_{2,k}|^2) \rightharpoonup \mu$ in the sense of measure,
iii) $|\tilde{w}_{1,k}(x,0)|^p|\tilde{w}_{2,k}(x,0)|^q \rightharpoonup \nu$ in the sense of measure,

and (6) holds with $\nu_k > 0$ and $\mu_k \ge C_{p,q}\mathscr{S}^0_{s,2_s^*}\nu_k^{2/2_s^*}$. It follows that

$$\lim_k \int_{\mathbb{R}^{N+1}_+} k_s y^{1-2s}(|\nabla \tilde{w}_{1,k}|^2 + |\nabla \tilde{w}_{2,k}|^2)\varphi dxdy = \int_{\mathbb{R}^{N+1}_+} \varphi\, d\mu, \quad \forall \varphi \in L^\infty \cap C(\mathbb{R}^{N+1}_+),$$

$$\lim_k \int_{\mathbb{R}^N} |\tilde{w}_{1,k}(x,0)|^p|\tilde{w}_{2,k}(x,0)|^q \varphi dx = \int_{\mathbb{R}^N} \varphi\, d\nu, \quad \forall \varphi \in L^\infty \cap C(\mathbb{R}^N).$$

In particular, we infer that

$$\lim_k \int_{\mathbb{R}^{N+1}_+} k_s y^{1-2s}(|\nabla \tilde{w}_{1,k}|^2 + |\nabla \tilde{w}_{2,k}|^2)dxdy = \mu(\mathbb{R}^{N+1}_+),$$

$$\lim_k \int_{\mathbb{R}^N} |\tilde{w}_{1,k}(x,0)|^p|\tilde{w}_{2,k}(x,0)|^q dx = \nu(\mathbb{R}^N).$$

Claim: $I \neq \emptyset$.

Verification: If $I = \emptyset$, we would have $\int_B |w_1(x,0)|^p |w_2(x,0)|^q dx = 1$ and

$$C_{p,q}\mathcal{S}^0_{s,2^*_s} = \lim_k \int_{C_B} k_s y^{1-2s}(|\nabla \tilde{w}_{1,k}|^2 + |\nabla \tilde{w}_{2,k}|^2) dxdy$$
$$\geq \int_{C_B} (k_s y^{1-2s}(|\nabla w_1|^2 + |\nabla w_2|^2)) dxdy,$$

yielding $C_{p,q}\mathcal{S}^0_{s,2^*_s} = \int_{C_B} k_s y^{1-2s}(|\nabla w_1|^2 + |\nabla w_2|^2) dxdy$, namely a contradiction to Corollary 2.3.

Claim: I contains only one point and $w_1 = w_2 = 0$.

Verification: We argue by contradiction and consider the following three cases:

i) $w_1 \neq 0$ and $w_2 \neq 0$;
ii) $w_1 \neq 0$ and $w_2 = 0$;
iii) $w_1 = 0$ and $w_2 \neq 0$.

In the case i), we have $\sum_{j \in J} \nu_j \in (0,1)$. Notice that

$$C_{p,q}\mathcal{S}^0_{s,2^*_s} = \lim_k \int_{C_B} k_s y^{1-2s}(|\nabla \tilde{w}_{1,k}|^2 + |\nabla \tilde{w}_{2,k}|^2) dxdy$$
$$\geq \int_{C_B} (k_s y^{1-2s}(|\nabla w_1|^2 + |\nabla w_2|^2)) dxdy + C_{p,q}\mathcal{S}^0_{s,2^*_s} \sum_{j \in I} \nu_j^{2/2^*_s},$$

as well as

$$1 = \nu(\mathbb{R}^N) = \int_B |w_1(x,0)|^p |w_2(x,0)|^q dx + \sum_{j \in I} \nu_j.$$

These facts imply that

$$\int_{C_B} k_s y^{1-2s}(|\nabla w_1|^2 + |\nabla w_2|^2) dxdy \leq C_{p,q}\mathcal{S}^0_{s,2^*_s}\Big(1 - \sum_{j \in I} \nu_j^{2/2^*_s}\Big)$$
$$\leq C_{p,q}\mathcal{S}^0_{s,2^*_s}\Big(1 - \sum_{j \in I} \nu_j\Big)^{2/2^*_s}$$
$$= \mathcal{S}^0_{s,p,q}\Big(\int_B |w_1(x,0)|^p |w_2(x,0)|^q dx\Big)^{2/2^*_s},$$

which is a contradiction. In the case ii) or iii), we have $\sum_{j \in J} \nu_j = 1$. Notice that

$$C_{p,q}\mathcal{S}^0_{s,2^*_s} \geq \int_{C_B} (k_s y^{1-2s}(|\nabla w_1|^2 + |\nabla w_2|^2)) dxdy + C_{p,q}\mathcal{S}^0_{s,2^*_s} \sum_{j \in I} \nu_j^{2/2^*_s}.$$

This implies, as above, that

$$\int_{C_B} k_s y^{1-2s}(|\nabla w_1|^2 + |\nabla w_2|^2)dxdy \le C_{p,q}\mathscr{S}^0_{s,2_s^*}\Big(1 - \sum_{j\in I}\nu_j\Big)^{2/2_s^*} = 0,$$

which is a contradiction. Then $w_1 = w_2 = 0$. We claim that J is singleton. Notice again

$$1 \ge \sum_{j\in I}\nu_j^{2/2_s^*} \ge \Big(\sum_{j\in I}\nu_j\Big)^{2/2_s^*} = 1,$$

so there is at most one $j^* \in I$ such that $\nu_{j^*} \neq 0$, proving the claim. Hence there exists $x_0 \in \overline{B}$ with

$$k_s y^{1-2s}(|\nabla \tilde{w}_{1,k}|^2 + |\nabla \tilde{w}_{2,k}|^2) \rightharpoonup \mu_0\delta_{x_0}, \quad |\tilde{w}_{1,k}(x,0)|^p|\tilde{w}_{2,k}(x,0)|^q \rightharpoonup \nu_0\delta_{x_0}, \tag{15}$$

in the sense of measure with $\mu_0 \ge C_{p,q}\mathscr{S}^0_{s,2_s^*}\nu_0^{2/2_s^*}$. Taking into account the relation (14) between $\tilde{w}_{i,k}$ and $w_{i,k}$ the same conclusion follows for the $w_{i,k}$.

Assume by contradiction that $x_0 \in B$. Then it follows $\mathrm{dist}(x_0, \partial E) = \sigma$, for $\sigma \in (0,1)$. Notice that $|w_{1,k}(x,0)|^{p_k} \le |w_{1,k}(x,0)|^p + o_k(1)$. By the concentration behavior of the sequence $|w_{1,k}(x,0)|^p|w_{2,k}(x,0)|^q$ stated in (15), there exists $\varphi \in L^\infty \cap C(\mathbb{R}^N)$ with $\varphi(x_0) = 0$ and

$$\begin{aligned}\int_{\mathbb{R}^N} &|w_{1,k}(x,0)|^p|w_{2,k}(x,0)|^q\chi_{B\setminus B(x_0,\sigma/2)}(x)dx\\ &\le \int_{\mathbb{R}^N}|w_{1,k}(x,0)|^p|w_{2,k}(x,0)|^q\varphi(x)dx = o_k(1).\end{aligned}$$

Since $\int_B |w_{2,k}(x,0)|^q dx \le C$ by the Sobolev inequality [5, formula (2.11)], then we conclude

$$\begin{aligned}\int_B |x|^\alpha|w_{1,k}(x,0)|^{p_k}|w_{2,k}(x,0)|^q dx &= \int_{B(x_0,\sigma/2)}|x|^\alpha|w_{1,k}(x,0)|^{p_k}|w_{2,k}(x,0)|^q dx\\ &+ \int_{\mathbb{R}^N}|x|^\alpha|w_{1,k}(x,0)|^{p_k}|w_{2,k}(x,0)|^q\chi_{B\setminus B(x_0,\sigma/2)}(x)dx\\ &\le (1-\sigma/2)^\alpha\int_{B(x_0,\sigma/2)}|w_{1,k}(x,0)|^{p_k}|w_{2,k}(x,0)|^q dx\\ &+ \int_{\mathbb{R}^N}|w_{1,k}(x,0)|^p|w_{2,k}(x,0)|^q\chi_{B\setminus B(x_0,\sigma/2)}(x)dx + o_k(1)\\ &\le \Lambda(\sigma)^{2_s^*/2}\Big(\int_B |w_{1,k}(x,0)|^{p_k}|w_{2,k}(x,0)|^q dx + o_k(1)\Big),\end{aligned} \tag{16}$$

where $\Lambda(\sigma) := (1-\sigma/2)^{2\alpha/2_s^*} \in (0,1)$. By formula (16), on account of by Lemma 2.2, it follows that

$$C_{p,q}\mathcal{S}^0_{s,2_s^*} = \lim_k \frac{\displaystyle\int_{C_B} k_s y^{1-2s}(|\nabla w_{1,k}|^2 + |\nabla w_{2,k}|^2)dxdy}{\left(\displaystyle\int_B |x|^\alpha |w_{1,k}(x,0)|^{p_k}|w_{2,k}(x,0)|^q dx\right)^{2/(p_k+q)}}$$

$$\geq \frac{1}{\Lambda(\sigma)} \lim_k \frac{\displaystyle\int_{C_B} k_s y^{1-2s}(|\nabla w_{1,k}|^2 + |\nabla w_{2,k}|^2)dxdy}{\left(\displaystyle\int_B |w_{1,k}(x,0)|^{p_k}|w_{2,k}(x,0)|^q dx\right)^{2/(p_k+q)}} > C_{p,q}\mathcal{S}^0_{s,2_s^*},$$

which is a contradiction, since $\Lambda^{-1}(\sigma) > 1$. The proof of Theorem 1.3 is complete.

4 Proof of Theorem 1.4

Let $(w_{1,\varepsilon}, w_{2,\varepsilon}) \in H$ be a nonnegative solution to (11). Then, up to a multiplicative constant depending upon the Lagrange multiplier, we may assume that $(w_{1,\varepsilon}, w_{2,\varepsilon})$ solves the system (12). In particular, identity (13) follows. Hence, from Lemma 2.2, we infer

$$\lim_{\varepsilon\to 0}\int_{C_B} k_s y^{1-2s}(|\nabla w_{1,\varepsilon}|^2 + |\nabla w_{2,\varepsilon}|^2)dxdy = \frac{(\mathcal{S}^0_{s,p,q}(C_B))^{\frac{N}{2s}}}{2^{\frac{N-2s}{2s}}} = \frac{(\mathcal{S}^0_{s,p,q}(\mathbb{R}^{N+1}_+))^{\frac{N}{2s}}}{2^{\frac{N-2s}{2s}}}. \tag{17}$$

We know that $(w_{1,\varepsilon}, w_{2,\varepsilon})$ is a solution of the system

$$\begin{cases} -\mathrm{div}(y^{1-2s}\nabla w_{1,\varepsilon}) = 0, \quad -\mathrm{div}(y^{1-2s}\nabla w_{2,\varepsilon}) = 0, & x \in C_B,\\ k_s y^{1-2s}\frac{\partial w_{1,\varepsilon}}{\partial \nu} = \frac{2p_\varepsilon}{p_\varepsilon+q}|x|^\alpha w_{1,\varepsilon}(x,0)^{p_\varepsilon-1}w_{2,\varepsilon}(x,0)^q, & x\in B,\\ k_s y^{1-2s}\frac{\partial w_{2,\varepsilon}}{\partial \nu} = \frac{2q}{p_\varepsilon+q}|x|^\alpha w_{1,\varepsilon}(x,0)^{p_\varepsilon}w_{2,\varepsilon}(x,0)^{q-1}, & x\in B,\\ w_{1,\varepsilon} = w_{2,\varepsilon} = 0, & x\in \partial_L C_B. \end{cases}$$

Then, we can assume $w_{i,\varepsilon} \in C^\tau(B)$, for some $\tau \in (0,1)$. There exist $x_{1,\varepsilon}, x_{2,\varepsilon} \in \overline{B}$ such that

$$M_{i,\varepsilon} = w_{i,\varepsilon}(x_{i,\varepsilon},0) = \sup_{(x,y)\in\overline{B}\times(0,\infty)} w_{i,\varepsilon}(x,y), \quad i=1,2. \tag{18}$$

In fact, let $x_{1,\varepsilon}, x_{2,\varepsilon} \in \overline{B}$ be such that

$$M_{i,\varepsilon} := \sup_{x\in\overline{B}} w_{i,\varepsilon}(x,0) = w_{i,\varepsilon}(x_{i,\varepsilon},0), \quad i=1,2.$$

Then the second equality in (18) holds, since we have the following maximum principle

Lemma 4.1. *$w_{i,\varepsilon}(x,y) \leq M_{i,\varepsilon}$ for a.e. $(x,y) \in B \times (0,\infty)$, for $i = 1,2$.*

Proof. Define $\tau_i(x,y) := (w_{i,\varepsilon}(x,y) - M_{i,\varepsilon})^+$ for $(x,y) \in B \times (0,\infty)$. Then, testing in (LS), we obtain

$$k_s \int_{C_B} y^{1-2s} |\nabla \tau_1(x,y)|^2 dxdy = \frac{2p_\varepsilon}{p_\varepsilon + q} \int_B |x|^\alpha w_{1,\varepsilon}(x,0)^{p_\varepsilon - 1} w_{2,\varepsilon}(x,0)^q \tau_1(x,0)dx = 0,$$

$$k_s \int_{C_B} y^{1-2s} |\nabla \tau_2(x,y)|^2 dxdy = \frac{2q}{p_\varepsilon + q} \int_B |x|^\alpha w_{1,\varepsilon}(x,0)^{p_\varepsilon} w_{2,\varepsilon}(x,0)^{q-1} \tau_2(x,0)dx = 0.$$

Then $\tau_i \equiv 0$ for $i = 1,2$, yielding he conclusion. □

Lemma 4.2. *For all $i = 1,2$, we have $M_{i,\varepsilon} \to +\infty$ as $p_\varepsilon + q \to 2_s^*$.*

Proof. Suppose by contradiction that there exist $C > 0$ and a sequence $\{\varepsilon_n\} \subset \mathbb{R}^+$ such that $p_{\varepsilon_n} + q \to 2_s^*$ and $M_{2,\varepsilon_n} \leq C$, for all $n \in \mathbb{N}$. Since (w_{i,ε_n}) is bounded in $H^1_{0,L}(C_B)$, up to a subsequence, by the conclusions of Theorem 1.3 we get $w_{i,\varepsilon_n} \rightharpoonup 0$ weakly in $H^1_{0,L}(C_B)$ and $w_{i,\varepsilon_n} \to 0$ in $L^r(B)$, for every $r < 2_s^*$. Then, from identity (13) and formula (17), there exists a positive constant σ independent of ε_n such that

$$\begin{aligned} 0 < \sigma &\leq \int_B |x|^\alpha |w_{1,\varepsilon_n}(x,0)|^{p_\varepsilon} |w_{2,\varepsilon_n}(x,0)|^q dx \\ &\leq \|w_{2,\varepsilon_n}(x,0)\|_\infty^q \|w_{1,\varepsilon_n}(x,0)\|_{L^{p_{\varepsilon_n}}(B)}^{p_{\varepsilon_n}} \leq Co_n(1), \end{aligned}$$

which yields a contradiction. □

Now we want to recall some general Pohožaev type identity. Consider the following system

$$(LG) \qquad \begin{cases} -\mathrm{div}(y^{1-2s}\nabla w_1) = 0, \text{ in } C_B = B \times (0,\infty), \\ -\mathrm{div}(y^{1-2s}\nabla w_2) = 0, \text{ in } C_B = B \times (0,\infty), \\ w_1 = w_2 = 0, \text{ on } \partial_L C_B = \partial B \times (0,\infty), \\ k_s y^{1-2s} \frac{\partial w_1}{\partial \nu} = C_1 w_1(x,0)^{p-1} w_2(x,0)^q, \text{ on } B \times \{0\}, \\ k_s y^{1-2s} \frac{\partial w_2}{\partial \nu} = C_2 w_1(x,0)^p w_2(x,0)^{q-1}, \text{ on } B \times \{0\}, \end{cases}$$

where $\frac{\partial}{\partial \nu}$ denotes the outward normal derivative, and ν is exterior normal vector to ∂B. For the scalar case, the next result was obtained in [7], while for the system we refer to [19].

Theorem 4.3. *Let $p + q = 2_s^*$. Then system (LG) does not admit any nontrivial nonnegative solution.*

The following nonexistence result is crucial for our argument. Consider the following problem

$$(LS)\qquad \begin{cases} -\mathrm{div}(y^{1-2s}\nabla w_1) = 0, \text{ in } \mathbb{R}^{N+1}_{++}, \\ -\mathrm{div}(y^{1-2s}\nabla w_2) = 0, \text{ in } \mathbb{R}^{N+1}_{++}, \\ w_1 = w_2 = 0, \text{ on } \{x_N = 0, y > 0\}, \\ k_s y^{1-2s}\frac{\partial w_1}{\partial \nu} = C_1 w_1(x,0)^{p-1} w_2(x,0)^q, \text{ on } \{x_N > 0, y = 0\}, \\ k_s y^{1-2s}\frac{\partial w_2}{\partial \nu} = C_2 w_1(x,0)^{p} w_2(x,0)^{q-1}, \text{ on } \{x_N > 0, y = 0\}, \end{cases}$$

where $C_1, C_2 > 0$, $\frac{\partial}{\partial \nu}$ denotes the outward normal derivative, $p + q = 2^*_s$ and

$$\mathbb{R}^{N+1}_{++} := \{(x_1, x_2, \dots, x_{N-1}, x_N, y) \in \mathbb{R}^{N+1} : x_N > 0, y > 0\}.$$

Proposition 4.4. *Let $w_1, w_2 \in H^1_{0,L}(\mathbb{R}^{N+1}_{++})$ be a bounded solution of (LS). Then $(w_1, w_2) = (0, 0)$.*

Proof. Since $(x, y)\cdot\nu = 0$ on $\partial\mathbb{R}^{N+1}_{++}$, one cannot apply directly Pohožaev identities. Whence, we use the Kelvin transformation as in [19, 21] to study a new system set in a ball. Let $w_i \in H^1_{0,L}(\mathbb{R}^{N+1}_{++})$ be a solution to system (*LS*). Then, the Kelvin transformation of w_i is defined by

$$\widetilde{w_i}(z) = |z|^{2s-N} w_i\Big(\frac{z}{|z|^2}\Big), \quad z \in \mathbb{R}^{N+1}_{++}.$$

and from [21, Proposition 2.6] we infer that $\widetilde{w_i}$ is also a solution to (*LS*). By [25, Corollary 2.1, Proposition 2.4], there exists $\gamma \in (0, 1)$ with $\widetilde{w_i}(z) \le C|z|^\gamma$, for $z \in B_1(0)$. Then there exists $C > 0$ such that

$$|w_i(z)| \le C(1 + |z|^2)^{-\frac{N-2s+\gamma}{2}}, \quad \text{for all } z \in \mathbb{R}^{N+1}_{++}. \tag{19}$$

Arguing as in [18], denote by $B_{\frac{1}{2}}(\frac{e_N}{2}) \subset \mathbb{R}^N$ the ball centered at $\frac{e_N}{2}$ with radius $\frac{1}{2}$. Define

$$v_i(z) := |z|^{2s-N} w_i\Big(-(e_N, 0) + \frac{z}{|z|^2}\Big), \quad \text{for all } z \in \overline{C_{B_{\frac{1}{2}}(\frac{e_N}{2})}} \setminus \{0\}.$$

By means of (19), for a positive constant C and for $|z|$ small enough, we have

$$v_i(z) \le C|z|^\gamma, \quad \text{for all } z \in \mathbb{R}^{N+1}_{++} \setminus \{0\}.$$

Therefore, we may extend v_i by 0 at 0. Then, as above, (v_1, v_2) is a weak solution of the system

$$(LSB)\qquad \begin{cases} -\mathrm{div}(y^{1-2s}\nabla v_1) = 0, \text{ in } C_{B_{\frac{1}{2}}(\frac{e_N}{2})}, \\ -\mathrm{div}(y^{1-2s}\nabla v_2) = 0, \text{ in } C_{B_{\frac{1}{2}}(\frac{e_N}{2})}, \\ v_1 = v_2 = 0, \text{ on } \partial_L C_{B_{\frac{1}{2}}(\frac{e_N}{2})}, \\ k_s y^{1-2s}\frac{\partial v_1}{\partial v} = C_1 v_1(x,0)^{p-1} v_2(x,0)^q, \text{ on } B_{\frac{1}{2}}(\frac{e_N}{2}) \times \{0\}, \\ k_s y^{1-2s}\frac{\partial v_2}{\partial v} = C_2 v_1(x,0)^{p} v_2(x,0)^{q-1}, \text{ on } B_{\frac{1}{2}}(\frac{e_N}{2}) \times \{0\}. \end{cases}$$

Now, applying Theorem 4.3 to system (*LSB*) we infer that $v_i = 0, i = 1, 2$, that is, $w_i = 0, i = 1, 2$. □

We are now ready to complete the proof. By Lemma 4.2, we may assume

$$M_{1,\varepsilon} = w_{1,\varepsilon}(x_{1,\varepsilon}, 0) = \sup_{(x,y)\in \bar{B}\times(0,\infty)} w_{1,\varepsilon}(x,y) \to +\infty,$$

We may assume $M_{1,\varepsilon} \geq M_{2,\varepsilon}$. Let $\lambda_\varepsilon > 0$ be such that

$$\lambda_\varepsilon^{\frac{N-2s}{2}} M_{1,\varepsilon} = 1, \qquad 0 \leq \lambda_\varepsilon^{\frac{N-2s}{2}} M_{2,\varepsilon} \leq 1,$$

where $\lambda_\varepsilon \to 0$, as $p_\varepsilon + q \to 2_s^*$. Define the scaled functions

$$\tilde{w}_{1,\varepsilon}(x,y) := \lambda_\varepsilon^{\frac{N-2s}{2}} w_{1,\varepsilon}(\lambda_\varepsilon x + x_{1,\varepsilon}, \lambda_\varepsilon y),$$

$$\tilde{w}_{2,\varepsilon}(x,y) := \lambda_\varepsilon^{\frac{N-2s}{2}} w_{2,\varepsilon}(\lambda_\varepsilon x + x_{1,\varepsilon}, \lambda_\varepsilon y),$$

$B_\varepsilon := \{x \in \mathbb{R}^N : \lambda_\varepsilon x + x_{1,\varepsilon} \in B_1(0)\}$ and $C_{B_\varepsilon} := B_\varepsilon \times (0,\infty)$. Then $(\tilde{w}_{1,\varepsilon}(x,y), \tilde{w}_{1,\varepsilon}(x,y))$ satisfies

$$\begin{cases} -\mathrm{div}(y^{1-2s}\nabla \tilde{w}_{1,\varepsilon}) = 0, \quad -\mathrm{div}(y^{1-2s}\nabla \tilde{w}_{2,\varepsilon}) = 0, \quad x \in C_{B_\varepsilon} \\ 0 < \tilde{w}_{1,\varepsilon} \leq 1, \ \ 0 < \tilde{w}_{2,\varepsilon} \leq 1, \ \ \tilde{w}_{1,\varepsilon}(0,0) = 1, \quad x \in C_{B_\varepsilon} \\ k_s y^{1-2s}\frac{\partial \tilde{w}_{1,\varepsilon}}{\partial v} = \frac{2p_\varepsilon}{p_\varepsilon+q}|\lambda_\varepsilon x + x_{1,\varepsilon}|^\alpha \lambda_\varepsilon^{N-\frac{N-2s}{2}(p_\varepsilon+q)} \tilde{w}_{1,\varepsilon}(x,0)^{p_\varepsilon-1}\tilde{w}_{2,\varepsilon}(x,0)^q, \quad x \in B_\varepsilon, \\ k_s y^{1-2s}\frac{\partial \tilde{w}_{2,\varepsilon}}{\partial v} = \frac{2q}{p_\varepsilon+q}|\lambda_\varepsilon x + x_{1,\varepsilon}|^\alpha \lambda_\varepsilon^{N-\frac{N-2s}{2}(p_\varepsilon+q)} \tilde{w}_{1,\varepsilon}(x,0)^{p_\varepsilon}\tilde{w}_{2,\varepsilon}(x,0)^{q-1}, \quad x \in B_\varepsilon, \\ \tilde{w}_{1,\varepsilon} = 0, \quad \tilde{w}_{2,\varepsilon} = 0, \ \ x \in \partial B_\varepsilon \cap \partial_L C_{B_\varepsilon}. \end{cases}$$

Suppose $x_{1,\varepsilon} \to x_0$ for some $x_0 \in \bar{B}_1(0)$. We claim that $x_0 \in \partial B_1(0)$. By contradiction, assume that $x_0 \in B_1(0)$ and let $d := \frac{1}{2}\mathrm{dist}(x_0, \partial B_1(0))$. Denote $\mathscr{B}(0,r) = \{z \in \mathbb{R}^{N+1} : |z| < r\}$. For $\varepsilon > 0$ small, both $\tilde{w}_{1,\varepsilon}$ and $\tilde{w}_{2,\varepsilon}$ are well defined in $\mathscr{B}(0, d/\lambda_\varepsilon) \cap \mathbb{R}_+^{N+1}$, and

$$\sup_{(x,y)\in\mathscr{B}(0,\frac{d}{\lambda_\varepsilon})\cap\mathbb{R}_+^{N+1}} \tilde{w}_{1,\varepsilon}(x,y) = \tilde{w}_{1,\varepsilon}(0,0) = 1, \qquad \sup_{(x,y)\in\mathscr{B}(0,\frac{d}{\lambda_\varepsilon})\cap\mathbb{R}_+^{N+1}} \tilde{w}_{2,\varepsilon}(x,y) \in (0,1].$$

Since $M_{1,\varepsilon} \to +\infty$, we have $0 \le \lambda_\varepsilon \le 1$, for $\varepsilon > 0$ small. Let

$$h(\varepsilon) := \lambda_\varepsilon^{N-\frac{N-2s}{2}(p_\varepsilon+q)} \quad \text{and} \quad h(0) := \lim_{p_\varepsilon+q\to 2_s^*} h(\varepsilon).$$

Hence, $0 \le h(\varepsilon) \le 1$. Three possibilities may occur, namely

(1) $h(0) = 0$,
(2) $h(0) = \beta \in (0,1)$,
(3) $h(0) = 1$.

We show that any of these cases yields a contradiction. We observe that, for any $R > 0$, $B_R(0) \subset B_{d/\lambda_\varepsilon}(0)$ for $\varepsilon > 0$ small enough. By Schauder estimates [7, 12, 14, 25] there are $C > 0$ and $0 < \vartheta < 1$ with

$$\|\tilde{w}_{1,\varepsilon}\|_{C^{0,\vartheta}(\mathscr{B}(0,2R)\cap\mathbb{R}_+^{N+1})} \le C, \quad \|\tilde{w}_{2,\varepsilon}\|_{C^{0,\vartheta}(\mathscr{B}(0,2R)\cap\mathbb{R}_+^{N+1})} \le C$$

for ε small enough. By Arzelà-Ascoli's Theorem, there exist subsequences $\{\tilde{w}_{i,\varepsilon_k}\}$ such that $\tilde{w}_{i,\varepsilon_k} \to w_i$ as $k \to \infty$, for $i = 1,2$, in $C^{0,\vartheta_0}_{\mathrm{loc}}$ for some $\vartheta_0 \in (0,\vartheta)$. Then, we derive that (w_1, w_2) satisfies

$$\begin{cases} -\mathrm{div}(y^{1-2s}\nabla w_1) = 0, \quad -\mathrm{div}(y^{1-2s}\nabla w_2) = 0, & \text{in } \mathbb{R}_+^{N+1}, \\ k_s y^{1-2s}\frac{\partial w_1}{\partial \nu} = \frac{2p}{2_s^*}|x_0|^\alpha h(0) w_1^{p-1}(x,0) w_2^q(x,0), & \text{on } \mathbb{R}^N \times \{0\}, \\ k_s y^{1-2s}\frac{\partial w_2}{\partial \nu} = \frac{2q}{2_s^*}|x_0|^\alpha h(0) w_1^{p}(x,0) w_2^{q-1}(x,0), & \text{on } \mathbb{R}^N \times \{0\}, \end{cases} \tag{20}$$

and $w_1(0,0) = 1$, $0 \le w_2 \le 1$. Moreover, $w_i \in H^1_{0,L}(\mathbb{R}_+^{N+1})$. If $x_0 = 0$ or $h(0) = 0$ or $w_2 = 0$, we have $w_1 \equiv 0$, which is impossible since $w_1(0,0) = 1$. Suppose $x_0 \ne 0$, $w_2 \not\equiv 0$ and $h(0) = \beta \in (0,1]$. Then

$$\begin{cases} -\mathrm{div}(y^{1-2s}\nabla w_1) = 0, \quad -\mathrm{div}(y^{1-2s}\nabla w_2) = 0, & \text{in } \mathbb{R}_+^{N+1}, \\ k_s y^{1-2s}\frac{\partial w_1}{\partial \nu} = \frac{2p}{2_s^*}\Lambda w_1^{p-1}(x,0) w_2^q(x,0), & \text{on } \mathbb{R}^N \times \{0\}, \\ k_s y^{1-2s}\frac{\partial w_2}{\partial \nu} = \frac{2q}{2_s^*}\Lambda w_1^{p}(x,0) w_2^{q-1}(x,0), & \text{on } \mathbb{R}^N \times \{0\}, \end{cases} \tag{21}$$

where $\Lambda = |x_0|^\alpha \beta \in (0,1)$. Setting

$$\bar{w}_1 := \Lambda^{\frac{1}{2_s^*-2}} w_1, \qquad \bar{w}_2 := \Lambda^{\frac{1}{2_s^*-2}} w_2,$$

we have $0 < \bar{w}_1 \le \Lambda^{\frac{1}{2_s^*-2}}$, $0 < \bar{w}_2 \le \Lambda^{\frac{1}{2_s^*-2}}$, $\bar{w}_1(0,0) = \Lambda^{\frac{1}{2_s^*-2}}$ and $(\bar{w}_1, \bar{w}_2)$ satisfies

$$\begin{cases} -\mathrm{div}(y^{1-2s}\nabla \bar{w}_1) = 0, \quad -\mathrm{div}(y^{1-2s}\nabla \bar{w}_2) = 0, & \text{in } \mathbb{R}^{N+1}_+, \\ k_s y^{1-2s}\frac{\partial \bar{w}_1}{\partial \nu} = \frac{2p}{2^*_s}\bar{w}_1^{p-1}(x,0)\bar{w}_2^q(x,0), & \text{on } \mathbb{R}^N \times \{0\}, \\ k_s y^{1-2s}\frac{\partial \bar{w}_2}{\partial \nu} = \frac{2q}{2^*_s}\bar{w}_1^{p}(x,0)\bar{w}_2^{q-1}(x,0), & \text{on } \mathbb{R}^N \times \{0\}, \end{cases} \tag{22}$$

Define $\mathscr{S} := \mathscr{S}^0_{s,p,q}(\mathbb{R}^{N+1}_+)$ and observe that

$$\int_{\mathbb{R}^{N+1}_+} k_s y^{1-2s}(|\nabla \bar{w}_1|^2 + |\nabla \bar{w}_2|^2)dxdy = 2\int_{\mathbb{R}^N} \bar{w}_1^p(x,0)\bar{w}_2^q(x,0)dx.$$

Then, by formula (17), we have

$$\begin{aligned} \mathscr{S}^{\frac{N}{2s}} &\le 2^{\frac{N-2s}{2s}} \int_{\mathbb{R}^{N+1}_+} k_s y^{1-2s}(|\nabla \bar{w}_1|^2 + |\nabla \bar{w}_2|^2)dxdy \\ &= \Lambda^{\frac{N-2s}{2s}} 2^{\frac{N-2s}{2s}} \int_{\mathbb{R}^{N+1}_+} k_s y^{1-2s}(|\nabla w_1|^2 + |\nabla w_2|^2)dxdy \\ &\le \Lambda^{\frac{N-2s}{2s}} \liminf_{\varepsilon\to 0} 2^{\frac{N-2s}{2s}} \int_{C_{B_\varepsilon}} k_s y^{1-2s}(|\nabla \tilde{w}_{1,\varepsilon}|^2 + |\nabla \tilde{w}_{2,\varepsilon}|^2)dxdy \\ &= \Lambda^{\frac{N-2s}{2s}} \liminf_{\varepsilon\to 0} 2^{\frac{N-2s}{2s}} \int_{C_B} k_s y^{1-2s}(|\nabla w_{1,\varepsilon}|^2 + |\nabla w_{2,\varepsilon}|^2)dxdy \\ &= \Lambda^{\frac{N-2s}{2s}} \mathscr{S}^{\frac{N}{2s}} < \mathscr{S}^{\frac{N}{2s}}, \end{aligned} \tag{23}$$

a contradiction. Then $x_0 \in \partial B_1(0)$. We can straighten ∂B in a neighborhood of x_0 by a non-singular C^1 change of coordinates. Let $x_N = \psi(x')$ be the equation of ∂B, where $x' = (x_1, x_2, \ldots, x_{N-1})$, $\psi \in C^1$. Define new coordinate system given by $z_i = x_i$ for $i = 1, \ldots, N-1$, $z_N = x_N - \psi(x')$ and $z_{N+1} = y$. Let $d_\varepsilon = \mathrm{dist}(x_\varepsilon, \partial B)$. For $p_\varepsilon + q \to 2^*_s$ as $\varepsilon \to 0$, $\tilde{w}_{i,\varepsilon}$ are well defined in $B(0, \frac{\delta}{\lambda_\varepsilon}) \cap \mathbb{R}^{N+1}_+ \cap \{z_N > -\frac{d_\varepsilon}{\lambda_\varepsilon}\}$ for some $\delta > 0$ small enough. Moreover

$$\sup_{B(0,\frac{\delta}{\lambda_\varepsilon})\cap\mathbb{R}^{N+1}_+\cap\{z_N>-\frac{d_\varepsilon}{\lambda_\varepsilon}\}} \tilde{w}_{1,\varepsilon}(x,y) = \tilde{w}_{1,\varepsilon}(0,0) = 1,$$

$$\sup_{B(0,\frac{\delta}{\lambda_\varepsilon})\cap\mathbb{R}^{N+1}_+\cap\{z_N>-\frac{d_\varepsilon}{\lambda_\varepsilon}\}} \tilde{w}_{2,\varepsilon}(x,y) \in (0,1].$$

We now have the following

Claim: $d_\varepsilon/\lambda_\varepsilon \to +\infty$ as $\varepsilon \to 0$.

Verification: Suppose by contradiction that $d_\varepsilon/\lambda_\varepsilon$ remains bounded and $d_\varepsilon/\lambda_\varepsilon \to s$ for some $s \ge 0$. By the previous argument, since $|x_0| = 1$, we get $\tilde{w}_{i,\varepsilon} \to \tilde{w}_i$ in $C^{0,\gamma}_{\mathrm{loc}}$, $\tilde{w}_1(0,0) = 1$ and

$$
\begin{cases}
\operatorname{div}(y^{1-2s}\nabla\tilde{w}_1) = 0, \text{ in } \{(z_1,\dots,z_N,z_{N+1}) : z_N > -s, z_{N+1} > 0\},\\
\operatorname{div}(y^{1-2s}\nabla\tilde{w}_2) = 0, \text{ in } \{(z_1,\dots,z_N,z_{N+1}) : z_N > -s, z_{N+1} > 0\},\\
k_s y^{1-2s}\frac{\partial\tilde{w}_1}{\partial\nu} = \Lambda\frac{2p}{2_s^*}\tilde{w}_1^{p-1}(x,0)\tilde{w}_2^q(x,0), \text{ on } \{(z_1,\dots,z_{N+1}) : z_N > -s, z_{N+1} = 0\},\\
k_s y^{1-2s}\frac{\partial\tilde{w}_2}{\partial\nu} = \Lambda\frac{2p}{2_s^*}\tilde{w}_1^{p}(x,0)\tilde{w}_2^{q-1}(x,0), \text{ on } \{(z_1,\dots,z_{N+1}) : z_N > -s, z_{N+1} = 0\},\\
\tilde{w}_1(z) = \tilde{w}_2(z) = 0, \text{ on } \{(z_1,\dots,z_N,z_{N+1}) : z_N = -s, z_{N+1} > 0\},\\
\tilde{w}_1(z), \tilde{w}_2(z) \in (0,1], \text{ in } \{(z_1,\dots,z_N,z_{N+1}) : z_N > -s, z_{N+1} > 0\}.
\end{cases}
\tag{24}
$$

By a translation, $(\tilde{w}_1, \tilde{w}_2)$ verifies

$$
\begin{cases}
\operatorname{div}(y^{1-2s}\nabla\tilde{w}_1) = 0, \quad \operatorname{div}(y^{1-2s}\nabla\tilde{w}_2) = 0, \text{ in } \mathbb{R}^{N+1}_{++},\\
k_s y^{1-2s}\frac{\partial\tilde{w}_1}{\partial\nu} = \Lambda\frac{2p}{2_s^*}\tilde{w}_1^{p-1}(x,0)\tilde{w}_2^q(x,0), \text{ on } \{(z_1,\dots,z_N,z_{N+1}) : z_N > 0, z_{N+1} = 0\},\\
k_s y^{1-2s}\frac{\partial\tilde{w}_2}{\partial\nu} = \Lambda\frac{2p}{2_s^*}\tilde{w}_1^{p}(x,0)\tilde{w}_2^{q-1}(x,0), \text{ on } \{(z_1,\dots,z_N,z_{N+1}) : z_N > 0, z_{N+1} = 0\},\\
\tilde{w}_1(z) = \tilde{w}_2(z) = 0, \text{ on} \{(z_1,\dots,z_N,z_{N+1}) : z_N = 0, z_{N+1} > 0\},\\
\tilde{w}_2(z) \in (0,1], \tilde{w}_1(0,\dots,s,0) = 1, \text{ in} \mathbb{R}^{N+1}_{++}.
\end{cases}
\tag{25}
$$

Since $\tilde{w}_i \in H^1_{0,L}(\mathbb{R}^{N+1}_{++})$, by Proposition 4.4, $(\tilde{w}_1, \tilde{w}_2) = (0,0)$, which violates $\tilde{w}_1(0,\dots,s,0) = 1$. Then the claim follows and C_{B_ε} converges to the entire $\mathbb{R}^{N+1}_{++}$ as $\varepsilon \to 0$.

Claim. $\Lambda = |x_0|h(0) = h(0) = 1$. We can assume $(\tilde{w}_{1,\varepsilon}, \tilde{w}_{2,\varepsilon}) \to (\tilde{w}_1, \tilde{w}_2)$, as $\varepsilon \to 0$, and $(\tilde{w}_1, \tilde{w}_2)$ satisfies

$$
\begin{cases}
\operatorname{div}(y^{1-2s}\nabla\tilde{w}_1) = 0, \quad \operatorname{div}(y^{1-2s}\nabla\tilde{w}_2) = 0, & \text{in } \mathbb{R}^{N+1}_{++},\\
k_s y^{1-2s}\frac{\partial\tilde{w}_1}{\partial\nu} = \Lambda\frac{2p}{2_s^*}\tilde{w}_1^{p-1}(x,0)\tilde{w}_2^q(x,0), & \text{on } \mathbb{R}^N \times \{0\},\\
k_s y^{1-2s}\frac{\partial\tilde{w}_2}{\partial\nu} = \Lambda\frac{2p}{2_s^*}\tilde{w}_1^{p}(x,0)\tilde{w}_2^{q-1}(x,0), & \text{on } \mathbb{R}^N \times \{0\},\\
\tilde{w}_1(z) = \tilde{w}_2(z) = 0, & \text{on } \{0\} \times (0,\infty),\\
\tilde{w}_i(z) \in (0,1], \tilde{w}_i(0,0) = 1, & \text{in } \mathbb{R}^{N+1}_{++}.
\end{cases}
\tag{26}
$$

If $\tilde{w}_2 \equiv 0$ or $0 \le \Lambda < 1$, we reach the contradiction either as in (23) or by Proposition 4.4. Hence, $\Lambda = 1$ and $\tilde{w}_2 \not\equiv 0$. This implies $M_{1,\varepsilon}^{-1}\tilde{w}_{2,\varepsilon}(\lambda_\varepsilon x + x_\varepsilon) \to v(x) \ne 0$, and then $1 \ge M_{1,\varepsilon}^{-1}M_{2,\varepsilon} \to \sigma > 0$ as $\varepsilon \to 0$, that is $M_{1,\varepsilon} = \mathcal{O}(1)M_{2,\varepsilon}$. This is (i) of Theorem 1.4.

Let $y_\varepsilon \in B_1(0)$ be such that $w_{2,\varepsilon}(y_\varepsilon) = \max_{B_1(0)} w_{2,\varepsilon}(y)$. We define $\tilde{w}_{2,\varepsilon}(x) = (\bar{\lambda}_\varepsilon)^{(N-2s)/2} w_{2,\varepsilon}(\bar{\lambda}_\varepsilon x + y_\varepsilon)$, where $\bar{\lambda}_\varepsilon^{(N-2s)/2} w_{2,\varepsilon}(y_\varepsilon) = 1$. Suppose $y_\varepsilon \to y_0$. Again, using a blow-up argument, we get $y_0 \in \partial B_1(0)$. Then, in light of Lemma 2.7, we have

$$
\tilde{w}_1(x,y) = a\mathscr{W}_1(x,y), \qquad \tilde{w}_2(x,y) = b\mathscr{W}_1(x,y)
$$

for some positive numbers a, b such that $a/b = \sqrt{p/q}$. Let $\tilde{v}_{i,\varepsilon} = \tilde{w}_{i,\varepsilon} - \tilde{w}_i$. Then $\tilde{v}_{i,\varepsilon} \rightharpoonup 0$ weakly in $H^1_{0,L}(C_\omega)$ for any $\omega \subset \mathbb{R}^{N+1}_+$ and

$$\begin{cases} \operatorname{div}(y^{1-2s}\nabla \tilde{v}_{1,\varepsilon}) = 0, \quad \operatorname{div}(y^{1-2s}\nabla \tilde{v}_{2,\varepsilon}) = 0, & \text{in } C_{B_\varepsilon}, \\ \kappa_{2s} y^{1-2s}\frac{\partial \tilde{v}_{1,\varepsilon}}{\partial \nu} = \frac{2p_\varepsilon}{p_\varepsilon+q} Q_\varepsilon \tilde{w}^{p_\varepsilon-1}_{1,\varepsilon}(x,0)\tilde{w}^q_{2,\varepsilon}(x,0) - \frac{p(N-2s)}{N} w_1^{p-1} w_2^q & \text{on } B_\varepsilon \times \{0\} \\ \kappa_{2s} y^{1-2s}\frac{\partial \tilde{v}_{2,\varepsilon}}{\partial \nu} = \frac{2q}{p_\varepsilon+q} Q_\varepsilon \tilde{w}^{p_\varepsilon}_{1,\varepsilon}(x,0)\tilde{w}^{q-1}_{2,\varepsilon}(x,0) - \frac{q(N-2s)}{N} w_1^{p} w_2^{q-1} & \text{on } B_\varepsilon \times \{0\} \\ \tilde{v}_{i,\varepsilon} = -\tilde{w}_i, & \text{on } \partial_L C_{B_\varepsilon}, \end{cases} \tag{27}$$

where we have set $Q_\varepsilon := |\lambda_\varepsilon x + x_{1,\varepsilon}|^\alpha h(\varepsilon)$. Multiplying the first equation in (27) by $\tilde{v}_{1,\varepsilon}$ and $\tilde{v}_{2,\varepsilon}$, respectively, integrating by parts, and applying Brézis-Lieb Lemma, as $p_\varepsilon + q \to 2^*_s$, we have

$$\begin{aligned}
&k_s \int_{C_{B_\varepsilon}} y^{1-2s}(|\nabla \tilde{v}_{1,\varepsilon}|^2 + |\nabla \tilde{v}_{2,\varepsilon}|^2)dxdy \\
&= \int_{B_\varepsilon} \Big(\frac{2p_\varepsilon}{p_\varepsilon+q} Q_\varepsilon \tilde{w}^{p_\varepsilon-1}_{1,\varepsilon}(x,0)\tilde{w}^q_{2,\varepsilon}(x,0) - \frac{p(N-2s)}{N}\tilde{w}_1^{p-1}(x,0)\tilde{w}_2^q(x,0)\Big)\tilde{v}_{1,\varepsilon}(x,0)dx \\
&- k_s \int_{\partial_L B_\varepsilon} y^{1-2s}\frac{\partial \tilde{v}_{1,\varepsilon}}{\partial \nu} w_1 dS \\
&+ \int_{B_\varepsilon} \Big(\frac{2q}{p_\varepsilon+q} Q_\varepsilon \tilde{w}^{p_\varepsilon}_{1,\varepsilon}(x,0)\tilde{w}^{q-1}_{2,\varepsilon}(x,0) - \frac{q(N-2s)}{N}\tilde{w}_1^{p}(x,0)\tilde{w}_2^{q-1}(x,0)\Big)\tilde{v}_{2,\varepsilon}(x,0)dx \\
&- k_s \int_{\partial_L B_\varepsilon} y^{1-2s}\frac{\partial \tilde{v}_{2,\varepsilon}}{\partial \nu} w_2 dS \\
&= \frac{p(N-2s)}{N}\int_{B_\varepsilon} Q_\varepsilon\Big(\tilde{w}^{p_\varepsilon-1}_{1,\varepsilon}(x,0)\tilde{w}^q_{2,\varepsilon}(x,0) - \tilde{w}_1^{p-1}(x,0)\tilde{w}_2^q(x,0)\Big)\tilde{v}_{1,\varepsilon}(x,0)dx \\
&- k_s \int_{\partial_L B_\varepsilon} y^{1-2s}\frac{\partial \tilde{v}_{1,\varepsilon}}{\partial \nu}\tilde{w}_1 dS + \frac{p(N-2s)}{N}\int_{B_\varepsilon}(Q_\varepsilon - 1)\tilde{w}_1^{p-1}(x,0)\tilde{w}_2^q(x,0)dx \\
&+ \frac{q(N-2s)}{N}\int_{B_\varepsilon} Q_\varepsilon\Big(\tilde{w}^{p_\varepsilon}_{1,\varepsilon}(x,0)\tilde{w}^{q-1}_{2,\varepsilon}(x,0) - \tilde{w}_1^{p}(x,0)\tilde{w}_2^{q-1}(x,0)\Big)\tilde{v}_{2,\varepsilon}(x,0)dx \\
&- k_s \int_{\partial_L B_\varepsilon} y^{1-2s}\frac{\partial \tilde{v}_{2,\varepsilon}}{\partial \nu}\tilde{w}_2 dS + \frac{q(N-2s)}{N}\int_{B_\varepsilon}(Q_\varepsilon - 1)\tilde{w}_1^{p}(x,0)\tilde{w}_2^{q-1}(x,0)dx + o_\varepsilon(1),
\end{aligned}$$

since $Q_\varepsilon \to 1$ as $\varepsilon \to 0$. Recalling that $\mathscr{W}_\varepsilon$ decay at infinity, we obtain

$$\begin{aligned}
&k_s \int_{C_{B_\varepsilon}} y^{1-2s}(|\nabla \tilde{v}_{1,\varepsilon}|^2 + |\nabla \tilde{v}_{2,\varepsilon}|^2)dxdy \\
&\qquad = \frac{p(N-2s)}{N}\int_{B_\varepsilon} Q_\varepsilon\Big((\tilde{v}_{1,\varepsilon} + \tilde{w}_1)^{p_\varepsilon-1}(x,0)(\tilde{v}_{2,\varepsilon} + \tilde{w}_2)^q(x,0)
\end{aligned}$$

$$- \tilde{w}_1^{p-1}(x,0)\tilde{w}_2^q(x,0)\Big)\tilde{v}_{1,\varepsilon}(x,0)dx$$

$$+ \frac{q(N-2s)}{N}\int_{B_\varepsilon} Q_\varepsilon\Big((\tilde{v}_{1,\varepsilon}+\tilde{w}_1)^{p_\varepsilon}(x,0)(\tilde{v}_{2,\varepsilon}+\tilde{w}_2)^{q-1}(x,0)$$

$$- \tilde{w}_1^{p}(x,0)\tilde{w}_2^{q-1}(x,0)\Big)\tilde{v}_{2,\varepsilon}(x,0)dx + o_\varepsilon(1).$$

Inserting $\tilde{w}_{i,\varepsilon} = \tilde{v}_{i,\varepsilon} = \tilde{w}_i$, and using the following inequalities (cf. [8, 34])

$$||a+b|^p - |a|^p - |b|^p - pab(|a|^{p-2} + |b|^{p-2}|)| \le C\begin{cases} |a||b|^{p-1} \text{ if } |a| \ge |b|, \\ |a|^{p-1}|b| \text{ if } |a| \le |b|, \\ \qquad\qquad 1 \le p \le 3, \end{cases}$$

$$||a+b|^p - |a|^p - |b|^p - pab(|a|^{p-2} + |b|^{p-2}|)| \le C(|a|^{p-2}|b|^2 + |a|^2|b|^{p-2}), \quad p \ge 3,$$

we infer that

$$k_s\int_{C_{B_\varepsilon}} y^{1-2s}(|\nabla\tilde{v}_{1,\varepsilon}|^2 + |\nabla\tilde{v}_{2,\varepsilon}|^2)dxdy = \frac{p(N-2s)}{N}\int_{B_\varepsilon} Q_\varepsilon \tilde{v}_{1,\varepsilon}^p(x,0)\tilde{v}_{2,\varepsilon}^q(x,0)dx$$

$$+\frac{q(N-2s)}{N}\int_{B_\varepsilon} Q_\varepsilon \tilde{v}_{1,\varepsilon}^p(x,0)\tilde{v}_{2,\varepsilon}^q(x,0)dx + o_\varepsilon(1) = 2\int_{B_\varepsilon} Q_\varepsilon \tilde{v}_{1,\varepsilon}^p(x,0)\tilde{v}_{2,\varepsilon}^q(x,0)dx + o_\varepsilon(1). \tag{28}$$

By definition of $\mathscr{S}^\alpha_{s,p_\varepsilon,q}$ and recalling that $\mathscr{S}^\alpha_{s,p_\varepsilon,q} = \mathscr{S} + o_\varepsilon(1)$, we get

$$k_s\int_{C_{B_\varepsilon}} y^{1-2s}(|\nabla\tilde{v}_{1,\varepsilon}|^2 + |\nabla\tilde{v}_{2,\varepsilon}|^2)dxdy$$

$$\ge \mathscr{S}\Big(\int_{B_\varepsilon} Q_\varepsilon \tilde{v}_{1,\varepsilon}^p(x,0)\tilde{v}_{2,\varepsilon}^q(x,0)dx\Big)^{\frac{2}{p_\varepsilon+q}} + o_\varepsilon(1).$$

Assume by contradiction that

$$\lim_{\varepsilon\to 0} k_s\int_{C_{B_\varepsilon}} y^{1-2s}(|\nabla\tilde{v}_{1,\varepsilon}|^2 + |\nabla\tilde{v}_{2,\varepsilon}|^2)dxdy = \rho > 0.$$

Then, we have

$$k_s\int_{C_{B_\varepsilon}} y^{1-2s}(|\nabla\tilde{v}_{1,\varepsilon}|^2 + |\nabla\tilde{v}_{2,\varepsilon}|^2)dxdy$$

$$= 2\int_{B_\varepsilon} Q_\varepsilon \tilde{v}_{1,\varepsilon}^p(x,0)\tilde{v}_{2,\varepsilon}^q(x,0)dx + o(1) \ge \frac{\mathscr{S}^{\frac{N}{2s}}}{2^{\frac{N-2s}{2s}}} + o_\varepsilon(1).$$

By using a Brézis-Lieb type Lemma and arguments similar to the ones above, we get

$$I(\tilde{w}_{1,\varepsilon}, \tilde{w}_{2,\varepsilon}) = \int_{C_{B_\varepsilon}} \frac{k_s}{2} y^{1-2s}(|\nabla \tilde{v}_{1,\varepsilon}|^2 + |\nabla \tilde{v}_{2,\varepsilon}|^2)dxdy$$

$$- \frac{2}{p_\varepsilon + q} \int_{B_\varepsilon} Q_\varepsilon \tilde{v}_{1,\varepsilon}(x,0)^p \tilde{v}_{2,\varepsilon}(x,0)^q dx + \int_{\mathbb{R}^{N+1}_+} \frac{k_s}{2} y^{1-2s}(a^2+b^2)|\nabla \mathscr{W}_1|^2 dxdy$$

$$- \frac{2}{p_\varepsilon + q} \int_{\mathbb{R}^N} |a\mathscr{W}_1(x,0)|^{p_\varepsilon} |b\mathscr{W}_1(x,0)|^q dx + o_\varepsilon(1) \geq \frac{2s}{N} \frac{\mathscr{S}^{\frac{N}{2s}}}{2^{\frac{N-2s}{2s}}} + o_\varepsilon(1).$$

On the other hand,

$$I(\tilde{w}_{1,\varepsilon}, \tilde{w}_{2,\varepsilon}) := \frac{k_s}{2} \int_{C_{B_\varepsilon}} y^{1-2s}(|\nabla \tilde{w}_{1,\varepsilon}|^2 + |\nabla \tilde{w}_{2,\varepsilon}|^2)dxdy$$

$$- \frac{2}{p+q} \int_{B_\varepsilon} Q_\varepsilon \tilde{w}^p_{1,\varepsilon}(x,0) \tilde{w}^q_{2,\varepsilon}(x,0)dx = \frac{s}{N} \frac{\mathscr{S}^{\frac{N}{2s}}}{2^{\frac{N-2s}{2s}}} + o_\varepsilon(1),$$

a contradiction. Hence $\rho = 0$, proving also Theorem 1.4(ii).

Acknowledgements O.H. Miyagaki was partially supported by CNPq/Brazil and CAPES/Brazil (Proc 2531/14-3). J. Yang was supported by NNSF of China, No:11271170; GAN PO 555 program of Jiangxi and NNSF of Jiangxi, No:2012BAB201008. This paper was completed while the second author was visiting the Department of Mathematics of the Rutgers University, whose hospitality he gratefully acknowledges. He wishes to thank Professor H. Brézis for invitation.

References

1. Alberti, G., Bouchitté, G., Seppecher, P.: Phase transition with the line-tension effect. Arch. Ration. Mech. Anal. **144**, 1–46 (1998)
2. Alves, C.O., de Morais Filho, D.C., Souto, M.A.S.: On systems of elliptic equations involving subcritical or critical Sobolev exponents. Nonlinear Anal. **42**, 771–787 (2000)
3. Bates, P.W.: On some nonlocal evolution equations arising in materials science. In: Nonlinear Dynamics and Evolution Equations, Fields Inst. Commun., vol. 48, pp. 13–52. American Mathematical Society, Providence (2006)
4. Badiale, M., Serra, E.: Multiplicity results for the supercritical Hénon equation. Adv. Nonlinear Stud. **4**, 453–467 (2004)
5. Barrios, B., Colorado, E., de Pablo, A., Sánchez, U.: On some critical problems for the fractional Laplacian operator. J. Differ. Equ. **252**, 613–6162 (2012)
6. Bianchi, G.: Non-existence of positive solutions to semilinear elliptic equations on $\mathbb{R}^n$ or $\mathbb{R}^n_+$ through the method of moving planes. Commun. Partial Differ. Equ. **22**, 1671–1690 (1997)
7. Brändle, C., Colorado, E., de Pablo, A., Sánchez, U.: A concave-convex elliptic problem involving the fractional Laplacian. Proc. R. Soc. Edinb. Sect. A **143**, 39–71 (2013)
8. Brézis, H., Nirenberg, L.: A minimization problem with critical exponent and non-zero data. In: Symmetric in Nature, Sc. Norm. Sup. Pisa, pp. 129–140 (1989)

9. Cabré, X., Sire, Y.: Nonlinear equations for fractional Laplacians I: regularity, maximum principles and Hamiltonian estimates. Ann. Inst. H. Poincaré **31**, 23–53 (2014)
10. Cabré, X., Tan, J.G.: Positive solutions of nonlinear problems involving the square root of the Laplacian. Adv. Math. **224**, 2052–2093 (2010)
11. Caffarelli, L., Roquejoffre, J.-M., Savin, O.: Non-local minimal surfaces. Commun. Pure Appl. Math. **63**, 1111–1144 (2010)
12. Caffarelli, L., Silvestre, L.: An extension problems related to the fractional Laplacian. Commun. Partial Differ. Equ. **32**, 1245–1260 (2007)
13. Cao, D.M., Peng, S.: The asymptotic behaviour of the ground state solutions for Hénon equation. J. Math. Anal. Appl. **278**, 1–17 (2003)
14. Capella, A., Dávila, J., Dupaigne, L., Sire, Y.: Regularity of radial extremal solutions for some non local semilnear equations. Commun. Partial Differ. Equ. **36**, 1353–1384 (2011)
15. Chen, G., Ni, W.-M., Zhou, J.: Algorithms and visualization for solutions of nonlinear elliptic equations. Int. J. Bifurcation Chaos Appl. Sci. Eng. **10**, 1565–1612 (2000)
16. Chen, W., Li, C., Ou, B.: Classification of solutions of an integral equation. Commun. Pure Appl. Math. **59**, 330–343 (2006)
17. Chen, W., Li, C.: Methods on Nonlinear Elliptic Equations, Aims Series on Differential Equations and Dynamical Systems, vol. 4. American Institute of Mathematical Science, Springfield (2010)
18. Chen, X.L., Yang, J.: Limiting behavior of solutions to an equation with the fractional Laplacian. Differ. Integr. Equ. **27**, 157–179 (2014)
19. Choi, W.: A priori bound for nonlinear elliptic equation and system involving a fractional laplacian, arXiv:1502.07913v1
20. Di Nezza, E., Palatucci, G., Valdinoci, E.: Hitchhiker's guide to the fractional Sobolev spaces. Bull. Sci. Math. **136**, 521–573 (2012)
21. Fall, M.M., Weth, T.: Nonexistence results for a class of fractional elliptic boundary value problems. J. Funct. Anal. **263**, 2205–2227 (2012)
22. Gidas, B., Ni, W.-M., Nirenberg, L.: Symmetry and related properties via the maximum principle. Commun. Math. Phys. **68**, 209–243 (1979)
23. Guo, Y., Liu, J.: Liouville type theorems for positive solutions of elliptic system in $\mathbb{R}^N$. Commun. Partial Differ. Equ. **33**, 263–284 (2008)
24. Hénon, M.: Numerical experiments on the stability of spheriocal stellar systems. Astron. Astrophys. **24**, 229–238 (1973)
25. Jin, T., Li, Y., Xiong, J.: On a fractional Nirenberg problem, part I: blow up analysis and compactness of solutions. J. Eur. Math. Soc. **16**, 1111–1171 (2014)
26. Kurzke, M.: A nonlocal singular perturbation problem with periodic well potential. ESAIM Control Optim. Calc. Var. **12**, 52–63 (2006)
27. Lions, P.L.: The concentration compactness principle in the calculus of variations. The limit case (Part 1 and Part 2). Rev. Mat. Iberoamewricana **1**, 145–201, 45–121 (1985)
28. Majda, A.J., Tabak, E.G.: A two-dimensional model for quasigeostrophic flow: comparison with the two-dimensional Euler flow. Phys. D **98**, 515–522 (1996)
29. Ni, W.M.: A nonlinear Dirichlet problem on the unit ball and its applications. Indiana Univ. Math. J. **31**, 801–807 (1982)
30. Silvestre, L.: Regularity of the obstacle problem for a fractional power of the Laplace operator. Commun. Pure Appl. Math. **60**, 67–112 (2007)
31. Smets, D., Su, J.B., Willem, M.: Non-radial ground states for the Hénon equation. Commun. Contemp. Math. **4**, 467–480 (2002)
32. Tan, J.: Positive solutions for non local elliptic problems. Discret. Cont. Dyn. Syst. **33**, 837–859 (2013)

33. Wang, Y., Yang, J.: Asymptotic behavior of ground state solution for Henon type systems. Electron. J. Differ. Equ. **116**, 1–14 (2010)
34. Wang, Y., Yang, J.: Existence and asymptotic behavior of solutions for Hénon type systems. Adv. Nonlinear Stud. **13**, 533–553 (2013)
35. Zhao, Y., Lei, Y.: Asymptotic behavior of positive solutions of a nonlinear integral system. Nonlinear Anal. **75**, 1989–1999 (2012)

Positive solutions for certain classes of fourth-order ordinary elliptic systems

J.M. do Ó, S. Lorca, J. Sánchez, and P. Ubilla

Mathematics Subject Classification (2010): 35J55, 35B45, 34B18

1 Introduction

In this article we study the existence of positive solutions for the following system involving fourth-order ordinary differential equations

J.M. do Ó (✉)
Departamento de Matemática, Universidade Federal da Paraíba, 58059-900, João Pessoa, Brazil
e-mail: jmbo@pq.cnpq.br

S. Lorca
Instituto de Alta Investigación, Universidad de Tarapacá, Casilla 7-D, Arica, Chile
e-mail: slorca@uta.cl

J. Sánchez
Departamento de Matemáticas, Universidad de La Serena, Avda. Cisternas 1200, La Serena, Chile
e-mail: jsanchez@userena.cl

P. Ubilla
Departamento de Matemática y C.C., Universidad de Santiago de Chile, Casilla 307, Correo 2, Santiago, Chile
e-mail: pedro.ubilla@usach.cl

A.N. Carvalho et al. (eds.), *Contributions to Nonlinear Elliptic Equations and Systems*, Progress in Nonlinear Differential Equations and Their Applications 86,
DOI 10.1007/978-3-319-19902-3_11

$$\begin{cases} u^{(4)}(t) = f(t, u(t), v(t)) & \text{for } 0 < t < 1, \\ v^{(4)}(t) = g(t, u(t), v(t)) & \text{for } 0 < t < 1, \\ u(0) = v(0) = u(1) = v(1) = 0, \\ u''(0) = v''(0) = u''(1) = v''(1) = 0 \end{cases} \tag{1}$$

where the nonlinearities f and g are continuous and may, in some sense, even be singular in the first variable at $t = 0$ and/or at $t = 1$. By applying well-known fixed point theorems in cones, we will establish the existence of positive solutions of (1) under the following two basic hypotheses on the nonlinearities $f,\ g : (0, 1) \times [0, +\infty)^2 \to [0, +\infty)$,

(h_1) (behaviour at zero)

$$\lim_{u+v\to 0+} \frac{f(s, u, v)}{u + v} = \lim_{u+v\to 0+} \frac{g(s, u, v)}{u + v} = 0 \quad \text{uniformly for } s \in (0, 1).$$

(h_2) (super-linearity) There exist constants $0 < \alpha_1' < \beta_1' < 1$ and $0 < \alpha_2' < \beta_2' < 1$ such that

$$\lim_{u+v\to +\infty} \frac{f(s, u, v)}{u + v} = +\infty \quad \text{uniformly for } s \in (\alpha_1', \beta_1')$$

or

$$\lim_{u+v\to +\infty} \frac{g(s, u, v)}{u + v} = +\infty \quad \text{uniformly for } s \in (\alpha_2', \beta_2').$$

Note that assumption (h_1) is the classical super-linearity condition at zero and the hypothesis (h_2) is a type of local super-linearity at $+\infty$.

Now we state the main result for (1).

Theorem 1.1. *Suppose f and g satisfy $(h_1) - (h_2)$. Then system (1) possesses at least one positive solution.*

The main purpose of this paper is two-fold. Firstly, we will perform an appropriated change of variables such that system (1) becomes a functional second order elliptic system. A second purpose is that with the help of this approach in combination with fixed point techniques we prove the existence of solution of system (1). We stress that this approach can be also applied to study the existence radial solutions for several classes of nonlocal elliptic systems defined on bounded annular domains or exterior domains.

For a study of fourth-order elliptic problems, see, for example, [1, 2, 7–11] and the references therein. We mention that scalar fourth-order boundary value problems describe the equilibrium state of an elastic beam which is supported simply at both ends.

We now give a brief description of the paper. In Section 2 we introduce the associated functional differential equations. Section 3 deals with the superlinear problems. Section 4 contains a well-known theorem used in Sections 5 to prove Theorem 1.1, which is our main result. In Section 6 we show the application to the systems involving fourth-order ordinary differential equations.

2 The associate functional differential equations

Now we show how a system such as (1) can be transformed into a class of systems of second-order ordinary functional differential equations. In fact, suppose (u, v) is a positive solution of system (1), and let $w = -u''$ and $z = -v''$. Using the Green's function

$$G(t,s) = \begin{cases} s(1-t) & \text{for } 0 \le s \le t \le 1\,, \\ t(1-s) & \text{for } 0 \le t \le s \le 1\,. \end{cases} \tag{2}$$

we can write

$$u(t) = -\int_0^1 G(t,s)\,u''(s)ds \quad \text{and} \quad v(t) = -\int_0^1 G(t,s)\,v''(s)ds,$$

and so, we may define the functional operators L_1 and L_2 given by

$$L_1(w,z)(t) = f\left(t, \int_0^1 G(t,s)w(s)ds, \int_0^1 G(t,s)z(s)ds\right) \quad \text{and}$$

$$L_2(w,z)(t) = g\left(t, \int_0^1 G(t,s)w(s)ds, \int_0^1 G(t,s)z(s)ds\right).$$

Thus (w, z) satisfies the following functional system:

$$\begin{cases} -w''(t) = L_1(w,z)(t) \text{ in } (0,1), \\ -z''(t) = L_2(w,z)(t) \text{ in } (0,1), \\ w(0) = z(0) = w(1) = z(1) = 0 \end{cases} \tag{S}$$

where the functionals $L_1, L_2 : (0,1) \times \mathscr{C}([0,1]) \times \mathscr{C}([0,1]) \longrightarrow [0,+\infty)$ are continuous and may, in some sense, even be singular at $t = 0$ and/or at $t = 1$. Thus, in the sections 3–6 we will concentrate studying systems like (S)

3 Locally superlinear nonlinearities

Consider $E := \mathscr{C}([0,1])$, the Banach space of continuous functions equipped with the usual norm

$$\|w\|_\infty = \max_{t\in[0,1]} |w(t)|\,.$$

Next we use the following notation

$$
\begin{aligned}
X &:= E \times E. \\
K &:= \{w \in E \ : \ w \text{ is concave} \quad \text{and} \quad w(0) = w(1) = 0\}. \\
C &:= K \times K \\
\mathscr{H} &:= \left\{ h \in \mathscr{C}((0,1), [0,+\infty)) : \int_0^1 s\,(1-s) h(s)\, ds < +\infty \right\}
\end{aligned}
$$

Note that K and C are cones, and X equipped with the norm $||(w,z)||_\infty = \|w\|_\infty + \|z\|_\infty$ is a Banach space. In what follows we assume that $(w,z) \in C$.

Suppose that the functionals $L_1,\, L_2 : (0,1) \times E \times E \longrightarrow [0,+\infty)$ are continuous and assume also that the following three hypotheses hold:

(H_1) For all $M > 0$, there exist $h_{1,M},\, h_{2,M} \in \mathscr{H}$, such that

$$L_i(w,z)(s) \leq h_{i,M}(s), \quad \text{for all } (w,z) \in C \text{ with } \|(w,z)\|_\infty \leq M \text{ and } s \in (0,1),$$

where $i = 1, 2$.

(H_2) For $i = 1, 2$, we have

$$\frac{L_i(w,z)(s)}{\|(w,z)\|_\infty} \to 0 \quad \text{as} \quad \begin{cases} \|(w,z)\|_\infty \to 0 \\ (w,z) \in C \end{cases} \quad \text{uniformly for } s \in (0,1).$$

(H_3) For some $i \in \{1,2\}$, there exist constants $0 < \alpha_i < \beta_i < 1$ such that

$$\frac{L_i(w,z)(s)}{\|(w,z)\|_\infty} \to +\infty \quad \text{as} \quad \begin{cases} \|(w,z)\|_\infty \to +\infty \\ (w,z) \in C \end{cases} \quad \text{uniformly for } s \in (\alpha_i, \beta_i).$$

We now state a result for the existence of a solution of system (S) under superlinear hypothesis at zero and a local superlinear hypothesis at $+\infty$.

Theorem 3.1. *Suppose L_1 and L_2 satisfy hypotheses (H_1) through (H_3). Then there exists a positive solution $(w,z) \in C$ of system (S).*

4 A fixed-point technique

Our approach is based on the following fixed-point theorem of cone expansion/compression type. We omit its proof. Readers who are interested in fixed-point theory in cones may consult, e.g., [3–6].

Theorem A. *Let X be a Banach space endowed with norm $\|\cdot\|$, and let $C \subset X$ be a cone in X. For $R > 0$, define $C_R = C \cap B[0,R]$, where $B[0,R]$ denotes the closed* *ball of radius R centered at the origin of X. Let r and R be real numbers satisfying* $0 < r < R$. *Assume that* $T : C \to C$ *is a completely continuous operator such that*

$$\|Tu\| < \|u\|, \text{ for all } u \in \partial C_r, \text{ and } \|Tu\| > \|u\|, \text{ for all } u \in \partial C_R \text{ or}$$
$$\|Tu\| > \|u\|, \text{ for all } u \in \partial C_r, \text{ and } \|Tu\| < \|u\|, \text{ for all } u \in \partial C_R$$

where $\partial C_R = \{u \in C : \|u\| = R\}$. *Then T has a fixed point* $u \in C$, *with* $r < \|u\| < R$.

Define the operator $T : X \to X$ by

$$T(w,z)(t) = (A(w,z)(t), B(w,z)(t)), \quad \text{for } 0 < t < 1$$

where

$$A(w,z)(t) = \int_0^1 G(t,s)L_1(w,z)(s)\,ds$$

and

$$B(w,z)(t) = \int_0^1 G(t,s)L_2(w,z)(s)\,ds\,.$$

Here $G(t,s)$ denotes the associated Green's function defined in (2). Observe that this function satisfies

$$G(t,s) \le s(1-s), \quad \text{for all } t,\, s \in [0,1]\,. \tag{3}$$

It is not difficult to see that system (S) is equivalent to the fixed-point equation

$$T(w,z) = (w,z)\,.$$

Thus the fixed points of the operator T correspond to the positive solutions of system (S). It follows from hypothesis (H_1) and inequality (3) that T is well defined.

Lemma 4.1. *T is completely continuous and* $T(C) \subset C$.

Proof. First, we show that A is continuous. In fact, let $(\phi_n, \psi_n) \subset X$ be such that

$$\|(\phi_n, \psi_n) - (\phi, \psi)\|_\infty \to 0 \quad \text{as} \quad n \to \infty\,. \tag{4}$$

Define the function

$$\zeta_n(\tau) = | L_1(\phi_n, \psi_n)(\tau) - L_1(\phi, \psi)(\tau) |\,.$$

For $\tau \in (0,1)$ and n sufficiently large, we have $\zeta_n(\tau) \to 0$ and $0 \le \zeta_n(\tau) \le 2h_{i,M}(\tau)$, where $M = \max\{\|\phi\|_\infty + \|\psi\|_\infty, \max_n\{\|\phi_n\|_\infty + \|\psi_n\|_\infty\}\}$. By (4), M is independent of n. Therefore,

$$|A(\phi_n,\psi_n)(t) - A(\phi,\psi)(t)| \le \int_0^1 \tau(1-\tau)\zeta_n(\tau)\,d\tau. \tag{5}$$

According to the Lebesgue dominated convergence theorem, we have

$$\|A(\phi_n,\psi_n) - A(\phi,\psi)\|_\infty \to 0 \quad \text{as} \quad n \to \infty$$

which implies that A is continuous. The proof for the operator B is analogous.

We next show that $T(C) \subset C$. Observe that

$$A(\phi,\psi)(t) = (1-t)\int_0^t \tau\, L_1(\phi,\psi)(\tau)\,d\tau + t\int_t^1 (1-\tau)L_1(\phi,\psi)(\tau)\,d\tau. \tag{6}$$

For $t \in (0,1)$, we have

$$(1-t)\int_0^t \tau\, L_1(\phi,\psi)(\tau)\,d\tau = \int_0^t (1-t)\tau L_1(\phi,\psi)(\tau)\,d\tau \le \int_0^1 \tau(1-\tau)h_{1,M}(\tau)\,d\tau.$$

Here M depends on both $\|\phi\|_\infty$ and $\|\psi\|_\infty$.

Let $\{t_n\}$ be a sequence such that $t_n \to 0^+$. For all $n \ge 1$, we have

$$\int_0^{t_n} (1-t_n)\tau\, L_1(\phi,\psi)(\tau)\,d\tau = \int_0^1 \chi_{[0,t_n]}(1-t_n)\tau\, L_1(\phi,\psi)(\tau)\,d\tau.$$

Define the function $l_n(\tau) = \chi_{[0,t_n]}(1-t_n)\tau\, L_1(\phi,\psi)(\tau)$. It is not difficult to see that $l_n(\tau) \to 0$ a. e. for $\tau \in (0,1)$ and that $0 \le l_n(\tau) \le \tau(1-\tau)h_{i,M}(\tau)$. Again, we have that M depends on both $\|\phi\|_\infty$ and $\|\psi\|_\infty$. According to the Lebesgue dominated convergence theorem, we have

$$\lim_{n\to+\infty} \int_0^{t_n} (1-t_n)\,\tau\, L_1(\phi,\psi)(\tau)\,d\tau = 0.$$

Similarly,

$$\lim_{n\to+\infty} \int_{t_n}^1 t_n\,(1-\tau)\, L_1(\phi,\psi)(\tau)\,d\tau = 0.$$

Taking the right-hand limit in (6) we obtain $A(\phi,\psi)(0) = 0$. Similarly, $A(\phi,\psi)(1) = 0$ when $t_n \to 1^-$.

A direct calculation shows that, for $(\phi, \psi) \in C$, the function $A(\phi, \psi)(s)$ is twice differentiable in (0,1), with the second derivative negative. The same holds for the function $B(\phi, \psi)(s)$. Hence, $T(C) \subset C$.

It remains to show that T is a completely continuous operator. Let $(\phi_n, \psi_n) \subset X$ be such that $\|(\phi_n, \psi_n)\| \leq M$. It follows that

$$\max_{t\in[0,1]} \{A(\phi_n, \psi_n)(t), B(\phi_n, \psi_n)(t)\} \leq \max_{i=1,2} \left\{ \int_0^1 \tau(1-\tau)h_{i,M}(\tau)\, d\tau \right\} < \infty .$$

Then $(T(\phi_n, \psi_n))$ is equibounded in X.

Let $(t_k) \subset [0, 1]$ be a sequence such that $t_k \to t_0^+$ (the case $t_k \to t_0^-$ is similar). By definition,

$$| A(\phi_n, \psi_n)(t_k) - A(\phi_n, \psi_n)(t_0) | \leq \int_{t_0}^{t_k} G_k(\tau)h_{i,M}(\tau)\, d\tau$$

where $G_k(\tau) =| G(t_k, \tau) - G(t_0, \tau)|$. According to the Lebesgue dominated convergence theorem, $A(\phi_n, \psi_n)(t_k)$ converges to $A(\phi_n, \psi_n)(t_0)$ uniformly in n as k tends to infinity. Therefore, $(A(\phi_n, \psi_n))$ is equicontinuous in [0, 1]. Analogously, $(B(\phi_n, \psi_n))$ is equicontinuous in [0, 1].

Finally, using Ascoli-Arzelà theorem, we see that the operator $T \quad X \to X$ is completely continuous. □

We next state an elementary property of concave functions that will be used in Section 6.

Lemma 4.2. *Suppose* $w \in K$. *Then, for all* $0 < \alpha < \beta < 1$, *we have*

$$\min_{t\in[t_0,t_1]} w(t) \geq \alpha(1-\beta)\, ||w||_\infty .$$

For any $0 < \alpha < \beta < 1$, we define the constant

$$\boxed{c_{\alpha,\beta} := \alpha(1-\beta)}$$

5 Proof of Theorem 3.1

Theorem 3.1 is an immediate consequence of Theorem A and the following two lemmas.

Lemma 5.1. *Suppose hypotheses* (H_1) *and* (H_3) *hold. Then there exists a* $\Theta > 0$ *so that, for all* $(w, z) \in \partial C_\Theta = \{(w, z) \in C : \|(w, z)\| = \Theta\}$, *we have*

$$||T(w, z)||_\infty > ||(w, z)||_\infty .$$

Proof. Without loss of generality, in (H_3) we may suppose that there exist constants $0 < \alpha_1 < \beta_1 < 1$ such that

$$\frac{L_1(w,z)(s)}{\|(w,z)\|_\infty} \to +\infty \quad \text{as} \quad \|(w,z)\|_\infty \to +\infty \quad \text{with} \quad (w,v) \in C,$$

uniformly for $s \in (\alpha_1, \beta_1)$. Thus for $\|(w,z)\|_\infty = \Theta$, with Θ sufficiently large, we may choose $\Upsilon > 0$ so that $\Upsilon \int_{\alpha_1}^{\beta_1} G(\frac{1}{2}, s)ds > 1$ and $L_1(w,z)(s) > \Upsilon\Theta$. Then

$$\begin{aligned} A(w,z)(1/2) &\geq \int_{\alpha_1}^{\beta_1} \frac{G(\frac{1}{2}, s)L_1(w,z)(s)}{\|(w,z)\|_\infty} \|(w,z)\|_\infty \, ds \\ &\geq \Upsilon \|(w,z)\|_\infty \int_{\alpha_1}^{\beta_1} G\left(\frac{1}{2}, s\right) ds > \|(w,z)\|_\infty . \end{aligned}$$

Therefore,

$$\|T(w,z)\|_\infty \geq A(w,z)(1/2) + B(w,z)(1/2) \geq A(w,z)(1/2) > \|(w,z)\|_\infty ,$$

for $\|(w,z)\|_\infty = \Theta$ and the proof is complete. □

Lemma 5.2. *Suppose that hypotheses* (H_1) *and* (H_2) *hold. Then there exists* $0 < \theta < \Theta$ *so that, for all* $(w,z) \in \partial C_\theta$*, we have*

$$||T(w,z)||_\infty < ||(w,z)||_\infty .$$

Proof. By hypothesis (H_2), for $(w,z) \in C$ with $\|(w,z)\| = \theta$ sufficiently small, we may choose a $\gamma > 0$ such that $2\gamma \int_0^1 s(1-s)\, ds < 1$ and $L_1(w,z)(s) < \gamma\,\theta$. Then

$$\begin{aligned} A(w,z)(t) &= \int_0^1 G(t,s)L_1(w,z)(s)\, ds \\ &\leq \int_0^1 s(1-s)\frac{L_1(w,z)(s)}{\|(w,z)\|_\infty} \|(w,z)\|_\infty \, ds \leq \|(w,z)\|_\infty \gamma \int_0^1 s(1-s)\, ds \end{aligned}$$

which implies that $A(w,z)(t) < \frac{\|(w,z)\|_\infty}{2}$. Analogously, $B(w,z)(t) < \frac{\|(w,z)\|_\infty}{2}$. Therefore,

$$\|T(w,z)\|_\infty = \|A(w,z)\|_\infty + \|B(w,z)\|_\infty < \|(w,z)\|_\infty \quad \text{for} \quad \|(w,z)\|_\infty = \theta,$$

and the lemma follows. □

6 Proof of Theorem 1.1

This will be an application of Theorem 3.1. For this purpose we will verify that assumptions $(h_1)-(h_2)$ imply that $(H_1)-(H_3)$ of Theorem 3.1 hold. Indeed, let us verify (H_3). We have

$$\frac{L_1(w,z)(t)}{\|(w,z)\|_\infty} =$$

$$\frac{f\left(t, \int_0^1 G(t,s)w(s)ds, \int_0^1 G(t,s)z(s)ds\right)}{\int_0^1 G(t,s)w(s)ds + \int_0^1 G(t,s)z(s)ds} \frac{\int_0^1 G(t,s)w(s)ds + \int_0^1 G(t,s)z(s)ds}{\|(w,z)\|_\infty} \geq$$

$$\frac{f\left(t, \int_0^1 G(t,s)w(s)ds, \int_0^1 G(t,s)z(s)ds\right)}{\int_0^1 G(t,s)w(s)ds + \int_0^1 G(t,s)z(s)ds} c_{\alpha_1',\beta_1'} \int_{\alpha_1'}^{\beta_1'} G(t,s)ds$$

by Lemma 4.2. It follows from hypothesis (h_2) that L_1 satisfies hypothesis (H_3). Analogously, L_1 satisfies hypothesis (H_2). Finally, it is easy to check that (H_1) is satisfied. Hence, we have a solution of system (1).

Acknowledgements J. M. do Ó: Research partially supported by the National Institute of Science and Technology of Mathematics INCT-Mat, CAPES-PROCAD and CNPq. S. Lorca: Research partially supported by FONDECYT grant 1120706. P. Ubilla: Research partially supported by FONDECYT grant 1120524.

References

1. Bai, Z., Wang, H.: On positive solutions of some nonlinear fourth-order beam equations. J. Math. Anal. Appl. **270**, 357–368 (2002)
2. Chu, J., O'Regan, D.: Positive solutions for regular and singular fourth–order boundary value problems. Commun. Appl. Anal. **10**, 185–199 (2006)
3. de Figueiredo, D.G.: Positive Solutions of Semilinear Elliptic Equations. Lecture Notes in Math, vol. 957, 34–87. Springer, Berlin (1982)
4. Deimling, K.: Nonlinear Functional Analysis. Springer, New York (1985)
5. Guo, D., Lakshmikantham, V.: Nonlinear Problems in Abstract Cones. Academinc Press, Orlando (1988)
6. Krasnosel'skii, M.A.: Positive Solutions of Operators Equations. Noordhoff, Groningen (1964)
7. Li, Y.: Positive solutions of fourth-order boundary value problems with two parameters. J. Math. Anal. Appl. **281**, 477–484 (2003)
8. Liu, Y.: Multiple Positive solutions of nonlinear singular boundary value problem for fourth-order equations. Appl. Math. Lett. **17**, 747–757 (2004)
9. Ma, R., Tisdell, C.: Positive solutions of singular sublinear fourth-order boundary value problems. Appl. Anal. **84**, 1199–1220 (2005)
10. Ma, T.F.: Positive solutions for a nonlocal fourth order equation of Kirchhoff type. Discret. Cont. Dyn. Syst.; Dynamical Systems and Differential Equations. In: Proceedings of the 6th AIMS International Conference, suppl., 694–703 (2007)
11. Minhós, F., Gyulov, T., Santos, A.I.: Existence and location result for a fourth order boundary value problem. Discrete Contin. Dyn. Syst. Suppl., 662–671 (2005)

Remarks on the behavior of the best Sobolev constants

Grey Ercole

1 Introduction

Let Ω be a bounded domain of $\mathbb{R}^N$, $N \geq 2$. It is well known that the Sobolev immersion $W_0^{1,p}(\Omega) \hookrightarrow L^q(\Omega)$ is compact if $1 \leq q < p^\star$, where $p^\star = Np/(N-p)$ if $1 < p < N$ and $p^\star = \infty$ if $p \geq N$. As a consequence of this fact, the infimum

$$\lambda_q(\Omega) := \inf\left\{\int_\Omega |\nabla u|^p \, dx : u \in W_0^{1,p}(\Omega) \text{ and } \int_\Omega |u|^q \, dx = 1\right\} \tag{1}$$

is achieved by at least one extremal function $u_q \in W_0^{1,p}(\Omega)$, which can be chosen nonnegative in Ω.

In the critical case, $q = p^\star$, the immersion $W_0^{1,p}(\Omega) \hookrightarrow L^{p^\star}(\Omega)$ remains continuous but no longer compact. Moreover, if $1 < p < N$, then $\lambda_{p^\star}(\Omega)$ coincides with the Sobolev constant:

$$\lambda_{p^\star}(\Omega) = S_p := \inf\left\{\int_{\mathbb{R}^N} |\nabla u|^p \, dx : u \in W_0^{1,p}(\mathbb{R}^N) \text{ and } \int_{\mathbb{R}^N} |u|^{p^\star} \, dx = 1\right\}. \tag{2}$$

It is also well known (see [6, 33]) that the Sobolev constant is explicitly given by

$$S_p := N\pi^{\frac{p}{2}} \left(\frac{N-p}{p-1}\right)^{p-1} \left(\frac{\Gamma(N/p)\Gamma(1+N-N/p)}{\Gamma(1+N/2)\Gamma(N)}\right)^{\frac{p}{N}}$$

G. Ercole (✉)
Departamento de Matemática, Universidade Federal de Minas Gerais, Belo Horizonte, MG 30.123-970, Brazil
e-mail: grey@mat.ufmg.br

A.N. Carvalho et al. (eds.), *Contributions to Nonlinear Elliptic Equations and Systems*, Progress in Nonlinear Differential Equations and Their Applications 86,
DOI 10.1007/978-3-319-19902-3_12

($\Gamma(t) = \int_0^\infty s^{t-1}e^{-s}ds$ is the Gamma Function) and it is achieved only at the functions of the form

$$w(x) = a\left(1 + b\,|x - x_0|^{\frac{p}{p-1}}\right)^{-\frac{N-p}{p}}, \quad x \in \mathbb{R}^N \tag{3}$$

for any $a \neq 0$, $b > 0$ and $x_0 \in \mathbb{R}^N$. This means that $\lambda_{p^\star}(\Omega)$ is not reached if Ω is a proper subset of $\mathbb{R}^N$.

The Euler–Lagrange equation associated with the minimization problem (1) is

$$-\Delta_p u = \lambda_q(\Omega)\,|u|^{q-2}\,u, \tag{4}$$

where $\Delta_p u := \operatorname{div}(|\nabla u|^{p-2}\,\nabla u)$. Thus, any extremal function of (1) must be a weak solution of this equation in $W_0^{1,p}(\Omega)$. Consequently, standard results imply that an extremal function of (1) does not change sign in Ω and also belongs to $C^{1,\alpha}(\overline{\Omega})$, for some $0 < \alpha < 1$. From now on, u_q denotes any positive extremal function. Thus, $u_q \in W_0^{1,p}(\Omega) \cap C^{1,\alpha}(\overline{\Omega})$ and satisfies

$$\|u_q\|_q = 1,\ \lambda_q(\Omega) = \|\nabla u_q\|_p^p \ \text{ and } \begin{cases} -\Delta_p u_q = \lambda_q(\Omega)u_q^{q-1} & \text{in } \Omega \\ u_q > 0 & \text{in } \Omega \\ u_q = 0 & \text{on } \partial\Omega, \end{cases} \tag{5}$$

where $\|\cdot\|_s$ stands for the standard norm of $L^s(\Omega)$, $s \geq 1$ (this notation is used through the text).

The particular values $\lambda_1(\Omega)$ and $\lambda_p(\Omega)$ have received much attention in the literature: $\lambda_1(\Omega) = \|\phi_p\|_1^{1-p}$, where ϕ_p is the p-torsion function, that is, the solution of the p-torsional creep problem (see [23])

$$\begin{cases} -\Delta_p u = 1 & \text{in } \Omega \\ u = 0 & \text{on } \partial\Omega, \end{cases}$$

whereas $\lambda_p(\Omega)$ is the first eigenvalue of the Dirichlet p-Laplacian, that is, the least λ such that the following Dirichlet problem

$$\begin{cases} -\Delta_p u = \lambda\,|u|^{p-2}\,u & \text{in } \Omega \\ u = 0 & \text{on } \partial\Omega \end{cases} \tag{6}$$

has a nontrivial weak solution.

It is interesting to notice that, as p is near to 1, they are closely related with the Cheeger constant $h(\Omega)$ by

$$\lim_{p\to 1^+} \lambda_p(\Omega) = h(\Omega) = \lim_{p\to 1^+} \lambda_1(\Omega). \tag{7}$$

We recall that the Cheeger constant of Ω is $h(\Omega) := \min_{E\subset\overline{\Omega}}(|\partial E| / |E|)$, where $|\partial E|$ and $|E|$ denote, respectively, the $(N-1)$-dimensional Lebesgue perimeter of ∂E in $\mathbb{R}^N$ and the N-dimensional Lebesgue volume of E, the quotients being evaluated among all smooth subsets $E \subset \overline{\Omega}$. The first equality in (7) was proved in [25] and the second in [9].

The first eigenvalue $\lambda_p(\Omega)$ is well studied in the literature and its properties are well known. Some of them are inherited by $\lambda_q(\Omega)$ depending on whether $1 < q < p$ or $p < q < p^\star$. For example, positive extremals functions of $\lambda_q(\Omega)$ are unique if $1 < q < p$ (see [21]). This property is not generically valid if $p < q < p^\star$: it is valid when Ω is a ball (see [2, 16, 24]), but not when Ω is an annulus (see [28]).

We find in the literature studies on the behavior of $\lambda_q(\Omega)$ when $q = p$ and p varies (see [20, 22, 25]) and also when p is fixed and q goes to $p^\star$ (see [1, 7, 29–31]) or to p (see [3, 12, 17]).

In the case $p = N > 1$, so that $p^\star = \infty$, the following asymptotic behavior, in which ω_N denotes the volume of the unit ball in $\mathbb{R}^N$, was proved in [31]

$$\lim_{q\to\infty} q^{N-1}\lambda_q(\Omega) = \frac{N^{2N-1}\omega_N}{(N-1)^{N-1}}e^{N-1}, \tag{8}$$

generalizing the result previously obtained by the same authors in [30], for $p = N = 2$. We note that (8) does not depend on Ω and also that it implies that $\lim_{q\to\infty}\lambda_q(\Omega) = 0$.

As regards to the case $p > N > 1$, we refer to [14] where the following result is proved for a ball B_R of radius $R > 0$

$$\lim_{q\to\infty} \lambda_q(B_R) = \frac{N\omega_N}{R^{p-N}}\left(\frac{p-N}{p-1}\right)^{p-1}. \tag{9}$$

We could not find in the literature any other result on the asymptotic behavior of $\lambda_q(\Omega)$ as $q \to \infty$ (and $p > N > 1$).

However, papers dealing with the behavior of the function $q \in [1, p^\star) \mapsto \lambda_q(\Omega)$ are quite rare. Let us refer to this function as the Best–Sobolev–Constant function, or simply, *BSC* function. As far as we are aware, the first paper dealing with the *BSC* function is [19], where the author proves that this function is continuous in the open interval $(1, p)$ and upper semi-continuous in the open interval $(p, p^\star)$ and still reports that it is decreasing provided that $|\Omega| \le 1$. We remark that this monotonicity is a particular case of the following fact: the function $q \mapsto |\Omega|^{\frac{p}{q}}\lambda_q(\Omega)$ is decreasing in the interval $[1, p^\star]$ if $1 < p < N$, and in the interval $[1, \infty)$ if $p \ge N$. This can be proved as a simple application of the Hölder inequality (see [12, Lemma 4.2]) and immediately implies that this function is of bounded variation in $[1, p^\star]$, if $1 < p < N$, and in closed intervals contained in $[1, \infty)$, if $p \ge N$. Of course, the same holds for the *BSC* function and

$$0 \le \lim_{q\to p^\star} \lambda_q(\Omega) = \lim_{q\to p^\star} |\Omega|^{-\frac{p}{q}} \inf_{1\le q<p^\star} |\Omega|^{\frac{p}{q}}\lambda_q(\Omega) < \infty.$$

We have started a study on the *BSC* function in [11]. In that paper we improved the results of [19] in the case $1 < p < N$, by arguing that the *BSC* function is of bounded variation in $[1, p^\star]$ and then proving that it is Lipschitz continuous in closed intervals contained in $[1, p^\star)$ and also that

$$\lim_{q\to p^\star} \lambda_q(\Omega) = \lambda_{p^\star}(\Omega)\,(=S_p). \tag{10}$$

Thus, by gathering these properties we concluded in [11] that the *BSC* function is absolutely continuous in the whole interval $[1, p^\star]$. We emphasize that our proofs given in that paper generalize easily for $p \geq N$, since in this case $\|u_q\|_\infty$ is bounded from above by a uniform constant with respect to q (see Lemma 3.1 in the next section). This fact does not occur if $1 < p < N$ and q approaches $p^\star$, what led us to derive suitable estimates to $\|u_q\|_\infty$ in [11] in order to extend the absolute continuity to the whole interval $[1, p^\star]$.

In [4] the authors extended the *BSC* function to the interval $(0, 1)$ and also proved Lipschitz continuity in compacts of $(0, p^\star]$ if $1 < p < N$ and of $(0, p^\star)$ if $p \geq N$.

More recently, in [13], by assuming $1 < p < N$, we have proved that the *BSC* function is continuously differentiable in the interval $[1, p]$ and, moreover, that it is α-Hölder continuous in $[1, p^\star]$, for any $\alpha \in (0, 1)$, provided that

$$\limsup_{q\to p^\star} (p^\star - q)\, \|u_q\|_\infty^\gamma < \infty, \quad \text{for some } \gamma > 0. \tag{11}$$

In the case where Ω is a ball the asymptotic behavior (11) was proved in [26], with $\gamma = \frac{p}{p-1}$, extending for $1 < p \neq 2$ a result already known for $p = 2 < N$ (see [5, 8] for balls and [18, 32] for a general bounded domain). The validity of an asymptotic behavior as (11) for a general bounded domain and $1 < p \neq 2$ appears to be still an open problem.

Thus, if $p = 2$ and Ω is a general bounded domain or if $1 < p < N$ and Ω is ball, our results in [13] guarantee that the *BSC* function belongs to $C^1([1, p]) \cap C^\alpha([1, p^\star])$ for any $\alpha \in (0, 1)$. Moreover, as we will see, our results also imply that the *BSC* function is continuously differentiable whenever uniqueness of positive extremal function holds. In particular, when Ω is ball and $p > 1$ the *BSC* function belongs to $C^1([1, p^\star))$.

In this paper we present and improve some of our results on the *BSC* function obtained in [13], by considering $p, N > 1$.

In Section 2 we make some definitions and fix the notation to be used in the sequel.

Then we prove, in Section 3, our main result: the *BSC* function satisfies an ordinary differential equation of first order in the Carathéodory sense. As a consequence we show that the derivative λ_q' of the *BSC* function at point $q \in [1, p^\star)$ exists if, and only if, a suitable functional $I_q : W_0^{1,p}(\Omega) \mapsto \mathbb{R}$, defined in Section 2, is constant on the set E_q of the positive extremal functions associated with q. We also show in Section 3 that the *BSC* function is semi-differentiable and that its one-sided derivatives λ'_{q-} and λ'_{q+} are given in terms of the extremal values of I_q on E_q, so that $\lambda'_{q+} \leq \lambda'_{q-}$. This result is new.

In Section 4 we use the results of the Section 3 to show that the *BSC* function is globally Lipschitz continuous, if $p \geq N > 1$ (which is also a new observation) and α-Hölder continuous, for any $\alpha \in (0,1)$, if $1 < p < N$. Some proofs given in [13] are slightly simplified in this paper.

2 Preliminaries

We recall that the *BSC* function is defined by

$$q \in [1, p^\star) \mapsto \lambda_q(\Omega) = \min \left\{ \|\nabla u\|_p^p : u \in W_0^{1,p}(\Omega) \text{ and } \|u\|_q = 1 \right\} \qquad \text{(BSC)}$$

and that it is differentiable almost everywhere. For the sake of simplicity we write λ_q instead of $\lambda_q(\Omega)$ and denote by λ'_q the derivative of the *BSC* function at a point $q \in [1, p^\star)$ where this function is differentiable.

We also define, for each $q \in [1, p^\star)$, the set

$$E_q := \left\{ u \in W_0^{1,p}(\Omega) : u > 0 \text{ in } \Omega, \; \|u\|_q = 1 \text{ and } \|\nabla u\|_p^p = \lambda_q(\Omega) \right\}$$

and the functional $I_q : W_0^{1,p}(\Omega) \to \mathbb{R}$ by

$$I_q(u) := \int_\Omega |u|^q \log |u| \, dx. \qquad (13)$$

We remark that E_q is precisely the set of positive extremal functions of (1). Thus, E_q is unitary if $1 \leq q \leq p$, but in the case $p < q < p^\star$ this set might have more than one element. As we have mentioned in the Introduction, for some particular bounded domains, as balls, the set E_q remains unitary when $p < q < p^\star$.

3 Differentiability

The following lemma follows from known estimates on positive solutions of the Lane-Emden problem

$$\begin{cases} -\Delta_p u = \lambda_q |u|^{q-2} u \text{ in } \Omega \\ u = 0 \qquad\qquad\quad \text{on } \partial\Omega \end{cases} \qquad (14)$$

combined with estimates of λ_q. We refer to [11, Lemma 5] for the case $1 < p < N$ and to [31, Theorem 1.1] to the case $p = N$. For the case $p > N$ this lemma follows immediately from Morrey's inequality

$$\|u\|_{C^{0,\alpha}(\overline{\Omega})} \leq C \|\nabla u\|_{L^p(\Omega)},$$

valid for all $u \in W^{1,p}(\Omega)$, where $\alpha := 1 - \frac{p}{N}$ and the constant C depends only on N, p and Ω.

Lemma 3.1. *There exists a positive constant C, depending only on N, p and Ω, such that*

(a) $\|u\|_\infty^{p^\star - q} \leq C$, *for all* $q \in [1, p^\star)$ *and* $u \in E_q$, *if* $1 < p < N$;
(b) $\|u\|_\infty \leq C$, *for all* $q \in [1, \infty)$ *and* $u \in E_q$, *if* $p \geq N > 1$.

Let us define the functions $\underline{f} : [1, p^\star) \to \mathbb{R}$ and $\overline{f} : [1, p^\star) \to \mathbb{R}$, respectively, by

$$\underline{f}(q) = \inf_{u \in E_q} I_q(u) \text{ and } \overline{f}(q) = \sup_{u \in E_q} I_q(u).$$

Proposition 3.2. *Let $q \in [1, p^\star)$. There exist $\underline{u}_q, \overline{u}_q \in E_q$ such that*

$$\underline{f}(q) = I_q(\underline{u}_q) \text{ and } \overline{f}(q) = I_q(\overline{u}_q). \tag{15}$$

Moreover, $\underline{f}$ is lower semi-continuous and $\overline{f}$ is upper semi-continuous.

Proof. Let $q \in [1, p^\star)$. If E_q has a finite number of elements, then (15) follows trivially. If not, let $\{u_n\} \subset E_q$ be a sequence such that $I_q(u_n) \to \underline{f}(q)$. It follows from Lemma 3.1 that $\|u_n\|_\infty \leq C$ for all $n \in \mathbb{N}$, where the positive constant C is uniform with respect to n. By applying standard regularity results (see [27]) we conclude that there exists a subsequence $\{u_{n_k}\}$ and a nonnegative function $\underline{u}_q \in C^1(\overline{\Omega}) \cap W_0^{1,p}(\Omega)$ such that: u_{n_k} converges in $C^1(\overline{\Omega})$ to $\underline{u}_q$, $\left\|\underline{u}_q\right\|_q = 1$ and $-\Delta_p \underline{u}_q = \lambda_q \underline{u}_q^{q-1}$ in Ω. These facts combined imply that $\underline{u}_q \in E_q$ and that $I_q(u_n) \to I_q(\underline{u}_q) = \underline{f}(q)$. Thus, we have proved the first equality in (15). The second equality in (15) follows analogously.

In order to prove that $\underline{f}$ is lower semi-continuous, let us take $q_n \to q \in [1, p^\star)$ and $\underline{u}_{q_n} \in E_{q_n}$ such that $\underline{f}(q_n) = I_{q_n}(\underline{u}_{q_n}) = \min_{u \in E_{q_n}} I_{q_n}(u)$. Then $-\Delta_p(\underline{u}_{q_n}) = \lambda_{q_n}(\underline{u}_{q_n})^{q_n - 1}$ in Ω. Since $q_n \to q$ and $q \in [1, p^\star)$, there exists a positive constant K such that $\left\|\underline{u}_{q_n}\right\|_\infty \leq K$ for all $n \in \mathbb{N}$, according to Lemma 3.1. Thus, we can use again standard regularity results to conclude that, up to a subsequence, $\underline{u}_{q_n}$ converges in $C^1(\overline{\Omega})$ to a nonnegative function $u \in C^1(\overline{\Omega}) \cap W_0^{1,p}(\Omega)$ satisfying $\|u\|_q = 1$. This fact, combined with the continuity of the *BSC* function, also implies that $-\Delta_p u = \lambda_q u^{q-1}$ in Ω. It follows that $u \in E_q$ and $\underline{f}(q_n) = I_{q_n}(\underline{u}_{q_n}) \to I_q(u) \geq \underline{f}(q)$. Since $\{q_n\}$ is an arbitrary sequence converging to q we conclude that $\underline{f}(q) \leq \liminf_{s \to q} \underline{f}(s)$. Analogously, we can prove that $\limsup_{s \to q} \overline{f}(s) \leq \overline{f}(q)$. □

We remark that if $q \in [1, p^\star)$ is such that $\underline{f}(q) = \overline{f}(q)$, then both $\underline{f}$ and $\overline{f}$ are continuous at q and I_q is constant on E_q.

Lemma 3.3. *We have*

$$\limsup_{s\to q^+} \frac{\lambda_q - \lambda_s}{q-s} \le -\frac{p}{q}\lambda_q I_q(\overline{u}_q), \quad \text{for all } q \in [1, p^\star) \tag{16}$$

and

$$\liminf_{s\to q^-} \frac{\lambda_q - \lambda_s}{q-s} \ge -\frac{p}{q}\lambda_q I_q(\underline{u}_q), \quad \text{for all } q \in (1, p^\star).$$

Proof. Let $q \in [1, p^\star)$. For each $s \in [1, p^\star)$, $s \neq q$, we have from (BSC) and (5) that

$$\left(\int_\Omega |u_q|^s\,dx\right)^{\frac{p}{s}} \lambda_s \le \int_\Omega |\nabla u_q|^p\,dx = \lambda_q, \tag{17}$$

where u_q is an arbitrary function in E_q. Hence, by choosing $\overline{u}_q$ in (17) we obtain

$$\begin{aligned}
\limsup_{s\to q^+} \frac{\lambda_q - \lambda_s}{q-s} &= \limsup_{s\to q^+} \frac{\lambda_s - \lambda_q}{s-q} \\
&\le \limsup_{s\to q^+} \lambda_s \frac{1 - \left(\int_\Omega |\overline{u}_q|^s\,dx\right)^{\frac{p}{s}}}{s-q} \\
&= -\lambda_q \frac{d}{ds}\left[\left(\int_\Omega |\overline{u}_q|^s\,dx\right)^{\frac{p}{s}}\right]_{s=q} = -\lambda_q \frac{p}{q} I_q(\overline{u}_q),
\end{aligned}$$

where we have used the continuity of the *BSC* function, L'Hôpital's rule and the fact that

$$\begin{aligned}
\frac{d}{ds}\left[\left(\int_\Omega |\overline{u}_q|^s\,dx\right)^{\frac{p}{s}}\right]_{s=q} &= \left(\int_\Omega |\overline{u}_q|^q\,dx\right)^{\frac{p}{q}} \exp\left(\frac{d}{ds}\left[\frac{p}{s}\log\left(\int_\Omega |\overline{u}_q|^s\,dx\right)\right]_{s=q}\right) \\
&= \frac{p}{q} I_q(\overline{u}_q).
\end{aligned}$$

Analogously, if $q \in (1, p^\star)$, we can choose $\underline{u}_q$ in (17) in order to find

$$\begin{aligned}
\liminf_{s\to q^-} \frac{\lambda_q - \lambda_s}{q-s} &\ge \liminf_{s\to q^-} \lambda_s \frac{\left(\int_\Omega |\underline{u}_q|^s\,dx\right)^{\frac{p}{s}} - 1}{q-s} \\
&= -\lambda_q \frac{d}{ds}\left[\left(\int_\Omega |\underline{u}_q|^s\,dx\right)^{\frac{p}{s}}\right]_{s=q} = -\lambda_q \frac{p}{q} I_q(\underline{u}_q).
\end{aligned}$$

□

Theorem 3.4. *Let $q \in (1, p^\star)$ be a point such that the derivative λ_q' exists. The functional I_q is constant on E_q and*

$$\lambda_q' + \frac{p}{q} I_q(u_q)\lambda_q = 0, \text{ for any choice of } u_q \in E_q. \qquad \text{(ODE)}$$

Proof. Let $u_q \in E_q$. It follows from Lemma 3.3 and (15) that

$$\lambda_q' = \limsup_{s \to q^+} \frac{\lambda_q - \lambda_s}{q - s} \le -\frac{p}{q}\lambda_q I_q(\overline{u}_q) \le -\frac{p}{q}\lambda_q I_q(u_q) \le -\frac{p}{q} I_q(\underline{u}_q)\lambda_q$$
$$\le \liminf_{s \to q^-} \frac{\lambda_q - \lambda_s}{q - s} = \lambda_q'.$$

□

Remark 3.5. The *BSC* function is a solution of (ODE) in the Carathéodory sense, that is, it is an absolutely continuous function that satisfies (ODE) almost everywhere.

Corollary 3.6. *The BSC function is semi-differentiable with*

$$\lambda_{q^+}' := \lim_{s \to q^+} \frac{\lambda_s - \lambda_q}{s - q} = -\frac{p}{q}\lambda_q \overline{f}(q), \text{ for all } q \in [1, p^\star) \qquad (18)$$

and

$$\lambda_{q^-}' := \lim_{s \to q^-} \frac{\lambda_s - \lambda_q}{s - q} = -\frac{p}{q}\lambda_q \underline{f}(q), \text{ for all } q \in (1, p^\star). \qquad (19)$$

Proof. Let us fix $q \in [1, p^\star)$. It follows from the absolute continuity of the *BSC* function that

$$\frac{\lambda_s - \lambda_q}{s - q} = \frac{1}{s - q}\int_q^s \lambda_t' dt, \text{ for all } s \in (q, p^\star).$$

Since λ_t' exists for almost all $t \in [q, s]$ we have from Theorem 3.4, with $u_q = \overline{u}_q$, that

$$\int_q^s \lambda_t' dt = -p \int_q^s \frac{\lambda_t}{t} \overline{f}(t) dt.$$

It follows that

$$\liminf_{s \to q^+} \frac{\lambda_s - \lambda_q}{s - q} = \liminf_{s \to q^+} \left(-\frac{p}{s - q}\int_q^s \frac{\lambda_t}{t}\overline{f}(t) dt \right) \ge -p \liminf_{s \to q^+} \frac{1}{s - q}\int_q^s \frac{\lambda_t}{t}\overline{f}(q) dt$$
$$= -\frac{p}{q}\lambda_q \overline{f}(q),$$

where the inequality comes from the upper semi-continuity of $\overline{f}$.

On the other hand, by (16) we have

$$\limsup_{s\to q^+} \frac{\lambda_q - \lambda_s}{q-s} \le -\frac{p}{q}\lambda_q I_q(\overline{u}_q) = -\frac{p}{q}\lambda_q \overline{f}(q),$$

which concludes the proof of (18). The proof of (19) is completely analogous. □

An immediate consequence of this last result is that $\lambda'_{q^+} \le \lambda'_{q^-}$. Another is stated in the sequel.

Corollary 3.7. *Suppose that I_q is constant on E_q for some $q \in [1, p^\star)$. The derivative of the BSC function exists at q and it is given by $\lambda'_q = -\frac{p}{q} I_q(u_q)\lambda_q$, where u_q is any function in E_q.*

Corollary 3.8. *The BSC function belongs to $C^1([1,p])$ if Ω is a general bounded domain and to $C^1([1,p^\star))$ if Ω is a ball.*

Proof. Let $J_\Omega := \{q \in [1,p^\star) : E_q \text{ is unitary}\}$. It is well known (and pointed out in the Introduction) that $[1,p] \subset J_\Omega$ in general, but when Ω is a ball we have that $J_\Omega = [1,p^\star)$. Therefore, if $q \in J_\Omega$, then I_q is constant on E_q. According to the Corollary 3.7, we can conclude that λ'_q exists at each $q \in J_\Omega$ and also that

$$\lambda'_q = -\lambda_q \frac{p}{q}\overline{f}(q) = -\lambda_q \frac{p}{q}\underline{f}(q).$$

Since the functions $\underline{f}$ and $\overline{f}$ coincide in J_Ω and are continuous in this set, we conclude that the derivative of the *BSC* function is continuous in J_Ω. □

Remark 3.9. E_q is also unitary if $p = 2$ and $q \in [2, 2+\epsilon]$, for some $\epsilon > 0$ (see [10]). Therefore, the *BSC* function belongs to $C^1([1, 2+\epsilon])$.

Remark 3.10. We can conclude from the results above that the differentiability of the *BSC* function at $q \in [1, p^\star)$ is equivalent to the property of the functional I_q to be constant on the set E_q. It is interesting to notice that this property holds for almost all $q \in [1, p^\star)$.

Now we prove a representation formula for the *BSC* function.

Corollary 3.11. *Let $1 \le q < p^\star$. Then*

$$\lambda_t = \lambda_1 \exp\left(-p\int_1^t \frac{I_s(u_s)}{s}ds\right) \quad \textit{for } \textbf{\textit{all}}\ t \in [1,q], \tag{20}$$

and, in the case $1 < p < N$, this representation formula also holds true if $q = p^\star$.

Proof. Since the *BSC* function is absolutely continuous in $[1,q]$ and its image is a closed interval $[a,b] \subset (0,\infty)$ the function $s \in [1,q] \mapsto \log \lambda_s$ is also absolutely continuous. Therefore,

$$\log(\lambda_t/\lambda_1) = \int_1^t (\log\lambda_s)'\,ds = -p\int_1^t \frac{I_s(u_s)}{s}ds \text{ for \textbf{all} } t \in [1,q] \tag{21}$$

where $u_s \in E_s$ for each $s \in [1,t)$ and the latter equality follows from (ODE). Now, (20) follows after exponentiation.

In the case $1 < p < N$ the first equality in (21) is also valid at $t = p^\star$, since the function $s \in [1,p^\star] \mapsto \log\lambda_s$ is absolutely continuous. Moreover, the second equality in this case also occurs because $(\log\lambda_s)' = \lambda_s'/\lambda_s = -pI_s(u_s)/s$ almost everywhere, according to (ODE). □

4 Hölder regularity

According to what we have seen in the Introduction the *BSC* function is Lipschitz continuous in any closed interval contained in $[1,p^\star)$. In this section we use the results of the Section 3 to show that the *BSC* function is globally Lipschitz continuous, if $p \geq N > 1$, and α-Hölder continuous, for any $\alpha \in (0,1)$, if $1 < p < N$.

For a bounded domain $\Omega \subset \mathbb{R}^N$ we can check that

$$\lambda_q(\Omega) = \lambda_q(\Omega_1)\,|\Omega|^{1-\frac{p}{N}-\frac{p}{q}} \tag{22}$$

where $\Omega_1 := \left\{x \in \mathbb{R}^N : x\,|\Omega|^{\frac{1}{N}} \in \Omega\right\}$ satisfies $|\Omega_1| = 1$. This allows us to assume, in this section, and without loss of generality for our study, that $|\Omega| = 1$. Under this condition the *BSC* function is decreasing and, moreover, $I_q(u) > 0$ for all $q \in [1,p^\star)$ and $u \in E_q$. This latter property comes from Jensen's inequality applied to the strictly convex and continuous function $\varphi : [0,\infty) \to \mathbb{R}$ given by $\varphi(\xi) = \xi\log\xi$, if $\xi > 0$ and $\varphi(0) = 0$. Indeed, we have

$$I_q(u) = \frac{1}{q}\int_\Omega \varphi(|u|^q)dx > \frac{|\Omega|}{q}\varphi(|\Omega|^{-1}\,\|u\|_q^q) = 0, \text{ if } \|u\|_q = 1 \text{ and } |\Omega| = 1. \tag{23}$$

We also observe that if $q \geq 1$ and $u \in E_q$, then

$$I_q(u) = \int_\Omega |u|^q \log|u|\,dx \leq \int_{u\geq 1} |u|^q \log|u|\,dx \leq \log\|u\|_\infty \int_\Omega |u|^q\,dx = \log\|u\|_\infty\,. \tag{24}$$

Theorem 4.1. *Suppose $p \geq N > 1$. The BSC function is globally Lipschitz continuous.*

Proof. It follows from (24) and Lemma 3.1 that if $t \geq 1$ and $u \in E_t$ then $0 < I_t(u) \leq \log C$, where C depends only on N, p and Ω. Therefore, if $1 \leq q_1 < q_2 < p^\star = \infty$, then

$$|\lambda_{q_2} - \lambda_{q_1}| = \lambda_{q_1} - \lambda_{q_2} = -\int_{q_1}^{q_2} \lambda_t' dt = p \int_{q_1}^{q_2} \frac{\lambda_t}{t} I_t(u_t) dt$$
$$\leq C_1(q_2 - q_1) = C_1 |q_2 - q_1|,$$

where $C_1 := \lambda_1 \log(C^p)$. □

We do not know if the *BSC* function is globally Lipschitz continuous in case $1 < p < N$. The great difficulty in obtaining such a result is to control $I_q(u_q)$, as $q \to p^\star$, since $\|u_q\|_\infty$ becomes unbounded. However, we can prove the following Hölder regularity result.

Theorem 4.2. *Suppose* $1 < p < N$ *and assume that*

$$\limsup_{q \to p^\star} (p^\star - q) \|u_q\|_\infty^\gamma < \infty, \quad \textit{for some } \gamma > 0. \tag{25}$$

The BSC function is α*-Hölder continuous, for any* $\alpha \in (0,1)$.

Proof. Let $\alpha \in (0,1)$. By hypothesis there exist $\sigma \in (1, p^\star)$ and a positive constant C such that

$$\|u_t\|_\infty \leq C(p^\star - t)^{-\frac{1}{\gamma}}, \quad \text{for all } t \in [\sigma, p^\star].$$

Since the *BSC* function is Lipschitz continuous in the interval $[1, \sigma]$, we need only to prove that this function is α-Hölder continuous in the interval $[\sigma, p^\star]$. Let $\sigma \leq q_1 < q_2 \leq p^\star$ and take $\beta > 0$ (to be chosen). Since $\log(y) \leq \beta^{-1} y^\beta$ for all $y \geq 1$, we obtain from (24) that

$$|\lambda_{q_2} - \lambda_{q_1}| = \lambda_{q_1} - \lambda_{q_2}$$
$$= -\int_{q_1}^{q_2} \lambda_t' dt = p \int_{q_1}^{q_2} \frac{\lambda_t}{t} I_t(u_t) dt \leq p\lambda_1 \int_{q_1}^{q_2} I_t(u_t) dt \leq \frac{p\lambda_1}{\beta} \int_{q_1}^{q_2} \|u_t\|_\infty^\beta \, dt.$$

Therefore,

$$|\lambda_{q_2} - \lambda_{q_1}| \leq \frac{p\lambda_1 C^\beta}{\beta} \int_{q_1}^{q_2} (p^\star - t)^{-\frac{\beta}{\gamma}} dt \leq \frac{p\lambda_1 C^\beta}{\beta} \int_{q_1}^{q_2} (q_2 - t)^{-\frac{\beta}{\gamma}} dt = \frac{p\lambda_1 C^\beta}{\beta\theta} (q_2 - q_1)^\theta,$$

where $\theta := 1 - \frac{\beta}{\gamma}$. By choosing $\beta = (1-\alpha)\gamma$ we obtain $\theta = \alpha$ and

$$|\lambda_{q_2} - \lambda_{q_1}| \leq \frac{p\lambda_1 C^{(1-\alpha)\gamma}}{\alpha(1-\alpha)\gamma} |q_2 - q_1|^\alpha.$$

□

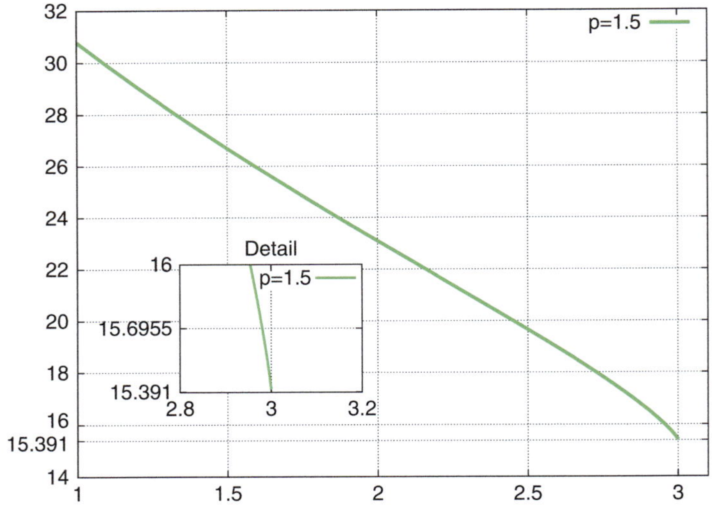

Fig. 1 Graph of the function $q \in [1, p^\star) \mapsto |B_1|^{\frac{p}{q}} \lambda_q(B_1)$ for $N = 3$ and $p = 1.5$. Note that $(1.5)^\star = 3$ and that $(4\pi/3)^{\frac{1.5}{3}} S_3 \approx 15.391$.

As mentioned in the Introduction, if $p = 2$ and Ω is an arbitrary bounded domain or if $1 < p \neq 2$ and Ω is ball, then (25) holds (with $\gamma = \frac{p}{p-1}$). Thus, in these situations the previous theorem implies that the *BSC* function is α-Hölder continuous, for any $\alpha \in (0, 1)$. If Ω is a ball, Corollary 3.8 also guarantees that the *BSC* function belongs to $C^1([1, p^\star))$.

We end this paper by stressing that the exact behavior of λ_q' as q approaches $p^\star$, in the case $1 < p < N$, is not known by now. We present in Figure 1 the graph of a numerical computation of the function $q \in [1, p^\star) \mapsto |B_1|^{\frac{p}{q}} \lambda_q(B_1)$, for $N = 3$ and $p = 1.5$, where B_1 is the unit ball. The graph has been implemented by using the method developed in [15] and appears to indicate that $\lambda_q'(B_1)$ might be unbounded.

References

1. Adimurthi, Grossi, M.: Asymptotic estimates for a two-dimensional problem with polynomial nonlinearity. Proc. Am. Math. Soc. **132**, 1013–1019 (2004)
2. Adimurthi, Yadava, S.L.: An elementary proof of the uniqueness of positive radial solutions of a quasilinear Dirichlet problem. Arch. Ration. Mech. Anal. **127**, 219–229 (1994)
3. Anello, G.: On the Dirichlet problem involving the equation $-\Delta_p u = \lambda u^{s-1}$. Nonlinear Anal. **70**, 2060–2066 (2009)
4. Anello, G., Faraci, F., Iannizzotto, A.: On a problem of Huang concerning best constants in Sobolev embeddings. Ann. Math. Pura Appl. **194**, 767–779 (2015)
5. Atkinson, F.V., Peletier, L.A.: Elliptic equations with nearly critical growth. J. Differ. Equat. **70**, 349–365 (1987)

6. Aubin, T.: Problèmes isopérimétriques et espaces de Sobolev. J. Differ. Geom. **11**, 573–598 (1976)
7. Benci, V., Cerami, G.: The effect of the domain topology on the number of positive solutions of nonlinear elliptic problems. Arch. Ration. Mech. Anal. **114**, 79–93 (1991)
8. Brezis, H., Peletier, L.A.: Asymptotics for elliptic equations involving critical growth. In: Colombini, F., Marino, A., Modica, L., Spagnolo, S. (eds.) Partial Differential Equations and the Calculus of Variations, vol. 1, pp. 149–192. Birkhäuser, Boston (1989)
9. Bueno, H., Ercole, G.: Solutions of the Cheeger problem via torsion functions. J. Math. Anal. Appl. **381**, 263–279 (2011)
10. Dancer, E.N.: Real analyticity and non-degeneracy. Math. Ann. **325**, 369–392 (2003)
11. Ercole, G.: Absolute continuity of the best Sobolev constant. J. Math. Anal. Appl. **404**, 420–428 (2013)
12. Ercole, G.: On the resonant Lane-Emden problem for the p-Laplacian. Commun. Contemp. Math. **16**, 1350033 (2014) [22 pp.]
13. Ercole, G.: Regularity results for the best-Sobolev-constant function. Ann. Math. Pura Appl. (2015) doi:10.1007/s10231-014-0425-3
14. Ercole, G., Pereira, G.A.: Asymptotics for the best Sobolev constants and their extremal functions arXiv:150600922 [math.AP]
15. Ercole, G., Espírito Santo, J.C., Martins, E.M.: A constructive method for obtaining the best Sobolev constants of the unit ball and their extremal functions. Submitted
16. Franzina, G., Lamberti, P.D.: Existence and uniqueness for a p-laplacian nonlinear eigenvalue problem. Electron. J. Differ. Equat. **26**, 10 pp. (2010)
17. Grumiau, C., Parini, E.: On the asymptotics of solutions of the Lane-Emden problem for the p-Laplacian. Arch. Math. **91**, 354–365 (2008)
18. Han, Z.: Asymptotic approach to singular solutions for nonlinear elliptic equations involving critical Sobolev exponent. Ann. Inst. Poincaré Anal. Non-linear **8**, 159–174 (1991)
19. Huang, Y.X.: A note on the asymptotic behavior of positive solutions for some elliptic equation. Nonlinear Anal. Theory Methods Appl. **29**, 533–537 (1997)
20. Huang, Y.X.: On the eigenvalues of the p-Laplacian with varying p. Proc. Am. Math. Soc. **125**, 3347–3354 (1997)
21. Idogawa, T., Otani, M.: The first eigenvalues of some abstract elliptic operators. Funkcial. Ekvac. **38**, 1–9 (1995)
22. Juutine, J., Lindqvist, P., Manfredi, J.: The ∞-eigenvalue problem. Arch. Ration. Mech. Anal. **148**, 89–105 (1999)
23. Kawohl, B.: On a family of torsional creep problems. J. Reine Angew. Math. **410**, 1–22 (1990)
24. Kawohl, B.: Symmetry results for functions yielding best constants in Sobolev-type inequalities. Discrete Contin. Dyn. Syst. **6**, 683–690 (2000)
25. Kawohl, B., Fridman, V.: Isoperimetric estimates for the first eigenvalue of the p-Laplace operator and the Cheeger constant. Comment. Math. Univ. Carol. **44**, 659–667 (2003)
26. Knaap, M.C., Peletier, L.A.: Quasilinear elliptic equations with nearly critical growth. Commun. Partial Differ. Equat. **14**, 1351–1383 (1989)
27. Lieberman, G.M.: Boundary regularity for solutions of degenerate elliptic equations. Nonlinear Anal. Theory Methods Appl. **12**, 1203–1219 (1988)
28. Nazarov, A.I.: The one-dimensional character of an extremum point of the Friedrichs inequality in spherical and plane layers. J. Math. Sci. **102**, 4473–4486 (2000)
29. Ren, X., Wei, J.: Single-point condensation and least-energy solutions. Proc. Am. Math. Soc. **124**, 111–120 (1996)
30. Ren, X., Wei, J.: On a two-dimensional elliptic problem with large exponent in nonlinearity. Trans. Am. Math. Soc. **343**, 749–763 (1994)
31. Ren, X., Wei, J.: Counting peaks of solutions to some quasilinear elliptic equations with large exponents. J. Differ. Equat. **117**, 28–55 (1995)
32. Rey, O.: Proof of two conjectures of H. Brezis and L. A. Peletier. Manus. Math. **65**, 19–37 (1989)
33. Talenti, G.: Best constant in Sobolev inequality. Ann. Mat. Pura Appl. **110**, 353–372 (1976)

Bubbling solutions to an anisotropic Hénon equation

Jorge Faya, Massimo Grossi, and Angela Pistoia

To Djairo de Figueiredo for his 80th birthday

1 Introduction

Hénon in 1973 in [10] introduced, as a model in the context of spherically symmetric clusters of stars, the Dirichlet problem

$$-\Delta u = |x|^{\alpha} u^{p} \text{ in } \Omega,\ u > 0 \text{ in } \Omega,\ u = 0 \text{ on } \partial\Omega, \tag{1}$$

where Ω is the unit ball in $\mathbb{R}^N$, with $N \geq 3$, the power α is positive and $p > 1$.

Hénon studied this equation numerically, for some definite values of α and p, but subsequent researches showed that the above problem exhibits very rich features from the functional–analytic point of view. In particular, various questions that arise quite naturally concerning existence, multiplicity and qualitative properties of solutions, have given the Hénon equation the role of a very interesting item in nonlinear analysis and critical point theory.

The first existence result is due to Ni, who in [12] introduced an *Hénon critical exponent* $p_\alpha := \frac{N+2+2\alpha}{N-2}$ and proved that for every $p \in (1, p_\alpha)$, problem (1) admits at least one *radial* solution. It is important to point out that p_α actually plays the same

J. Faya (✉)
Centro de Modelamiento Matemático, Universidad de Chile, Blanco Encalada 2120, Santiago, Chile
e-mail: jfaya@dim.uchile.cl

M. Grossi
Dipartimento di Matematica, Università di Roma La Sapienza, P.le A. Moro 2, 00185 Roma, Italy
e-mail: massimo.grossi@uniroma1.it

A. Pistoia
Dipartimento di Metodi e Modelli Matematici, Universitá di Roma "La Sapienza", via Antonio Scarpa 16, 00161 Roma, Italy
e-mail: pistoia@dmmm.uniroma1.it

A.N. Carvalho et al. (eds.), *Contributions to Nonlinear Elliptic Equations and Systems*, Progress in Nonlinear Differential Equations and Their Applications 86,
DOI 10.1007/978-3-319-19902-3_13

role of the usual critical Sobolev exponent, since standard arguments based on the Pohozaev identity yield that problem (1) has no solution if $p \geq p_\alpha$.

The existence of nonradial solutions was firstly obtained by Smets, Su, and Willem, who studied the *ground state* solutions associated with (1). In [15] they proved, among other results, that for $p \in (1, \frac{N-2}{N+2})$ no ground state is radial provided α is large enough. Successively, Byeon and Wang in [2, 3] described the symmetry breaking, asymptotics and single point concentration profile at the boundary of the ground state as α goes to $+\infty$. In view of the above considerations, existence of nonradial solutions could be expected also for larger values of p. When p is almost critical, i.e. $p = \frac{N+2}{N-2} - \varepsilon$, Pistoia and Serra in [13] proved that problem (1) admits an arbitrarily large number of nonradial solutions provided ε is positive and small enough. On the other hand, when p is critical, i.e. $p = \frac{N+2}{N-2}$ Serra in [14] proved that problem (1) has a nonradial solution for α large and Wei and Yan in [16] constructed infinitely many solutions to (1) for any α positive.

The supercritical case, i.e. $p > \frac{N+2}{N-2}$ is much more delicate and only a few results have been obtained. Ni in [12] proved that a radial solution to (1) exists for any $p \in (1, p_\alpha)$. Recently, Dos Santos and Pacella in [5] found a solution to (1) which concentrate along spheres as α approaches $+\infty$ for some suitable $2m$-dimensional balls. When the ball is replaced by a more general domain Ω, the only existence result was obtained by Gladiali and Grossi in [6] and by Gladiali, Grossi, and Neves in [7] when the exponent $p + 1$ is close enough to the *Hénon critical exponent* p_α. They found a solution to the problem

$$\begin{cases} -\Delta u = |x - \xi|^\alpha u^{p_\alpha - 1 - \varepsilon} & \text{in } \Omega, \\ u > 0 & \text{in } \Omega, \\ u = 0 & \text{on } \partial\Omega, \end{cases} \tag{2}$$

in a smooth bounded domain $\Omega \subset \mathbb{R}^N$ where $\xi \in \Omega$, provided ε is a positive small parameter and α is different from an even integer. In particular, that solution blows up at the point ξ as ε goes to zero. If Ω is the ball and ξ is the center of Ω, this is nothing but the radial solution found by Ni.

It is interesting to ask what happens when the Laplace operator Δu is replaced by a more general elliptic operator $\mathscr{L}u = \operatorname{div}(a(x)\nabla u)$. Up to our knowledge there are no results about this problem. In particular, in this paper we are interested in solving the problem

$$\begin{cases} -\operatorname{div}(a(x)\nabla u) = c_\alpha a(x)|x - \xi|^\alpha u^{p_\alpha - 1 \pm \varepsilon} & \text{in } \Omega, \\ u > 0 & \text{in } \Omega, \\ u = 0 & \text{on } \partial\Omega, \end{cases} \tag{3}$$

where Ω is a smooth bounded domain in $\mathbb{R}^N$, $N \geq 3$, $\xi \in \Omega$, the function $a \in C^2(\overline{\Omega})$ is strictly positive on $\overline{\Omega}$ and ε is a positive and small parameter. Here $c_\alpha := (N + \alpha)(N - 2)$. Our existence result reads as follows.

Theorem 1.1. *Let $N \geq 6$. Assume that $\alpha > 0$ is a positive real number which is not an even integer.*

Assume $\xi \in \Omega$ is a critical point of a such that $\Delta a(\xi) \neq 0$.

(i) *If $\Delta a(\xi) < 0$, then, for ε small enough, there exists a solution u_ε to problem*

$$\begin{cases} -\mathrm{div}(a(x)\nabla u) = c_\alpha a(x)|x-\xi|^\alpha u^{p_\alpha - \varepsilon} & \text{in } \Omega, \\ u > 0 & \text{in } \Omega, \\ u = 0 & \text{on } \partial\Omega, \end{cases} \tag{4}$$

which blows up at ξ as ε goes to zero.

(ii) *If $\Delta a(\xi) > 0$, then, for ε small enough, there exists a solution u_ε to problem*

$$\begin{cases} -\mathrm{div}(a(x)\nabla u) = c_\alpha a(x)|x-\xi|^\alpha u^{p_\alpha + \varepsilon} & \text{in } \Omega, \\ u > 0 & \text{in } \Omega, \\ u = 0 & \text{on } \partial\Omega, \end{cases} \tag{5}$$

which blows up at ξ as ε goes to zero.

The proof of our result relies on a well-known Ljapunov–Schmidt procedure which is completely carried out in Section 3. The necessary background is introduced in Section 2.

2 Preliminaries

The main ingredients in building the solutions u_ε of Theorem 3 are the *bubbles of order α*

$$U_\lambda(x) := \frac{\lambda^{\frac{N-2}{2}}}{(1+\lambda^{2+\alpha}|x-\xi|^{2+\alpha})^{\frac{N-2}{2+\alpha}}}, \qquad \lambda > 0,$$

which are all the solutions of the problem

$$\begin{cases} -\Delta U = c_\alpha |x-\xi|^\alpha U^{p_\alpha} & \text{in } \mathbb{R}^N, \\ U > 0 & \text{in } \mathbb{R}^N, \\ U \in D^{1,2}(\mathbb{R}^N), \end{cases}$$

where $N \geq 3, \alpha > 0, p_\alpha = \frac{N+2+2\alpha}{N-2}, c_\alpha = (N+\alpha)(N-2)$, $D^{1,2}(\mathbb{R}^N) = \{u \in L^{2^*}(\mathbb{R}^N) : |\nabla u| \in L^2(\mathbb{R}^N)\}$ and $2^* = \frac{2N}{N-2}$.

We are looking for a solution of problem (3) of the form $u_{\varepsilon,\lambda} = PU_\lambda + \phi_{\varepsilon,\lambda}$. Here Pu denotes the projection onto $H_0^1(\Omega)$, namely if $u \in D^{1,2}(\mathbb{R}^N)$ then Pu is the unique solution to the problem

$$-\Delta(Pu) = -\Delta u \text{ in } \Omega, \quad Pu = 0 \text{ on } \partial\Omega. \tag{6}$$

The concentration parameter λ is chosen so that

$$\lambda_\varepsilon := d\varepsilon^{-1} \text{ for some } d > 0. \tag{7}$$

Finally, the function $\phi_{\varepsilon,\lambda}$ is a higher order term which satisfies the orthogonality condition

$$\int_\Omega a(x)\nabla\phi_{\varepsilon,\lambda}\nabla PZ_\lambda(x)dx = 0.$$

Here PZ_λ is the projection of the function

$$Z_\lambda(x) := \frac{\partial U_\lambda}{\partial\lambda}(x) = \frac{N-2}{2}\lambda^{\frac{N-4}{2}} \frac{1-\lambda^{2+\alpha}|x-\xi|^{2+\alpha}}{(1+\lambda^{2+\alpha}|x-\xi|^{2+\alpha})^{\frac{N+\alpha}{2+\alpha}}}. \tag{8}$$

Let us recall the following result proved in [6, 7].

Theorem 2.1. *Let α be a positive real number which is not an even integer. Then, the function Z_λ defined in* (8) *is the unique, up to a constant, solution of the problem*

$$\begin{cases} -\Delta Z = c_\alpha p_\alpha |x-\xi|^\alpha U_\lambda^{p_\alpha - 1} Z \text{ in } \mathbb{R}^N, \\ Z \in D^{1,2}(\mathbb{R}^N). \end{cases} \tag{9}$$

Remark 2.2. In the case when $\alpha = 2(k-1)$ for some integer k, we have that the space of solutions to problem (9) has dimension $\frac{(N+2k-2)(N+k-3)!}{(N-2)!k!}$ and it is spanned by functions of the form

$$\frac{Y_k}{(1+|x-\xi|^{2+\alpha})^{\frac{N+\alpha}{2+\alpha}}}$$

where the functions Y_k form a basis for the space of all homogeneous harmonic polynomials of degree k in $\mathbb{R}^N$.

It is necessary to recall some useful estimates which involve the projection of the bubble U_λ and its derivative $\partial_\lambda U_\lambda$, whose proof is given in [6].

Proposition 2.3. *Let Ω be a bounded smooth domain in $\mathbb{R}^N$, $N \geq 3$, and $\xi \in \Omega$. Then*

1. $PU_\lambda(x) = U_\lambda - \frac{\omega_N(N-2)}{\lambda^{\frac{N-2}{2}}}H(x,\xi) + O\left(\lambda^{-\frac{N+2+2\alpha}{2}}\right)$.

2. $PZ_\lambda(x) = Z_\lambda(x) + \frac{(N-2)^2}{2}\omega_N \frac{1}{\lambda^{\frac{N}{2}}} H(x,\xi) + O(\lambda^{-\frac{N+4+2\alpha}{2}})$.
3. $\nabla PZ_\lambda(x) = \nabla Z_\lambda(x) + O(\lambda^{-\frac{N}{2}})$, *with* $\nabla Z_\lambda(x) = -\frac{N-2}{2}\lambda^{\frac{N+2\alpha}{2}}(x-\xi)|x-\xi|^\alpha \frac{(N+2+2\alpha)-(N-2)\lambda^{2+\alpha}|x-\xi|^{2+\alpha}}{\left(1+\lambda^{2+\alpha}|x-\xi|^{2+\alpha}\right)^{\frac{N+\alpha}{2+\alpha}+1}}$.

Here $G(x,y)$ is Green's function of the Laplace operator in Ω with zero Dirichlet boundary condition and $H(x,y)$ is its regular part, i.e.

$$G(x,y) = \frac{1}{\omega_N(N-2)|x-y|^{N-2}} - H(x,y)$$

where ω_N is the area of the unit sphere in $\mathbb{R}^N$.

Moreover it is also useful to recall the following results.

Lemma 2.4. *For any $\alpha > 0$, it holds true that*

$$\int_{\mathbb{R}^N} |z|^\alpha \frac{(1-|z|^{2+\alpha})}{(1+|z|^{2+\alpha})^{\frac{N+\alpha}{2N+2+3\alpha}}} dz = 0, \tag{10}$$

$$\int_{\mathbb{R}^N} \frac{|z|^\alpha(1-|z|^{2+\alpha})}{(1+|z|^{2+\alpha})^{\frac{2N+2+3\alpha}{2+\alpha}}} \log(1+|z|^{2+\alpha}) dz < 0, \tag{11}$$

$$\int_{\mathbb{R}^N} \frac{|z|^{2+\alpha}(1-|z|^{2+\alpha})}{(1+|z|^{2+\alpha})^{\frac{2(N+\alpha)}{2+\alpha}}} dz < 0. \tag{12}$$

Proof. See Lemma 3.7 in [6] for the proof of equations (10) and (11). To show (12) note that

$$\int_{\mathbb{R}^N} \frac{|z|^{2+\alpha}(1-|z|^{2+\alpha})}{(1+|z|^{2+\alpha})^{\frac{2(N+\alpha)}{2+\alpha}}} dz = \omega_N \int_0^1 \frac{|\rho|^{2+\alpha+N-1}(1-|\rho|^{2+\alpha})}{(1+|\rho|^{2+\alpha})^{\frac{2(N+\alpha)}{2+\alpha}}} d\rho + \omega_N \int_1^\infty \frac{|\rho|^{2+\alpha+N-1}(1-|\rho|^{2+\alpha})}{(1+|\rho|^{2+\alpha})^{\frac{2(N+\alpha)}{2+\alpha}}} d\rho$$

On the one hand, setting $\rho = \frac{1}{t}$, we obtain

$$\int_1^\infty \frac{|\rho|^{2+\alpha+N-1}(1-|\rho|^{2+\alpha})}{(1+|\rho|^{2+\alpha})^{\frac{2(N+\alpha)}{2+\alpha}}} d\rho = -\int_0^1 \frac{(\frac{1}{t})^{N+\alpha+1}[(\frac{1}{t})^{2+\alpha}(t^{2+\alpha}-1)](-t^{-2})}{[(\frac{1}{t})^{2+\alpha}(t^{2+\alpha}+1)]^{\frac{2(N+\alpha)}{2+\alpha}}} dt = -\int_0^1 \frac{t^{N-5}(1-t^{2+\alpha})}{(1+t^{2+\alpha})^{\frac{2(N+\alpha)}{2+\alpha}}} dt. \tag{13}$$

On the other hand, since $N+1+\alpha > N-5$, it is clear that

$$t^{N+1+\alpha} < t^{N-5} \text{ for every } t \in (0,1)$$

and, using that $(1-t^{2+\alpha}) > 0$ for every $t \in (0,1)$, we obtain

$$\int_0^1 \frac{t^{N+1+\alpha}(1-t^{2+\alpha})}{(1+t^{2+\alpha})^{\frac{2(N+\alpha)}{2+\alpha}}}dt < \int_0^1 \frac{t^{N-5}(1-t^{2+\alpha})}{(1+t^{2+\alpha})^{\frac{2(N+\alpha)}{2+\alpha}}}dt$$

i.e.

$$\int_0^1 \frac{t^{N+1+\alpha}(1-t^{2+\alpha})}{(1+t^{2+\alpha})^{\frac{2(N+\alpha)}{2+\alpha}}}dt - \int_0^1 \frac{t^{N-5}(1-t^{2+\alpha})}{(1+t^{2+\alpha})^{\frac{2(N+\alpha)}{2+\alpha}}}dt < 0 \tag{14}$$

Therefore, using equations (13) and (14), we obtain

$$\begin{aligned}\int_{\mathbb{R}^N} \frac{|z|^{2+\alpha}(1-|z|^{2+\alpha})}{(1+|z|^{2+\alpha})^{\frac{2(N+\alpha)}{2+\alpha}}}dz &= \omega_N \int_0^1 \frac{|\rho|^{2+\alpha+N-1}(1-|\rho|^{2+\alpha})}{(1+|\rho|^{2+\alpha})^{\frac{2(N+\alpha)}{2+\alpha}}}d\rho \\ &\quad + \omega_N \int_1^\infty \frac{|\rho|^{2+\alpha+N-1}(1-|\rho|^{2+\alpha})}{(1+|\rho|^{2+\alpha})^{\frac{2(N+\alpha)}{2+\alpha}}}d\rho \\ &= \omega_N\Big(\int_0^1 \frac{t^{N+1+\alpha}(1-t^{2+\alpha})}{(1+t^{2+\alpha})^{\frac{2(N+\alpha)}{2+\alpha}}}dt - \int_0^1 \frac{t^{N-5}(1-t^{2+\alpha})}{(1+t^{2+\alpha})^{\frac{2(N+\alpha)}{2+\alpha}}}dt\Big) < 0.\end{aligned}$$

□

Lemma 2.5. *For any $\alpha > 0$ and $\theta > 0$ we have*

$$\int_{\mathbb{R}^N} \frac{|y|^\alpha}{|x-y|^{N-2}} \frac{1}{(1+|y|^{2+\alpha})^\theta}dy \le \begin{cases} \frac{C}{(1+|x|)^{(2+\alpha)(\theta-1)}} & \text{if } N+\alpha-\theta(2+\alpha) > 0, \\ \frac{C}{(1+|x|)^{N-2}} \log|x| & \text{if } N+\alpha-\theta(2+\alpha) = 0, \\ \frac{C}{(1+|x|)^{N-2}} & \text{if } N+\alpha-\theta(2+\alpha) < 0. \end{cases} \tag{15}$$

and

$$\int_{\mathbb{R}^N} \frac{1}{|x-y|^{N-2}} \frac{1}{(1+|y|^{2+\alpha})^\theta}dy \le \begin{cases} \frac{C}{(1+|x|)^{(2+\alpha)\theta-2}} & \text{if } N-\theta(2+\alpha) > 0, \\ \frac{C}{(1+|x|)^{N-2}} \log|x| & \text{if } N-\theta(2+\alpha) = 0, \\ \frac{C}{(1+|x|)^{N-2}} & \text{if } N-\theta(2+\alpha) < 0. \end{cases} \tag{16}$$

Proof. See Lemma 3.8 and Corollary 3.9 in [6]. □

3 The finite-dimensional reduction

3.1 The linear theory

In order to perform the finite-dimensional reduction we have to study the following linear auxiliary problem.

For any $\lambda > 0$ large enough, given a function $\psi \in L^2(\Omega)$ with $\int\limits_\Omega \psi PZ_\lambda dx = 0$ let us find a function $\phi_\lambda \in H_0^1(\Omega)$ and a real number c_λ solutions to the linear problem

$$(L_\lambda) \quad \begin{cases} \mathrm{div}(a(x)\nabla\phi) + a(x)p_\alpha|x-\xi|^\alpha (PU_\lambda)^{p_\alpha-1}\phi = \psi - c_\lambda \mathrm{div}(a(x)\nabla PZ_\lambda) & \text{in } \Omega, \\ \phi = 0 & \text{on } \partial\Omega, \\ \int\limits_\Omega a(x)\nabla\phi\nabla PZ_\lambda dx = 0. \end{cases}$$

In this section we will show that problem (L_λ) is uniquely solvable in certain appropriate norms, provided that λ is large enough. To do this, we consider the following weighted L^∞-norms. For functions ϕ, ψ defined on Ω, we define the norms

$$\| \phi \|_* = \sup_{y\in\Omega} \left(\frac{\lambda^{\frac{N-2}{2}}}{(1+\lambda^{2+\alpha}|y-\xi|^{2+\alpha})^{\frac{N-2}{2(2+\alpha)}}} \right)^{-1} |\phi(y)|, \tag{17}$$

and

$$\| \psi \|_{**} = \sup_{y\in\Omega} \left(\frac{\lambda^{\frac{N+2}{2}}}{(1+\lambda^{2+\alpha}|y-\xi|^{2+\alpha})^{\frac{N+2}{2(2+\alpha)}}} \right)^{-1} |\psi(y)|.$$

Proposition 3.1. *There exist $\lambda_0 > 0$ and $C > 0$, not depending on λ, such that, for any $\lambda \geq \lambda_0$ and for any $\psi \in L^2(\Omega)$ with $\int\limits_\Omega \psi PZ_\lambda dx = 0$, problem (L_λ) admits a unique solution $\phi_\lambda := L_\lambda(\psi)$ for some $c_\lambda \in \mathbb{R}$. In addition*

$$\|L_\lambda(\psi)\|_* \leq C\|\psi\|_{**}. \tag{18}$$

Proof. The proof follows using the same argument as in the proof of Proposition 4.1 in [4]. In particular it relies on the following result: *Let (λ_n) be a sequence such that $\lambda_n \to \infty$ as $n \to \infty$. For any $n \in \mathbb{N}$, let $c_n \in \mathbb{R}$ and ϕ_n be a solution to problem (L_{λ_n}) for ψ_n. If $\| \psi_n \|_{**} \to 0$, then $\| \phi_n \|_* \to 0$.* For the reader convenience, we give here the proof which is similar to that of Lemma 4.1 in [6].

Proceeding by contradiction, let us suppose that there exist sequences (λ_n), (c_n), (ϕ_n) and (ψ_n) with the following properties:

- $\lambda_n \to \infty$,
- the function ϕ_n is a solution to problem (L_{λ_n}) for ψ_n,
- $\| \psi_n \|_{**} \to 0$,
- there exist $c > 0$ such that $\| \phi_n \|_* \geq c > 0$ for every $n \in \mathbb{N}$.

We may also assume that $\| \phi_n \|_* = 1$. In the following C will denote some positive constant independent of n, not necessarily the same one.

Step 1: We will show that $c_n \to 0$. We multiply problem (L_{λ_n}) by PZ_{λ_n} and we get

$$\int_\Omega a(x)\nabla\phi_n \nabla PZ_{\lambda_n} dx + \int_\Omega a(x)p_\alpha |x-\xi|^\alpha (PU_{\lambda_n})^{p_\alpha - 1}\phi_n PZ_{\lambda_n}$$

$$= \int_\Omega \psi_n PZ_{\lambda_n} dx + c_n \int_\Omega a(x)|\nabla PZ_{\lambda_n}|^2 dx \tag{19}$$

Now, by Proposition 2.3 we easily get

$$\int_\Omega a(x)|\nabla PZ_{\lambda_n}|^2 = \lambda_n^{-2}\left(a(\xi)\int_{\mathbb{R}^N} |\nabla Z_1|^2 dy + o(1)\right), \tag{20}$$

$$\int_\Omega \psi_n PZ_{\lambda_n} dx = 0, \tag{21}$$

$$\int_\Omega a(x)\nabla\phi_n \nabla PZ_{\lambda_n} dx + \int_\Omega a(x)p_\alpha |x-\xi|^\alpha (PU_{\lambda_n})^{p_\alpha - 1}\phi_n PZ_{\lambda_n} dx$$

$$= \int_\Omega a(x)p_\alpha |x-\xi|^\alpha \left[(PU_{\lambda_n})^{p_\alpha-1} - (U_{\lambda_n})^{p_\alpha-1}\right] Z_{\lambda_n}\phi_n dx$$

$$+ \int_\Omega a(x)p_\alpha |x-\xi|^\alpha (U_{\lambda_n})^{p_\alpha-1}\left(PZ_{\lambda_n} - Z_{\lambda_n}\right)\phi_n dx$$

$$- \int_\Omega \nabla a(x)\nabla PZ_{\lambda_n}\phi_n dx = o\left(\lambda_n^{-2}\right). \tag{22}$$

Indeed

$$\int_\Omega \nabla a(x)\nabla PZ_{\lambda_n}\phi_n dx$$

$$= O\left(\|\phi_n\|_* \int_\Omega |\nabla a(x)|\lambda_n^{N-2} \frac{(\lambda_n|x-\xi|)^{\alpha+1}}{\left(1+(\lambda_n|x-\xi|)^{\alpha+2}\right)^{\frac{2}{N+\alpha}+\alpha+\frac{N+2}{2(2-\alpha)}}} dx \right)$$

$$= O\left(\lambda_n^{-2} \int_{\mathbb{R}^N} |\nabla a(\lambda_n^{-1}x+\xi)| \frac{|y|^{\alpha+1}}{(1+|y|^{\alpha+2})^{\frac{N+\alpha}{2+\alpha}+\frac{N+2}{2(2+\alpha)}}} dy \right) = O\left(\lambda_n^{-3}\right),$$

since $\nabla a(\xi) = 0$. Moreover

$$\int_\Omega a(x)p_\alpha|x-\xi|^\alpha (U_{\lambda_n})^{p_\alpha-1}\left(PZ_{\lambda_n}-Z_{\lambda_n}\right)\phi_n dx$$

$$\leq C\|\phi_n\|_* \left(\int_\Omega |x-\xi|^\alpha \frac{\lambda_n^{2+\alpha+\frac{N-2}{2}}}{\left(1+\lambda_n^{2+\alpha}|x-\xi|^{2+\alpha}\right)^{\frac{N+2+2\alpha}{2+\alpha}}} (\lambda_n^{-\frac{N}{2}}\omega_N H(x,p))dx \right)$$

$$\leq C\|\phi_n\|_*\lambda_n^{1-N} \left(\int_{\Omega_n} \frac{|y|^\alpha}{(1+|y|^{2+\alpha})^{\frac{N+6+4\alpha}{2+\alpha}}} dy \right),$$

where

$$\Omega_n := \{y \in \mathbb{R}^N : \frac{1}{\lambda_n}y+\xi \in \Omega\}.$$

Finally, if $p_\alpha \leq 2$, i.e. $N-2\alpha-6 \geq 0$

$$\int_\Omega a(x)p_\alpha|x-\xi|^\alpha \left[(PU_{\lambda_n})^{p_\alpha-1}-(U_{\lambda_n})^{p_\alpha-1}\right] Z_{\lambda_n}\phi_n$$

$$\leq C\left(\int_\Omega |x-\xi|^\alpha (PU_{\lambda_n})^{p_\alpha-2}(PU_{\lambda_n}-U_{\lambda_n})Z_{\lambda_n}\phi_n \right)$$

$$\leq \|\phi_n\|_* C \int_\Omega \left\{ |x-\xi|^\alpha \left(\frac{\lambda_n^{\frac{N-2}{2}}}{(1+\lambda^{2+\alpha}|x-\xi|^{2+\alpha})^{\frac{N-2}{2+\alpha}}} \right)^{\frac{-N+6+2\alpha}{N-2}} \right. \times$$

$$\left(\frac{\lambda_n^{\frac{N-2}{2}}}{(1+\lambda_n^{2+\alpha}|x-\xi|^{2+\alpha})^{\frac{N-2}{2(2+\alpha)}}}\right)\left[\frac{\omega_N(N-2)}{\lambda_n^{\frac{N-2}{2}}}H(x)+O\left(\lambda_n^{-\frac{N+2+2\alpha}{2}}\right)\right]\times$$

$$\left.\left((\frac{N-2}{2})\lambda_n^{\frac{N-4}{2}}\frac{1-\lambda_n^{2+\alpha}|x-\xi|^{2+\alpha}}{(1+\lambda_n^{2+\alpha}|x-\xi|^{2+\alpha})^{\frac{N+\alpha}{2+\alpha}}}\right)\right\}dx$$

$$\leq C\lambda_n^{-N+1}\int_{\Omega_n}\frac{|y|^\alpha(1-|y|^{2+\alpha})}{(1+\lambda_n^{2+\alpha}|y|^{2+\alpha})^{\frac{N+10+6\alpha}{2(2+\alpha)}}}dy$$

and if $p_\alpha > 2$, i.e. $N-2\alpha-6<0$

$$\int\limits_\Omega a(x)p_\alpha|x-\xi|^\alpha\left[(PU_{\lambda_n})^{p_\alpha-1}-(U_{\lambda_n})^{p_\alpha-1}\right]Z_{\lambda_n}\phi_n dx$$

$$\leq C\left(\int\limits_\Omega |x-\xi|^\alpha(PU_{\lambda_n})^{p_\alpha-2}(PU_{\lambda_n}-U_{\lambda_n})Z_{\lambda_n}\phi_n dx\right)$$

$$+C\left(\int\limits_\Omega |x-\xi|^\alpha(PU_{\lambda_n}-U_{\lambda_n})^{p_\alpha-1}Z_{\lambda_n}\phi_n dx\right)$$

$$\leq C\int_{\Omega_n}\frac{\lambda_n^{-N+1}|y|^\alpha(1-|y|^{2+\alpha})}{(1+\lambda_n^{2+\alpha}|y|^{2+\alpha})^{\frac{N+10+6\alpha}{2(2+\alpha)}}}dy$$

$$+C\|\phi_n\|_*\left(\int\limits_\Omega \lambda_n^{-(2+\alpha)+\frac{N-4}{2}+\frac{N-2}{2}}|x-\xi|^\alpha\frac{(1-\lambda_n^{2+\alpha}|x-\xi|^{2+\alpha})}{\left(1+(\lambda_n|x-\xi|)^{\alpha+2}\right)^{\frac{3N-2+2\alpha}{2(2+\alpha)}}}dx\right)$$

$$\leq C\int_{\Omega_n}\frac{\lambda_n^{-N+1}|y|^\alpha(1-|y|^{2+\alpha})}{(1+\lambda_n^{2+\alpha}|y|^{2+\alpha})^{\frac{N+10+6\alpha}{2(2+\alpha)}}}dy$$

$$+C\left(\int_\Omega \lambda_n^{-(2+\alpha)+\frac{N-4}{2}+\frac{N-2}{2}}|x-\xi|^\alpha\frac{(1-\lambda_n^{2+\alpha}|x-\xi|^{2+\alpha})}{\left(1+(\lambda_n|x-\xi|)^{\alpha+2}\right)^{\frac{3N-2+2\alpha}{2(2+\alpha)}}}dx\right)$$

$$\leq C\lambda_n^{-N+1}\int_{\Omega_n}\frac{|y|^\alpha(1-|y|^{2+\alpha})}{(1+\lambda_n^{2+\alpha}|y|^{2+\alpha})^{\frac{N+10+6\alpha}{2(2+\alpha)}}}dy$$

$$+C\lambda_n^{-(5+2\alpha)}\int_{\Omega_n}\frac{|y|^\alpha(1-|y|^{2+\alpha})}{(1+|y|^{\alpha+2})^{\frac{3N-2+2\alpha}{2(2+\alpha)}}}dy.$$

Here we used the crucial estimate that for any $a \geq 0$ and $b \in \mathbb{R}$

$$||a+b|^q - a^q| \leq \begin{cases} c(q) \min\left\{|b|^q, a^{q-1}|b|\right\} & \text{if } 0 < q \leq 1, \\ c(q)\left(|a|^{q-1}|b| + |b|^q\right) & \text{if } q > 1. \end{cases} \tag{23}$$

Therefore by (19), (20), (21), and (22) we immediately get that $c_n = o(1)$ and the claim is proved.

Step 2: We will show that there exist $b > 0$ and $R > 0$ such that

$$0 < b \leq \left\| \lambda_n^{-\frac{N-2}{2}} |\phi_n| \right\|_{L^\infty(B_{\frac{R}{\lambda_n}}(\xi))} \quad \text{for all } n \in \mathbb{N}. \tag{24}$$

Let us denote by $G_a(x,y)$ the Green function for the uniformly elliptic operator $M_a(f) = -\mathrm{div}(a\nabla f)$ in Ω with zero Dirichlet boundary condition. It is known (see, e.g., [9, Theorem 1.1]) that the function $G_a(x,y)$ satisfies

$$0 \leq G_a(x,y) \leq \frac{C}{|x-y|^{N-2}} \text{ for all } x, y \in \Omega.$$

Therefore, we have that

$$\begin{aligned}
|\phi_n(x)| &= \Big| \int_\Omega G_a(x,y)\left(-a(y)c_\alpha|y-\xi|^\alpha p_\alpha (PU_{\lambda_n})^{p_\alpha-1}\phi_n(y)\right. \\
&\qquad \left. +\psi_n(y) - c_n \mathrm{div}\left(a(y)PZ_{\lambda_n}(y)\right)\right) dy \Big| \\
&\leq \int_\Omega \frac{C}{|x-y|^{N-2}} a(y)c_\alpha|x-\xi|^\alpha p_\alpha (PU_{\lambda_n})^{p_\alpha-1}|\phi_n(y)|dy + \int_\Omega \frac{C}{|x-y|^{N-2}}|\psi_n(y)|dy \\
&\quad + \int_\Omega \frac{C}{|x-y|^{N-2}} a(y)c_\alpha|x-\xi|^\alpha p_\alpha (PU_{\lambda_n})^{p_\alpha-1}|Z_{\lambda_n}(y)|dy \\
&\quad + \int_\Omega \frac{C}{|x-y|^{N-2}}|\nabla a(y)||\nabla Z_{\lambda_n}(y)|dy.
\end{aligned}$$

We will denote by

$$\begin{aligned}
A_1^n(x) &:= \int_\Omega \frac{|\psi_n(y)|}{|x-y|^{N-2}} dy, \\
A_2^n(x) &:= \int_\Omega p_\alpha a(y)c_\alpha|y-\xi|^\alpha PU_{\lambda_n}^{(p_\alpha-1)}|\phi_n(y)| \frac{1}{|x-y|^{N-2}} dy, \\
A_3^n(x) &:= \int_\Omega p_\alpha a(y)c_\alpha|y-\xi|^\alpha PU_{\lambda_n}^{(p_\alpha-1)}|Z_{\lambda_n}(y)| \frac{1}{|x-y|^{N-2}} dy, \\
A_4^n(x) &:= \int_\Omega \frac{C}{|x-y|^{N-2}}|\nabla a(y)||\nabla Z_{\lambda_n}(y)|dy
\end{aligned}$$

and, as before,

$$\Omega_n = \{z \in \mathbb{R}^N : \frac{1}{\lambda_n} z + \xi \in \Omega\}.$$

Firstly, we have that

$$\begin{aligned}
A_1^n(x) &= \int_\Omega \frac{|\psi_n(y)|}{|x-y|^{N-2}} dy \\
&\leq \|\psi_n\|_{**} \int_\Omega \frac{1}{|x-y|^{N-2}} \left[\frac{\lambda_n^{(N+2)/2}}{(1+\lambda_n^{2+\alpha}|y-\xi|^{2+\alpha})^{\frac{N+2}{2(2+\alpha)}}} \right] dy \\
&\leq \|\psi_n\|_{**} \int_{\Omega_n} \frac{1}{|\lambda_n(x-\xi)-z|^{N-2}} \frac{\lambda_n^{(N-2)/2}}{(1+|z|^{2+\alpha})^{\frac{N+2}{2(2+\alpha)}}} dz \\
&\leq C \|\psi_n\|_{**} \frac{\lambda_n^{(N-2)/2}}{(1+|\lambda_n(x-\xi)|)^{\frac{N-2}{2}}},
\end{aligned}$$

the last inequality follows from (16) in Lemma 2.5. Therefore

$$\begin{aligned}
&A_1^n(x) \left(\frac{\lambda_n^{(N-2)/2}}{(1+\lambda_n^{2+\alpha}|x-\xi|^{2+\alpha})^{\frac{N-2}{2(2+\alpha)}}} \right)^{-1} \\
&\leq C \|\psi_n\|_{**} \frac{\lambda_n^{(N-2)/2}}{(1+|\lambda_n(x-\xi)|)^{\frac{N-2}{2}}} \left(\frac{\lambda_n^{(N-2)/2}}{(1+\lambda_n^{2+\alpha}|x-\xi|^{2+\alpha})^{\frac{N-2}{2(2+\alpha)}}} \right)^{-1} \\
&\leq C \|\psi_n\|_{**} \frac{(1+\lambda_n^{2+\alpha}|x-\xi|^{2+\alpha})^{\frac{N-2}{2(2+\alpha)}}}{(1+|\lambda_n(x-\xi)|)^{\frac{N-2}{2}}} \\
&\leq C \|\psi_n\|_{**}.
\end{aligned} \tag{25}$$

On the other hand, using again Lemma 2.5, we obtain

$$\begin{aligned}
A_2^n(x) &= \int_\Omega p_\alpha a(y) c_\alpha |y-\xi|^\alpha PU_{\lambda_n}^{(p_\alpha-1)} |\phi_n(y)| \frac{1}{|x-y|^{N-2}} dy \\
&\leq C\|\phi_n\|_* \int_\Omega c_\alpha |y-\xi|^\alpha \frac{1}{|x-y|^{N-2}} \left(\frac{\lambda_n^{\frac{N-2}{2}}}{(1+\lambda_n^{2+\alpha}|y-\xi|^{2+\alpha})^{\frac{N-2}{2+\alpha}}} \right)^{(p_\alpha-1)} \times \\
&\qquad \times \left(\frac{\lambda_n^{\frac{N-2}{2}}}{(1+\lambda_n^{2+\alpha}|y-\xi|^{2+\alpha})^{\frac{N-2}{2(2+\alpha)}}} \right) dy
\end{aligned}$$

$$
\begin{aligned}
&\le C\|\phi_n\|_* \int_\Omega \lambda_n^{\frac{N+2+2\alpha}{2}} \frac{|y-\xi|^\alpha}{|x-y|^{N-2}} \frac{1}{(1+\lambda_n^{2+\alpha}|y-\xi|^{2+\alpha})^{\frac{N+6+4c}{2(2+\alpha)}}} dy \\
&\le C\|\phi_n\|_* \lambda_n^{\frac{N-2}{2}} \int_{\Omega_n} \frac{|z|^\alpha}{|\lambda_n(x-\xi)-z|^{N-2}} \frac{1}{(1+|z|^{2+\alpha})^{\frac{N+6+4\alpha}{2(2+\alpha)}}} dz \\
&\le C\|\phi_n\|_* \lambda_n^{\frac{N-2}{2}} \frac{1}{(1+\lambda_n|x-\xi|)^\beta},
\end{aligned}
$$

where $\beta = \min\{\frac{N+2+2\alpha}{2}, N-2\}$. This implies that

$$
\begin{aligned}
&A_2^n(x) \left(\frac{\lambda_n^{\frac{N-2}{2}}}{(1+\lambda_n^{2+\alpha}|x-\xi|^{2+\alpha})^{\frac{N-2}{2(2+\alpha)}}} \right)^{-1} \\
&\le C\,\|\phi_n\|_*\, \lambda_n^{\frac{N-2}{2}} \frac{1}{(1+\lambda_n|x-\xi|)^\beta} \left(\frac{\lambda_n^{\frac{N-2}{2}}}{(1+\lambda_n^{2+\alpha}|x-\xi|^{2+\alpha})^{\frac{N-2}{2(2+\alpha)}}} \right)^{-1} \qquad (26)\\
&\le C\,\|\phi_n\|_* \frac{1}{(1+\lambda_n|x-\xi|)^{\beta-\frac{(N-2)}{2}}}.
\end{aligned}
$$

Moreover, we have that

$$
\begin{aligned}
A_3^n(x) &\le C \int_\Omega p_\alpha a(y) c_\alpha |y-\xi|^\alpha P U_{\lambda_n}^{(p_\alpha-1)} |Z_{\lambda_n}(y)| \frac{1}{|x-y|^{N-2}} dy \\
&\le C \int_\Omega |y-\xi|^\alpha \Big(\frac{\lambda_n^{2+\alpha}}{(1+\lambda_n^{2+\alpha}|y-\xi|^{2+\alpha})^2} \Big) \frac{\lambda_n^{\frac{N-4}{2}} |1-\lambda_n^{2+\alpha}|y-\xi|^{2+\alpha}|}{(1+\lambda_n^{2+\alpha}|y-\xi|^{2+\alpha})^{\frac{N+\alpha}{2+\alpha}}} \frac{1}{|z-y|^{N-2}} dy \\
&\le C \int_\Omega |y-\xi|^\alpha \frac{\lambda_n^{\frac{N}{2}+\alpha} |1-\lambda_n^{2+\alpha}|y-\xi|^{2+\alpha}|}{(1+\lambda_n^{2+\alpha}|y-\xi|^{2+\alpha})^{\frac{N+4+3\alpha}{2+\alpha}}} \frac{1}{|x-y|^{N-2}} dy \\
&\le C \int_\Omega \lambda_n^{\frac{N}{2}-2} \frac{|z|^\alpha (1-|z|^{2+\alpha})}{(1+|z|^{2+\alpha})^{\frac{N+4+3\alpha}{2+\alpha}}} \frac{1}{|x-y|^{N-2}} dy \\
&\le C \int_\Omega |y-\xi|^\alpha \frac{\lambda_n^{\frac{N}{2}+\alpha} |1-\lambda_n^{2+\alpha}|y-\xi|^{2+\alpha}|}{(1+\lambda_n^{2+\alpha}|y-\xi|^{2+\alpha})^{\frac{N+4+3\alpha}{2+\alpha}}} \frac{1}{|x-y|^{N-2}} dy \\
&\le C \int_{\Omega_n} \lambda_n^{\frac{N}{2}-2} \frac{|z|^\alpha}{(1+|z|^{2+\alpha})^{\frac{N+2+2\alpha}{2+\alpha}}} \frac{1}{|z-\lambda_n(x-\xi)|^{N-2}} dz \le C \frac{\lambda_n^{\frac{N}{2}-2}}{(1+\lambda_n|x-\xi|)^{N-2}}
\end{aligned}
$$

and so

$$A_3^n(x)\left(\frac{\lambda_n^{\frac{N-2}{2}}}{(1+\lambda_n^{2+\alpha}|x-\xi|^{2+\alpha})^{\frac{N-2}{2(2+\alpha)}}}\right)^{-1}$$

$$\leq C\frac{\lambda_n^{\frac{N}{2}-2}}{(1+\lambda_n|x-\xi|)^{N-2}}\left(\frac{\lambda_n^{(N-2)/2}}{(1+\lambda_n^{2+\alpha}|x-\xi|^{2+\alpha})^{\frac{N-2}{2(2+\alpha)}}}\right)^{-1} \tag{27}$$

$$\leq C\frac{\lambda_n^{-1}}{(1+\lambda_n|x-\xi|)^{\frac{N-2}{2}}}.$$

Finally, taking into account that $\nabla a(\xi)=0$ we have that

$$A_4^n(x)\leq C\int_\Omega\frac{1}{|x-y|^{N-2}}|\nabla a(y)|\lambda_n^{\frac{N-2}{2}}\frac{(\lambda_n|y-\xi|)^{\alpha+1}}{\left(1+(\lambda_n|y-\xi|)^{\alpha+2}\right)^{\frac{N+\alpha}{2+\alpha}}}dy$$

$$\leq C\int_{\Omega_n}\lambda_n^{\frac{N-8}{2}}\frac{1}{|z-\lambda_n(x-\xi)|^{N-2}}\frac{|z|^{\alpha+2}}{(1+|z|^{\alpha+2})^{\frac{N+\alpha}{2+\alpha}}}dz$$

$$\leq C\lambda_n^{\frac{N-8}{2}}\frac{1}{(1+|\lambda_n(x-\xi)|)^{N-4}}$$

and so

$$A_4^n(x)\left(\frac{\lambda_n^{\frac{N-2}{2}}}{(1+\lambda_n^{2+\alpha}|x-\xi|^{2+\alpha})^{\frac{N-2}{2(2+\alpha)}}}\right)^{-1}$$

$$\leq C\frac{\lambda_n^{\frac{N-8}{2}}}{(1+\lambda_n|x-\xi|)^{N-4}}\left(\frac{\lambda_n^{(N-2)/2}}{(1+\lambda_n^{2+\alpha}|x-\xi|^{2+\alpha})^{\frac{N-2}{2(2+\alpha)}}}\right)^{-1}$$

$$\leq C\frac{\lambda_n^{-3}}{(1+\lambda_n|x-\xi|)^{\frac{N-6}{2}}}. \tag{28}$$

Thus, using equations (25), (26), (27), and (28), we obtain that

$$|\phi_n(x)|\left(\frac{\lambda_n^{\frac{N-2}{2}}}{(1+\lambda_n^{2+\alpha}|x-\xi|^{2+\alpha})^{\frac{N-2}{2(2+\alpha)}}}\right)^{-1}$$

$$\leq (A_1^n(x)+A_2^n(x)+A_3^n(x)+A_4^n(x))\left(\frac{\lambda_n^{(N-2)/2}}{(1+\lambda_n^{2+\alpha}|x-\xi|^{2+\alpha})^{\frac{N-2}{2(2+\alpha)}}}\right)^{-1}$$

$$\leq C\left(\|\phi_n\|_*\frac{1}{(1+\lambda_n|x-\xi|)^{\beta-\frac{(N-2)}{2}}}+\|\psi_n\|_{**}\right.$$
$$\left.+\frac{\lambda_n^{-1}}{(1+\lambda_n|x-\xi|)^{\frac{N-2}{2}}}+\frac{\lambda_n^{-3}}{(1+\lambda_n|x-\xi|)^{\frac{N-6}{2}}}\right). \tag{29}$$

Since $\|\phi_n\|_* = 1$, we have that for every $n \in \mathbb{N}$ there exists $y_n \in \Omega$ such that

$$|\phi_n(y_n)|\left(\frac{\lambda_n^{\frac{N-2}{2}}}{(1+\lambda_n^{2+\alpha}|y_n-\xi|^{2+\alpha})^{\frac{N-2}{2(2+\alpha)}}}\right)^{-1}=1.$$

and by (29) we deduce that that there exists $R > 0$ such that

$$\lambda_n|y_n-\xi|<R \quad \text{for every} \ \ n\in\mathbb{N},$$

hence

$$y_n \in B_{\frac{R}{\lambda_n}}(\xi).$$

This yields

$$\begin{aligned}1&=\sup_{y\in\Omega}\left(\frac{\lambda_n^{\frac{N-2}{2}}}{(1+\lambda_n^{2+\alpha}|y-\xi|^{2+\alpha})^{\frac{N-2}{2(2+\alpha)}}}\right)^{-1}|\phi(y)|\\&\leq\sup_{y\in B_{\frac{R}{\lambda_n}}(\xi)}\left(\frac{\lambda_n^{\frac{N-2}{2}}}{(1+\lambda_n^{2+\alpha}|y-\xi|^{2+\alpha})^{\frac{N-2}{2(2+\alpha)}}}\right)^{-1}|\phi(y)|\\&\leq(1+R^{2+\alpha})^{\frac{N-2}{2(2+\alpha)}}\left\|\lambda_n^{-\frac{N-2}{2}}|\phi_n|\right\|_{L^\infty(B_{\frac{R}{\lambda_n}}(\xi))}\end{aligned}$$

which gives claim (24).
Step 3. We prove that a contradiction arises.
Let

$$\widetilde{\phi_n}(y):=\lambda_n^{-\frac{N-2}{2}}\phi_n(\frac{1}{\lambda_n}y+\xi) \text{ and } a_n(y):=a(\frac{1}{\lambda_n}y+\xi),\ y\in\Omega_n.$$

Then, we have that

$$\mathrm{div}(a_n\nabla\widetilde{\phi_n}) + a_n|y|^\alpha p_\alpha\left(\lambda_n^{-\frac{N-2}{2}}PU_{\lambda_n}(\frac{y}{\lambda_n}+\xi)\right)^{p_\alpha-1}\widetilde{\phi_n}(y) = \lambda_n^{-\frac{N+2}{2}}\psi_n(\frac{y}{\lambda_n}+\xi)$$
$$+\lambda_n^{-\frac{N+2}{2}}c_n\left[\nabla a(\frac{y}{\lambda_n}+\xi)(\nabla PZ_{\lambda_n})(\frac{y}{\lambda_n}+\xi)+a(\frac{y}{\lambda_n}+\xi)(\Delta PZ_{\lambda_n})(\frac{y}{\lambda_n}+\xi)\right] \quad (30)$$

in Ω_n. Note that, using that $\|\phi_n\|_* = 1$, we get

$$|\tilde{\phi}_n(y)| = |\lambda_n^{-\frac{N-2}{2}}\phi_n\left(\frac{y}{\lambda_n}+\xi\right)| \leq \frac{1}{(1+|y-\xi|^{2+\alpha})^{\frac{N-2}{2(2+\alpha)}}} \leq 1. \quad (31)$$

On the other hand, since $\|\psi_n\|_{**} = o(1)$, we have that

$$|\lambda_n^{-\frac{N+2}{2}}\psi_n\left(\frac{y}{\lambda_n}+\xi\right)| = \frac{o(1)}{(1+|y-\xi|^{2+\alpha})^{\frac{N+2}{2(2+\alpha)}}} = o(1).$$

Moreover, straightforward computations show that

$$\lambda_n^{-\frac{N+2}{2}}c_n\nabla a(\frac{y}{\lambda_n}+\xi)(\nabla PZ_{\lambda_n})(\frac{y}{\lambda_n}+\xi) = O\left(\frac{1}{\lambda_n^3}\frac{|y|^{\alpha+2}}{(1+|y|^{2+\alpha})^{\frac{N+\alpha}{\alpha+2}}}\right)$$
$$= O\left(\frac{1}{\lambda_n^3}\right) = o(1)$$

and

$$\lambda_n^{-\frac{N+2}{2}}c_n a(\frac{y}{\lambda_n}+\xi)(\Delta PZ_{\lambda_n})(\frac{y}{\lambda_n}+\xi) = O\left(\frac{1}{\lambda_n}\frac{|y|^{\alpha}}{(1+|y|^{2+\alpha})^{\frac{N+2\alpha+2}{\alpha+2}}}\right)$$
$$= O\left(\frac{1}{\lambda_n}\right) = o(1).$$

This implies that the right side of the equation (30) converges uniformly at zero on compact sets on $\mathbb{R}^N$. The elliptic theory implies that $(\tilde{\phi}_n)$ has a subsequence, which we still denote in the same way, such that $\tilde{\phi}_n \to \tilde{\phi}$ uniformly on every compact set (for more details see, e.g., [11]). Therefore we can pass into the limit into equation (30) to obtain

$$-a(\xi)\Delta\tilde{\phi} = a(\xi)c_\alpha p_\alpha|y|^\alpha U_{1,0}^{p_\alpha-1}\tilde{\phi} \quad \text{in } \mathbb{R}^N,$$

i.e, the function $\tilde{\phi}$ satisfies the equation

$$-\Delta\tilde{\phi} = c_\alpha p_\alpha |y|^\alpha U_{1,0}^{p_\alpha-1}\tilde{\phi} \quad \text{in } \mathbb{R}^N, \tag{32}$$

and, using equation (24), we have

$$0 < b \leq \|\tilde{\phi}\|_{L^\infty(B_R(0))}, \tag{33}$$

which implies that $\tilde{\phi} \neq 0$.

If we proceed exactly as in Lemma 4.1 in [6], then one can show that the function $\tilde{\phi}$ belongs to $D^{1,2}(\mathbb{R}^N)$ and therefore it satisfies that

$$\begin{cases} -\Delta\tilde{\phi} = c_\alpha p_\alpha |y|^\alpha U_{1,0}^{p_\alpha-1}\tilde{\phi} \text{ in } \mathbb{R}^N, \\ \tilde{\phi} \in D^{1,2}(\mathbb{R}^N). \end{cases} \tag{34}$$

On the other hand, since ϕ_n is orthogonal to PZ_{λ_n}, we have

$$\begin{aligned} 0 &= \int_\Omega a(x)\nabla\phi_n \cdot \nabla PZ_{\lambda_n} = -\int_\Omega a(x)\phi_n \Delta PZ_{\lambda_n} - \int_\Omega \phi_n \nabla a(x)\cdot\nabla PZ_{\lambda_n} \\ &= c_\alpha p_\alpha \int_\Omega |x-\xi|^\alpha U_{\lambda_n}^{p_\alpha-1}\frac{\partial U_{\lambda_n}}{\partial\lambda_n}\phi_n dx + o(1) \\ &= c_\alpha p_\alpha \frac{N-2}{2}\lambda_n^{-1}\int_{\Omega_n}|y|^\alpha \frac{1-|y|^{2+\alpha}}{(1+|y|^{2+\alpha})^{\frac{N+\alpha}{2+\alpha}+2}}\tilde{\phi}_n(y)dy + o(1). \end{aligned}$$

Here we used the fact that by scaling $\int_\Omega \phi_n \nabla a(x)\cdot\nabla PZ_{\lambda_n} = o(1)$ since $\nabla a(\xi) = 0$. Since equation (31) guaranties that $|\widetilde{\phi_n}| \leq 1$ in Ω_n, we can pass to the limit to show that

$$0 = \int_{\mathbb{R}^N} |y|^\alpha U_{1,0}^{p_\alpha-1} Z\tilde{\phi}(y)dy = \int_{\mathbb{R}^N} \nabla Z(y)\cdot\nabla\tilde{\phi}(y)dy$$

where $Z := \frac{(1-|z|^{2+\alpha})}{(1+|z|^{2+\alpha})^{\frac{N+\alpha}{2+\alpha}}}$. This implies that the function $\tilde{\phi}$ is not a multiple of Z.

This is a contradiction, since Theorem 2.1 shows that Z is the unique, up to a constant, solution to problem (34). This concludes the proof. □

3.2 *The one-dimensional reduction*

We will reduce problem (3) to a 1-dimensional one.

Proposition 3.2. *There exist λ_0 and ε_0 such that for every $\varepsilon \in (0, \varepsilon_0)$ and for every $\lambda > \lambda_0$ there exists a unique function $\phi_{\lambda,\varepsilon} \in L^\infty(\Omega) \cap H_0^1(\Omega)$ which is a solution of*

$$\begin{cases} -\text{div}\,(a(x)\,(PU_\lambda + \phi_{\varepsilon,\lambda})) = a(x)c_\alpha|x-\xi|^\alpha[(PU_\lambda+\phi_{\varepsilon,\lambda})^+]^{p_\alpha\pm\varepsilon} \\ \qquad\qquad\qquad\qquad\qquad\qquad +c_{\lambda,\varepsilon}\text{div}\,(a(x)\,(\nabla PZ_\lambda)) \quad in\ \Omega, \\ \phi_{\varepsilon,\lambda} = 0 \quad on\ \partial\Omega, \\ \int_\Omega a(x)\nabla\phi_{\lambda,\varepsilon}\nabla PZ_\lambda = 0. \end{cases} \tag{35}$$

for some $c_{\lambda,\varepsilon} \in \mathbb{R}$. *Moreover, it satisfies*

$$\|\phi_{\lambda,\varepsilon}\|_* \le c\left(\varepsilon + \frac{1}{\lambda^2}\right). \tag{36}$$

for some positive constant c.

Proof. The proof is completely analogous to that of Proposition 4.3 of [6], we outline the proof for reader's convenience.

Let us consider the space

$$X^* := \{f \in L^\infty(\Omega) : \|f\|_* < \infty\},$$

and the operator $S_{\varepsilon,\lambda} : X^* \mapsto X^*$ defined by

$$S_{\varepsilon,\lambda}(\phi) := -L_\lambda^{-1}\left\{c_\alpha|x-\xi|^\alpha a(x)\left[((PU_\lambda+\phi)^+)^{p_\alpha\pm\varepsilon} - U_\lambda^{p_\alpha} - p_\alpha PU_\alpha^{p_\alpha-1}\phi\right]\right.$$
$$\left. -\nabla a(x)\cdot\nabla PU_\lambda\right\}.$$

We remark that a function ϕ satisfies that $S_{\varepsilon,\lambda}(\phi) = \phi$ if and only if ϕ is a solution to problem (35). So we are led to show that the operator $S_{\varepsilon,\lambda}$ is a contraction on a suitable ball. First we will show that, if ε is sufficiently small and λ is large enough, then

$$S_{\varepsilon,\lambda} : B_{\varepsilon,\lambda} \to B_{\varepsilon,\lambda} \tag{37}$$

where $B_{\varepsilon,\lambda} = \{f \in X^* : \|f\|_* < \rho\left(\varepsilon + \lambda^{-2}\right)\}$.

Using property (18), we obtain

$$\|S_{\varepsilon,\lambda}(\phi)\|_* \le \|c_\alpha|x-\xi|^\alpha a(x)\left[((PU_\lambda+\phi)^+)^{p_\alpha\pm\varepsilon} - U_\lambda^{p_\alpha} - p_\alpha(PU_\lambda)^{p_\alpha-1}\phi\right]\|_{**}$$
$$+\|-\nabla a(x)\cdot\nabla PU_\lambda\|_{**}$$
$$\le \|N^1_{\lambda,\varepsilon}(\phi)\|_{**} + \|N^2_{\lambda,\varepsilon}(\phi)\|_{**} + \|N^3_{\lambda,\varepsilon}\|_{**} + \|N^4_{\lambda,\varepsilon}\|_{**} + \|N^5_{\lambda,\varepsilon}\|_{**} + \|N^6_{\lambda,\varepsilon}\|_{**}$$

where

$$N^1_{\lambda,\varepsilon}(\phi) := c_\alpha|x-\xi|^\alpha a(x)(((PU_\lambda+\phi)^+)^{p_\alpha\pm\varepsilon} - (PU_\lambda)^{p_\alpha\pm\varepsilon}$$
$$-(p_\alpha\pm\varepsilon)(PU_\lambda)^{(p_\alpha\pm\varepsilon-1)}\phi)$$
$$N^2_{\lambda,\varepsilon}(\phi) := c_\alpha|x-\xi|^\alpha a(x)((p_\alpha\pm\varepsilon)(PU_\lambda)^{(p_\alpha\pm\varepsilon-1)}\phi - p_\alpha(PU_\lambda)^{(p_\alpha-1)}\phi)$$

$$N^3_{\lambda,\varepsilon} := c_\alpha|x-\xi|^\alpha a(x)((PU_\lambda)^{(p_\alpha\pm\varepsilon)} - (PU_\lambda)^{p_\alpha})$$
$$N^4_{\lambda,\varepsilon} := c_\alpha|x-\xi|^\alpha a(x)\left[(PU_\lambda)^{p_\alpha} - U_\lambda^{p_\alpha}\right]$$
$$N^5_{\lambda,\varepsilon} := \nabla a(x)\cdot\nabla PU_\lambda.$$

Arguing exactly as in [6], we obtain the following estimates

$$\|N^1_{\lambda,\varepsilon}(\phi)\|_{**} \le \begin{cases} C\|\phi\|_*^{p_\alpha\pm\varepsilon} & \text{if } p_\alpha \le 2 \\ C\|\phi\|_*^{p_\alpha\pm\varepsilon} + C\|\phi\|_*^2 & \text{if } p_\alpha > 2. \end{cases}$$

and that

$$\|N^2_{\lambda,\varepsilon}(\phi)\|_{**} \le C\varepsilon\|\phi\|_*$$
$$\|N^3_{\lambda,\varepsilon}\|_{**} \le c_1\varepsilon$$
$$\|N^4_{\lambda,\varepsilon}\|_{**} \le c_2\lambda^{-k}, \text{ where } k = \frac{N+2+2\alpha}{2}.$$

We estimate $\|N^5_{\lambda,\varepsilon}\|_{**}$. First, using that

$$-\Delta(PU_\lambda) = -\Delta U_\lambda \text{ in } \Omega, \quad \text{and} \quad PU_\lambda = 0 \text{ on } \partial\Omega, \tag{38}$$

we obtain that

$$\nabla PU_\lambda(x) = \int_\Omega \nabla_x\left(\frac{1}{\omega_N(N-2)|x-y|^{N-2}} - H(x,y)\right) c_\alpha|y-\xi|^\alpha U_\lambda^{p_\alpha}(y)dy.$$

Moreover, we have

$$\left|\nabla_x\left(\frac{1}{|x-y|^{N-2}}\right)\right| \le \frac{C}{|x-y|^{N-1}}$$

and that (see, e.g., [1] or [8]).

$$|\nabla_x(H(x,y))| \le \frac{C}{|x-y|^{N-1}}.$$

This implies that

$$\begin{aligned}|\nabla PU_\lambda(x)| &\le C\int_\Omega \frac{c_\alpha|y-\xi|^\alpha U_\lambda^{p_\alpha}(y)}{|x-y|^{N-1}}dy \\ &\le C\lambda^{\frac{N}{2}}\int_{\Omega_\lambda} \frac{|z|^\alpha}{(1+|z|^{2+\alpha})^{\frac{n+2+2\alpha}{2+\alpha}}}\frac{1}{(|\lambda(x-\xi)-z|^{N-1})}dz\end{aligned}$$

Arguing as in [16] we can show that

$$\int_{\mathbb{R}^N} \frac{|z|^\alpha}{(1+|z|^{2+\alpha})^{\frac{N+2+2\alpha}{2+\alpha}}} \frac{1}{(|\lambda(x-\xi)-z|^{N-1})} dz \le \frac{C}{(1+|\lambda(x-\xi)|)^{N-1}}. \tag{39}$$

Therefore

$$\begin{aligned} |\nabla a(x)\cdot\nabla PU_\lambda(x)| \frac{\left(1+\lambda^{2+\alpha}|x-\xi|^{2+\alpha}\right)^{\frac{N+2}{2(2+\alpha)}}}{\lambda^{\frac{N+2}{2}}} &\le c\frac{1}{\lambda}\frac{|\nabla a(x)|}{(1+\lambda|x-\xi|)^{\frac{N-4}{2}}} \\ &\le c\frac{1}{\lambda}\frac{|x-\xi|}{(1+\lambda|x-\xi|)^{\frac{N-4}{2}}} \le c\frac{1}{\lambda^2}, \end{aligned}$$

because $\nabla a(\xi) = 0$ and $N \ge 6$. Finally, we get

$$\|N^5_{\lambda,\varepsilon}\|_{**} \le c\lambda^{-2}.$$

Summing up all the above information, we obtain that

$$\begin{aligned} \|S_{\varepsilon,\lambda}(\phi)\|_* &\le c(\|\phi\|_*^{p_\alpha\pm\varepsilon} + \|\phi\|_*^2 + \varepsilon\|\phi\|_* + \varepsilon + \lambda^{-k} + \lambda^{-2}) \\ &\le \rho\left(\varepsilon + \lambda^{-2}\right) \end{aligned}$$

provided ρ is large enough, ε is small enough, and λ is large enough. This shows (37).

Moreover, arguing in a standard way (see, for example, Proposition 4.3 of [6]) we show that, if ρ is large enough, ε is small enough, and λ is large enough, then

$$\|S_{\varepsilon,\lambda}(\phi_1) - S_{\varepsilon,\lambda}(\phi_2)\|_* \le \ell\|\phi_1 - \phi_2\|$$

for some constant $\ell < 1$, namely $S_{\varepsilon,\lambda}$ is a contraction map. This concludes the proof. □

3.3 The one-dimensional problem

We will study the reduced one-dimensional problem.

Proposition 3.3. *It holds true that in the sub-critical case*

$$c_{d_\varepsilon,\varepsilon}\int_\Omega a(x)|\nabla PZ_{\lambda_\varepsilon}|^2 := \varepsilon^2\left[+d_\varepsilon a(\xi)A(\alpha) + d_\varepsilon^2 B(\alpha)\Delta a(\xi) + o(1)\right]$$

and in the super-critical case

$$c_{d_\varepsilon,\varepsilon}\int_\Omega a(x)|\nabla PZ_{\lambda_\varepsilon}|^2 := \varepsilon^2\left[-d_\varepsilon a(\xi)A(\alpha)+d_\varepsilon^2 B(\alpha)\Delta a(\xi)+o(1)\right]$$

Here $A(\alpha)$ and $B(\alpha)$ are positive constants.

Proof. The proof is analogous to that of Lemma 5.1 of [6]. We sketch it here for reader's convenience. We write

$$\begin{aligned}
c_{d_\varepsilon,\varepsilon}&\int_\Omega a(x)\nabla PZ_{\lambda_\varepsilon}\cdot\nabla PZ_{\lambda_\varepsilon}dx\\
&=\int_\Omega a(x)\nabla(PU_{\lambda_\varepsilon}+\phi_{\varepsilon,\lambda_\varepsilon})\cdot\nabla PZ_{\lambda_\varepsilon}dx\\
&\quad-\int_\Omega a(x)c_\alpha|x-\xi|^\alpha\left[(PU_{\lambda_\varepsilon}+\phi_{\varepsilon,\lambda_\varepsilon})^+\right]^{p_\alpha\pm\varepsilon}PZ_{\lambda_\varepsilon}dx\\
&=\int_\Omega a(x)c_\alpha|x-\xi|^\alpha U_{\lambda_\varepsilon}^{p_\alpha}PZ_{\lambda_\varepsilon}dx-\int_\Omega(\nabla a(x)\cdot\nabla PU_{\lambda_\varepsilon})PZ_{\lambda_\varepsilon}dx\\
&\quad-\int_\Omega a(x)c_\alpha|x-\xi|^\alpha\left[(PU_{\lambda_\varepsilon}+\phi_{\varepsilon,\lambda_\varepsilon})^+\right]^{p_\alpha\pm\varepsilon}PZ_{\lambda_\varepsilon}dx\\
&=\int_\Omega a(x)c_\alpha|x-\xi|^\alpha\left(U_{\lambda_\varepsilon}^{p_\alpha}-U_{\lambda_\varepsilon}^{p_\alpha\pm\varepsilon}\right)PZ_{\lambda_\varepsilon}dx\\
&\quad+\int_\Omega a(x)c_\alpha|x-\xi|^\alpha\left(U_{\lambda_\varepsilon}^{p_\alpha\pm\varepsilon}-PU_{\lambda_\varepsilon}^{p_\alpha\pm\varepsilon}\right)PZ_{\lambda_\varepsilon}dx\\
&\quad+\int_\Omega a(x)c_\alpha|x-\xi|^\alpha\left(PU_{\lambda_\varepsilon}^{p_\alpha\pm\varepsilon}-\left[(PU_{\lambda_\varepsilon}+\phi_{\varepsilon,\lambda_\varepsilon})^+\right]^{p_\alpha\pm\varepsilon}\right)PZ_{\lambda_\varepsilon}dx\\
&\quad-\int_\Omega(\nabla a(x)\cdot\nabla PU_{\lambda_\varepsilon})PZ_{\lambda_\varepsilon}dx=I_1+I_2+I_3+I_4,
\end{aligned}$$

where

$$\begin{aligned}
I_1&:=\int_\Omega a(x)c_\alpha|x-\xi|^\alpha\left(U_{\lambda_\varepsilon}^{p_\alpha}-U_{\lambda_\varepsilon}^{p_\alpha\pm\varepsilon}\right)PZ_{\lambda_\varepsilon}dx,\\
I_2&:=\int_\Omega a(x)c_\alpha|x-\xi|^\alpha\left(U_{\lambda_\varepsilon}^{p_\alpha\pm\varepsilon}-PU_{\lambda_\varepsilon}^{p_\alpha\pm\varepsilon}\right)PZ_{\lambda_\varepsilon}dx\\
I_3&:=\int_\Omega a(x)c_\alpha|x-\xi|^\alpha\left(PU_{\lambda_\varepsilon}^{p_\alpha\pm\varepsilon}-\left[(PU_{\lambda_\varepsilon}+\phi_{\varepsilon,\lambda_\varepsilon})^+\right]^{p_\alpha\pm\varepsilon}\right)PZ_{\lambda_\varepsilon}dx,\\
I_4&:=-\int_\Omega(\nabla a(x)\cdot\nabla PU_{\lambda_\varepsilon})PZ_{\lambda_\varepsilon}dx.
\end{aligned}$$

We will set

$$\Omega_{\lambda_\varepsilon}:=\{x\in\mathbb{R}^N:\frac{1}{\lambda_\varepsilon}x+\xi\in\Omega\}.$$

Step 1: We have that

$$I_1 = \varepsilon\lambda_\varepsilon^{-1}\left(A(\alpha) + o(1)\right), \ \textit{with}\ A(\alpha) > 0. \tag{40}$$

First, we have

$$I_1 = I_{11} + I_{12} \tag{41}$$

where

$$I_{11} = \int_\Omega a(x)c_\alpha|x-\xi|^\alpha \left(U_{\lambda_\varepsilon}^{p_\alpha} - U_{\lambda_\varepsilon}^{p_\alpha \pm \varepsilon} \pm \varepsilon U_{\lambda_\varepsilon}^{p_\alpha} log(U_{\lambda_\varepsilon})\right) PZ_{\lambda_\varepsilon} dx,$$
$$I_{12} = \int_\Omega a(x)c_\alpha|x-\xi|^\alpha \varepsilon U_{\lambda_\varepsilon}^{p_\alpha} log(U_{\lambda_\varepsilon}) PZ_{\lambda_\varepsilon} dx.$$

An application of the Mean Value Theorem shows that

$$U_{\lambda_\varepsilon}^{p_\alpha} - U_{\lambda_\varepsilon}^{p_\alpha \pm \varepsilon} \pm \varepsilon U_{\lambda_\varepsilon}^{p_\alpha} log(U_{\lambda_\varepsilon}) = -\frac{\varepsilon^2}{2}(logU_{\lambda_\varepsilon})^2 U_{\lambda_\varepsilon}^{p_\alpha - \theta_x\varepsilon}$$

for some $\theta_x \in [0, 1]$. Therefore

$$\begin{aligned}
|I_{11}| &\leq C\frac{\varepsilon^2}{2}\int_\Omega c_\alpha|x-\xi|^\alpha (logU_{\lambda_\varepsilon})^2 U_{\lambda_\varepsilon}^{p_\alpha-\theta_x\varepsilon}|PZ_{\lambda_\varepsilon}|dx \\
&= C\varepsilon^2\lambda_\varepsilon^{N-1+\alpha}\int_\Omega \left[\frac{|x-\xi|^\alpha}{(1+\lambda_\varepsilon^{2+\alpha}|x-\xi|^{2+\alpha})^{\frac{N+2+2\alpha}{2+\alpha}-\theta_x\varepsilon\frac{N-2}{2+\alpha}}}\frac{1}{(1+\lambda_\varepsilon^{2+\alpha}|x-\xi|^{2+\alpha})^{\frac{N-2}{2+\alpha}}}\right] \\
&\qquad \times\left(\frac{N-2}{2}\log(\lambda_\varepsilon) - \frac{N-2}{2+\alpha}\log(1+\lambda_\varepsilon^{2+\alpha}|x-\xi|^{2+\alpha})\right)^2 dx \\
&\leq C\varepsilon^2\lambda_\varepsilon^{-1}\int_{\Omega_{\lambda_\varepsilon}}\left[\frac{|z|^\alpha}{(1+|z|^{2+\alpha})^{\frac{N+2+2\alpha}{2+\alpha}-\theta_x\varepsilon\frac{N-2}{2+\alpha}}}\frac{1}{(1+\lambda_\varepsilon^{2+\alpha}|x-\xi|^{2+\alpha})^{\frac{N-2}{2+\alpha}}}\right] \\
&\qquad \times\left(\frac{N-2}{2}\log(\lambda_\varepsilon) - \frac{N-2}{2+\alpha}\log(1+|z|^{2+\alpha})\right)^2 dx = o(\varepsilon\lambda_\varepsilon^{-1}).
\end{aligned}$$

On the other hand, we have that

$$\begin{aligned}
I_{12} &= \int_\Omega a(x)c_\alpha|x-\xi|^\alpha\varepsilon U_{\lambda_\varepsilon}^{p_\alpha}\log(U_{\lambda_\varepsilon})PZ_{\lambda_\varepsilon}dx \\
&= \varepsilon\int_\Omega \left\{ a(x)c_\alpha|x-\xi|^\alpha\left(\frac{\lambda_\varepsilon^{\frac{N-2}{2}}}{(1+\lambda_\varepsilon^{2+\alpha}|x-\xi|^{2+\alpha})^{\frac{N-2}{2+\alpha}}}\right)^{p_\alpha}\right.
\end{aligned}$$

$$\left(\log\left(\lambda_\varepsilon^{\frac{N-2}{2}}\right) - \log\left((1+\lambda_\varepsilon^{2+\alpha}|x-\xi|^{2+\alpha})^{\frac{N-2}{2+\alpha}}\right)\right) PZ_{\lambda_\varepsilon}\Big\} dx$$

$$= \varepsilon\lambda_\varepsilon^{\frac{N+2+2\alpha}{2}} \frac{N-2}{2}\log(\lambda_\varepsilon)\int_\Omega \frac{a(x)c_\alpha|x-\xi|^\alpha}{(1+\lambda_\varepsilon^{2+\alpha}|x-\xi|^{2+\alpha})^{\frac{N+2+2\alpha}{2+\alpha}}} PZ_{\lambda_\varepsilon} dx$$

$$-\varepsilon\lambda_\varepsilon^{\frac{N+2+2\alpha}{2}} \frac{N-2}{2+\alpha}\int_\Omega \frac{a(x)c_\alpha|x-\xi|^\alpha}{(1+\lambda_\varepsilon^{2+\alpha}|x-\xi|^{2+\alpha})^{\frac{N+2+2\alpha}{2+\alpha}}} \log(1+\lambda_\varepsilon^{2+\alpha}|x-\xi|^{2+\alpha}) PZ_{\lambda_\varepsilon} dx$$

that is,

$$I_{12} = \varepsilon\lambda_\varepsilon^{\frac{N+2+2\alpha}{2}} \frac{N-2}{2}\log(\lambda_\varepsilon)\int_\Omega \frac{a(x)c_\alpha|x-\xi|^\alpha}{(1+\lambda_\varepsilon^{2+\alpha}|x-\xi|^{2+\alpha})^{\frac{N+2+2\alpha}{2+\alpha}}} Z_{\lambda_\varepsilon} dx$$

$$-\varepsilon\lambda_\varepsilon^{\frac{N+2+2\alpha}{2}} \frac{N-2}{2+\alpha}\int_\Omega \frac{a(x)c_\alpha|x-\xi|^\alpha}{(1+\lambda_\varepsilon^{2+\alpha}|x-\xi|^{2+\alpha})^{\frac{N+2+2\alpha}{2+\alpha}}} \log(1+\lambda_\varepsilon^{2+\alpha}|x-\xi|^{2+\alpha}) Z_{\lambda_\varepsilon} dx$$

$$+\varepsilon\lambda_\varepsilon^{\frac{N+2+2\alpha}{2}} \frac{N-2}{2}\log(\lambda_\varepsilon)\int_\Omega \frac{a(x)c_\alpha|x-\xi|^\alpha}{(1+\lambda_\varepsilon^{2+\alpha}|x-\xi|^{2+\alpha})^{\frac{N+2+2\alpha}{2+\alpha}}} [PZ_{\lambda_\varepsilon} - Z_{\lambda_\varepsilon}] dx$$

$$-\varepsilon\lambda_\varepsilon^{\frac{N+2+2\alpha}{2}} \frac{N-2}{2+\alpha}\int_\Omega \left\{ \frac{a(x)c_\alpha|x-\xi|^\alpha}{(1+\lambda_\varepsilon^{2+\alpha}|x-\xi|^{2+\alpha})^{\frac{N+2+2\alpha}{2+\alpha}}} \times \right.$$

$$\left. \times \log(1+\lambda_\varepsilon^{2+\alpha}|x-\xi|^{2+\alpha}) [PZ_{\lambda_\varepsilon} - Z_{\lambda_\varepsilon}]\right\} dx = A_1 + A_2 + A_3 + A_4.$$

First of all, we have

$$A_1 = \left(\frac{N-2}{2}\right)^2 \varepsilon\lambda_\varepsilon^{N-1+\alpha}\log(\lambda_\varepsilon)\int_\Omega \frac{a(x)c_\alpha|x-\xi|^\alpha(1-\lambda_\varepsilon^{2+\alpha}|x-\xi|^{2+\alpha})}{(1+\lambda_\varepsilon^{2+\alpha}|x-\xi|^{2+\alpha})^{\frac{2N+2+3\alpha}{2-\alpha}}} dx$$

$$= \left(\frac{N-2}{2}\right)^2 c_\alpha\varepsilon\lambda_\varepsilon^{-1}\log(\lambda_\varepsilon)\int_{\Omega_{\lambda_\varepsilon}} \frac{a(\frac{1}{\lambda_\varepsilon}z+\xi)|z|^\alpha(1-|z|^{2+\alpha})}{(1+|z|^{2+\alpha})^{\frac{2N+2+3\alpha}{2+\alpha}}} dz$$

$$= \left(\frac{N-2}{2}\right)^2 c_\alpha\varepsilon\lambda_\varepsilon^{-1}\log(\lambda_\varepsilon)\int_{\mathbb{R}^N} \frac{a(\frac{1}{\lambda_\varepsilon}z+\xi)|z|^\alpha(1-|z|^{2+\alpha})}{(1+|z|^{2+\alpha})^{\frac{2N+2+3\alpha}{2+\alpha}}} dz$$

$$+ \left(\frac{N-2}{2}\right)^2 c_\alpha\varepsilon\lambda_\varepsilon^{-1}\log(\lambda_\varepsilon)\int_{\mathbb{R}^N\setminus\Omega_{\lambda_\varepsilon}} \frac{a(\frac{1}{\lambda_\varepsilon}z+\xi)|z|^\alpha(1-|z|^{2+\alpha})}{(1+|z|^{2+\alpha})^{\frac{2N+2+3\alpha}{2+\alpha}}} dz$$

$$= \left(\frac{N-2}{2}\right)^2 c_\alpha\varepsilon\lambda_\varepsilon^{-1}\log(\lambda_\varepsilon)\int_{\mathbb{R}^N} \left[a(\frac{1}{\lambda_\varepsilon}z+\xi)-a(\xi)\right] \frac{|z|^\alpha(1-|z|^{2+\alpha})}{(1+|z|^{2+\alpha})^{\frac{2N+2+3\alpha}{2+\alpha}}} dz$$

$$+ \left(\frac{N-2}{2}\right)^2 c_\alpha\varepsilon\lambda_\varepsilon^{-1}\log(\lambda_\varepsilon)\int_{\mathbb{R}^N\setminus\Omega_{\lambda_\varepsilon}} \frac{a(\frac{1}{\lambda_\varepsilon}z+\xi)|z|^\alpha(1-|z|^{2+\alpha})}{(1+|z|^{2+\alpha})^{\frac{2N+2+3\alpha}{2+\alpha}}} dz$$

$$= A_{11} + A_{12} = o(\varepsilon\lambda_\varepsilon^{-1}),$$

because

$$\int_{\mathbb{R}^N} \frac{|z|^\alpha(1-|z|^{2+\alpha})}{(1+|z|^{2+\alpha})^{\frac{2N+2+3\alpha}{2+\alpha}}}dz = 0.$$

Moreover

$$|A_{11}| \le c\left|\int_{\mathbb{R}^N} \lambda_\varepsilon^{-2}\frac{|z|^{2+\alpha}(1-|z|^{2+\alpha})}{(1+|z|^{2+\alpha})^{\frac{2N+2+3\alpha}{2+\alpha}}}dz\right| \le c\lambda_\varepsilon^{-2},$$

because $\nabla a(\xi) = 0$, and

$$|A_{12}| \le \left|\int_{\mathbb{R}^N\setminus\Omega_{\lambda_\varepsilon}} \frac{a(\frac{1}{\lambda_\varepsilon}z+\xi)|z|^\alpha(1-|z|^{2+\alpha})}{(1+|z|^{2+\alpha})^{\frac{2N+2+3\alpha}{2+\alpha}}}dz\right| \le c\lambda_\varepsilon^{-(N+\alpha)}.$$

In addition, we have that

$$\begin{aligned}
A_2 &= \mp\varepsilon\lambda_\varepsilon^{\frac{N+2+2\alpha}{2}}\frac{N-2}{2+\alpha}\int_\Omega\left\{\frac{a(x)c_\alpha|x-\xi|^\alpha}{(1+\lambda_\varepsilon^{2+\alpha}|x-\xi|^{2+\alpha})^{\frac{N+2+2\alpha}{2+\alpha}}}\times\right.\\
&\quad\left.\times\log(1+\lambda_\varepsilon^{2+\alpha}|x-\xi|^{2+\alpha})Z_{\lambda_\varepsilon}\right\}dx\\
&= \mp\varepsilon\lambda_\varepsilon^{N+\alpha-1}\frac{(N-2)^2}{2(2+\alpha)}\int_\Omega\left\{\frac{a(x)c_\alpha|x-\xi|^\alpha(1-\lambda_\varepsilon^{2+\alpha}|x-\xi|^{2+\alpha})}{(1+\lambda_\varepsilon^{2+\alpha}|x-\xi|^{2+\alpha})^{\frac{2N+2+3\alpha}{2+\alpha}}}\times\right.\\
&\quad\left.\times\log(1+\lambda_\varepsilon^{2+\alpha}|x-\xi|^{2+\alpha})\right\}dx\\
&= \mp c_\alpha\varepsilon\lambda_\varepsilon^{-1}\frac{(N-2)^2}{2(2+\alpha)}\int_{\Omega_{\lambda_\varepsilon}}\frac{a(\frac{1}{\lambda_\varepsilon}z+\xi)|z|^\alpha(1-|z|^{2+\alpha})}{(1+|z|^{2+\alpha})^{\frac{2N+2+3\alpha}{2+\alpha}}}\log(1+|z|^{2+\alpha})dz\\
&= \mp c_\alpha\varepsilon\lambda_\varepsilon^{-1}\frac{(N-2)^2}{2(2+\alpha)}\int_{\mathbb{R}^N}\frac{a(\frac{1}{\lambda_\varepsilon}z+\xi)|z|^\alpha(1-|z|^{2+\alpha})}{(1+|z|^{2+\alpha})^{\frac{2N+2+3\alpha}{2+\alpha}}}\log(1+|z|^{2+\alpha})dz\\
&\quad\pm c_\alpha\varepsilon\lambda_\varepsilon^{-1}\frac{(N-2)^2}{2(2+\alpha)}\int_{\mathbb{R}^N\setminus\Omega_{\lambda_\varepsilon}}\frac{a(\frac{1}{\lambda_\varepsilon}z+\xi)|z|^\alpha(1-|z|^{2+\alpha})}{(1+|z|^{2+\alpha})^{\frac{2N+2+3\alpha}{2+\alpha}}}\log(1+|z|^{2+\alpha})dz\\
&= \mp c_\alpha\varepsilon\lambda_\varepsilon^{-1}(1+o(1))\frac{(N-2)^2}{2(2+\alpha)}\int_{\mathbb{R}^N}\frac{a(\xi)|z|^\alpha(1-|z|^{2+\alpha})}{(1+|z|^{2+\alpha})^{\frac{2N+2+3\alpha}{2+\alpha}}}\log(1+|z|^{2+\alpha})dz\\
&\quad\pm c_\alpha\varepsilon\lambda_\varepsilon^{-1}\frac{(N-2)^2}{2(2+\alpha)}\int_{\mathbb{R}^N\setminus\Omega_{\lambda_\varepsilon}}\frac{a(\frac{1}{\lambda_\varepsilon}z+p)|z|^\alpha(1-|z|^{2+\alpha})}{(1+|z|^{2+\alpha})^{\frac{2N+2+3\alpha}{2+\alpha}}}\log(1+|z|^{2+\alpha})dz
\end{aligned}$$

and since

$$\left| c_\alpha \varepsilon \lambda_\varepsilon^{-1} \frac{(N-2)^2}{2(2+\alpha)} \int_{\mathbb{R}^N \setminus \Omega_{\lambda_\varepsilon}} \frac{a(\frac{1}{\lambda_\varepsilon} z + p)|z|^\alpha (1-|z|^{2+\alpha})}{(1+|z|^{2+\alpha})^{\frac{2N+2+3\alpha}{2+\alpha}}} \log(1+ \ z|^{2+\alpha}) dz \right|$$

$$\le c_\alpha \varepsilon \lambda_\varepsilon^{-1} \frac{(N-2)^2}{2(2+\alpha)} O\left(\int_{\mathbb{R}^N \setminus \Omega_{\lambda_\varepsilon}} \frac{|z|^\alpha (1-|z|^{2+\alpha})}{(1+|z|^{2+\alpha})^{\frac{2N+2+3\alpha}{2+\alpha}}} \log(1+|z|^{2-\alpha}) dz \right)$$

$$= o(\varepsilon \lambda_\varepsilon^{-1})$$

we get

$$A_2 = \varepsilon \lambda_\varepsilon^{-1} \left(\pm a(\xi) A(\alpha) + o(1) \right) \tag{42}$$

where (see (11))

$$A(\alpha) := -c_\alpha \frac{(N-2)^2}{2(2+\alpha)} \int_{\mathbb{R}^N} \frac{|z|^\alpha (1-|z|^{2+\alpha})}{(1+|z|^{2+\alpha})^{\frac{2N+2+3\alpha}{2+\alpha}}} \log(1+|z|^{2+\alpha}) dz > 0. \tag{43}$$

On the other hand, using that $|PZ_{\lambda_\varepsilon} - Z_{\lambda_\varepsilon}| < C\lambda_\varepsilon^{-\frac{N}{2}}$, one can show that

$$\begin{aligned} |A_3| &\le C\varepsilon \lambda_\varepsilon^{1-N} \log(\lambda_\varepsilon) \int_{\mathbb{R}^N} \frac{|z|^\alpha}{(1+|z|^{2+\alpha})^{\frac{N+2+2\alpha}{2+\alpha}}} dz \\ &= o(\varepsilon \lambda_\varepsilon^{-1}) \end{aligned} \tag{44}$$

and

$$\begin{aligned} |A_4| &\le C\varepsilon \lambda_\varepsilon^{1-N} \int_{\mathbb{R}^N} \frac{|z|^\alpha}{(1+|z|^{2+\alpha})^{\frac{N+2+2\alpha}{2+\alpha}}} \log(1+|z|^{2+\alpha}) dz \\ &= o(\varepsilon \lambda_\varepsilon^{-1}). \end{aligned} \tag{45}$$

Finally, collecting all the previous estimates, the proof of Step 1 is completed.
Step 2: It holds true that

$$I_2 = o(\lambda_\varepsilon^{-2}).$$

We write

$$I_2 := I_{21} + I_{22},$$

where

$$I_{21} := \int_{\Omega} \{a(x)c_\alpha|x-\xi|^\alpha \times \\ \times \left(U_{\lambda_\varepsilon}^{p_\alpha \pm \varepsilon} - (U_{\lambda_\varepsilon} + R_{\lambda_\varepsilon})^{p_\alpha \pm \varepsilon} + (p_\alpha \pm \varepsilon) U_{\lambda_\varepsilon}^{p_\alpha \pm \varepsilon - 1} R_{\lambda_\varepsilon}\right) PZ_{\lambda_\varepsilon}\Big\} \, dx$$

and

$$I_{22} := -\int_{\Omega} a(x)c_\alpha|x-\xi|^\alpha (p_\alpha \pm \varepsilon) U_{\lambda_\varepsilon}^{p_\alpha \pm \varepsilon - 1} R_{\lambda_\varepsilon} PZ_{\lambda_\varepsilon} dx.$$

Following exactly the same argument in [6], a direct computation shows that

$$|I_{21}| \leq \begin{cases} C\int_\Omega |x-\xi|^\alpha |R_{\lambda_\varepsilon}|^{p_\alpha \pm \varepsilon} |PZ_{\lambda_\varepsilon}| & \text{if } p_\alpha \leq 2 \\ C\int_\Omega |x-\xi|^\alpha \left(|R_{\lambda_\varepsilon}|^{p_\alpha \pm \varepsilon} + U_{\lambda_\varepsilon}^{p_\alpha - 2 \pm \varepsilon} |R_{\lambda_\varepsilon}|^2\right) |PZ_{\lambda_\varepsilon}| & \text{if } p_\alpha > 2, \end{cases}$$

and that

$$I_{21} = o(\lambda_\varepsilon^{-2}).$$

On the other hand, we have that

$$\begin{aligned} I_{22} &= (p_\alpha \pm \varepsilon)\lambda_\varepsilon^{2+\alpha \pm \varepsilon \frac{N-2}{2}} \int_\Omega \Bigg\{ \frac{a(x)c_\alpha|x-\xi|^\alpha}{(1+\lambda_\varepsilon^{2+\alpha}|x-\xi|^{2+\alpha})^{2 \pm \varepsilon \frac{N-2}{2+\alpha}}} \times \\ &\times \left(-\omega_N (N-2)\lambda_\varepsilon^{-\frac{N-2}{2}} H(x) + f_{\lambda_\varepsilon}(x)\right)\left(Z_{\lambda_\varepsilon} + \frac{(N-2)^2}{2}\omega_N \lambda_\varepsilon^{-\frac{N}{2}} H(x) + N_{\lambda_\varepsilon}(x)\right)\Bigg\} \, dx \\ &= (1+o(1))c_\alpha (N-2) p_\alpha \omega_N \lambda_\varepsilon^{\alpha - \frac{N-6}{2}} \int_\Omega \frac{a(x)|x-\xi|^\alpha}{(1+\lambda_\varepsilon^{2+\alpha}|x-\xi|^{2+\alpha})^{2\pm\varepsilon\frac{N-2}{2+\alpha}}} H(x) Z_{\lambda_\varepsilon} dx \\ &= (1+o(1))c_\alpha \frac{(N-2)^2}{2} p_\alpha \omega_N \lambda_\varepsilon^{1+\alpha} \int_\Omega \frac{a(x)|x-\xi|^\alpha (1-\lambda_\varepsilon^{2+\alpha}|x-\xi|^{2+\alpha})}{(1+\lambda_\varepsilon^{2+\alpha}|x-\xi|^{2+\alpha})^{2\pm\varepsilon\frac{N-2}{2+\alpha}+\frac{N+\alpha}{2+\alpha}}} H(x) dx \\ &= (1+o(1))c_\alpha \frac{(N-2)^2}{2} p_\alpha \omega_N \lambda_\varepsilon^{1-N} \int_{\Omega_{\lambda_\varepsilon}} \frac{a(\frac{1}{\lambda_\varepsilon}z+\xi)|z|^\alpha(1-|z|^{2+\alpha})}{(1+|z|^{2+\alpha})^{2\pm\varepsilon\frac{N-2}{2+\alpha}+\frac{N+\alpha}{2+\alpha}}} H(\frac{1}{\lambda_\varepsilon}z+\xi) dz \\ &= o(\lambda_\varepsilon^{-2}). \end{aligned}$$

The proof of Step 2 is complete.

Step 3: We have that

$$I_3(\varepsilon) = o(\lambda_\varepsilon^{-2}).$$

Following the same argument as in Lemma 5.1 of [6], we have that

$$I_3(\varepsilon) = \begin{cases} O(\lambda_\varepsilon^{-1}\|\phi_\varepsilon\|_*^2) + O(\varepsilon\lambda_\varepsilon^{-1}\log(\varepsilon)\|\phi_\varepsilon\|_*) & \text{if } p_\alpha \le 2 \\ O(\lambda_\varepsilon^{-1}(\|\phi_\varepsilon\|_*^{p_\alpha \pm \varepsilon} + \|\phi_\varepsilon\|_*^2)) + O(\varepsilon\lambda_\varepsilon^{-1}\log(\varepsilon)\|\phi_\varepsilon\|_*) & \text{if } p_\alpha > 2, \end{cases}$$

which proves the claim, since $\lambda_\varepsilon \sim \varepsilon$ and $\|\phi_{\varepsilon,\lambda_\varepsilon}\|_* = O(\varepsilon)$.
Step 4: We have that

$$I_4(\varepsilon) = \lambda_\varepsilon^{-2}(-\Delta a(\xi)B(\alpha) + o(1)) \text{ with } B(\alpha) > 0.$$

We have

$$I_4 := -\int_\Omega (\nabla a(x)\cdot\nabla PU_{\lambda_\varepsilon})\,PZ_{\lambda_\varepsilon}dx = -\int_\Omega (\nabla a(x)\cdot\nabla U_{\lambda_\varepsilon})\,Z_{\lambda_\varepsilon}dx$$
$$-\int_\Omega (\nabla a(x)\cdot\nabla U_{\lambda_\varepsilon})\,(PZ_{\lambda_\varepsilon} - Z_{\lambda_\varepsilon})\,dx - \int_\Omega (\nabla a(x)\cdot\nabla\,(PU_{\lambda_\varepsilon} - U_{\lambda_\varepsilon})\,PZ_{\lambda_\varepsilon}dx.$$

The leading term is

$$\begin{aligned}
&-\int_\Omega (\nabla a(x)\cdot\nabla U_{\lambda_\varepsilon})\,Z_{\lambda_\varepsilon}dx \\
&= -\lambda_\varepsilon^{N-1+\alpha}\int_\Omega (\nabla a(x)\cdot(x-\xi))\,\frac{|x-\xi|^\alpha(1-\lambda_\varepsilon^{2+\alpha}|x-\xi|^{2+\alpha})}{(1+\lambda_\varepsilon^{2+\alpha}|x-\xi|^{2+\alpha})^{\frac{2(N+\alpha)}{2+\alpha}}}dx \\
&= -\lambda_\varepsilon^{-2}\int_{\Omega_{\lambda_\varepsilon}} \nabla a\left(\frac{y}{\lambda}+\xi\right)\cdot y\frac{|y|^\alpha(1-|y|^{2+\alpha})}{(1+|y|^{2+\alpha})^{\frac{2(N+\alpha)}{2+\alpha}}}dy \\
&= -\lambda_\varepsilon^{-2}\int_{\Omega_{\lambda_\varepsilon}} \sum_{i,j=1}^N \partial_{ij}a(\xi)y_iy_j\frac{|y|^\alpha(1-|y|^{2+\alpha})}{(1+|y|^{2+\alpha})^{\frac{2(N+\alpha)}{2+\alpha}}}dy + O\left(\lambda_\varepsilon^{-3}\right) \\
&= +\lambda_\varepsilon^{-2}\Delta a(\xi)\frac{1}{N}\left(-\int_{\mathbb{R}^N}\frac{|y|^{2+\alpha}(1-|y|^{2+\alpha})}{(1+|y|^{2+\alpha})^{\frac{2(N+\alpha)}{2+\alpha}}}dy\right) + O\left(\lambda_\varepsilon^{-3}\right) \\
&= \lambda_\varepsilon^{-2}\Delta a(\xi)B(\alpha) + O\left(\lambda_\varepsilon^{-3}\right)
\end{aligned}$$

where (see (12))

$$B(\alpha) := -\frac{1}{N}\int_{\mathbb{R}^N} \frac{|y|^{2+\alpha}(1-|y|^{2+\alpha})}{(1+|y|^{2+\alpha})^{\frac{2(N+\alpha)}{2+\alpha}}}dy > 0. \tag{46}$$

Collecting all the previous estimates, the proof of Proposition 3.3 is completed. □

Proof (Proof of Theorem 1.1). It is clear that we have to find a positive real number d_ε such that the number $c_{\lambda_\varepsilon,\varepsilon}$ in (35) is zero. By our construction, taking also into account the maximum principle, it will immediately follow that the function $u_\varepsilon = PU_{\lambda_\varepsilon} + \phi_{\varepsilon,\lambda_\varepsilon}$ is a solution of problem (3). On the other hand, it is easy to check that if $\Delta a(\xi) < 0$ there exists $d_\varepsilon > 0$ such that $c_{d_\varepsilon,\varepsilon} = 0$ in the sub-critical case and so problem (4) has a solution and if $\Delta a(\xi) > 0$ there exists $d_\varepsilon > 0$ such that $c_{d_\varepsilon,\varepsilon} = 0$ in the super-critical case and so problem (5) has a solution. The proof of Theorem 1.1 is completed. □

Acknowledgements M. Grossi and A. Pistoia are supported by PRIN-2009-WRJ3W7 grant. J. Faya is supported by Fondecyt postdoctoral grant 3150172 (Chile).

References

1. Ackermann, N., Clapp, M., Pistoia, A.: Boundary clustered layers near the higher critical exponents. J. Differ. Equat. **245**, 4168–4193 (2013)
2. Byeon, J., Wang, Z.: On the Hénon equation: asymptotic profile of ground states II. J. Differ. Equat. **216**(1), 78–108 (2005)
3. Byeon, J., Wang, Z.: On the Hénon equation: asymptotic profile of ground states. I. Ann. Inst. H. Poincaré Anal. Non Linéaire **23**(6), 803–828 (2006)
4. del Pino, M., Felmer, P., Musso, M.: Two-bubble solutions in the super-critical Bahri-Coron's problem. Calc. Var. Partial Differ. Equat. **16**, 113–145 (2003)
5. Dos Santos, E.M., Pacella, F.: Hénon type equations and concentration on spheres. arXiv:1407.6581, Indiana Univ. Math. Journal (to appear)
6. Gladiali, F., Grossi, M.: Supercritical elliptic problem with nonautonomous nonlinearities. J. Differ. Equat. **253**(9), 2616–2645 (2012)
7. Gladiali, F., Grossi, M., Neves, S.: Nonradial solutions for the Hénon equation in $\mathbb{R}^N$. Adv. Math. **249**, 1–36 (2013)
8. Gomez, S.M.: On a singular nonlinear elliptic problem. SIAM J. Math. Anal. **17**, 1359–1369 (1986)
9. Grüter, M., Widman, K.O.: The Green function for uniformly elliptic equations. Manuscripta Math. **37**(3), 303–342 (1982)
10. Hénon, M.: Numerical experiments on the stability of spherical stellar systems. Astron. Astrophys. **24**, 229–238 (1973)
11. Miranda, C.: Partial Differential Equations of Elliptic Type. Springer, New York (1970)
12. Ni, W.M.: A Nonlinear Dirichlet problem on the unit ball and its applications. Indiana Univ. Math. J. **31**, 801–807 (1982)
13. Pistoia, A., Serra, E.: Multi-peak solutions for the Hénon equation with slightly subcritical growth. Math. Z. **23**, 75–97 (2007)
14. Serra, E.: Non radial positive solutions for the Hénon equation with critical growth. Calc. Var. Partial Differ. Equat. **23**(3), 301–326 (2005)

15. Smets, D., Su, J., Willem, M.: Non radial ground states for the Hénon equation. Commun. Contemp. Math. **4**, 467–480 (2002)
16. Wei, J., Yan, S.: Infinitely many nonradial solutions for the Hénon equation with critical growth. Rev. Mat. Iberoam. **29**, 997–1020 (2013)

The sub-supersolution method for Kirchhoff systems: applications

Giovany M. Figueiredo and Antonio Suárez

Mathematics Subject Classification (2010): 35J35, 35J65,49J40

1 Introduction

In this note we study the existence of solutions of a nonlinear Kirchhoff system

$$\begin{cases} -M_1(\|u_1\|^2)\Delta u_1 = f_1(x,u_1,u_2) & \text{in } \Omega, \\ -M_2(\|u_2\|^2)\Delta u_2 = f_2(x,u_1,u_2) & \text{in } \Omega, \\ u_1 = u_2 = 0 & \text{on } \partial\Omega, \end{cases} \tag{1}$$

where $\Omega \subset \mathbb{R}^N$, $N \geq 1$, is a regular and bounded domain,

$$\|u\|^2 := \int_\Omega |\nabla u|^2 dx, \qquad \text{for } u \in H_0^1(\Omega),$$

G.M. Figueiredo (✉)
Faculdade de Matemática, Universidade Federal do Pará,
CEP 66075-110 Belém, PA, Brazil
e-mail: giovany@ufpa.br

A. Suárez
Faculdade de Matemáticas, Departamento de Ecuaciones Diferenciales y Análisis Numérico,
University de Sevilla, C/Tarfia s/n, 41012 Sevilla, Spain
e-mail: suarez@us.es

A.N. Carvalho et al. (eds.), *Contributions to Nonlinear Elliptic Equations and Systems*, Progress in Nonlinear Differential Equations and Their Applications 86,
DOI 10.1007/978-3-319-19902-3_14

M_i, $i = 1, 2$ are continuous functions verifying

$$(M) \qquad M_i : \mathbb{R}_+ \mapsto \mathbb{R}_+ \quad \text{and } \exists m_0 > 0 \text{ such that } M_i(t) \geq m_0 > 0 \ \forall t \in \mathbb{R}_+,$$

and $f_i \in C(\overline{\Omega} \times \mathbb{R}^2)$. We assume (M) along the paper.

Basically, in our knowledge, similar systems to (1) have been analyzed in several papers. In [3, 4, 6, 8, 10] and the references therein, variational methods have been applied to prove existence and multiplicity of positive solutions for systems as (1). In [1] and [2] the sub-supersolution method has been used to prove the existence of solution with M_i increasing and bounded from above and below for positive constants, that is, there exist positive constants $0 < m_i \leq m_i^\infty < \infty$ such that

$$0 < m_i \leq M_i(t) \leq m_i^\infty < \infty \quad i = 1, 2, \quad \forall t \geq 0.$$

However, in both papers the authors use a comparison principle (see, for instance, Lemma 2.1 in [1]) which seems not to be correct, see [5].

In this paper, we prove that the sub-supersolution method works for system (1), when the sub-supersolution is defined in an appropriate way, see Theorem 3.3. Indeed, in this case, the definition of sub-supersolution depends on the monotony of the nonlinear reaction term (in a similar way to the local problems, see, for instance, [9]) and on the functions M_i. In order to prove this result, we transform our Kirchhoff system (1) into another with general non-local term depending only on the unknown variable u_i but not the $\|u_i\|^2$. So, as a consequence, we establish a very general sub-supersolution method for a large class of systems with nonlinear and non-local terms (see Theorem 2.2).

The paper is organized as follows. In Section 2 we show that the sub-supersolution method works for general non-local systems. In Section 3, under very general conditions on M_i, we transform our system (1) into a non-local systems, and apply the method of Section 2. Section 4 is devoted to apply our method for different particular systems.

2 The sub-super method for non-local systems

First of all we show that the sub-supersolution method works well for non-local systems of the following type

$$\begin{cases} -\Delta u_1 = g_1(x, u_1, u_2, B_1(u_1), B_2(u_2), C_1(u_1, u_2)) & \text{in } \Omega, \\ -\Delta u_2 = g_2(x, u_1, u_2, B_1(u_1), B_2(u_2), C_2(u_1, u_2)) & \text{in } \Omega, \\ u_1 = u_2 = 0 & \text{on } \partial\Omega, \end{cases} \tag{2}$$

where $g_i : \Omega \times \mathbb{R}^5 \mapsto \mathbb{R}$ is a continuous function, $B_i : L^\infty(\Omega) \mapsto \mathbb{R}$, $C_i : (L^\infty(\Omega))^2 \mapsto \mathbb{R}$ are continuous operators. Given $w \leq z$ a.e. in Ω, we denote by

$$[w, z] := \{u : w(x) \leq u(x) \leq z(x) \quad \text{a.e. } x \in \Omega\}.$$

Definition 2.1. We say that the pair $(\underline{u}_1, \overline{u}_1)$, $(\underline{u}_2, \overline{u}_2)$, with $\underline{u}_i, \overline{u}_i \in H^1(\Omega) \cap L^\infty(\Omega)$, is a pair of sub-supersolution of (2) if

1. $\underline{u}_i \leq \overline{u}_i$ in Ω and $\underline{u}_i \leq 0 \leq \overline{u}_i$ on $\partial\Omega$ for $i = 1, 2$,
2. $$-\Delta \underline{u}_1 - g_1(x, \underline{u}_1, v, B_1(u), B_2(v), C_1(u, v)) \leq 0$$ $$-\Delta \overline{u}_1 - g_1(x, \overline{u}_1, v, B_1(u), B_2(v), C_1(u, v)) \geq 0$$ in the weak sense for all $(u, v) \in [\underline{u}_1, \overline{u}_1] \times [\underline{u}_2, \overline{u}_2]$.
3. $$-\Delta \underline{u}_2 - g_2(x, u, \underline{u}_2, B_1(u), B_2(v), C_2(u, v)) \leq 0$$ $$-\Delta \overline{u}_2 - g_2(x, u, \overline{u}_2, B_1(u), B_2(v), C_2(u, v)) \geq 0$$ in the weak sense for all $(u, v) \in [\underline{u}_1, \overline{u}_1] \times [\underline{u}_2, \overline{u}_2]$.

The main result in this section is:

Theorem 2.2. *Assume that there exists a pair of sub-supersolution of (2) in the sense of Definition 2.1. Then, there exists a solution $(u_1, u_2) \in (H_0^1(\Omega) \cap L^\infty(\Omega))^2$ of (2) such that $u_i \in [\underline{u}_i, \overline{u}_i]$, $i = 1, 2$.*

Proof. For $i = 1, 2$, define the truncation operators

$$T_i u(x) := \begin{cases} \overline{u}_i(x) \text{ if } u(x) \geq \overline{u}_i(x), \\ u(x) \ \text{ if } \underline{u}_i(x) \leq u(x) \leq \overline{u}_i(x), \\ \underline{u}_i(x) \text{ if } u(x) \leq \underline{u}_i(x), \end{cases} \tag{3}$$

and the Nemytskii operators $F_i : (L^\infty(\Omega))^2 \mapsto L^\infty(\Omega)$ given by

$$F_i(u_1, u_2)(x) := g_i(x, T_1(u_1)(x), T_2(u_2)(x), B_1(T_1(u_1)), B_2(T_2(u_2)), C_i(T_1(u_1), T_2(u_2))).$$

It is clear that F_i is continuous and bounded, because there exists $M > 0$ such that

$$\|F_i(u_1, u_2)\|_\infty \leq M \quad \text{for all } u_1, u_2 \in L^\infty(\Omega).$$

Consider the problem

$$\begin{cases} -\Delta w_1 = F_1(u_1, u_2) \text{ in } \Omega, \\ -\Delta w_2 = F_2(u_1, u_2) \text{ in } \Omega, \\ w_1 = w_2 = 0 \qquad \text{on } \partial\Omega. \end{cases} \tag{4}$$

We can define the operator $\mathscr{T}$ by $(u_1, u_2) \mapsto (w_1, w_2) := \mathscr{T}(u_1, u_2)$ being (w_1, w_2) the unique solution of (4). It is clear that $\mathscr{T}$ is well defined, it is a compact operator and $\mathscr{T}(B_M) \subset B_M$ for some $M > 0$, where B_M denotes the ball in $(L^\infty(\Omega))^2$ centered in $(0,0)$ and radius M. Hence, by the Schauder Fixed Point Theorem there exists $(u_1, u_2) \in (L^\infty(\Omega))^2$ such that $(u_1, u_2) = \mathscr{T}(u_1, u_2)$, and then

$$\begin{cases} -\Delta u_1 = F_1(u_1, u_2) \text{ in } \Omega, \\ -\Delta u_2 = F_2(u_1, u_2) \text{ in } \Omega, \\ u_1 = u_2 = 0 \qquad\quad \text{on } \partial\Omega. \end{cases} \tag{5}$$

Now, we show that $u_i \in [\underline{u}_i, \overline{u}_i]$, which implies that (u_1, u_2) is solution of (2). Let us show that

$$u_1 \leq \overline{u}_1 \quad \text{in } \Omega,$$

the other inequalities can be proved similarly. Indeed, in the definition of supersolution of $\overline{u}_1$ we can take $u = T_1(u_1)$, $v = T_2(u_2)$ and then,

$$-\Delta\overline{u}_1 \geq g_1(x, \overline{u}_1, T_2(u_2), B_1(T_1(u_1)), B_2(T_2(u_2)), C_1(T_1(u_1), T_2(u_2))),$$

and so, denoting $z := \overline{u}_1 - u_1$ we get

$$\begin{aligned} -\Delta z &\geq g_1(x, \overline{u}_1, T_2(u_2), B_1(T_1(u_1)), B_2(T_2(u_2)), C_1(T_1(u_1), T_2(u_2))) - F(u_1, u_1) \\ &= g_1(x, \overline{u}_1, T_2(u_2), B_1(T_1(u_1)), B_2(T_2(u_2)), C_1(T_1(u_1), T_2(u_2))) \\ &\quad - g_1(x, T_1(u_1)(x), T_2(u_2)(x), B_1(T_1(u_1)), B_2(T_2(u_2)), C_1(T_1(u_1), T_2(u_2))). \end{aligned}$$

Now, multiplying by $(\overline{u}_1 - u_1)^-$ we obtain

$$\int_\Omega |\nabla(\overline{u}_1 - u_1)^-|^2 \leq 0,$$

whence we conclude the result. □

3 The sub-supersolution for Kirchhoff systems

First, we are going to transform (1) into a nonlocal system as (2). Indeed, define

$$N_i(t) := M_i(t)t$$

and assume that N_i is invertible, and so define

$$G_i(t) = N_i^{-1}(t).$$

Finally, define the non-local operators $\mathscr{R}_i : (L^\infty(\Omega))^2 \mapsto \mathbb{R}$ by

$$\mathscr{R}_i(u_1, u_2) = M_i\left(G_i\left(\int_\Omega f_i(x, u_1, u_2)u_i\right)\right).$$

Lemma 3.1. *Assume that*

(N) $\quad N_i,\ i = 1, 2$ *are invertible.*

Then, (1) is equivalent to

$$\begin{cases} -\Delta u_1 = F_1(x, u_1, u_2, C_1(u_1, u_2)) \text{ in } \Omega, \\ -\Delta u_2 = F_2(x, u_1, u_2, C_2(u_1, u_2)) \text{ in } \Omega, \\ u_1 = u_2 = 0 \qquad\qquad\qquad\qquad\quad \text{on } \partial\Omega, \end{cases} \tag{6}$$

where

$$C_i(u_1, u_2) = \mathscr{R}_i(u_1, u_2), \quad F_i(x, t_1, t_2, r) = \frac{f_i(x, t_1, t_2)}{r},\ i = 1, 2.$$

Proof. Assume that (u_1, u_2) is solution of (1). Multiplying (1) by u_i and integrating, we get

$$M_i(\|u_i\|^2)\|u_i\|^2 = \int_\Omega f_i(x, u_1, u_2)u_i,$$

and then,

$$\|u_i\|^2 = G_i\left(\int_\Omega f_i(x, u_1, u_2)u_i\right) \Longrightarrow M_i(\|u_i\|^2) = \mathscr{R}_i(u_1, u_2).$$

By (M), $\mathscr{R}_i(u_1, u_2) \geq m_0$ and then we can divide by $\mathscr{R}_i(u_1, u_2)$. Hence, we conclude that (u_1, u_2) is solution of (6).

Reciprocally, if (u_1, u_2) is solution of (6), then multiplying by u_i we obtain

$$\|u_i\|^2 = \frac{\int_\Omega f_i(x, u_1, u_2)u_i}{\mathscr{R}_i(u_1, u_2)} = \frac{\int_\Omega f_i(x, u_1, u_2)u_i}{M_i(G_i(\int_\Omega f_i(x, u_1, u_2)u_i)} = G_i\left(\int_\Omega f_i(x, u_1, u_2)u_i\right),$$

where we have used that $N_i \circ G_i(t) = t$, that is $M_i(G_i(t))G_i(t) = t$. Applying M_i in that above equality we get

$$M_i(\|u_i\|^2) = \mathscr{R}_i(u_1, u_2),$$

and so (u_1, u_2) is solution of (1). This completes the proof. □

As a consequence of this result and Theorem 2.2, we have the following results.

Definition 3.2. We say that the pair $(\underline{u}_1, \overline{u}_1)$, $(\underline{u}_2, \overline{u}_2)$, with $\underline{u}_i, \overline{u}_i \in H^1(\Omega) \cap L^\infty(\Omega)$, is a pair of sub-supersolution of (1) if

1. $\underline{u}_i \leq \overline{u}_i$ in Ω and $\underline{u}_i \leq 0 \leq \overline{u}_i$ on $\partial\Omega$ for $i = 1, 2$,
2. $$-\mathscr{R}_1(u, v)\Delta \underline{u}_1 - f_1(x, \underline{u}_1, v) \leq 0 \leq -\mathscr{R}_1(u, v)\Delta \overline{u}_1 - f_1(x, \overline{u}_1, v)$$
   in the weak sense for all $(u, v) \in [\underline{u}_1, \overline{u}_1] \times [\underline{u}_2, \overline{u}_2]$.
3. $$-\mathscr{R}_2(u, v)\Delta \underline{u}_2 - f_2(x, u, \underline{u}_2) \leq 0 \leq -\mathscr{R}_2(u, v)\Delta \overline{u}_2 - f_2(x, u, \overline{u}_2)$$
   in the weak sense for all $(u, v) \in [\underline{u}_1, \overline{u}_1] \times [\underline{u}_2, \overline{u}_2]$.

Theorem 3.3. *Assume (M) and (N). If there exists a pair of sub-supersolution of (6) in the sense of Definition 3.2, then there exists a solution* (u_1, u_2) *of (1) such that* $(u_1, u_2) \in [\underline{u}_1, \overline{u}_1] \times [\underline{u}_2, \overline{u}_2]$.

Remark 3.4. Observe that if M_i is increasing, then it verifies (N).

4 Applications

4.1 Non-local Lotka–Volterra models

Consider the classical diffusive Lotka–Volterra model with non-local interaction

$$\begin{cases} -\Delta u_1 = u_1(\lambda - u_1 - b \int_\Omega u_2) & \text{in } \Omega, \\ -\Delta u_2 = u_2(\mu - u_2 - c \int_\Omega u_1) & \text{in } \Omega, \\ u_1 = u_2 = 0 & \text{on } \partial\Omega, \end{cases} \tag{7}$$

where $\lambda, \mu \in \mathbb{R}$ and $b, c \in \mathbb{R}$. Here, u_1 and u_2 denote two species inhabiting in Ω, the habitat, which is surrounded by inhospitable areas. Here, λ and μ represent the intrinsic growth rates of each species, and b, c the interaction rates between the species: if both b and c are positive numbers the species compete, if both are negative they cooperate and finally in the case $b > 0$ and $c < 0$, u_1 denotes the prey and u_2 the predator. The main novelty in (7) is that this interaction is non-local, that is, the interaction between both species at the point $x \in \Omega$ depends on not only the value at x but also the value to the entire domain Ω, see [7].

In order to enunciate the main result, we need introduce some notation. Denote by $\varphi > 0$ the eigenfunction associated with λ_1, the principal eigenvalue of the $-\Delta$ under Dirichlet boundary conditions, such that $\|\varphi\|_\infty = 1$. It is well known that the classical logistic equation

$$\begin{cases} -\Delta w = w(\gamma - w) & \text{in } \Omega, \\ w = 0 & \text{on } \partial\Omega, \end{cases} \tag{8}$$

possesses a unique positive solution if and only if $\gamma > \lambda_1$. In such case, the positive solution is unique. We denote it by θ_γ. We prolong the definition of $\theta_\gamma \equiv 0$ when $\gamma \le \lambda_1$. It is well known that $\gamma \mapsto \theta_\gamma$ is increasing in γ and that $\theta_\gamma \le \gamma$.

Theorem 4.1. *1. Assume that $b, c > 0$. Then, (7) possesses at least a positive solution if*

$$\lambda - b\int_\Omega \theta_\mu > \lambda_1 \quad \textit{and} \quad \mu - c\int_\Omega \theta_\lambda > \lambda_1. \tag{9}$$

2. Assume that $b, c < 0$ and $bc|\Omega|^2 < 1$. Then, (7) possesses at least a positive solution if (λ, μ) verifies condition (9).

3. Assume $b > 0$, $c < 0$ and

$$\lambda - b|\Omega|(\mu + c\int_\Omega \theta_\lambda) > \lambda_1 \quad \textit{and} \quad \mu > \lambda_1. \tag{10}$$

Proof. 1. We can take as pair of sub-supersolution

$$(\underline{u}_1, \overline{u}_1) = (\theta_{\lambda - b\int_\Omega \theta_\mu}, \theta_\lambda), \quad (\underline{u}_2, \overline{u}_2) = (\theta_{\mu - c\int_\Omega \theta_\lambda}, \theta_\mu)$$

First, observe that $\underline{u}_1 \le \overline{u}_1$ and $\underline{u}_2 \le \overline{u}_2$ in Ω. Now, we have to verify four inequalities. Let us only check two of them:

$$-\Delta \underline{u}_1 \le \underline{u}_1(\lambda - \underline{u}_1 - b\int_\Omega \overline{u}_2), \quad -\Delta \overline{u}_1 \ge \overline{u}_1(\lambda - \overline{u}_1 - b\int_\Omega \underline{u}_2).$$

Observe that

$$\begin{aligned} -\Delta \underline{u}_1 &= -\Delta \theta_{\lambda - b\int_\Omega \theta_\mu} = \theta_{\lambda - b\int_\Omega \theta_\mu}(\lambda - b\int_\Omega \theta_\mu - \theta_{\lambda - b\int_\Omega \theta_\mu}) \\ &= \underline{u}_1(\lambda - \underline{u}_1 - b\int_\Omega \overline{u}_2). \end{aligned}$$

On the other hand,

$$-\Delta \overline{u}_1 = -\Delta \theta_\lambda = \theta_\lambda(\lambda - \theta_\lambda) \ge \theta_\lambda(\lambda - \theta_\lambda - b\int_\Omega \underline{u}_2) = \overline{u}_1(\lambda - \overline{u} - b\int_\Omega \underline{u}_2).$$

This completes the first paragraph.

2. In this case, take

$$(\underline{u}_1, \overline{u}_1) = (\theta_{\lambda - b\int_\Omega \theta_\mu}, M), \quad (\underline{u}_2, \overline{u}_2) = (\theta_{\mu - c\int_\Omega \theta_\lambda}, N),$$

where M, N are positive constants verifying

$$M \geq \lambda - bN|\Omega| \quad \text{and} \quad N \geq \mu - cM|\Omega|,$$

which exist because $bc|\Omega|^2 < 1$.
We prove now that they are sub-supersolutions. Again we only show two inequalities:

$$-\Delta \underline{u}_1 \leq \underline{u}_1(\lambda - \underline{u}_1 - b\int_\Omega \underline{u}_2), \quad -\Delta \overline{u}_1 \geq \overline{u}_1(\lambda - \overline{u}_1 - b\int_\Omega \overline{u}_2).$$

The first inequality is equivalent to

$$\theta_\mu \leq \theta_{\mu - c\int_\Omega \theta_\lambda},$$

and the second one to

$$0 \geq \lambda - M - bN|\Omega|.$$

Taking M and N large we get both inequalities and $\underline{u}_1 \leq \overline{u}_1$ and $\underline{u}_2 \leq \overline{u}_2$.

3. Take in this case

$$(\underline{u}_1, \overline{u}_1) = (\varepsilon\varphi, \theta_\lambda), \quad (\underline{u}_2, \overline{u}_2) = (\theta_\mu, N),$$

with $\varepsilon, N > 0$ to choose. Observe that N has to verify that $N \geq \mu - c\int_\Omega \overline{u}_1$, and so, we can take

$$N = \mu - c\int_\Omega \theta_\lambda.$$

It is clear that $\overline{u}_1$ and $\underline{u}_2$ verify the inequalities. Finally, we consider $\underline{u}_1$. It has to verify that

$$\lambda_1 \leq \lambda - \varepsilon\varphi - bN|\Omega|,$$

so, if $\lambda - bN|\Omega| > \lambda_1$ we can take ε small enough that the above inequality holds and $\underline{u}_1 \leq \overline{u}_1$. Finally, observe that since $\theta_\mu \leq \mu < N$ we get that $\underline{u}_2 \leq \overline{u}_2$.

□

4.2 Kirchhoff systems

Along this section, we assume that M_i verifies (M) and (N). We present different applications of Theorem 3.3. First, we study a system with concave nonlinearities

$$\begin{cases} -M_1(\|u_1\|^2)\Delta u_1 = \lambda u_1^{q_1} + u_2^{q_2} & \text{in } \Omega, \\ -M_2(\|u_2\|^2)\Delta u_2 = \mu u_2^{p_2} + u_1^{p_1} & \text{in } \Omega, \\ u_1 = u_2 = 0 & \text{on } \partial\Omega, \end{cases} \tag{11}$$

where $\lambda, \mu \in \mathbb{R}$ and $0 < q_i, p_i < 1$.

Theorem 4.2. *Assume that $\lambda, \mu > 0$. Then, there exists a positive solution of (11).*

Proof. We are going to build again a pair of sub-supersolution. Denote also by e the unique positive solution of

$$\begin{cases} -\Delta e = 1 & \text{in } \Omega, \\ e = 0 & \text{on } \partial\Omega. \end{cases} \tag{12}$$

We show that $(\underline{u}_1, \overline{u}_1) = (\varepsilon_1\varphi, K_1 e)$ and $(\underline{u}_2, \overline{u}_2) = (\varepsilon_2\varphi, K_2 e)$ is a pair of sub-supersolution of (11) taking the positive constants $\varepsilon_1, \varepsilon_2, K_1$ and K_2 in an appropriate way. We start with $\overline{u}_1$. We need to verify that

$$-\mathscr{R}_1(u,v)\Delta\overline{u}_1 \geq \lambda\overline{u}_1^{q_1} + \overline{u}_2^{q_2}, \quad \forall (u,v) \in [\underline{u}_1, \overline{u}_1] \times [\underline{u}_2, \overline{u}_2]$$

Using (M), it suffices to show that

$$K_1 m_0 \geq \lambda K_1^{q_1}\|e\|_\infty^{q_1} + K_2^{q_2}\|e\|_\infty^{q_2}.$$

Similarly for $\overline{u}_2$,

$$K_2 m_0 \geq \mu K_2^{p_2}\|e\|_\infty^{p_2} + K_1^{p_1}\|e\|_\infty^{p_1}.$$

Fix, K_1 and K_2 verifying above inequalities. Now, we study $\underline{u}_1$ and $\underline{u}_2$. They have to verify

$$\begin{aligned} \mathscr{R}_1(u,v)\lambda_1\varepsilon_1\varphi &\leq \lambda(\varepsilon_1\varphi)^{q_1} + (\varepsilon_2\varphi)^{q_2}, \quad \forall (u,v) \in [\underline{u}_1, \overline{u}_1] \times [\underline{u}_2, \overline{u}_2], \\ \mathscr{R}_2(u,v)\lambda_1\varepsilon_2\varphi &\leq \mu(\varepsilon_2\varphi)^{p_2} + (\varepsilon_1\varphi)^{p_1}, \quad \forall (u,v) \in [\underline{u}_1, \overline{u}_1] \times [\underline{u}_2, \overline{u}_2]. \end{aligned}$$

Since $\mathscr{R}_i$ is bounded in $[0, \overline{u}_1] \times [0, \overline{u}_2]$, it is clear that we can take ε_1 and ε_2 small enough, and we conclude the result. □

Finally, we consider the competition Kirchhoff model with local nonlinearities

$$\begin{cases} -M_1(\|u_1\|^2)\Delta u_1 = u_1(\lambda - u_1 - bu_2) & \text{in } \Omega, \\ -M_2(\|u_2\|^2)\Delta u_2 = u_2(\mu - u_2 - cu_1) & \text{in } \Omega, \\ u_1 = u_2 = 0 & \text{on } \partial\Omega, \end{cases} \tag{13}$$

where $\lambda, \mu \in \mathbb{R}$ and $0 < b, c$. The meaning of the parameters was given at the beginning of this section.

Theorem 4.3. *Assume that there exist positive constants m_i^∞, $i = 1, 2$, such that $M_i \leq m_i^\infty$, and*

$$\lambda > b\mu + \lambda_1 m_1^\infty \quad \textit{and} \quad \mu > c\lambda + \lambda_1 m_2^\infty.$$

Then, there exists a positive solution of (13).

Proof. We show that

$$(\underline{u}_1, \overline{u}_1) = (\varepsilon_1\varphi, M_1) \quad \text{and} \quad (\underline{u}_2, \overline{u}_2) = (\varepsilon_2\varphi, M_2)$$

is a pair of sub-supersolution of (13) taking positive constants $\varepsilon_1, \varepsilon_2$ small $M_1 = \lambda$, $M_2 = \mu$. Indeed, $\overline{u}_1$ is supersolution if

$$0 \geq \lambda - M_1 - b\varepsilon_2\varphi,$$

which is true for $M_1 = \lambda$.

Consider now $\underline{u}_1$. The function $\underline{u}_1 = \varepsilon_1\varphi$ is subsolution provided of

$$\mathscr{R}_1(u, v)\lambda_1\varepsilon_1\varphi \leq (\varepsilon_1\varphi)(\lambda - \varepsilon_1\varphi - b\mu), \qquad \forall (u, v) \in [\underline{u}_1, \overline{u}_1] \times [\underline{u}_2, \overline{u}_2],$$

for which it suffices $\lambda > b\mu + \lambda_1 m_1^\infty$. Analogously for $\underline{u}_2$ and $\overline{u}_2$. □

Acknowledgements The authors have been partially supported for the following projects: CNPQ-Proc. 400426/2013-7 (GF and AS), CNPQ-Proc. 301242/2011-9 (GF), and Ministerio de Economía y Competitividad under grant MTM2012-31304 (AS).

References

1. Chung, N.T.: An existence result for a class of Kirchhoff type systems via sub and supersolutions method. Appl. Math. Lett. **35**, 95–101 (2014)
2. Chung, N.T., Afrouzi, G.A.: Existence of positive solutions for a class of nonlocal elliptic systems with multiple parameter. Matematicki Vesnik (2014, to appear)
3. Corrêa, F.J.S.A., Nascimento, R.G.: On a nonlocal elliptic system of p-Kirchhoff-type under Neumann boundary condition. Math. Comput. Model. **49**(3–4), 598–604 (2009)
4. Figueiredo, G.M.: On a nonlocal system with critical growth. In: Zographopoulos, N. (ed.) Recent Trends on Nonlinear Elliptic System. International Scientific Press, Greece (2012)

5. Figueiredo, G.M., Suárez, A.: Some remarks on the comparison principle in Kirchhoff equations. Submitted
6. Khademloo, S., Valipour, E., Babakhani, A.: Multiplicity of positive solutions for a second order elliptic system of Kirchhoff type Abstr. Appl. Anal. 2014, Art. ID 280130 9 pp.
7. Leman, H., Méléard, S., Mirrahimi, S.: Influence of a spatial structure on the long time behavior of a competitive Lotka-Volterra type system. Discrete Contin. Dyn. Syst. Ser. B **20**(2), 469–493 (2015)
8. Li, Q., Yang, Z.: Existence of positive solutions for a quasilinear elliptic systems of p-Kirchhoff type. Differ. Equat. Appl. **6**(1), 73–80 (2014)
9. Pao, C.V.: Nonlinear Parabolic and Elliptic Equations. Plenum Press, New York (1992)
10. Zhang, Z., Suny, Y.: Existence and multiplicity of solutions for nonlocal systems with Kirchhoff type. to appear in Acta. Math. Applica. Sinica, (2015)

Nodal structure of the solution of a cooperative elliptic system

J. Fleckinger, J.-P. Gossez, and F. de Thélin

A Djairo, avec toute notre amitié et notre admiration

Mathematics Subject Classification (2010): 35J57, 35B50, 35B09

1 Introduction

This paper is concerned with the study of the nodal structure of the (weak) solution $U = (u_i) \in (H_0^1(\Omega))^{n\times 1}$ of the following elliptic system

$$(S_F) \qquad -\Delta U = AU + \mu U + F \; in \; \Omega, \;\; U|_{\partial\Omega} = 0.$$

Here Ω is a smooth bounded domain in $\mathbb{R}^N$, $N \geq 1$, $A = (a_{ij}) \in \mathbb{R}^{n\times n}$ is a cooperative matrix (i.e. $a_{ij} \geq 0$ for $i \neq j$), $F = (f_i)$ is given in some space $(L^p(\Omega))^{n\times 1}$, and μ is a real parameter.

We are interested in the nodal structure of $U = U_\mu$ when the parameter μ varies near a real eigenvalue of the associated homogeneous system (S_0). These eigenvalues are of the form $\mu_{t,l} := \lambda_t - \xi_l$ where λ_t is an eigenvalue of $-\Delta$ on $H_0^1(\Omega)$ and ξ_l is a real eigenvalue of the matrix A (cf. Proposition 2.1). The general

J. Fleckinger
Institut de Mathématique (UMR CNRS 5219), CEREMATH-UT1, Université de Toulouse, 31042 Toulouse Cedex, France
e-mail: jfleck@math.cnrs.fr

J.-P. Gossez (✉)
Département de Mathématique, C.P. 214, Université Libre de Bruxelles, 1050 Bruxelles, Belgium
e-mail: gossez@ulb.ac.be

F. de Thélin
Institut de Mathématique (UMR CNRS 5219), Université de Toulouse, 31062 Toulouse Cédex, France
e-mail: francoisdethelin@yahoo.fr

A.N. Carvalho et al. (eds.), *Contributions to Nonlinear Elliptic Equations and Systems*, Progress in Nonlinear Differential Equations and Their Applications 86,
DOI 10.1007/978-3-319-19902-3_15

objective is to show that under suitable assumptions, if μ is sufficiently close to one $\mu_{s,k}$, then the nodal structure of U is similar to that of an eigenfunction ϕ_s associated with λ_s.

We are also interested in the change of sign of U when the parameter μ crosses $\mu_{s,k}$. The most classical situation of such a change of sign is that of the maximum and antimaximum principles for the single equation

$$-\Delta u = \mu u + f \text{ in } \Omega, \;\; u|_{\partial\Omega} = 0 \tag{1}$$

when the parameter μ crosses the first eigenvalue λ_1 (cf. [1, 3]). As we shall see, in the case of a system, the algebraic multiplicity of ξ_k plays a role and may prevent such a change of sign, even near the first eigenvalue $\mu_{1,1}$ of (S_0) (cf. Remark 3.6, Example 3.7 and Remark 3.8).

Let us consider for a moment Problem (1) and let $\hat{\lambda}$ be a (higher) eigenvalue of $-\Delta$ on $H_0^1(\Omega)$, with $\hat{\phi}$ an associated eigenfunction. Under suitable assumptions on f and on the nodal domains of $\hat{\phi}$, it was shown in [9] that if μ is sufficiently close to $\hat{\lambda}$, then the solution u of (1) has the same number of nodal domains as $\hat{\phi}$ and the nodal domains of u appear as small perturbations of those of $\hat{\phi}$; moreover, if $\hat{\phi}$ is positive in one of its nodal domains, then u is positive (resp. negative) in the corresponding nodal domain of u when $\mu < \hat{\lambda}$ (resp $\mu > \hat{\lambda}$).

The main result of this paper (cf. Theorem 3.1) provides an extension of the above results of [9] to systems. As already mentioned the question of the change of sign of the solution for μ crossing an eigenvalue leads to a new phenomena proper to systems.

Several works have been devoted to the study of the sign of the solution of a system like (S_F), for instance [5–8, 13],... In Theorem 2.1 of [6], it is shown that if $\mu < \mu_{1,1}$, then $f_i \geq 0, f_i \not\equiv 0$ for all i implies $u_i \geq 0, u_i \not\equiv 0$ for all i. In Theorem 2.4 of [8], it is shown (for A strictly cooperative, i.e. $a_{ij} > 0$ for $i \neq j$), that if $f_i \geq 0$ for all i with $f_i \not\equiv 0$ for at least one i, then $u_i < 0$ for all i provided $\mu > \mu_{1,1}$ is sufficiently close to $\mu_{1,1}$. These two results are versions of the classical maximum and antimaximum principles for the system (S_F).

The contribution of Theorem 3.1 below with respect to the above two results is twofold. First we obtain a variant of them (cf. Corollary 3.4) in the direction initiated in [1] for the case of Equation (1) (in [1] the classical sign hypothesis on f is weakened into a sign hypothesis on the Fourier coefficient $\int_\Omega f\phi_1$). Second we consider higher eigenvalues $\mu_{s,k}$. It is of course in this context of higher eigenvalues that the notion of nodal domain really plays a role.

It is worth observing that while most references for systems deal with solutions having a sign, we are dealing here with sign changing solutions.

The plan of the paper is the following. Some preliminaries are given in Section 2, our main result is stated and commented in Section 3, and all the proofs are contained in Section 4. The notion of nonsingular M-matrix (cf. [2]) plays an important role in these proofs.

To simplify notations, we will abbreviate $\mathbb{R}^{n\times 1}$ into $\mathbb{R}^n$, $(L^2(\Omega))^{n\times 1}$ into $(L^2)^n$, ... Moreover all vectors are understood to be column vectors and inequalities involving vectors or matrices are understood to be componentwise.

2 Preliminaries

Let $A = (a_{ij}) \in \mathbb{R}^{n\times n}$ be a *cooperative* matrix, i.e. $a_{ij} \geq 0$ for $i \neq j$. We denote by ξ_l the real eigenvalues of A, written in a non-increasing order and repeated according to multiplicity: $\xi_1 \geq \ldots \geq \xi_l \geq \ldots \geq \xi_{l_0}$, where clearly $l_0 \leq n$. We denote by $X_l \in \mathbb{R}^n$ the associated independent eigenvectors:

$$AX_l = \xi_l X_l.$$

Note that the existence of ξ_1 (with $X_1 \geq 0$) follows by applying the finite dimensional version of the Krein–Rutman theorem to $A + \tau I$ with τ sufficiently large (cf. p.6 of [2]).

The domain $\Omega \subset \mathbb{R}^N$ is assumed to be bounded and of class $\mathscr{C}^2$. As usual we denote by $(0 <)\lambda_1 < \lambda_2 \leq \ldots \leq \lambda_t \leq \ldots$ the eigenvalues of $-\Delta$ on $H_0^1(\Omega)$, repeated according to multiplicity, and by (ϕ_t) an orthonormal basis in $L^2(\Omega)$ of associated eigenfunctions. We choose $\phi_1 > 0$ in Ω.

A real number μ is called an *eigenvalue of System* (S_0) if there exists a nonzero $U \in (H_0^1)^n$ satisfying

$$(S_0) \qquad -\Delta U = AU + \mu U \text{ in } \Omega, \;\; U|_{\partial\Omega} = 0.$$

Proposition 2.1. *The real eigenvalues of* (S_0) *are the numbers*

$$\mu_{t,l} := \lambda_t - \xi_l, \text{ for } t = 1, 2, \ldots \text{ and } l = 1, \ldots, l_0.$$

Associated eigenvectors are given by $U = X_l \phi_t$.

The (easy) proof of Proposition 2.1 will be given in Section 4. The first (smallest) eigenvalue of (S_0) is thus $\mu_{1,1}$. Note that cooperativeness is not required in Proposition 2.1.

In the following, the parameter μ will always be assumed to satisfy:

$$(H_1) \qquad \mu \neq \mu_{t,l} \text{ for all } t = 1, 2, \ldots. \text{ and } l = 1, \ldots, l_0.$$

Proposition 2.2. *Let A be cooperative and take μ satisfying (H_1). Then, for any $F \in (L^2)^n$, System (S_F) has a unique solution U in $(H_0^1)^n \cap (H^2)^n$.*

The conclusion of Proposition 2.2 is probably known, although we could not find an explicit reference. For completeness a proof is included in Section 4.

3 Main Result

From now on we fix an eigenvalue λ_s, an eigenvalue ξ_k, as well as one eigenfunction ϕ_s associated with λ_s. We consider the eigenvalue $\mu_{s,k} = \lambda_s - \xi_k$ of (S_0).

Our assumptions will bear on the eigenvalue $\mu_{s,k}$ (cf. (H_2)), on the nodal domains of ϕ_s (cf. (H_3)), and on the right-hand side F (cf. (H_4)):

(H_2) If $\mu_{t,l} = \mu_{s,k}$ for some t, l, then $\lambda_t = \lambda_s$ and $\xi_l = \xi_k$. Equivalently $\lambda_t \neq \lambda_s$ or $\xi_l \neq \xi_k$ implies $\mu_{t,l} \neq \mu_{s,k}$.

(H_3) ϕ_s has q nodal domains $\Omega^1, \ldots, \Omega^r, \ldots, \Omega^q$ which enjoy the following two properties:

- (P_1) each Ω^r satisfies at each $x \in \partial\Omega^r$ the interior ball condition,
- (P_2) for σ sufficiently small, say $0 < \sigma < \sigma_1$, each Ω^r_σ is connected, where $\Omega^r_\sigma := \{x \in \Omega^r : dist(x, \partial\Omega^r) > \sigma\}$.

(H_2) can be looked as some sort of uniqueness for $\mu_{s,k}$ with respect to the decomposition $\mu_{s,k} = \lambda_s - \xi_k$. (H_3) is a regularity property of the nodal domains of ϕ_s. Various comments and references relative to (H_3) are given in [9]. See also Remarks 3.2 and 3.3.

To state our assumption on F we need the following notations relative to λ_s:

Notations (s_1, s_2). Denote by s_1 the largest t such that $\lambda_t < \lambda_s$ and by s_2 the smallest t such that $\lambda_t > \lambda_s$. Obviously if $s = 1$, then s_1 does not exist and $s_2 = 2$.

(H_4) F is given in $(L^p)^n$ with $p > N$ and $p \geq 2$, and can be written as

$$F = \sum_{t=1}^{s_1} Z^t \phi_t + Z\phi_s + \hat{F}, \tag{2}$$

where $Z^t = \int_\Omega F\phi_t$, $Z = \int_\Omega F\phi_s$ are Fourier coefficients of F, and $\hat{F}$ has all its components orthogonal to the eigenspaces associated with $\lambda_1, \ldots, \lambda_{s_1}, \lambda_s$.

The decomposition (2) is analogous to that considered on p.820 of [9]. It is not a restriction on F when λ_s is simple.

To state our main result we need the following notations relative to ξ_k and F.

Notations (m_k, n_k). Denote by m_k the algebraic multiplicity of ξ_k and by n_k the number of eigenvalues ξ_l of A with $\xi_l > \xi_k$, taking into account multiplicity.

Notation ($M_{s,k,i}$). Let $B_1, \ldots, B_n$ be the columns of the matrix $\xi_k I - A$. For $i = 1, \ldots, n$, we write

$$M_{s,k,i} := det(B_1, \ldots, B_{i-1}, Z, B_{i+1}, \ldots, B_n),$$

where Z is as above the Fourier coefficient $\int_\Omega F\phi_s$.

Finally introduce $\sigma_2 > 0$ such that

$$dist(\bar{\Omega}^r, \bar{\Omega}^{r'}) \geq 4\sigma_2 \quad \text{if } \bar{\Omega}^r \cap \bar{\Omega}^{r'} = \emptyset,$$

$$dist(\bar{\Omega}^r, \partial\Omega) \geq 4\sigma_2 \quad \text{if } \bar{\Omega}^r \cap \partial\Omega = \emptyset.$$

We thus consider System (S_F) with μ close to $\mu_{s,k}$. Let $U = U_\mu \in (H_0^1)^n$ be its (unique) solution (cf. Proposition 2.2). By standard regularity, $U \in (W^{2,p})^n \subset (\mathscr{C}^1(\bar{\Omega}))^n$.

Theorem 3.1. *Let A be a cooperative matrix and let λ_s, ξ_k and ϕ_s be fixed. Consider $\mu_{s,k}$. Assume (H_1), (H_2), (H_3) and (H_4). Assume also that for some i_0:*

$$M_{s,k,i_0} > 0. \tag{3}$$

Take $\sigma > 0$ with $\sigma < \sigma_1$ and $\sigma < \sigma_2$. Then there exists $\delta = \delta(F,\sigma) > 0$ such that

(i) if $\mu_{s,k} - \delta < \mu < \mu_{s,k}$, then the component u_{i_0} of U has exactly q nodal domains $\mathscr{O}^1, \ldots, \mathscr{O}^r, \ldots \mathscr{O}^q$, which enjoy the following three properties:

- *(i_1) $\Omega^r_\sigma \subset \mathscr{O}^r \subset \tilde{\Omega}^r_\sigma$ for $1 \leq r \leq q$, where $\tilde{\Omega}^r_\sigma := \{x \in \Omega : dist(x, \Omega^r) < \sigma\}$,*
- *(i_2) $(-1)^{n_k} u_{i_0}(x)\phi_s(x) > 0$ for all $x \in \mathscr{O}^r \cap \Omega^r$ and any $1 \leq r \leq q$,*
- *(i_3) if $\bar{\mathscr{O}}^r \cap \bar{\mathscr{O}}^{r'} \neq \emptyset$ with $r \neq r'$, then $u_{i_0}(x)u_{i_0}(y) < 0$ for all $x \in \mathscr{O}^r, y \in \mathscr{O}^{r'}$,*

(ii) if $\mu_{s,k} < \mu < \mu_{s,k} + \delta$, then the same conclusion as in (i) above holds, with the only change that in (i_2) one now has $(-1)^{n_k}(-1)^{m_k} u_{i_0}(x)\phi_s(x) > 0$ for all $x \in \mathscr{O}^r \cap \Omega^r$.

Finally if assumption (3) is replaced by $M_{s,k,i_0} < 0$, the same conclusion holds with the only change that > 0 should be replaced by < 0 in (i_2) (both in case (i) $\mu < \mu_{s,k}$ and in case (ii) $\mu > \mu_{s,k}$).

We thus see that, for μ close to $\mu_{s,k}$, u_{i_0} looks like ϕ_s in the sense that it has the same number of nodal domains and that each $\mathscr{O}^r$ appears as a small perturbation of the corresponding Ω^r (cf. (i_1)), with the same or opposite sign for u_{i_0} and ϕ_s on the intersection (cf. (i_2)). Moreover a change of sign occurs when going from one $\mathscr{O}^r$ to an neighbouring $\mathscr{O}^{r'}$ (cf.(i_3)). We will discuss later the question of the change of sign of u_{i_0} when μ crosses $\mu_{s,k}$ (cf. Remark 3.6, Example 3.7 and Remark 3.8).

Theorem 3.1 provides an extension to systems of the results obtained in [9] for the case of a single equation.

Remark 3.2. Assumptions (H_3) and (H_4) are trivially satisfied for $N = 1$ (i.e. for a system of ODEs). Assumption (H_3) is also satisfied if Ω^r is of class $\mathscr{C}^2$. This latter fact can be verified by a construction similar to that given on p.355 of [12].

Remark 3.3. Suppose Ω is the open unit disk in $\mathbb{R}^2$. Assumption (H_3) is satisfied for $s = 1, 6, 15, \ldots$ (cf. [4, 9]) and Theorem 3.1 applies. For $s = 2$, assumption (H_3) is not satisfied. One can however conjecture that the conclusion of Theorem 3.1 still holds under the additional assumption $F \in (W^{1,p})^n$. See [10] for the case of a single equation. For $s = 4$, assumption (H_3) is not satisfied and the conclusion of Theorem 3.1 does not hold anymore, even in the case of a single equation (cf. [11]).

In the particular case where $s = 1$, the conclusion of Theorem 3.1 reads as follows:

Corollary 3.4. *Let A be cooperative and fix one ξ_k. Consider the eigenvalue $\mu_{1,k} = \lambda_1 - \xi_k$. Assume (H_1), (H_2) and let $F \in (L^p)^n$ with $p > N$ and $p \geq 2$. Suppose $M_{1,k,i_0} > 0$ for some i_0. Then there exists $\delta = \delta(F) > 0$ such that*

(i) If $\mu_{1,k} - \delta < \mu < \mu_{1,k}$, then $(-1)^{n_k} u_{i_0} > 0$ in Ω,
(ii) If $\mu_{1,k} < \mu < \mu_{1,k} + \delta$, then $(-1)^{n_k}(-1)^{m_k} u_{i_0} > 0$ in Ω.

The following example illustrates Corollary 3.4 for $k = 1$ and $k = 2$. It also allows some comparison with the results of [6] and [8] mentioned in the introduction.

Example 3.5. Suppose $n = 2$ and A strictly cooperative:

$$A = \begin{pmatrix} a_{11} & a_{12} \\ a_{21} & a_{22} \end{pmatrix}.$$

A simple calculation shows that A has 2 distinct real eigenvalues $\xi_1 > \xi_2$ which verify $\xi_2 < a_{11} < \xi_1$ and $\xi_2 < a_{22} < \xi_1$. So $m_1 = m_2 = 1$ and $n_1 = 0$, $n_2 = 1$. Write $z_1 := \int_\Omega f_1 \phi_1 > 0$ and $z_2 := \int_\Omega f_2 \phi_1 > 0$. We first consider the case where μ remains close to $\mu_{1,1}$. Assuming $z_1 \geq 0$ and $z_2 > 0$ (or $z_1 > 0$ and $z_2 \geq 0$), we obtain

$$M_{1,1,1} = z_1(\xi_1 - a_{22}) + a_{12} z_2 > 0,$$

$$M_{1,1,2} = z_2(\xi_1 - a_{11}) + a_{21} z_1 > 0.$$

Theorem 3.1 thus yields the existence of $\delta > 0$ such that u_1 and u_2 are > 0 in Ω for $\mu_{1,1} - \delta < \mu < \mu_{1,1}$, and u_1 and u_2 are < 0 in Ω for $\mu_{1,1} < \mu < \mu_{1,1} + \delta$. These results are versions of the maximum and antimaximum principles in the line of [1], i.e. with assumptions bearing on the Fourier coefficients of F. We consider now the situation where μ remains close to $\mu_{1,2}$ ($> \mu_{1,1}$). Assuming $z_1 < 0$ and $z_2 \geq 0$ (or $z_1 \leq 0$ and $z_2 > 0$), we obtain

$$M_{1,2,1} = z_1(\xi_2 - a_{22}) + a_{12} z_2 > 0,$$

$$M_{1,2,2} = z_2(\xi_2 - a_{11}) + a_{21} z_1 < 0.$$

Corollary 3.4 thus yields the existence of $\delta > 0$ such that $u_1 < 0$ in Ω and $u_2 > 0$ in Ω for $\mu_{1,2} - \delta < \mu < \mu_{1,2}$, and $u_1 > 0$ in Ω and $u_2 < 0$ in Ω for $\mu_{1,2} < \mu < \mu_{1,2} + \delta$.

Remark 3.6. One observes in the preceding example a change of sign of the solution when μ crosses $\mu_{1,1}$ or $\mu_{1,2}$. On the contrary, when the multiplicity of ξ_s is even, no change of sign occurs when μ crosses $\mu_{s,k}$. The following example illustrates this fact.

Example 3.7. Consider the cooperative matrix:

$$A = \begin{pmatrix} \xi & 1 \\ 0 & \xi \end{pmatrix}.$$

Here $\xi_1 = \xi_2 = \xi$, $m_1 = m_2 = 2$ and $n_1 = n_2 = 0$. We consider the situation where μ remains close to $\mu_{1,1} = \mu_{1,2}$. Write again $z_i = \int_\Omega f_i \phi_1$. Assuming $z_2 > 0$ we find $M_{1,1,1} = z_2 > 0$. Corollary 3.4 yields that u_1 is > 0 in Ω for $\mu_{1,1} - \delta < \mu < \mu_{1,1}$ and also for $\mu_{1,1} < \mu < \mu_{1,1} + \delta$. Corollary 3.4 does not apply to the second component u_2 since $M_{1,1,2} = 0$ for any value of z_1 and z_2.

Remark 3.8. A direct argument can be used to derive the conclusion in the preceding example, with in addition some information about u_2. Indeed the system reads

$$\begin{cases} -\Delta u_1 = (\xi + \mu)u_1 + u_2 + f_1, \\ -\Delta u_2 = (\xi + \mu)u_2 + f_2. \end{cases}$$

Assume as above $z_2 > 0$. Consider first the case $\mu < \mu_{1,1}$, *i.e.* $\xi + \mu < \lambda_1$. Applying the maximum principle of [1] to the second equation, we get $u_2 > 0$ for μ near $\mu_{1,1}$, with in addition $u_2 \to +\infty$ as $\mu \to \mu_{1,1}$. It follows that the same maximum principle can be applied to the first equation to get $u_1 > 0$ for μ near $\mu_{1,1}$. Consider now the case $\mu > \mu_{1,1}$, *i.e.* $\xi + \mu > \lambda_1$. A similar argument using twice the antimaximum principle of [1] yields first $u_2 < 0$ and then $u_1 > 0$ for μ near $\mu_{1,1}$. A change of sign occurs for u_2 when μ crosses $\mu_{1,1}$, but not for u_1 (as already observed in the preceding example).

4 Proofs

We start with the

Proof of Proposition 2.1. We first verify that $\lambda_t - \xi_l$ is an eigenvalue of (S_0). Putting $U = X_l \phi_t$, this follows directly from the relations

$$-\Delta U = -\Delta(X_l \phi_t) = X_l \lambda_t \phi_t,$$

$$AU = A(X_l \phi_t) = (AX_l)\phi_t = \xi_l X_l \phi_t.$$

Conversely, let μ be an eigenvalue of (S_0), with $U = (u_i)$ an associated eigenvector. So

$$-\Delta u_i = \Sigma_j a_{ij} u_j + \mu u_i, \tag{4}$$

where $u_j \not\equiv 0$ for at least one j. Take such a j, choose t such that $\int_\Omega u_j \phi_t \neq 0$ and write $Y = (y_i)$ with $y_i = \int_\Omega u_i \phi_t$. Multiplying (4) by ϕ_t and integrating, one obtains

$$\lambda_t y_i = \Sigma_j a_{ij} y_j + \mu y_i,$$

i.e. $(\lambda_t - \mu)Y = AY$. This shows that $(\lambda_t - \mu)$ is an eigenvalue of A. □

The notion of nonsingular M-matrix will enter the proofs of Proposition 2.2 and Theorem 3.1. We recall that a matrix $B \in \mathbb{R}^{n \times n}$ is a *nonsingular M-matrix* if it can be written in the form $B = \sigma I - C$ with $C \geq 0$ and $\sigma > \rho(C)$, where $\rho(C)$ denotes the spectral radius of the matrix C. We list below 3 of the 50 equivalent conditions given in [2] (p.132 to p.138) for a matrix $B = (b_{ij})$, with $b_{ij} \leq 0$ for $i \neq j$, to be a nonsingular M-matrix:

(A1) All the principal minors of B are positive.
(D16) Every real eigenvalue of B is positive.
(N39) B is monotone in the sense that for $X \in \mathbb{R}^n$, $BX \geq 0$ implies $X \geq 0$.

Lemma 4.1. *Suppose A cooperative. Then there exists σ_0 such that for $\sigma \geq \sigma_0$, $(\sigma I - A)$ is a non singular M-matrix. Moreover*

$$\sigma_0 > \xi_1 \geq \xi_2 \geq \ldots . \geq \xi_{l_0}. \tag{5}$$

Proof. Take τ sufficiently large so that $B = \tau I + A$ is ≥ 0. Take $\sigma_0 > \rho(B) - \tau$. It follows that for $\sigma \geq \sigma_0$, $(\sigma I - A) = (\sigma + \tau)I - B$ with $\sigma + \tau > \rho(B)$. Consequently $(\sigma I - A)$ is a nonsingular M-matrix. Moreover, by (D16), all real eigenvalues of $(\sigma_0 I - A)$ are positive. So $\sigma_0 - \xi_1 > 0$, which implies (5). □

The following algebraic system will appear repeatedly in the proofs:

$$(\mathfrak{T}_{t,Z}) \qquad ((\lambda_t - \mu)I - A)X = Z$$

where $X, Z \in \mathbb{R}^n$. By (H_1) it has a unique solution X for any given Z.

Let us finally introduce the following decomposition of the spaces $(L^2)^n$ and $(H_0^1)^n$:

Notations $(L, L^\perp, H, H^\perp)$. For a given s_0, we consider in $(L^2)^n$ (resp. $(H_0^1)^n$) the finite dimensional spaces L (resp. H) generated by $Y\phi_t$ where Y varies in $\mathbb{R}^n$ and $t = 1, \ldots, s_0 - 1$. We call $L^\perp$ (resp. $H^\perp$) its orthogonal complement in $(L^2)^n$ (resp. $(H_0^1)^n$). Orthogonality is understood in the usual componentwise sense.

We are now ready to start the

Proof of Proposition 2.2. Fix μ satisfying (H_1). Uniqueness in (S_F) is obvious. To prove existence, we take σ_0 from Lemma 4.1 and choose s_0 such that $\lambda_{s_0} - \mu \geq \sigma_0$, which is clearly possible since $\lambda_t \to +\infty$ as $t \to +\infty$. We will use the spaces $L^\perp$ and $H^\perp$ associated with s_0 in the above notations.

Claim. For any given $G \in L^\perp$, System (S_G) has a unique solution $V \in H^\perp$.

Proof of the Claim. We proceed by approximation in finite dimensional spaces. Let $L_r^\perp$ (resp. $H_r^\perp$) be the subspace of $L^\perp$ (resp. $H^\perp$) generated by $Y\phi_t$ where Y varies in $\mathbb{R}^n$ and $t = s_0, \ldots, s_0 + r$. Let $G^r \in L_r^\perp$ be the orthogonal projection of G onto $L_r^\perp$. We thus have $\|G^r\|_{(L^2)^n} \leq \|G\|_{(L^2)^n}$ and $G^r \to G$ in $(L^2)^n$ as $r \to +\infty$.

We first fix r and seek a solution $V^r \in H_r^\perp$ of the system

$$(S_{G^r}) \qquad -\Delta V^r = AV^r + \mu V^r + G^r \text{ in } \Omega, \ \ V^r|_{\partial\Omega} = 0.$$

Multiplying by ϕ_t for $t = s_0, \ldots, s_0 + r$, one sees that solving (S_{G^r}) is equivalent to solving the $r + 1$ algebraic systems

$$((\lambda_t - \mu)I - A)\, X = \int_\Omega G^r \phi_t$$

for $t = s_0, \ldots, s_0 + r$. Each of the latter systems is of the form $(\mathfrak{T}_{t,Z})$ and so has a unique solution.

We write $V^r = (v_i^r)$ the solution of (S_{G^r}). We also write $G^r = (g_i^r)$. Multiplying (S_{G^r}) by V^r, integrating and using the variational characterization of λ_{s_0} as well as $a_{ij} \geq 0$ for $i \neq j$, we obtain, for $i = 1, \ldots, n$

$$\lambda_{s_0} \int_\Omega (v_i^r)^2 \leq \int_\Omega |\nabla v_i^r)|^2 = \Sigma_{j=1}^n a_{ij} \int_\Omega v_i^r v_j^r + \mu \int_\Omega (v_i^r)^2 + \int_\Omega g_i^r v_i^r$$

$$\leq \Sigma_{j \neq i} a_{ij} \|v_i^r\|_{L^2} \|v_j^r\|_{L^2} + (a_{ii} + \mu)\|v_i^r\|_{L^2}^2 + \|g_i^r\|_{L^2} \|v_i^r\|_{L^2}.$$

Dividing by $\|v_i^r\|_{L^2}$ (in case it is $\neq 0$) and denoting by $\tilde{V}^r$ (resp. $\tilde{G}^r$) the vector with components $\|v_i^r\|_{L^2}$ (resp. $\|g_i^r\|_{L^2}$), we derive

$$((\lambda_{s_0} - \mu)I - A)\, \tilde{V}^r \leq \tilde{G}^r.$$

By the choice of s_0 and Lemma 4.1, $((\lambda_{s_0} - \mu)I - A)$ is a nonsingular M-matrix and consequently, by $(N39)$, one deduces

$$0 \leq \tilde{V}^r \leq ((\lambda_{s_0} - \mu)I - A)^{-1}\, \tilde{G}^r.$$

It follows

$$\|\tilde{V}^r\| \leq \|\, ((\lambda_{s_0} - \mu)I - A)^{-1}\, \| \, \|\tilde{G}^r\|,$$

where $\|.\|$ denotes the euclidian norm of a vector and of a matrix. Hence, for some constant C_1 independent of r

$$\|V^r\|_{(L^2)^n} \leq C_1\|G^r\|_{(L^2)^n} \leq C_1\|G\|_{(L^2)^n},$$

and by the regularity properties of the Laplacian,

$$\|V^r\|_{(H^2)^n} \leq C_2\|G\|_{(L^2)^n},$$

with another constant C_2. A standard compactness argument then yields that when $r \to +\infty$, for a subsequence, V^r converges in $(H_0^1)^n$ to some $V \in H^\perp$ which is a solution of (S_G). This concludes the proof of the claim.

To conclude the proof of Proposition 2.2 we take $F \in (L^2)^n$ and write $F = G + G^\perp$ where $G \in L$ and $G^\perp \in L^\perp$. Consider the two systems (S_G) and $(S_{G^\perp})$. System $(S_{G^\perp})$ has a unique solution $V \in H^\perp$ by the above claim. Multiplying (S_G) by ϕ_t for $t = 1, \ldots, s_0 - 1$ and integrating, one sees as above that solving (S_G) is equivalent to solving $(s_0 - 1)$ algebraic systems of the form $(\mathfrak{T}_{t,Z})$ where $Z = \int_\Omega G\phi_t$. We obtain in this way a unique solution W to System (S_G). Finally $U = W + V$ yields a solution to (S_F). □

We now turn to the proof of Theorem 3.1. Three preliminary lemmas will be needed, whose objective is to get estimates independent of μ.

We consider σ_0 as given by Lemma 4.1, introduce some constant K to be specified later and chose s_0 such that

$$\lambda_{s_0} \geq \sigma_0 + K. \tag{6}$$

Lemma 4.2. *(i) There exists $\epsilon > 0$ such that in the factorization*

$$det((\lambda_t - \mu)I - A) = (\lambda_t - \mu - \xi_1)\ldots.(\lambda_t - \mu - \xi_{l_0})Q(\lambda_t - \mu), \tag{7}$$

one has $Q(\lambda_t - \mu) \geq \epsilon$ for all $t = 1, 2, \ldots$ and all μ in $\mathbb{R}$.

(ii) If $\mu < K$ and s_0 satisfies (6), then $((\lambda_{s_0} - \mu)I - A)$ is a nonsingular M-matrix and one has

$$det((\lambda_{s_0} - \mu)I - A) \geq (\sigma_0 - \xi_1)^{l_0}\epsilon > 0, \tag{8}$$

Proof. (i) The determinant in (7) is the characteristic polynomial of A computed in $(\lambda_t - \mu)$. The quotient polynomial Q thus involves only nonreal eigenvalues and consequently does not vanish on $\mathbb{R}$.

(ii) By (6), for $\mu < K$, $\lambda_{s_0} - \mu \geq \sigma_0 + K - \mu \geq \sigma_0$, and consequently, by Lemma 4.1, $((\lambda_{s_0} - \mu)I - A)$ is a nonsingular M-matrix. Moreover, using (5), one has for $l = 1, \ldots, l_0$,

$$\lambda_{s_0} - \mu - \xi_l \geq \sigma_0 + K - \mu - \xi_l \geq \sigma_0 - \xi_1 > 0.$$

Replacing in (7) yields (8). □

We now consider the decomposition $(L, L^{\perp}, H, H^{\perp})$ of the spaces $(L^2)^n$ and $(H_0^1)^n$ associated with s_0, s_0 verifying as before (6). If $\mu < K$, then $\lambda_{s_0} \geq \sigma_0 + K > \sigma_0 + \mu$, and consequently the claim in the proof of Proposition 2.2 applies: given $G \in L^{\perp}$, there exists a unique $V = V_{\mu} \in H^{\perp}$ solution of the system

$$(S_{G^{\perp}}) \qquad -\Delta V = AV + \mu V + G \text{ in } \Omega, \quad V|_{\partial\Omega} = 0.$$

Lemma 4.3. *Suppose $G \in L^{\perp} \cap (L^p)^n$ with $p > N$ and $p \geq 2$. Let K' be a constant with $K' < K$. Then there exists a constant C, independent of μ, such that*

$$\|V_{\mu}\|_{(\mathscr{C}^1(\overline{\Omega}))^n} \leq C\|G\|_{(L^p)^n}$$

for all μ with $K' < \mu < K$.

Proof. Write $V = (v_i)$ and $G = (g_i)$. Multiplying (S_G) by V, integrating, and using the variational characterization of λ_{s_0} as well as $a_{ij} \geq 0$ for $i \neq j$, one can proceed exactly as in the proof of the above-mentioned claim to obtain

$$((\lambda_{s_0} - \mu)I - A)\,\tilde{V} \leq \tilde{G},$$

where $\tilde{V}$ (resp. $\tilde{G}$) denotes the vector in $\mathbb{R}^n$ with components $\|v_i\|_{L^2}$ (resp. $\|g_i\|_{L^2}$). By part (*ii*) of Lemma 4.2, $((\lambda_{s_0} - \mu)I - A)$ is a nonsingular M-matrix, and so by (N39),

$$0 \leq \tilde{V} \leq ((\lambda_{s_0} - \mu)I - A)^{-1}\,\tilde{G}.$$

This implies

$$\|\tilde{V}\| \leq \|\,((\lambda_{s_0} - \mu)I - A)^{-1}\,\|\,\|\tilde{G}\|,$$

where $\|.\|$ denotes the euclidian norm of a vector or a matrix. By (8), $\|\,((\lambda_{s_0} - \mu)I - A)^{-1}\,\|$ is bounded independently of μ for $K' < \mu < K$. Consequently there exists a constant C_1 such that

$$\|V_{\mu}\|_{(L^2)^n} \leq C_1\|G\|_{(L^2)^n},$$

for $K' < \mu < K$. The conclusion of the lemma now follows by combining the Sobolev imbedding theorem and standard regularity properties of the Laplacian. □

In the last lemma μ remains near $\mu_{s,k}$ and assumption (H_2) is used, as well as the notations s_1 and s_2.

Lemma 4.4. *Take t with either $t \leq s_1$ or $t \geq s_2$. Then there exists $\delta_t > 0$ such that if $|\mu - \mu_{s,k}| < \delta_t$, then*

$$|det((\lambda_t - \mu)I - A)| \geq (\delta_t)^{l_0}\epsilon > 0, \tag{9}$$

where ϵ is defined in Lemma 4.2.

Proof. Put

$$\delta_t := \frac{1}{2}min\{|\mu_{t,l} - \mu_{s,k}|, \, l = 1, \ldots, l_0\}.$$

Since $\lambda_t \neq \lambda_s$, assumption (H_2) implies $\delta_t > 0$. Now for $|\mu - \mu_{s,k}| < \delta_t$, one has, for $l = 1, \ldots, l_0$,

$$\begin{aligned} |\lambda_t - \mu - \xi_l| &= |\mu_{t,l} - \mu_{s,k} + \mu_{s,k} - \mu| \\ &\geq |\mu_{t,l} - \mu_{s,k}| - \delta_t \geq \delta_t. \end{aligned}$$

The conclusion (9) now follows from (7). □

We are now ready to start the

Proof of Theorem 3.1. By (H_4),

$$F = \sum_{t=1}^{s_1} F^t + Z\phi_s + \hat{F},$$

where $F^t = Z^t\phi_t$ with $Z^t = \int_\Omega F\phi_t$, $Z = \int_\Omega F\phi_s$ and $\hat{F}$ has all its components orthogonal to the eigenspaces associated with $\lambda_1, \ldots, \lambda_{s_1}, \lambda_s$. The solution $U = U_\mu$ of (S_F) reads

$$U = \sum_{t=1}^{s_1} U^t + X\phi_s + \hat{U}$$

where U^t satisfies (S_{F^t}), $X\phi_s$ satisfies $(S_{Z\phi_s})$ and $\hat{U}$ satisfies $(S_{\hat{F}})$.

We start estimating the component i_0 of $X\phi_s$ as $\mu \to \mu_{s,k}$. Write $X = (x_i) = (x_i(\mu))$.

Claim 1. One has

$$(-1)^{n_k} \, (sgn(\mu_{s,k} - \mu))^{m_k} \, x_{i_0}(\mu) \to +\infty$$

as $\mu \to \mu_{s,k}$.

Proof of Claim 1. It will be convenient to write $\mu = \mu_{s,k} + \eta$. Multiplying (S_F) by ϕ_s and integrating, one sees that X solves

$$(\mathfrak{T}_{s,Z}) \qquad ((\lambda_s - \mu)I - A)X = ((\xi_k - \eta)I - A)X = Z.$$

Denote by $B_1^\eta, \ldots, B_n^\eta$ the column vectors of the matrix $(\xi_k - \eta)I - A$. By Cramer formula the ith component of X is $x_i = x_i(\eta) = M_i^\eta / D^\eta$, where

$$M_i^\eta = det(B_1^\eta, \ldots, B_{i-1}^\eta, Z, B_{i+1}^\eta, \ldots, B_n^\eta)$$

and

$$D^\eta = det((\xi_k - \eta)I - A).$$

When $\eta \to 0$, $M_i^\eta \to M_i^0$. Note that M_i^0 is in fact $M_{s,k,i}$ as introduced in the notation before the statement of Theorem 3.1. Consequently, by assumption (3), $M_i^0 > 0$. On the other hand, $D^\eta \to D^0 = 0$ when $\eta \to 0$. We now investigate the sign of D^η for $\eta \neq 0$. By (7), one has

$$D^\eta = \Pi_{l=1}^{l_0} (\xi_k - \xi_l - \eta)\, Q(\xi_k - \eta).$$

$Q(\xi_k - \eta)$ is $\geq \epsilon > 0$ for all η as seen in Lemma 4.2. For $0 < |\eta|$ small, the sign of $(\xi_k - \xi_l - \eta)$ is that of $(\xi_k - \xi_l)$ in case $\xi_l \neq \xi_k$, and that of $(-\eta)$ in case $\xi_l = \xi_k$. Consequently, for $0 < |\eta|$ small, D^η has the sign of $(-1)^{n_k}(-\eta)^{m_k}$. The conclusion of Claim 1 follows.

We will now get bounds on the remaining parts of the solution U_μ as $\mu \to \mu_{s,k}$. From now on, we assume $K' < \mu < \mu_{s_2,k}$, where K' is a constant $< \mu_{s,k}$. The preliminary lemmas will be used with $K = \mu_{s_2,k}$. Pick s_0 such that $\lambda_{s_0} \geq \sigma_0 + \mu_{s_2,k}$. Note that by (5), $\lambda_{s_0} > \xi_1 + \lambda_{s_2} - \xi_k \geq \lambda_{s_2}$. We consider the decomposition $(L, L^\perp, H, H^\perp)$ of $(L^2)^n$ and $(H_0^1)^n$ associated with s_0 and write F as

$$F = \sum_{l=1}^{s_1} F^t + Z\phi_s + \sum_{t=s_2}^{s_0-1} F^t + F^\perp$$

where $F^\perp \in L^\perp$. As before $F^t = Z^t \phi_t$ with $Z^t = \int_\Omega F\phi_t$. The solution U is now written as

$$U = \sum_{t=1}^{s_1} U^t + X\phi_s + \sum_{t=s_2}^{s_0-1} U^t + U^\perp,$$

where U^t satisfies (S_{F^t}) for $t = 1, \ldots, s_1$ and $t = s_2, \ldots, s_0 - 1$, $X\phi_s$ satisfies $(S_{Z\phi_s})$ and $U^\perp$ satisfies $(S_{F^\perp})$. All these components U of course depend on μ.

Claim 2. $\|U^\perp\|_{\mathscr{C}^1(\overline{\Omega})}$ is bounded independently of μ for $K' < \mu < \mu_{s_2,k}$.

Proof of Claim 2. This is a direct consequence of Lemma 4.3.

Claim 3. There exists $\delta > 0$ such that for each $t = 1, \dots, s_1$ and $t = s_2, \dots, s_0 - 1$, $\|U^t\|_{\mathscr{C}^1(\bar{\Omega})}$ is bounded independently of μ for $|\mu - \mu_{s,k}| < \delta$.

Proof of Claim 3. Fix t as indicated in Claim 3. $U^t = X^t\phi_t$ satisfies (S_{F^t}) and so, multiplying by ϕ_t and integrating, one sees that X^t satisfies the algebraic system $(\mathfrak{T}_{t,Z^t})$. It follows from Cramer formula and Lemma 4.4 that for $|\mu - \mu_{s,k}| < \delta_t$, X^t is bounded independently of μ. Setting $\delta = min\{\delta_1, \dots, \delta_{s_1}, \delta_{s_2} \dots, \delta_{s_0-1}\}$ yields the conclusion of Claim 3.

We are now ready to conclude the proof of Theorem 3.1. Combining Claims 1, 2 and 3, we see that when $\mu \to \mu_{s,k}$,

$$u_{i_0} = (-1)^{n_k} (sgn(\mu_{s,k} - \mu))^{m_k} \alpha(\mu)\phi_s + \tilde{u}_\mu, \tag{10}$$

where $\alpha(\mu) \to +\infty$ and $\tilde{u}_\mu$ remains bounded in $\mathscr{C}^1(\bar{\Omega})$. Relation (10) is the key observation which allows to derive the conclusion of Theorem 3.1. Roughly speaking it shows that the first term in (10) is the dominating term when $\mu \to \mu_{s,k}$: for μ close to $\mu_{s,k}$, u_{i_0} looks like $(-1)^{n_k}(sgn(\mu_{s,k} - \mu))^{m_k}\phi_s$. Reaching the precise statements of Theorem 3.1 about the nodal domains of u_{i_0} can then be carried out by following word by word the arguments developed in the proofs of Theorems 2.1 and 2.2 of [9] in the case $N = 1$ and of Theorem 3.1 of [9] when $N \geq 2$. It is at this latter stage that assumption (H_3) plays a role. □

References

1. Arcoya, D., Gamez, J.: Bifurcation theory and related problems: antimaximum principle and resonance. Commun. Part. Differ. Equ. **26**, 1879–1911 (2001)
2. Berman, A., Plemmons, R.J.: Nonnegative Matrices in the Mathematical Sciences. SIAM, Philadelphia (1994)
3. Clément, P., Peletier, L.A.: An antimaximum principle for second order elliptic operators. J. Differ. Equ. **34**, 218–229 (1979)
4. Courant, R., Hilbert, D.: Methods of Mathematical Physics. Intersciences, New York (1953)
5. de Figueiredo, D.G., Mitidieri, E.: A maximum principle for an elliptic system and applications to semilinear problems. SIAM J. Math. Anal. **17**, 836–849 (1986)
6. de Figueiredo, D.G., Mitidieri, E.: Maximum principles for linear elliptic systems. Rend. Istit. Mat. Univ. Trieste 22(1-2), 36-66 (1990/1992)
7. Fleckinger, J., Hernández, J., de Thélin, F.: On maximum principles and existence of positive solutions for some cooperative elliptic systems. Differ. Int. Equ. **8**, 69–85 (1995)
8. Fleckinger, J., Gossez, J.-P., Takác, P., de Thélin, F.: Nonexistence of Solutions and an antimaximum principle for cooperative systems with the p-Laplacian. Math. Nachr. **194**, 49–78 (1998)
9. Fleckinger, J., Gossez, J.-P., de Thélin, F.: Maximum and antimaximum principles: beyond the first eigenvalue. Differ. Int. Equ. **22**, 815–828 (2009)
10. Fleckinger, J., Gossez, J.-P., de Thélin, F.: Maximum and antimaximum principles near the second eigenvalue. Differ. Int. Equ. **24**, 389–400 (2011)

11. Fleckinger, J., Gossez, J.-P., de Thélin, F.: Nodal domains for an elliptic problem with the spectral parameter near the fourth eigenvalue. Differ. Int. Equ. **25**, 1189–1202 (2012)
12. Gilbarg, D., Trudinger, N.S.: Elliptic Partial Differential Equations of Second Order. Springer, New York (1998)
13. Protter, M.H., Weinberger, H.: Maximum Principles in Differential Equations. Prentice Hall, New York (1967)

Nonquadraticity condition on superlinear problems

Marcelo F. Furtado and Edcarlos D. Silva

Dedicated to Prof. Djairo de Figueiredo on the occasion of his 80th birthday

Mathematics Subject Classification (2010): 35J20, 35J60

1 Introduction

In this paper we consider the nonlinear elliptic equation

$$(P) \qquad \begin{cases} -\Delta u = f(x,u) & \text{in } \Omega, \\ u = 0 & \text{on } \partial\Omega, \end{cases}$$

where $\Omega \subseteq \mathbb{R}^N, N \geq 3$, is a bounded smooth domain and $f \in C(\Omega \times \mathbb{R}, \mathbb{R})$ is subcritical and superlinear in the following sense:

(f_0) there exist $a_1 > 0$ and $p \in (2, 2^*)$ such that

$$|f(x,s)| \leq a_1(1 + |s|^{p-1}), \quad \text{for any } (x,s) \in \Omega \times \mathbb{R}.$$

(f_1) for $F(x,s) := \int_0^s f(x,\tau)d\tau$, uniformly in $x \in \Omega$ there holds

M.F. Furtado (✉)
Departamento de Matemática, Universidade de Brasília, 70910-900 Brasília, Distrito Federal, Brazil
e-mail: mfurtado@unb.br

E.D. Silva
Universidade Federal de Goiás, Instituto de Matemática e Estatística, 74001-970 Goiânia, State of Goiás, Brazil
e-mail: edcarlos@mat.ufg.br

A.N. Carvalho et al. (eds.), *Contributions to Nonlinear Elliptic Equations and Systems*, Progress in Nonlinear Differential Equations and Their Applications 86,
DOI 10.1007/978-3-319-19902-3_16

$$\lim_{|s|\to\infty} \frac{F(x,s)}{s^2} = +\infty.$$

The weak solutions of the problem are precisely the critical points of the C^1-functional

$$I(u) := \frac{1}{2}\int_\Omega |\nabla u|^2 \mathrm{d}x - \int_\Omega F(x,u)\mathrm{d}x, \quad u \in H_0^1(\Omega).$$

and therefore we can use all the machinery of the Critical Point Theory. This theory is based on the existence of a linking structure and deformation lemmas [1, 2, 15, 19, 21]. In general, to be able to derive such deformation results, it is supposed that the functional satisfies some compactness condition. We use here the well-known Cerami condition, which reads as: the functional I satisfies the Cerami condition at level $c \in \mathbb{R}$ ($(\mathrm{Ce})_c$ for short) if any sequence $(u_n) \subset H_0^1(\Omega)$ such that $I(u_n) \to c$ and $\|I'(u_n)\|_{H_0^1(\Omega)'}(1 + \|u_n\|) \to 0$ has a convergent subsequence.

In order to get compactness we shall assume the following condition (see [3])

(NQ) setting $H(x,s) := f(x,s)s - 2F(x,s)$, we have that

$$\lim_{|s|\to\infty} H(x,s) = +\infty, \quad \text{uniformly for } x \in \Omega.$$

The behaviour of the nonlinearity at the origin will be done by the condition

(f_2) there exists $K_0 \in L^t(\Omega)$, $t > N/2$, with nontrivial positive part such that

$$\lim_{s\to 0} \frac{2F(x,s)}{s^2} = K_0(x), \quad \text{uniformly for } x \in \Omega.$$

As it is well known (see deFigueiredo [4]), under this condition the weighted eigenvalue problem

$$(LP) \qquad -\Delta u = \lambda K_0(x) u, \quad u \in H_0^1(\Omega)$$

has an increasing sequence of eigenvalues $(\lambda_j(K_0))_{j\in\mathbb{N}}$ with $\lambda_1(K_0) > 0$.

We establish the existence of one weak solution by assuming a crossing condition at the origin. Related conditions on weighted eigenvalue problems have already appeared in the paper of deFigueiredo and Massabó [5] (see also [6]). Our main result can be stated as follows:

Theorem 1.1. *Suppose that f satisfies (f_0), (f_1) and (NQ). If f also satisfies (f_2) with*

$$\lambda_m(K_0) < 1 < \lambda_{m+1}(K_0),$$

then the problem (P) has at least one nonzero solutions.

We emphasize that our existence result works without the well-known Ambrosetti-Rabinowitz condition [1]. It reads as: there exist $\theta > 2$, $R > 0$ such that

$$(AR) \qquad \theta F(x,s) \le sf(x,s), \quad x \in \Omega,\ |s| \ge R.$$

The main role of (AR) condition is to ensure the boundedness of the Palais-Smale sequences for I. It is not hard to verify that it implies that $F(x,s) \ge c_1|s|^\theta - c_2$ for any $x \in \Omega, t \in \mathbb{R}$, in such a way that F goes to infinity at least like $|s|^\theta$. We observe that the condition (f_1) is a more natural superlinear condition. Indeed, there are many superlinear functions which do not behave like $|s|^\theta$, $\theta > 2$, at infinity. For instance, we can take $f(s) = \lambda s + s\log(1+|s|)$, with $\lambda > 0$, and easily conclude that it does not verify (AR), but (NQ) holds. Actually, we can verify that hypotheses $(f_0) - (f_2)$ are also satisfied with $K_0 \equiv 1$ and $\lambda_j(K_0) = \lambda_j \in \sigma(-\Delta, H_0^1(\Omega)), \lambda \in (\lambda_m, \lambda_{m+1})$ for some fixed $m \in \mathbb{N}$. Thus our results extend and complement earlier results on superlinear problems.

As it will be clear from the proofs the ideas presented here can be used to handle with other settings of conditions on f. We could consider the case $\lambda_1(K_0) > 1$ as an application of the classical Mountain Pass theorem. If $K_0 \equiv 0$, the same ideas of the proof provide a weak solution. Also we can deal with the existence of multiple solutions under some symmetry assumptions (see [8] for instance). The main point here is to guarantee compactness. We show that the condition (NQ), introduced by Costa-Magalhães [3], is a powerful tool. More precisely, we prove that (f_0) and (NQ) are sufficient to prove that the functional I satisfies the Cerami condition, this being the main novelty of this work.

Semilinear superlinear elliptic problems have been considered during the past forty years, see [3, 7, 10–13, 21]. In all these works some condition on the nonlinearity ensure some kind of compactness condition. For example, in [7] de Figueiredo *et al.* considered superlinear elliptic problems such as (P) which satisfies the following conditions

(FLN_1) there exist $\theta \in (0, 2^*)$ such that

$$\limsup_{|s|\to\infty} \frac{f(x,s)s - \theta F(x,s)}{s^2 f(s)^{2/N}} \le 0, \quad \text{uniformly for } x \in \Omega.$$

(FLN_2) for each $x \in \Omega$, the function $s \mapsto f(x,s)/s^{2^*-1}$ is nonincreasing.

They proved that problem (P) admits at least one positive solution using topological methods. Posteriorly, Jeanjean [9] considered the problem (P) requiring convexity in s for the function $H(x,s)$ defined in (NQ). We also mention the papers [18, 20] and references therein for similar results. Superlinear elliptic problems have also been studied under monotonicity conditions for the function $s \mapsto f(x,s)/s$, for $|s| \ge R$ (see [14]). In other works [10, 11] monotonicity was imposed on $H(x,\cdot)$ (see also [16–18]).

Here we do not assume any kind of monotonicity or convexity on the nonlinear therm *f* nor in the function *H* defined in (*NQ*). Hence, our results complement and/or extended the aforementioned works.

The paper has just one more section, where we present the variational setting of the problem and prove Theorem 1.1. Throughout the paper we suppose that the function *f* satisfies (f_0). For save notation, we write only $\int_\Omega g$ instead of $\int_\Omega g(x)\mathrm{d}x$. For any $1 \le t < \infty$, $|g|_t$ denotes the norm in $L^t(\Omega)$.

2 Proof of the main theorem

We denote by H the Hilbert space $H_0^1(\Omega)$ endowed with the norm

$$\|u\|^2 = \left(\int |\nabla u|^2\right)^{1/2}, \quad \text{for any } u \in H.$$

By the Sobolev theorem we know that, for any $2 \le \sigma \le 2^*$ fixed, the embedding $H \hookrightarrow L^\sigma(\Omega)$ is continuous and therefore we can find $S_\sigma > 0$ such that

$$\int |u|^\sigma \le S_\sigma \|u\|^\sigma. \tag{1}$$

If $\sigma < 2^*$, the Rellich-Kondrachov theorem implies that the above embedding is also compact.

As quoted in the introduction, the linear problem (*LP*) has a sequence of eigenvalues $(\lambda_j(K_0))_{j\in\mathbb{N}}$ with $\lambda_1(K_0) > 0$. If we denote by φ_j the eigenfunction associated with $\lambda_j(K_0)$, we set

$$V := \mathrm{span}\{\varphi_1, \ldots, \varphi_m\}, \quad W := V^\perp.$$

and write H as being $H = V \oplus W$. The following variational inequalities hold

$$\|u\|^2 \le \lambda_m(K_0) \int K_0(x)u^2, \ \ \forall\, u \in V, \tag{2}$$

and

$$\|u\|^2 \ge \lambda_{m+1}(K_0) \int K_0(x)u^2, \ \ \forall\, u \in W. \tag{3}$$

As a consequence of our assumption at the origin we have the following:

Lemma 2.1. *Suppose that f satisfies* (f_0) *and* (f_2) *with* $\lambda_m(A_0) < 1 < \lambda_{m+1}(A_0)$. *Then I has a local link at the origin, i.e.,*

(i) *there exists* $\rho_1 > 0$ *such that* $I(z) \leq 0$, *for all* $z \in V \cap B_{\rho_1}(0)$,
(ii) *there exists* $\rho_2 > 0$ *such that* $I(z) > 0$, *for all nonzero* $z \in W \cap B_{\rho_2}(0)$.

Proof. Given $\varepsilon > 0$, we can use (f_0) and (f_2) to obtain $A_\varepsilon > 0$ such that

$$\frac{1}{2}K_0(x)s^2 - \frac{\varepsilon}{2}|s|^2 - A_\varepsilon|s|^p \leq F(x,s) \leq \frac{1}{2}K_0(x)s^2 + \frac{\varepsilon}{2}|s|^2 + A_\varepsilon|s|^p, \tag{4}$$

for any $(x,s) \in \Omega \times \mathbb{R}$. By taking $\varepsilon > 0$ sufficiently small we can use (4), (2), (1) and $\lambda_m(A_0) < 1$ to obtain

$$\begin{aligned} I(u) &\leq \frac{1}{2}\|u\|^2 - \frac{1}{2}\int K_0(x)u^2 + \frac{\varepsilon}{2}\int |u|^2 + A_\varepsilon \int |u|^p \\ &\leq \frac{1}{2}\left(1 - \frac{1}{\lambda_m(A_0)} + \varepsilon S_2\right)\|u\|^2 + A_\varepsilon S_p\|u\|^p \leq \left(\frac{\kappa}{2} + A_\varepsilon S_p\|u\|^{p-2}\right)\|u\|^2 \end{aligned}$$

for some $\kappa < 0$ and for all $u \in V$. Hence the condition (i) holds for $\rho_1 := \left(-\kappa/2A_\varepsilon S_p\right)^{1/(p-2)} > 0$.

In order to verify (ii), we choose $\varepsilon > 0$ small and use (4), (3), (1) and $\lambda_{m+1}(A_0) > 1$, to get

$$\begin{aligned} I(u) &\geq \frac{1}{2}\left(1 - \frac{1}{\lambda_{m+1}(A_0)} - \varepsilon S_2\right)\|u\|^2 - A_\varepsilon S_p\|u\|^p \\ &\geq \left(\frac{\mu}{2} + A_\varepsilon S_p\|z\|^{p-2}\right)\|u\|^2, \end{aligned}$$

for some $\mu > 0$ and for all $u \in W$. As before, we can check that (ii) holds for $\rho_2 := \left(\mu/2A_\varepsilon S_p\right)^{1/(p-2)} > 0$. The lemma is proved. □

We are now ready to prove our main theorem.

Proof of Theorem 1.1. According to the last lemma the functional I has a local linking at the origin. For any given $k \in \mathbb{N}$, let $H_k \subset H$ be a k-dimensional subspace. Since all the norms in H_k are equivalent, there exists $c_1 > 0$ such that $\|u|^2 \leq c_1 \int u^2$ for any $u \in H_k$. Given $M > (2/c_1)$, it follows from (f_1) that $F(x,s) \geq Ms^2 - c_2$ for any $x \in \Omega$ and $s \in \mathbb{R}$. Hence,

$$I(u) \leq \frac{1}{2}\left(1 - \frac{2M}{c_2}\right)\|u\|^2 + c_1|\Omega|,$$

and we conclude that $I(u) \to -\infty$ as $\|u\| \to +\infty$, $u \in H_k$. Moreover, by (f_0), we can easily see that I maps bounded sets into bounded sets.

The above consideration shows that the functional I satisfies all the geometric condition of the Local Linking Theorem proved by Li and Willem in [13, Theorem 2]. Hence, if we can prove that I satisfies the Cerami condition, this last theorem provides a nonzero critical point for I. Here we mention that Theorem 2 in [13] is stated for a Palais-Smale type condition. However, as it is well known (see [2]), the deformation lemma used in [13] also holds for the Cerami condition.

It remains to check that I satisfies the Cerami condition. Let $(u_n) \subset H$ be such that

$$I(u_n) \to c, \quad \|I'(u_n)\|_{H'}(1 + \|u_n\|) \to 0,$$

where $c \in \mathbb{R}$. Since f has subcritical growth it suffices to prove that (u_n) is bounded.

Arguing by contradiction we suppose that, along a subsequence, $\|u_n\| \to +\infty$ as $n \to +\infty$. For each $n \in \mathbb{N}$, let $t_n \in [0, 1]$ be such that

$$I(t_n u_n) = \max_{t \in [0,1]} I(tu_n). \tag{5}$$

Setting $v_n := u_n/\|u_n\|$ we obtain $v \in H$ such that, along a subsequence,

$$\begin{cases} v_n \rightharpoonup v \text{ weakly in } H, \\ v_n \to v \text{ strongly in } L^q(\Omega), \ \text{for any } 1 \le q < 2^*, \\ v_n(x) \to v(x). \end{cases} \tag{6}$$

In what follows we prove that $v \neq 0$. Indeed, suppose by contradiction that $v = 0$. Then it follows from (f_0) and the strong convergence in (6) that $\int F(x, \sqrt{4m} v_n) \to 0$, as $n \to +\infty$, for any fixed $m > 0$. Since we may suppose that $\sqrt{4m} < \|u_n\|$, the definition of t_n in (5) provides

$$I(t_n u_n) \ge I\left(\frac{\sqrt{4m}}{\|u_n\|} u_n\right) = 2m - \int F(x, \sqrt{4m} v_n) \ge m > 0, \tag{7}$$

for any $n \ge n_0$, where $n_0 \in \mathbb{N}$ depends only on m.

We look for a contradiction by considering two cases:

Case 1: along a subsequence, $t_n < (2/\|u_n\|)$

In this case we use condition (f_0) and the Sobolev embeddings to obtain $c_1, c_2 > 0$ such that

$$\left| \int H(x, t_n u_n) \right| \le c_1 t_n \|u_n\| + c_2 t_n^p \|u_n\|^p \le 2c_1 + c_2 2^p = c_3.$$

If $t_n > 0$, it follows from $I'(t_n u_n)(t_n u_n) = 0$ that

$$0 = t_n^2 \|u_n\|^2 - \int f(x, t_n u_n)(t_n u_n) = 2I(t_n u_n) - \int H(x, t_n u_n),$$

and therefore

$$I(t_n u_n) = \frac{1}{2}\int H(x, t_n u_n) \le \frac{c_3}{2}.$$

The above inequality also holds if $t_n = 0$, and therefore we obtain a contradiction with (7), since the number $m > 0$ in that expression is arbitrary. Hence, the case 1 cannot occurs.

It remains to discard the

Case 2: along a subsequence, $t_n \ge (2/\|u_n\|)$

We fix $\gamma > 0$ in such a way that

$$3\gamma|\Omega| > 4, \tag{8}$$

where $|\Omega|$ stands for the Lebesgue measure of Ω. In view of (NQ) we can obtain $s_0 > 0$ such that $H(x,s) \ge \gamma$ for any $x \in \Omega$, $|s| \ge s_0$. On the other hand, since H has a subcritical growth, we have that $H(x,s) \ge -C|s|$ for any $x \in \Omega$, $|s| \le s_1$, where $s_1 > 0$ is small.

We consider the nonnegative cut-off function $\psi_\varepsilon : \mathbb{R} \to \mathbb{R}$ given by

$$\psi_\varepsilon(s) = \begin{cases} e^{-\varepsilon/s^2}, & \text{if } s \neq 0,\\ 0, & \text{if } s = 0,\end{cases}$$

with $\varepsilon > 0$ free for now. We mention that ψ_ε is smooth and

$$\lim_{s\to 0}\psi_\varepsilon(s) = \lim_{s\to 0}\psi_\varepsilon'(s) = 0.$$

These limits, (f_0) and the continuity of H provide $C_{\gamma,\varepsilon} > 0$ such that

$$H(x,s) \ge \gamma\psi_\varepsilon(s) - C_{\gamma,\varepsilon}|s|, \quad \text{for any } (x,s) \in \Omega\times\mathbb{R}.$$

Given $0 < s < t$, we can use the above inequality and the definition of H to get

$$\begin{aligned}\frac{I(tu_n)}{t^2\|u_n\|^2} - \frac{I(su_n)}{s^2\|u_n\|^2} &= -\int_\Omega\int_s^t \frac{d}{d\tau}\left(\frac{F(x,\tau u_n)}{\tau^2\|u_n\|^2}\right) d\tau\, dx\\ &= -\int_\Omega\int_s^t \frac{H(x,\tau u_n)}{\tau^3\|u_n\|^2}\, d\tau\, dx\\ &\le \int_\Omega\int_s^t \left(\frac{C_{\gamma,\varepsilon}}{\|u_n\|}\frac{|u_n|}{\|u_n\|}\tau^{-2} - \frac{\gamma\psi_\varepsilon(\tau u_n)}{\|u_n\|^2}\tau^{-3}\right) d\tau\, dx\end{aligned}$$

from which it follows that

$$\frac{I(tu_n)}{t^2\|u_n\|^2} \le \frac{I(su_n)}{s^2\|u_n\|^2} + C_{\gamma,\varepsilon}\frac{|v_n|_1}{s\|u_n\|} - \gamma\int_\Omega\int_s^t \frac{\psi_\varepsilon(|\tau u_n|)}{\|u_n\|^2}\tau^{-3} d\tau\, dx.$$

We now set

$$s = s_n = \frac{1}{\|u_n\|} < \frac{2}{\|u_n\|} \le t_n.$$

Since $\int_{s_n}^{t_n} \tau^{-3}\mathrm{d}\tau = (1/2)(\|u_n\|^2 - t_n^{-2})$ we have that

$$\begin{aligned}\frac{I(t_n u_n)}{t_n^2\|u_n\|^2} &\le I(v_n) + C_{\gamma,\varepsilon}|v_n|_1 - \frac{\gamma|\Omega|}{2}\left(1 - \frac{1}{t_n^2\|u_n\|^2}\right) + \gamma A_n \\ &\le B_\gamma + C_{\gamma,\varepsilon}|v_n|_1 - \int F(x, v_n) + \gamma A_n,\end{aligned} \tag{9}$$

with

$$A_n = \int_{s_n}^{t_n}\int_\Omega \frac{1-\psi_\varepsilon(|\tau u_n|)}{\|u_n\|^2}\tau^{-3}\mathrm{d}x\,\mathrm{d}\tau \ge 0$$

and

$$B_\gamma = \frac{1}{2}\left(1 - \frac{3}{4}\gamma|\Omega|\right) < 0,$$

where we have used (8) in the last inequality.

We shall verify in a few moments that, uniformly in $n \in \mathbb{N}$, the following limit holds

$$\lim_{\varepsilon\to 0}\int_{s_n}^{t_n}\int_\Omega \frac{1-\psi_\varepsilon(|\tau u_n|)}{\|u_n\|^2}\tau^{-3}\mathrm{d}x\,\mathrm{d}\tau = 0. \tag{10}$$

If this is true, we can choose $\varepsilon > 0$ in such a way that $\gamma A_n < -B_\gamma/2$, for all $n \in \mathbb{N}$. Since we are supposing that $v = 0$, it follows from (6) and (f_0) that $|v_n|_1 = o_n(1)$ and $\int F(x, v_n) = o_n(1)$, as $n \to +\infty$. Hence, we can take the limit in (9) to obtain

$$\limsup_{n\to+\infty}\frac{I(t_n u_n)}{t_n^2\|u_n\|^2} \le B_\gamma - \frac{B_\gamma}{2} = \frac{B_\gamma}{2} < 0,$$

and therefore $I(t_n u_n) < 0$, for n large, contradicting (7) again.

We proceed now with the proof that the limit in (10) is uniform. We start by considering $\delta > 0$ and splitting the term A_n into two integrals

$$\int_{s_n}^{t_n}\int_\Omega \frac{1-\psi_\varepsilon(|\tau u_n|)}{\|u_n\|^2}\tau^{-3}\mathrm{d}x\,\mathrm{d}\tau = \int_{s_n}^{t_n}\int_{|\tau u_n|\ge\delta}(\cdots) + \int_{s_n}^{t_n}\int_{|\tau u_n|<\delta}(\cdots).$$

In order to save notation we call $A^+_{n,\delta}$ the first integral on the right-hand side above and $A^-_{n,\delta}$ the second one. It suffices to show that these quantities go to 0, uniformly in n, as $\varepsilon \to 0$.

Since ψ_ε is nondecreasing we have that

$$\begin{aligned} A^+_{n,\delta} &\leq \frac{1-e^{-\varepsilon/\delta^2}}{\delta\|u_n\|^2}\int_{s_n}^{t_n}\int_{|\tau u_n|\geq\delta}|\tau u_n|\tau^{-3}\mathrm{d}x\,\mathrm{d}\tau \\ &\leq \frac{1-e^{-\varepsilon/\delta^2}}{\delta\|u_n\|}\left(\frac{1}{s_n}-\frac{1}{t_n}\right)\int_\Omega\frac{|u_n|}{\|u_n\|}\leq\left(\frac{1-e^{-\varepsilon/\delta^2}}{\delta}\right)|v_n|\,, \end{aligned}$$

since $s_n\|u_n\| = 1$. Recalling that $(|v_n|_1)$ is uniformly bounded, we conclude that the limit $\lim_{\varepsilon\to 0} A^+_{n,\delta} = 0$ is uniform.

The calculations for $A^-_{n,\delta}$ are more involved. We first notice that, for each $|s| \leq \delta$ fixed, the function $\varepsilon \mapsto \psi_\varepsilon(s)$ is smooth. Hence, it follows from Taylor's Theorem that, for $h(s) = s^{-2}e^{-\varepsilon/s^2}$, there holds

$$1-\psi_\varepsilon(s) = \varepsilon s^{-2}e^{-\varepsilon/s^2} + r(\varepsilon,s) = \varepsilon\left(h(s)+\frac{r(\varepsilon,s)}{\varepsilon}\right) \leq \varepsilon(h(s)+1),$$

since the continuous remainder term r is such that $\lim_{\varepsilon\to 0} r(\varepsilon,s)/\varepsilon = 0$ uniformly in the compact set $|s| \leq \delta$. By applying Taylor's Theorem again we get for $|s| \leq \delta$,

$$h(s) = h(0) + h'(0)s + r_1(\varepsilon,s) = r_1(\varepsilon,s),$$

with $r_1(\varepsilon,s) = o(|s|)$ as $s \to 0$ uniformly in $\varepsilon \in (0,1]$. Thus, we conclude that, if $\delta > 0$ is small,

$$1-\psi_\varepsilon(s) \leq \varepsilon(1+|s|), \quad \text{for any } |s| \leq \delta.$$

The above inequality and the definition of $A^-_{n,\delta}$ provide

$$\begin{aligned} A^-_{n,\delta} &= \int_{s_n}^{t_n}\int_{|\tau u_n|<\delta}\frac{1-\psi_\varepsilon(|\tau u_n|)}{\|u_n\|^2}\tau^{-3}\mathrm{d}x\,\mathrm{d}\tau \\ &\leq \varepsilon\int_{s_n}^{t_n}\int_\Omega\frac{\tau^{-3}}{\|u_n\|^2}\mathrm{d}x\,\mathrm{d}\tau + \varepsilon\int_{s_n}^{t_n}\int_\Omega\frac{|u_n|}{\|u_n\|^2}\tau^{-2}\mathrm{d}x\,\mathrm{d}\tau \\ &= \varepsilon\frac{|\Omega|}{2}\left(1-\frac{1}{t_n^2\|u_n\|^2}\right)+\frac{\varepsilon}{\|u_n\|}\left(1-\frac{1}{t_n\|u_n\|}\right)\int_\Omega|v_n|\mathrm{d}x \leq \varepsilon\left(\frac{|\Omega|}{2}+|v_n|_1\right), \end{aligned}$$

since we may assume that $\|u_n\| > 1$. This implies that, uniformly in n, there holds $\lim_{\varepsilon\to 0} A^-_{n,\delta} = 0$. This finishes the proof that the weak limit v is nonzero.

After proving that $v \neq 0$ we can prove the theorem in the following way: the set $\tilde{\Omega} := \{x \in \Omega : v(x) \neq 0\}$ has positive measure. Moreover, since $\|u_n\| \to +\infty$, we have that $|u_n(x)| \to +\infty$ a.e. in $\tilde{\Omega}$. Thus, the continuity of H, Fatou's Lemma and (NQ) provide

$$\begin{aligned} 2c &= \lim_{n\to+\infty} (2I(u_n) - I'(u_n)u_n) \\ &\geq \text{meas}(\Omega \setminus \tilde{\Omega}) \cdot \min_{\overline{\Omega}\times\mathbb{R}} H + \int_{\tilde{\Omega}} \liminf_{n\to+\infty} H(x, u_n) = +\infty, \end{aligned}$$

which is a contradiction. Hence, we have that (u_n) is bounded and the theorem is proved. □

Acknowledgements The authors were partially supported by CNPq/Brazil under the grants 307327/2013-2 and 211623/2013-0, respectively.

References

1. Ambrosetti, A., Rabinowitz, P.: Dual variational methods in critical point theory and applications. J. Funct. Anal. **14**, 349–381 (1973)
2. Bartolo, P., Benci, V., Fortunato, D.: Abstract critical point theorems and applications to some nonlinear problems with strong resonance at infinity. Nonlinear Anal. **7**, 981–1012 (1983)
3. Costa, D.G., Magalhães, C.A.: Variational elliptic problems which are nonquadratic at infinity. Nonlinear Anal. **23**, 1401–1412 (1994)
4. de Figueiredo, D.G.: Positive solutions of semilinear elliptic problems. In: Differential Equations (S ao Paulo, 1981). Lecture Notes in Mathematics, vol. 957, pp. 34–87. Springer, New York (1982)
5. de Figueiredo, D.G., Massabó, I.: Semilinear elliptic equations with the primitive of the nonlinearity interacting with the first eigenvalue. J. Math. Anal. Appl. **156**, 381–394 (1991)
6. de Figueiredo, D.G., Miyagaki, O.H.: Semilinear elliptic equations with the primitive of the nonlinearity away from the spectrum. Nonlinear Anal. **17**, 1201–1219 (1991)
7. de Figueiredo, D.G., Lions, P.L., Nussbaum, R.D.: A priori estimates and existence of positive solutions of semilinear elliptic equations. J. Math. Pures Appl. **9**(61, 1), 41–63 (1982)
8. Furtado, M.F., Silva, E.D.: Superlinear elliptic problems under the nonquadriticty condition at infinity. Proc. Roc. Soc. Edinburgh Sect. A. (to appear)
9. Jeanjean, L.: On the existence of bounded Palais-Smale sequences and a application to Landemann-Lazer type problem set $\mathbb{R}^N$. Proc. Roc. Soc. Edinb. Sect. A **129**, 797–809 (1999)
10. Liu, S.: On superlinear problems without Ambrosetti-Rabinowitz condition. Nonlinar Anal. **73**, 788–795 (2010)
11. Li, G., Wang, C.: The existence of a nontrivial solution to a nonlinear elliptic problem of linking type without the Ambrosetti-Rabinowitz condition. Ann. Acad. Sci. Fenn. Math. **36**, 461–480 (2011)
12. Liu, Z., Wang, Z.Q.: On the Ambrosetti-Rabinowitz superlinear condition. Adv. Nonlinear Studies **4**, 653–574 (2004)
13. Li, S., Willem, M.: Applications of local linking to critical point theory. J. Math. Anal. Appl. **189**, 6–32 (1995)
14. Miyagaki, O.H., Souto, M.A.S.: Supelinear problems without Ambrosetti-Rabinowitz growth condition . J. Differ. Equ. **245**, 3628–3638 (2008)

15. Rabinowitz, P.H.: Minimax Methods in Critical Point Theory with Applications to Differential Equations. C.B.M.S. Regional Conference Series Mathematics, vol. 65. American Mathematical Society, Providence (1986)
16. Schechter, M.: Superlinear elliptic boundary value problems. Manuscripta Math. **86**, 253–265 (1995)
17. Schechter, M., Zou, W.: Double linking theorem and multiple solutions. J. Funct. Anal. **205**, 37–61 (2003)
18. Schechter, M., Zou, W.: Superlinear problems. Pacific J. Math. **214**, 145–160 (2004)
19. Struwe, M.: Variational Methods-Applications to Nonlinear Partial Differential Equations and Hamiltonian Systems. Springer, Berlin (1990)
20. Wang, Z.Q.: On a supelinear ellitic equation. Anal. Inst. H. Poincaré Anal. Nonlinéare **8**, 43–57 (1991)
21. Willem, M.: Minimax Theorems. Birkhäuser, Basel (1996)

Radial solutions of a Neumann problem coming from a burglary model

M. Garcia-Huidobro, R. Manásevich, and J. Mawhin

To Djairo de Figueiredo, for his 80th birthday anniversary, with friendship and gratitude

Mathematics Subject Classification (2010): 35J57, 34B15, 47H11

1 The problem

Let $\Omega \subset \mathbb{R}^n$ be a bounded open domain with smooth boundary, $\eta > 0$, $A^0 > 0$, $\psi > 0$ and $\omega > 0$. The Neumann problem on Ω

$$\begin{aligned} \eta\Delta(A - A^0) - A + A^0 + \psi NA(1 - A) &= 0 \text{ in } \Omega, \\ \Delta N - \nabla \cdot \left(\frac{2N}{A}\nabla A\right) - \omega^2(N - 1) &= 0 \text{ in } \Omega, \\ \frac{\partial A}{\partial \nu} = \frac{\partial N}{\partial \nu} &= 0 \text{ on } \partial\Omega \end{aligned} \tag{1}$$

M. Garcia-Huidobro
Facultad de Matematicas, Pontificia Universidad Catolica de Chile,
Casilla 306 Correo 22, Santiago, Chile
e-mail: mgarcia@mat.puc.cl

R. Manásevich
DIM & CMM, FCFM, Universidad de Chile, Casilla 170 Correo 3,
Santiago, Chile
e-mail: manasevi@dim.uchile.cl

J. Mawhin (✉)
Université Catholique de Louvain, IRMP, chemin du cyclotron, 2,
1348 Louvain-la-Neuve, Belgium
e-mail: jean.mawhin@uclouvain.be

A.N. Carvalho et al. (eds.), *Contributions to Nonlinear Elliptic Equations and Systems*, Progress in Nonlinear Differential Equations and Their Applications 86,
DOI 10.1007/978-3-319-19902-3_17

arises in some mathematical modeling for burglary of houses (see [5] and related papers [1, 2, 6–8]). In most of these models A represents attractiveness for a house to be burgled and N density of burglars, thus the restrictions $A > 0$ and $N > 0$ appear as natural. When η, A^0, ψ are positive constants, system (1) admits the unique positive solution

$$A = (2\psi)^{-1}\left[\psi - 1 + \sqrt{(\psi-1)^2 + 4A^0\psi}\right], \quad N = 1.$$

A natural question is to know if a positive (non constant) solution still exists when η, A^0 and ψ may depend upon x.

The question was positively answered in [4] in the case where $n = 1$ for more general problems of the form

$$\begin{aligned} \eta(x)[A - A^0(x)]'' - A + A^0 + Nf(x, A) &= 0 \text{ in } (0, L), \\ N'' + [g(x, A, A')N]' - \omega^2(N-1) &= 0 \text{ in } (0, L) \\ A'(0) = A'(L) = N'(0) &= N'(L) = 0 \end{aligned}$$

where the continuous function η, the C^2 function A^0 are positive, and the continuous functions f and g satisfy suitable conditions.

The aim of this paper is to consider the existence of positive radial solutions $A = A(r)$, $N = N(r)$ (where as is customary we have set $|x| = r$) in the annulus of radii $0 < l < L$

$$\Omega = \{x \in \mathbb{R}^n : l < |x| < L\}$$

for systems of the form

$$\begin{aligned} \eta(r)\Delta[A - A^0(r)] - A + A^0(r) + Nf(r, A) &= 0 \text{ in } \Omega, \\ \Delta N + \nabla \cdot [N\nabla h(A)] - \omega^2(N-1) &= 0 \text{ in } \Omega, \\ \frac{\partial A}{\partial \nu} = \frac{\partial N}{\partial \nu} &= 0 \text{ on } \partial\Omega, \end{aligned} \tag{2}$$

where $\eta : [l, L] \to \mathbb{R}$ is continuous and positive, $A^0 : [l, L] \to \mathbb{R}$ is of class C^2 and positive, $h : \mathbb{R} \to \mathbb{R}$ is of class C^2, $\omega > 0$, and $f : [l, L] \times \mathbb{R} \to \mathbb{R}$ continuous satisfies condition (f_1) or (f_2) mentioned below.

This type of domain is not unrealistic in burglary problems because, for $n = 2$, the annulus Ω corresponds to a city with circular shape, a choice made almost since the beginning of cities in the history of mankind. Indeed, a round architectural shape has both aesthetic and practical advantages.

In a classical way, problem (2) is reduced to the Neumann problem on (l, L) for an ordinary differential system

$$\eta(r)[r^{n-1}(A - A^0(r))']' + r^{n-1}[-A + A^0(r) + Nf(r, A)] = 0 \text{ in } (l, L),$$

$$(r^{n-1}N')' + \{r^{n-1}[h(A)]'N\}' - r^{n-1}\omega^2(N-1) = 0 \text{ in } (l,L),$$
$$A'(l) = A'(L) = N'(l) = N'(L) = 0.$$

and we will study indeed the more general problem

$$\eta(r)[r^{n-1}(A - A^0(r))']' + r^{n-1}[-A + A^0(r) + Nf(r,A)] = 0 \text{ in } (l,L), \tag{3}$$

$$(r^{n-1}N')' + [r^{n-1}g(r,A,A')N]' - r^{n-1}\omega^2(N-1) = 0 \text{ in } (l,L), \tag{4}$$

$$A'(l) = A'(L) = N'(l) = N'(L) = 0. \tag{5}$$

for a suitable class of C^1 functions $g : [l,L]\times\mathbb{R}\times\mathbb{R} \to \mathbb{R}$. A solution of (3)–(4)–(5) is a couple of real functions $(A,N) \in C^2([0,T]) \times C^2([0,T])$ such that $A(r) > 0$ for all $r \in [l,L]$, which satisfies the system and the boundary conditions. We are interested in *positive solutions* of this problem, i.e. in solutions (A,N) such that $A(r) > 0$ and $N(r) > 0$ for all $r \in [l,L]$.

The corresponding problem on a ball $B(0,R)$ remains open when $n \geq 2$, and we will indicate the obstacles in the course of the paper.

We write $\mathbb{R}^+ = \{x \in \mathbb{R} : x > 0\}$, $\mathbb{R}_0^+ = \{x \in \mathbb{R} : x \geq 0\}$, and for a function $u : (l,L) \to \mathbb{R}$, we set

$$osc\ u = \sup_{r\in(l,L)} u(r) - \inf_{r\in(l,L)} u(r).$$

Define the continuous function $B^0 : [l,L] \to \mathbb{R}$ by

$$B^0(r) = A^0(r) - \eta(r)\left[(A^0)''(r) + \frac{n-1}{r}(A^0)'(r)\right], \tag{6}$$

and introduce the following assumptions upon f and g.

(f_1) *there exists $R >$ osc A^0 such that, for all $r \in [l,L]$,*

$$f(r,A) \geq 0 \text{ when } 0 \leq A \leq R \text{ and } f(r,A) \leq 0 \text{ when } A \geq R.$$

(f_2) *B^0 is a positive function and there exists $R > 0$ such that, for all $r \in [l,L]$,*

$$f(r,A) \geq 0 \text{ when } 0 \leq A \leq R \text{ and } f(r,A) \leq 0 \text{ when } A \geq R.$$

(g) $g(l,A,0) = 0$ *and* $g(L,A,0) = 0$ *for all* $A \in \mathbb{R}^+$.

The main result of the paper is that *if f satisfies Assumption (f_1) or Assumption (f_2) and if g satisfies Assumption (g), then problem (3)–(4)–(5) has at least one positive solution*. In particular, *if f satisfies Assumption (f_1) or Assumption (f_2), then problem (2) has at least one positive radial solution.*

Like in [4], we approach this question using Leray-Schauder degree [3]. The requested *a priori* estimates are obtained through a combination of pointwise estimates based upon maximum or minimum properties and of L^1-estimates. The reduction to a fixed point problem takes into account the specificities of the system.

2 The homotopy

Let us associate with (3)–(4)–(5) the homotopy, with $\lambda \in [0,1]$,

$$\begin{aligned} -\eta(r)[r^{n-1}(A-A^0(r))']' + r^{n-1}[A-A^0(r)] &= \lambda r^{n-1} N f(r,A), \\ A'(l) = A'(L) &= 0, \end{aligned} \tag{7}$$

$$\begin{aligned} -(r^{n-1}N')' + r^{n-1}\omega^2 N &= r^{n-1}\omega^2 + \lambda[r^{n-1}g(r,A,A')N]', \\ N'(l) = N'(L) &= 0. \end{aligned} \tag{8}$$

For $\lambda = 1$, (7)–(8) reduces to (3)–(4)–(5), and for $\lambda = 0$, (7)–(8) reduces to the non-homogeneous decoupled linear system

$$-\eta(r)[r^{n-1}(A-A^0(r))']' + r^{n-1}[A-A^0(r)] = 0, \quad A'(l) = A'(L) = 0, \tag{9}$$

$$-(r^{n-1}N')' + r^{n-1}\omega^2 N = r^{n-1}\omega^2, \quad N'(l) = N'(L) = 0. \tag{10}$$

3 The *a priori* estimates

For any continuous function $B : [l,L] \to \mathbb{R}$, we write for simplicity

$$\min B := \min_{[l,L]} B, \quad \max B := \max_{[l,L]} B.$$

Lemma 3.1. *If (A,N) is any possible solution of (7)–(8) for some $\lambda \in [0,1]$, then*

$$\int_l^L N(r)\, r^{n-1} dr = \frac{L^n - l^n}{n}. \tag{11}$$

In particular, if N is nonnegative,

$$\int_l^L N(r)\,dr \le \frac{L^n - l^n}{nl^{n-1}}. \tag{12}$$

Proof. Integration of both members of (8) over $[l, L]$, and use of the boundary conditions and Assumption (g) give (11). Inequality (12) follows from

$$l^{n-1}\int_l^L N(r)\,dr \le \int_l^L N(r)r^{n-1}\,dr.$$

□

Remark 3.2. In the case where Ω is the ball $B(0, L)$, inequality (12) does not hold, which creates difficulties in getting the further a priori estimates.

Lemma 3.1 implies that, for any possible solution (A, N) of (7)–(8) for some $\lambda \in [0, 1]$, N must take values greater or equal to one. We now show that N cannot have a minimum equal to zero.

Lemma 3.3. *If (A, N) is any possible solution of (7)–(8) for some $\lambda \in [0, 1]$, then either* $\min N < 0$ *or* $\min N > 0$. *In particular, is N is nonnegative,* $\min N > 0$.

Proof. If N reaches a minimum equal to zero at some $\xi \in [l, L]$, then $N'(\xi) = 0$ and

$$\begin{aligned}0 \le \xi^{n-1}N''(\xi) &= -(n-1)\xi^{n-2}N'(\xi) - \lambda N'(\xi)\{\xi^{n-1}g[\xi, A(\xi), A'(\xi)]\}\\ &- \lambda N(\xi)[r^{n-1}g(r, A, A')]'(\xi) + \xi^{n-1}\omega^2 N(\xi) - \xi^{n-1}\omega^2 = -\xi^{n-1}\omega^2 < 0,\end{aligned}$$

a contradiction. □

Lemma 3.4. *If Assumption (f_1) holds and if (A, N) is any possible positive solution of (7)–(8) for some $\lambda \in [0, 1]$, then, for all $x \in [l, L]$,*

$$\begin{aligned}0 < A_{0,1} &:= \min\{\min A^0, R - osc\, A^0\}\\ &\le A(r) \le \max\{\max A^0, R + osc\, A^0\} := A_{1,1}.\end{aligned} \tag{13}$$

Proof. Let (A, N) be a possible positive solution of (7–8) for some $\lambda \in [0, 1]$. Assume that $A - A^0$ has a maximum greater or equal to $R - \min A^0$ at some $\xi \in [l, L]$. Then $A(\xi) \ge R - \min A^0 + A^0(\xi) \ge R$. Now, , using Assumption (f_1),

$$\begin{aligned}0 \ge \eta(\xi)\xi^{n-1}(A - A^0)''(\xi) &= -(n-1)\eta(\xi)\xi^{n-2}(A - A^0)'(\xi)\\ &+ \xi^{n-1}[A(\xi) - A^0(\xi)] - \lambda\xi^{n-1}N(\xi)f(\xi, A(\xi))\\ &\ge \xi^{n-1}[A(\xi) - A^0(\xi)],\end{aligned}$$

so that, for all $r \in [l, L]$,

$$A(r) - A^0(r) \leq A(\xi) - A^0(\xi) \leq 0,$$

and hence.

$$\max A \leq \max A^0.$$

Similarly, if $A - A^0$ reaches a minimum lower or equal to $R - \max A^0$ at some $\xi \in [l, L]$, then $A(\xi) \leq R - (\max A^0 - A^0(\xi)) \leq R$ and, using Assumption (f_1) and proceeding like above we get, for all $r \in [l, L]$,

$$A(r) \geq A^0(r)$$

and hence

$$\min A \geq \min A^0.$$

The result follows easily. □

Lemma 3.5. *If Assumption (f_2) holds and if (A, N) is any possible positive solution of (7)–(8) for some $\lambda \in [0, 1]$, then, for all $r \in [l, L]$,*

$$\begin{aligned} 0 < A_{0,2} &:= \min\{\min B^0, R\} \\ &\leq A(r) \leq \max\{\max B^0, R\} := A_{1,2}. \end{aligned} \tag{14}$$

Proof. Let (A, N) be a possible positive solution of (7–8) for some $\lambda \in [0, 1]$. Assume that A has a maximum greater or equal to R at some $\xi \in [l, L]$. Then, using Assumption (f_2) and definition (6),

$$\begin{aligned} 0 \geq \eta(\xi)\xi^{n-1}A''(\xi) &= \eta(\xi)\xi^{n-1}(A^0)''(\xi) - (n-1)\eta(\xi)\xi^{n-2}A'(\xi) \\ &+ (n-1)\eta(\xi)\xi^{n-2}(A^0)'(\xi) + \xi^{n-1}[A(\xi) - A^0(\xi)] - \lambda\xi^{n-1}N(\xi)f(\xi, A(\xi)) \\ &\geq \xi^{n-1}\eta(\xi)(A^0)''(\xi) + (n-1)\eta(\xi)\xi^{n-2}(A^0)'(\xi) + \xi^{n-1}[A(\xi) - A^0(\xi)] \\ &= \xi^{n-1}[A(\xi) - B^0(\xi)], \end{aligned}$$

so that

$$\max A = A(\xi) \leq B^0(\xi) \leq \max B^0.$$

Similarly, if A reaches a minimum lower or equal to R at some $\xi \in [l, L]$, then, using Assumption (f_2),

$$\min A = A(\xi) \geq B^0(\xi) \geq \min B^0.$$

The result follows easily. □

From now on, we, respectively, write A_0 and A_1 for $A_{0,1}$ and $A_{1,1}$ or for $A_{0,2}$ and $A_{1,2}$ depending upon the Assumption made upon f. We set

$$\max|f| := \max_{[l,L]\times[A_0,A_1]} |f|.$$

Lemma 3.6. *If (A,N) is any possible positive solution of (7)–(8) for some $\lambda \in [0,1]$, then, for all $r \in [l,L]$,*

$$|A'(r)| \le \frac{1}{l^{n-1}} \int_l^L \left|[s^{n-1}(A^0)'(s)]'\right| ds + \frac{L^n - l^n}{nl^{n-1}\min\eta}\left(A_1 + \max A^0 + \max|f|\right) := A_2. \tag{15}$$

Proof. Let (A,N) be a possible positive solution of (7)–(8) for some $\lambda \in [0,1]$. It follows from equation (7) that, for all $x \in [l,L]$,

$$|[(r^{n-1}A'(r))']| \le |[(r^{n-1}(A^0)'(r))']| + \frac{r^{n-1}}{\eta(r)}\left[A(r) + A^0(r) + N(r)|f(r,A(r))|\right],$$

and hence, by integration, using the boundary conditions, (11) and (13) or (14)

$$l^{n-1}|A'(r)| \le r^{n-1}|A'(r)| \le \left|\int_l^r [s^{n-1}A'(s)]'\,ds\right| \le \int_l^r \left|[s^{n-1}A'(s)]'\right| ds$$
$$\le \int_l^L \left|[s^{n-1}(A^0)'(s)]'\right| ds + \frac{L^n - l^n}{n\min\eta}\left[A_1 + \max A^0 + \max|f|\right].$$

□

We set

$$\max|g| = \max_{[l,L]\times[A_0,A_1]\times[-A_2,A_2]} |g|.$$

Lemma 3.7. *If (A,N) is any possible positive solution of (7)–(8) for some $\lambda \in [0,1]$, then, for all $x \in [l,L]$,*

$$N(r) \le 1 + \frac{L^n - l^n}{nl^{n-1}}\left[2\omega^2(L-l) + \max|g|\right] := A_3. \tag{16}$$

Proof. Integrating equation (8) and using the boundary conditions and Assumption (g), we get, for all $r \in [l,L]$,

$$r^{n-1}N'(r) = \omega^2\int_l^r N(s)\,s^{n-1}\,ds - \omega^2\frac{r^n - l^n}{n} - \lambda g(r,A(r),A'(r))N(r))r^{n-1},$$

and hence, using (11),

$$l^{n-1}|N'(r)| \le r^{n-1}|N'(r)| \\ \le 2\omega^2 \frac{L^n - l^n}{n} + |g(r, A(r), A'(r))| N(r) r^{n-1}. \quad (17)$$

Because of (11), there exists $\xi \in [l, L]$ such that $N(\xi) = 1$, which, together with (17) gives, for all $x \in [l, L]$,

$$\begin{aligned} |N(r) - 1| = |N(r) - N(\xi)| &= \left| \int_\xi^r N'(s)\, ds \right| \le \int_l^L |N'(s)|\, ds \\ &\le 2\omega^2 \frac{(L-l)(L^n - l^n)}{n l^{n-1}} + \frac{1}{l^{n-1}} \max |g| \int_l^L N(r) r^{n-1}\, dr \\ &= \frac{L^n - l^n}{n l^{n-1}} \left[2\omega^2 (L - l) + \max |g| \right], \end{aligned}$$

and the result follows easily. □

4 The associated linear system

Lemma 4.1. *For every continuous* $f : [l, L] \to \mathbb{R}$*, the problem*

$$-\eta(r)(r^{n-1}A')' + r^{n-1}A = r^{n-1}f(r) \ \text{ in } \ (l, L), \quad A'(l) = A'(L) = 0 \quad (18)$$

has unique solution $A \in C^2([l, L])$*, and there exists a constant* $C > 0$ *such that*

$$\|A\|_\infty + \|A'\|_\infty \le C\|f\|_\infty. \quad (19)$$

Furthermore, if $f(r) > 0$ *for all* $r \in [l, L]$*, then* $A(r) > 0$ *for all* $r \in [l, L]$*.*

Proof. Consider the homogeneous problem associated to (18)

$$-\eta(r)(r^{n-1}C')' + r^{n-1}C = 0 \ \text{ in } \ (l, L), \quad C'(l) = C'(L) = 0.$$

If C reaches a positive maximum at some $\xi \in [l, L]$, then $C'(\xi) = 0$ and

$$0 \ge \eta(\xi)\xi^{n-1}C''(\xi) = \xi^{n-1}C(\xi) > 0$$

a contradiction. Similarly if C reaches a negative maximum at some $\xi \in [l, L]$. Thus $C \equiv 0$ is the unique solution. As the homogeneous problem only has the trivial solution, standard linear theory implies that (18) has a unique solution satisfying (19). Furthermore, if $f(r) > 0$ for all $r \in [l, L]$ and A reaches a nonpositive minimum at $\xi \in [l, L]$, then $A''(\xi) \geq 0$ and

$$0 \leq \eta(\xi)\xi^{n-1}A''(\xi) = \xi^{n-1}A(\xi) - f(\xi) < 0,$$

a contradiction. □

Lemma 4.2. *Problem (10) has the unique solution $N \equiv 1$.*

Proof. By a similar argument as in Lemma 4.1, the corresponding homogeneous problem

$$-(r^{n-1}N')' + r^{n-1}\omega^2 N = 0 \text{ in } (l, L), \quad N'(l) = N'(L) = 0$$

has only the trivial solution. Consequently, problem (10) is uniquely solvable, and $N \equiv 1$ is a solution. □

The following standard result (see for example [9], p. 207–211) will be used later.

Lemma 4.3. *For any continuous $f : [l, L] \to \mathbb{R}$, the problem*

$$-(r^{n-1}N')' + r^{n-1}\omega^2 N = r^{n-1}f(r) \text{ in } (l, L), \quad N'(l) = N'(L) = 0$$

has a unique solution N given by

$$N(r) = \int_0^L G(r, s)f(s)\, ds,$$

where the Green function $G : [l, L]^2 \to \mathbb{R}$ is defined by

$$G(x, y) = \begin{cases} C^{-1}N_1(r)N_2(s) \text{ if } l \leq r \leq s \leq L \\ C^{-1}N_1(s)N_2(r) \text{ if } l \leq s < r \leq L \end{cases}$$

with N_1 (resp. N_2) a solution of

$$-(r^{n-1}N')' + r^{n-1}\omega^2 N = 0$$

such that $N_1'(l) = 0$ (resp. $N_2'(L) = 0$), and $C \neq 0$ the constant such that

$$r^{n-1}[N_1(r)N_2'(r) - N_2(r)N_1'(r)] = C \quad (l \leq r \leq L).$$

Remark 4.4. It is not clear if a result like Lemma 4.3 is available on $(0, L)$.

5 The fixed point reduction

It follows from Lemma 4.1 that the linear operator $K : C([l,L]) \to C^1([l,L])$ which to any $f \in C([l,L])$ associates the unique solution A of (18) is continuous and takes positive functions into positive functions.

Let E denote the Banach space $E = C^1([l,L]) \times C([l,L])$ with the usual norm $\|(A,N)\|_E = \|A\|_\infty + \|A'\|_\infty + \|N\|_\infty$, and let us define the function $\rho : [l,L] \to \mathbb{R}$ by $\rho(r) = r^{n-1}$, and the operator

$$\mathscr{F} : \{(A,U,\lambda) \in E \times [0,1] \mid A > 0, N > 0\} \mapsto E,$$

by

$$\mathscr{F}(A,N,\lambda) = \begin{pmatrix} K\{-\eta[\rho(A^0)']' + \rho A^0 + \lambda\rho N f(\cdot,A)\}, \\ -\lambda \int_l^L \frac{\partial}{\partial s} G(\cdot,s)\rho(s) g(s,A(s),A'(s))N(s)\,ds + \omega^2 \int_l^L G(\cdot,s)\rho(s)\,ds \end{pmatrix}.$$

Lemma 5.1. *For any fixed $\lambda \in [0,1]$, (A,N) is a positive solution of problem (7)–(8) if and only if (A,N) is a fixed point of $\mathscr{F}$.*

Proof. Clearly, using Lemma 4.1, (7) is equivalent to

$$A = K\{-\eta[(\rho(A^0)']' + \rho A^0 + \lambda\rho N f(\cdot,A)\}.$$

On the other hand, using Lemma 4.3, boundary conditions and Assumption (g), (8) is equivalent to

$$\begin{aligned} N(r) &= \int_l^L G(r,s)\left[\left(\lambda\rho(s)g(s,A(s),A'(s))N(s)\right)' + \omega^2\rho(s)\right] ds \\ &= \lambda \int_l^L \frac{\partial}{\partial s}\left[G(r,s)\rho(s)g(s,A(s),A'(s))N(s)\right] ds \\ &\quad - \lambda \int_l^L \frac{\partial}{\partial s} G(r,s)\rho(s)g(s,A(s),A'(s))N(s)\,ds + \omega^2 \int_l^L G(r,s)\rho(s)\,ds \\ &= -\lambda \int_l^L \frac{\partial}{\partial s} G(r,s)g(s,A(s),A'(s))N(s)\rho(s)\,ds + \omega^2 \int_l^L G(r,s)\rho(s)\,ds. \end{aligned}$$

□

6 The existence theorem and applications

We can now state and prove our existence theorem.

Theorem 6.1. *Let $L > l > 0, \omega > 0$, let $\eta : [l, L] \to \mathbb{R}$ be a continuous positive function, $A^0 : [l, L] \to \mathbb{R}$ a positive function of class C^2. If the continuous functions $f : [l, L] \times \mathbb{R}^+ \to \mathbb{R}$ and $g : [l, L] \times \mathbb{R}^+ \times \mathbb{R} \to \mathbb{R}$ satisfy either conditions (f_1) and (g), or conditions (f_2) and (g), then problem (3)–(5) has at least one positive solution.*

Proof. In order to apply Leray-Schauder's continuation theorem to the operator $\mathscr{F}$, we fix $0 < R_0 < A_0, R_1 > A_1, R_2 > A_2, R_3 > A_3$ and we consider the open bounded set $\Omega \subset E$ defined by

$$\Omega = \{(A, N) \in E : R_0 < A(r) < R_1,\ \|A'\|_\infty < R_2,\ 0 < N(r) < R_3\ \ (r \in [l, L])\}.$$

The continuity and compactness of $\mathscr{F}$ on $\overline{\Omega}$ are proved in a standard way, using Ascoli-Arzela's theorem.

It follows from Lemmas 3.1 to 3.7 and Lemma 5.1 that, for any $\lambda \in [0, 1]$ and any possible fixed point (A, N) of $\mathscr{F}$, one has $(A, N) \notin \partial\Omega$. Indeed, any possible solution in $\overline{\Omega}$ belongs to Ω. Consequently, using the homotopy invariance of the Leray-Schauder degree d_{LS}, we obtain

$$d_{LS}[I - \mathscr{F}(\cdot, 1), \Omega, 0] = d_{LS}[I - \mathscr{F}(\cdot, 0), \Omega, 0].$$

But

$$I - \mathscr{F}(\cdot, 0) = I - \left(K\{-\eta[(\rho(A^0)']' + \rho A^0\},\ \omega^2 \int_l^L G(\cdot, s)\rho(s)\, ds\right),$$

and since $\left(K\{-\eta[\rho(A^0)']' + \rho A^0\}, \omega^2 \int_l^L G(\cdot, s)\rho(s)\, ds\right) \in \Omega$, we obtain that

$$d_{LS}[I - \mathscr{F}(\cdot, 0), \Omega, 0] = 1.$$

Consequently

$$d_{LS}[I - \mathscr{F}(\cdot, 1), \Omega, 0] = 1$$

and the existence property of Leray-Schauder degree implies that $\mathscr{F}(\cdot, 1)$ has at least one fixed point in Ω. Hence, using Lemma 5.1 with $\lambda = 1$, problem (3)–(5) has at least one solution in Ω. □

A direct consequence of Theorem 6.1 for problem (2) is the following

Corollary 6.2. *Let $L > l > 0, \omega > 0$, let $\eta : [l, L] \to \mathbb{R}$ be a continuous positive function, $A^0 : [l, L] \to \mathbb{R}$ a positive function of class C^2 and $h : \mathbb{R} \to \mathbb{R}$ of function*

of class C^2. If the continuous function $f : [l, L] \times \mathbb{R}^+ \to \mathbb{R}$ satisfies either condition (f_1) or condition (f_2), then problem (2) has at least one positive radial solution.

Applied to the model problem (1) with Ω an annulus of radii $0 < l < L$, Corollary 6.2 provides the following existence results.

Corollary 6.3. *Let $L > l > 0$, $\omega > 0$, let $\eta : [l, L] \to \mathbb{R}$ and $\psi : [l, L] \to \mathbb{R}$ be continuous positive functions, and $A^0 : [l, L] \to \mathbb{R}$ be a positive function of class C^2. If either osc $A^0 < 1$ or $B^0 > 0$, with B^0 defined in (6), then problem (1) has at least one positive radial solution.*

Acknowledgements The three authors were partially supported by FONDECYT GRANT 1110268. RM was also partially supported by Basal-CMM-Conicyt, Milenio grant-P05-004F, and Anillo grant ACT 87-Conicyt.

References

1. Berestycki, H., Nadal, J.-P.: Self-organised critical hot spots of criminal activity. European J. Appl. Math. **21**, 371–399 (2010)
2. Cantrell, S., Cosner, C., Manásevich, R.: Global bifurcation of solutions for crime modeling equations. SIAM J. Math. Anal. **44**, 1340–1358 (2012)
3. Deimling, K.: Nonlinear Functional Analysis. Springer, Berlin (1985)
4. Garcia-Huidobro, M., Manásevich, R., Mawhin, J.: Existence of solutions for a 1-D boundary value problem coming from a model for burglary. Nonlinear Anal. Real World Appl. **14**, 1939–1946 (2013)
5. Pitcher, A.B.: Adding police to a mathematical model of burglary. Eur. J. Appl. Math. **21**, 401–419 (2010)
6. Rodriguez, N., Bertozzi, A.L.: Local existence and uniqueness of solutions to a PDE model for criminal behavior. Math. Models Methods Appl. Sci. **20**, 1425–1457 (2010)
7. Short, M.B., D'Orsogna, M.R., Pasour, V.B., Tita, G.E., Brantingham, P.J., Bertozzi, A.L., Chayes, L.B.: A statistical model of criminal behavior. Math. Models Methods Appl. Sci. **18**, 1249–1267 (2008)
8. Short, M.B., Bertozzi, A.L., Brantingham, P.J.: Nonlinear patterns in urban crime: hotspots, bifurcations, and suppression. SIAM J. Appl. Dyn. Syst. **9**, 462–483 (2010)
9. Stakgold, I.: Green's Functions and Boundary Value Problems. Wiley Interscience, New York (1998)

Modeling suspension bridges through the von Kármán quasilinear plate equations

Filippo Gazzola and Yongda Wang

Dedicated to Djairo Guedes de Figueiredo, on the occasion of his 80th birthday

Mathematics Subject Classification (2010): 35G60, 74B20, 74K20, 35A15

1 Introduction and motivations: nonlinear behavior of suspension bridges

The purposes of the present paper are to set up a nonlinear model to describe the static behavior of a suspension bridge and to study possible multiplicity of the equilibrium positions. We view the deck of the bridge as a long narrow rectangular thin plate, hinged on its short edges where the bridge is supported by the ground, and free on its long edges. Let L denote its length and 2ℓ denote its width; a realistic assumption is that $2\ell \cong \frac{L}{100}$.

The rectangular plate resists to transverse loads exclusively by means of bending. The flexural properties of a plate strongly depend on its thickness, which we denote by d, compared with its width 2ℓ and its length L. We assume here that $2\ell < L$ so that d is to be compared with 2ℓ. From Ventsel-Krauthammer [41, § 1.1] we learn that plates may be classified according to the ratio $2\ell/d$:

- if $2\ell \leq 8d$, we have a thick plate and the analysis of these plates includes all the components of stresses, strains, and displacements as for solid three-dimensional bodies;
- if $8d \leq 2\ell \leq 80d$, we have a thin plate which may behave in both linear and nonlinear regime according to how large is the ratio between its deflection and its thickness d;

F. Gazzola (✉) • Y. Wang
Dipartimento di Matematica, Politecnico di Milano, Milano, Italy
e-mail: filippo.gazzola@polimi.it; yongda.wang@polimi.it

A.N. Carvalho et al. (eds.), *Contributions to Nonlinear Elliptic Equations and Systems*, Progress in Nonlinear Differential Equations and Their Applications 86,
DOI 10.1007/978-3-319-19902-3_18

- if $2\ell \geq 80d$, the plate behaves like a membrane and lacks of flexural rigidity. Let us now turn to a particular suspension bridge. The main span of the collapsed Tacoma Narrows Bridge [2, 39] had the measures

$$L = 2800\,\text{ft.}\,, \qquad 2\ell = 39\,\text{ft.}\,, \qquad d = 4\,\text{ft.}\,, \tag{1}$$

see p.11 and Drawings 2 and 3 in [2]. Therefore, $2\ell/d = 9.75$ and

the deck of the Tacoma Narrows Bridge may be considered as a thin plate.

It is clear that modern suspension bridges with their stiffening trusses are more similar to thick plates.

Which theory (linear or nonlinear) models a thin plate depends on the magnitude W of its maximal deflection. If we denote again by d its thickness, two cases may occur, according to Ventsel-Krauthammer [41, § 1.1]:

- if $W/d \leq 0.2$, the plate is classified as stiff: these plates carry loads two dimensionally, mostly by internal bending, twisting moments and by transverse shear forces;
- if $W/d \geq 0.3$, the plate is classified as flexible: in this case, the deflections will be accompanied by stretching of the surface.

A fundamental feature of stiff plates is that the equation of static equilibrium for a plate element may be set up for an original (undeformed) configuration of the plate: in this case, a linear theory describes with sufficient accuracy the behavior of the plate. Flexible plates behave somehow in between membranes and stiff plates: when $W \gg d$ the membrane action is dominant and the flexural stress can be neglected compared with the membrane stress: in this case, a linear theory is not enough to describe accurately the behavior of the plate and one has to stick to nonlinear theories.

According to Scott [38, pp. 49–51] (see also [2, p. 60] and the video [39]), the Board of Engineers stated that under pure longitudinal oscillations ... *the lateral deflection of the center bridge was not measured but did not appear excessive, perhaps four times the width of the yellow center line (about 2 ft.)* while, after the appearance of the torsional oscillation, ... *the roadway was twisting almost* 45° *from the horizontal, with one side lurching 8.5 m. above the other*. This means that it was W =2 ft. during the vertical oscillations without torsion and W =14 ft. when the torsional oscillation appeared at the Tacoma Narrows Bridge. In view of (1), we then have $W/d = 0.5$ under pure longitudinal oscillations and $W/d = 3.5$ in presence of torsional oscillations. The conclusion is that

the Tacoma Narrows Bridge oscillated in a nonlinear regime.

This was already known to civil engineers about half a century ago (see, e.g., [35]) although the difficulties in tackling nonlinear models prevented a systematic study of the nonlinear regimes. In recent years, the necessity of nonlinear models became even more evident [13, 20, 26, 33] and the progress of tools in nonlinear analysis and in numerics gives the chance to obtain responses from nonlinear models. Which nonlinear model should be used is questionable. For two different models of "nonlinear degenerate bridges" a structural instability has been recently highlighted in [3, 7], both numerically and theoretically: it is shown that the torsional instability has a structural origin and not a mere aerodynamic justification as usually assumed in engineering literature, see [34, Section 12] and [36, 37]. By "degenerate" bridge we mean that the deck is not modeled through a full plate as in actual bridges.

A first interesting linear plate theory is due to Kirchhoff [22] in 1850, but it was only 60 years later (in 1910) that von Kármán [43] suggested a two-dimensional system in order to describe large deformations of a thin plate. This theory was considered a breakthrough in several scientific communities, including in the National Advisory Committee for Aeronautics, an American federal agency during the 19th century: the purpose of this agency was to undertake, to promote, and to institutionalize aeronautical research and the von Kármán equations were studied for a comparison between theoretical and experimental results, see [29, 30]. In his report, Levy [29] writes that *In the design of thin plates that bend under lateral and edge loading, formulas based on the Kirchhoff theory which neglects stretching and shearing in the middle surface are quite satisfactory provided that the deflections are small compared with the thickness. If deflections are of the same order as the thickness, the Kirchhoff theory may yield results that are considerably in error and a more rigorous theory that takes account of deformations in the middle surface should therefore be applied. The fundamental equations for the more exact theory have been derived by von Kármán.*

In order to describe its structural behavior, in this paper we view the bridge deck as a plate subject to the restoring force due to the hangers and behaving nonlinearly: we adapt the quasilinear von Kármán [43] model to a suspension bridge. In spite of the fact that this model received severe criticisms about its physical soundness (see [40, pp. 601–602]), many authors have studied the von Kármán system, see our incomplete bibliography. In particular, Ciarlet [15] provides an important justification of the von Kármán equations. He makes an asymptotic expansion with respect to the thickness of a three-dimensional class of elastic plates under suitable loads. He then shows that the leading term of the expansion solves a system of equations equivalent to those of von Kármán. Davet [17] pursues further and proves that the von Kármán equations may be justified by asymptotic expansion methods starting from very general 3-dimensional constitutive laws.

Following the setting in [19] (see also [1, 44, 45]), we consider a thin and narrow rectangular plate Ω where the two short edges are assumed to be hinged whereas the two long edges are assumed to be free. The plate is subject to three actions:

- normal dead and live loads acting orthogonally on the plate;
- edge loading, also called buckling loads, namely compressive forces along its edges;
- the restoring force due to the hangers, which acts in a neighborhood of the long edges.

The simplest action is the first one: the dead load is the structural weight whereas the live load may be a wind gust or some vehicle going through the bridge. As already pointed out by von Kármán [43], large edge loading may yield buckling, that is, the plate may deflect out of its plane when these forces reach a certain magnitude. The edge loading is called prestressing in engineering literature, see [32]. This was mathematically modeled by Berger [8] with a suitable nonlocal term and tackled with variational methods in a recent paper [1] which shows that large prestressing leads to buckling, that is, multiplicity of solutions of the corresponding equation. The critical buckling load may be computed by finding the smallest eigenvalue of an associated linear problem.

An important contribution of Berger-Fife [10] reduces the von Kármán system to a variational problem and tackles it with critical point and bifurcation theories (we point out that there are two different authors named Berger in our references). Subsequently, Berger [9] made a full analysis of the unloaded clamped plate problem (Dirichlet boundary conditions) which is somehow the simplest one but does not model the physical situation of a bridge. The loaded clamped plate was analyzed in [23, 24] where existence and possible nonuniqueness results were obtained. Different boundary conditions for the hinged plate (named after Navier) and for free boundaries were then analyzed with the same tools by Berger-Fife [11]. Since free edges of the plate are considered, this last paper is of particular interest for our purposes. As clearly stated by Ciarlet [15, p.353] the boundary conditions for the Airy function *are often left fairly vague in the literature*; we take them in a "dual form", that is, more restrictions for the edges yield less restrictions for the Airy function and vice versa.

We adapt here these plate models to a suspension bridge. The main novelties are that the function representing the vertical displacement of the rectangular plate Ω satisfies a mixed hinged and free boundary conditions and that the restoring force due to the hangers is taken into account. It is well known [18] that nonlinear elliptic systems are fairly delicate to tackle with variational methods. The model describing a suspension bridge involves a fourth order quasilinear elliptic system and this brings further difficulties, in particular in the definition of the action functional. We start by setting in full detail the linear theory which enables us to determine the critical prestressing values leading to buckling and to the multiplicity of solutions. Then we analyze the problem with normal dead loads but no restoring force and we obtain results in the spirit of [9, 10]. Finally, we introduce the restoring force due to the hangers and we prove existence and multiplicity of the equilibrium positions.

2 Functional framework and the quasilinear equations

2.1 Elastic energies of a plate

The bending energy of the plate Ω involves curvatures of the surface. Let κ_1 and κ_2 denote the principal curvatures of the graph of the (smooth) function u representing the vertical displacement of the plate in the downwards direction, then the Kirchhoff model [22] for the bending energy of a deformed plate Ω of thickness $d > 0$ is

$$\mathbb{E}_B(u) = \frac{E\,d^3}{12(1-\sigma^2)} \int_\Omega \left(\frac{\kappa_1^2}{2} + \frac{\kappa_2^2}{2} + \sigma \kappa_1 \kappa_2 \right) dxdy \tag{2}$$

where σ is the Poisson ratio defined by $\sigma = \frac{\lambda}{2(\lambda+\mu)}$ and E is the Young modulus defined by $E = 2\mu(1+\sigma)$, with the so-called Lamé constants λ, μ that depend on the material. For physical reasons it holds that $\mu > 0$ and usually $\lambda > 0$ so that

$$0 < \sigma < \frac{1}{2}. \tag{3}$$

For small deformations the terms in (2) are taken as approximations being purely quadratic with respect to the second order derivatives of u. More precisely, for small deformations u, one has

$$(\kappa_1 + \kappa_2)^2 \approx (\Delta u)^2\,, \quad \kappa_1 \kappa_2 \approx \det(D^2 u) = u_{xx} u_{yy} - u_{xy}^2\,, \tag{4}$$

and therefore

$$\frac{\kappa_1^2}{2} + \frac{\kappa_2^2}{2} + \sigma \kappa_1 \kappa_2 \approx \frac{1}{2}(\Delta u)^2 + (\sigma - 1)\det(D^2 u).$$

Then, if f denotes the external vertical load (including both dead and live loads) acting on the plate Ω and if u is the corresponding (small) vertical displacement of the plate, by (2) we have that the total energy $\mathbb{E}_T$ of the plate becomes

$$\begin{aligned} \mathbb{E}_T(u) &= \mathbb{E}_B(u) - \int_\Omega fu\,dxdy \qquad (5) \\ &= \frac{E\,d^3}{12(1-\sigma^2)} \int_\Omega \left(\frac{1}{2}(\Delta u)^2 - (1-\sigma)\det(D^2 u) \right) dxdy - \int_\Omega fu\,dxdy. \end{aligned}$$

Note that the "quadratic" functional $\mathbb{E}_B(u)$ is positive whenever $|\sigma| < 1$, a condition which is ensured by (3).

If large deformations are involved, one does not have a linear strain-displacement relation resulting in (4). For a plate of uniform thickness $d > 0$, one assumes that the plate has a middle surface midway between its parallel faces that, in equilibrium, occupies the region Ω in the plane $z = 0$. Let $w = w(x, y)$, $v = v(x, y)$ $u = u(x, y)$ denote the components (respectively in the x, y, z directions) of the displacement vector of the particle of the middle surface which, when the plate is in equilibrium, occupies the position $(x, y) \in \Omega$: u is the component in the vertical z-direction which is related to bending while w and v are the in-plane stretching components. For large deformations of Ω there is a coupling between u and (w, v). In order to describe it, we compute the stretching in the x and y directions (see, e.g., [41, (7.80)]):

$$\varepsilon_x = \sqrt{1+2w_x+u_x^2}-1 \approx w_x+\frac{u_x^2}{2}\,, \quad \varepsilon_y = \sqrt{1+2v_y+u_y^2}-1 \approx v_y+\frac{u_y^2}{2} \tag{6}$$

where the approximation is due to the fact that, compared to unity, all the components are small in the horizontal directions x and y. One can also compute the shear strain (see, e.g., [41, (7.81)]):

$$\gamma_{xy} \approx w_y + v_x + u_x u_y\,. \tag{7}$$

Finally, it is convenient to introduce the so-called stress resultants which are the integrals of suitable components of the strain tensor (see e.g. [27, (1.22)]), namely,

$$N^x = \frac{Ed}{1-\sigma^2}\left(w_x+\sigma v_y+\frac{1}{2}u_x^2+\frac{\sigma}{2}u_y^2\right),\ N^y = \frac{Ed}{1-\sigma^2}\left(v_y+\sigma w_x+\frac{1}{2}u_y^2+\frac{\sigma}{2}u_x^2\right),$$

$$N^{xy} = \frac{Ed}{2(1+\sigma)}\left(w_y+v_x+u_xu_y\right)\,, \tag{8}$$

so that

$$\varepsilon_x = \frac{N^x-\sigma N^y}{Ed}\,, \quad \varepsilon_y = \frac{N^y-\sigma N^x}{Ed}\,, \quad \gamma_{xy} = \frac{2(1+\sigma)}{Ed}N^{xy}\,.$$

We are now in a position to define the energy functional. The first term $\mathbb{E}_T(u)$ of the energy is due to pure bending and to external loads and was already computed in (5). For large deformations, one needs to consider also the interaction with the stretching components v and w and the total energy reads (see [28, (1.7)])

$$J(u,v,w) = \mathbb{E}_T(u) + \frac{E\,d}{2(1-\sigma^2)}\int_\Omega \left(\varepsilon_x^2+\varepsilon_y^2+2\sigma\,\varepsilon_x\varepsilon_y+\frac{1-\sigma}{2}\,\gamma_{xy}^2\right)dxdy \tag{9}$$

which has to be compared with (5). In view of (6)–(7) the additional term $I := J - \mathbb{E}_T$ may also be written as

$$I(u,v,w) = \frac{E\,d}{2(1-\sigma^2)}\int_\Omega \left\{\left(w_x+\frac{u_x^2}{2}\right)^2+\left(v_y+\frac{u_y^2}{2}\right)^2\right.$$
$$\left.+2\sigma\left(w_x+\frac{u_x^2}{2}\right)\left(v_y+\frac{u_y^2}{2}\right)\right\}dxdy + \frac{E\,d}{4(1+\sigma)}\int_\Omega (w_y+v_x+u_xu_y)^2dxdy\,.$$

The next step is to derive the equations and boundary conditions which characterise the critical points of J; this will be done in the two following subsections.

2.2 The Euler-Lagrange equation

Let L denote the length of the plate Ω and 2ℓ denote its width with $2\ell \cong \frac{L}{100}$. In order to simplify the Fourier series expansions we take $L = \pi$ so that, in the sequel,

$$\Omega = (0,\pi) \times (-\ell,\ell) \subset \mathbb{R}^2 \qquad \text{(with } \ell \ll \pi\text{)}.$$

The natural functional space where to set up the problem is

$$H^2_*(\Omega) := \left\{ w \in H^2(\Omega);\ w = 0 \text{ on } \{0,\pi\} \times (-\ell,\ell) \right\}.$$

We also define

$$\mathscr{H}_*(\Omega) := \text{ the dual space of } H^2_*(\Omega)$$

and we denote by $\langle\cdot,\cdot\rangle$ the corresponding duality. Since we are in the plane, $H^2(\Omega) \subset C^0(\overline{\Omega})$ so that the condition on $\{0,\pi\} \times (-\ell,\ell)$ introduced in the definition of $H^2_*(\Omega)$ makes sense. On the space $H^2(\Omega)$ we define the Monge-Ampère operator

$$[\phi,\psi] := \phi_{xx}\psi_{yy} + \phi_{yy}\psi_{xx} - 2\phi_{xy}\psi_{xy} \qquad \forall \phi,\psi \in H^2(\Omega) \tag{10}$$

so that, in particular, $[\phi,\phi] = 2\det(D^2\phi)$ where $D^2\phi$ is the Hessian matrix of ϕ.

As pointed out in [19, Lemma 4.1], $H^2_*(\Omega)$ is a Hilbert space when endowed with the scalar product

$$(u,v)_{H^2_*(\Omega)} := \int_\Omega \Big(\Delta u \Delta v - (1-\sigma)[u,v] \Big)\, dxdy\,.$$

The corresponding norm then reads

$$\|u\|_{H^2_*(\Omega)} := \left(\int_\Omega \Big(|\Delta u|^2 - (1-\sigma)[u,u] \Big)\, dxdy \right)^{1/2}.$$

The unique minimizer u of the convex functional $\mathbb{E}_T$ in (5) over the space $H^2_*(\Omega)$ satisfies the Euler-Lagrange equation (see e.g. [21])

$$\frac{E\,d^3}{12(1-\sigma^2)}\, \Delta^2 u = f(x,y) \qquad \text{in } \Omega\,. \tag{11}$$

On the other hand, the Euler-Lagrange equation for the energy J in (9) characterizes the critical points of J: we need to compute the variation δJ of J and to find triples (u,v,w) such that

$$\langle \delta J(u,v,w),(\phi,\psi,\xi)\rangle = \lim_{t\to 0}\frac{J(u+t\phi,v+t\psi,w+t\xi)-J(u,v,w)}{t}=0$$

for all $\phi,\psi,\xi\in C_c^\infty(\Omega)$. After replacing N^x, N^y, N^{xy}, see (8), this yields

$$\frac{E d^3}{12(1-\sigma^2)}\int_\Omega \big(\Delta u\Delta\phi+(\sigma-1)[u,\phi]\big)\,dxdy$$

$$+\int_\Omega \big((N^x u_x+N^{xy}u_y)\phi_x+(N^y u_y+N^{xy}u_x)\phi_y\big)\,dxdy=\int_\Omega f\phi\,dxdy \quad \forall\phi\in C_c^\infty(\Omega)$$

$$\int_\Omega \big(N^y\psi_y+N^{xy}\psi_x\big)\,dxdy=0 \qquad \forall\psi\in C_c^\infty(\Omega)$$

$$\int_\Omega \big(N^x\xi_x+N^{xy}\xi_y\big)\,dxdy=0 \qquad \forall\xi\in C_c^\infty(\Omega).$$

Thanks to some integration by parts and by arbitrariness of the test functions, we may rewrite the above identities in strong form

$$\begin{array}{c}\frac{E d^3}{12(1-\sigma^2)}\Delta^2 u-(N^x u_x+N^{xy}u_y)_x-(N^y u_y+N^{xy}u_x)_y=f \qquad \text{in } \Omega\,,\\ N^y_y+N^{xy}_x=0\,,\quad N^x_x+N^{xy}_y=0 \qquad \text{in } \Omega\,.\end{array} \tag{12}$$

The last two equations in (12) show that there exists a function Φ (called Airy stress function), unique up to an affine function, such that

$$\Phi_{yy}=N^x,\quad \Phi_{xx}=N^y,\quad \Phi_{xy}=-N^{xy}\,. \tag{13}$$

Then, after some tedious computations, by using the Monge-Ampère operator (10) and by normalizing the coefficients, the system (12) may be written as

$$\begin{cases}\Delta^2\Phi=-[u,u] & \text{in } \Omega\\ \Delta^2 u=[\Phi,u]+f & \text{in } \Omega\,.\end{cases} \tag{14}$$

In a plate subjected to compressive forces along its edges, one should consider a prestressing constraint which may lead to buckling. Then the system (14) becomes

$$\begin{cases}\Delta^2\Phi=-[u,u] & \text{in } \Omega\\ \Delta^2 u=[\Phi,u]+f+\lambda[F,u] & \text{in } \Omega\,.\end{cases} \tag{15}$$

The term $\lambda[F,u]$ in the right-hand side of (15) represents the boundary stress. The parameter $\lambda\ge 0$ measures the magnitude of the compressive forces acting on $\partial\Omega$ while the smooth function F satisfies

$$F \in C^4(\overline{\Omega}), \quad \Delta^2 F = 0 \text{ in } \Omega, \quad F_{xx} = F_{xy} = 0 \text{ on } (0,\pi) \times \{\pm\ell\}, \tag{16}$$

see [11, pp. 228–229]: the term λF represents the stress function in the plate resulting from the applied force if the plate were artificially prevented from deflecting and the boundary constraints in (16) physically mean that no external stresses are applied on the free edges of the plate. Following Knightly-Sather [25], we take

$$F(x,y) = \frac{\ell^2 - y^2}{2} \qquad \text{so that} \quad [F,u] = -u_{xx}. \tag{17}$$

Therefore, (15) becomes

$$\begin{cases} \Delta^2 \Phi = -[u,u] & \text{in } \Omega \\ \Delta^2 u = [\Phi,u] + f - \lambda u_{xx} & \text{in } \Omega. \end{cases} \tag{18}$$

2.3 Boundary conditions

We now determine the boundary conditions to be associated with (18). In literature these equations are usually considered under Dirichlet boundary conditions, see [16, § 1.5] and [42, p.514]. But since we aim to model a suspension bridge, these conditions are not the correct ones. Following [19] (see also [1, 44]) we view the deck of a suspension bridge as a long narrow rectangular thin plate hinged at its two opposite short edges and free on the remaining two long edges.

Let us first consider the two short edges $\{0\} \times (-\ell,\ell)$ and $\{\pi\} \times (-\ell,\ell)$. Due to the connection with the ground, u is assumed to be hinged there and hence it satisfies the Navier boundary conditions:

$$u = u_{xx} = 0 \qquad \text{on } \{0,\pi\} \times (-\ell,\ell). \tag{19}$$

In this case, Ventsel-Krauthammer [41, Example 7.4] suggest that $N^x = v = 0$ on $\{0,\pi\} \times (-\ell,\ell)$. In view of (8) this yields

$$0 = w_x + \sigma v_y + \frac{1}{2}u_x^2 + \frac{\sigma}{2}u_y^2 = w_x + \frac{1}{2}u_x^2 = \frac{Ed}{(1-\sigma^2)\sigma}N^x$$

where the condition $u_y = 0$ comes from the first of (19). In turn, by (13) this implies that $\Phi_{xx} = 0$ on $\{0,\pi\} \times (-\ell,\ell)$. For the second boundary condition we recall that $N^x = 0$ so that, by (13), also $\Phi_{yy} = 0$: since the Airy function Φ is defined up to the addition of an affine function, we may take $\Phi = 0$. Summarizing, we also have

$$\Phi = \Phi_{xx} = 0 \qquad \text{on } \{0,\pi\} \times (-\ell,\ell). \tag{20}$$

On the long edges $(0,\pi)\times\{\pm\ell\}$ the plate is free, which results in

$$u_{yy}+\sigma u_{xx}=u_{yyy}+(2-\sigma)u_{xxy}=0 \qquad \text{on } (0,\pi)\times\{\pm\ell\}, \tag{21}$$

see, e.g., [41, (2.40)] or [19]. Note that here the boundary conditions do not depend on λ. For the Airy stress function Φ, we follow the usual Dirichlet boundary condition on $(0,\pi)\times\{\pm\ell\}$, see [10, 11]. Then

$$\Phi=\Phi_y=0 \qquad \text{on } (0,\pi)\times\{\pm\ell\}. \tag{22}$$

These boundary conditions suggest to introduce the following subspace of $H^2_*(\Omega)$

$$H^2_{**}(\Omega):=\{u\in H^2_*(\Omega): u=u_y=0 \text{ on } (0,\pi)\times\{\pm\ell\}\},$$

which is a Hilbert space when endowed with the scalar product and norm

$$(u,v)_{H^2_{**}(\Omega)}:=\int_\Omega \Delta u\Delta v\,, \qquad \|u\|_{H^2_{**}(\Omega)}:=\left(\int_\Omega|\Delta u|^2\right)^{1/2}.$$

We denote the dual space of $H^2_{**}(\Omega)$ by $\mathscr{H}_{**}(\Omega)$.

2.4 The quasilinear von Kármán equations modeling suspension bridges

By putting together the Euler-Lagrange equation (18) and the boundary conditions (19)–(22) we obtain the system

$$\begin{cases} \Delta^2\Phi=-[u,u] & \text{in } \Omega\\ \Delta^2 u=[\Phi,u]+f-\lambda u_{xx} & \text{in } \Omega\\ u=\Phi=u_{xx}=\Phi_{xx}=0 & \text{on } \{0,\pi\}\times(-\ell,\ell)\\ u_{yy}+\sigma u_{xx}=u_{yyy}+(2-\sigma)u_{xxy}=0 & \text{on } (0,\pi)\times\{\pm\ell\}\\ \Phi=\Phi_y=0 & \text{on } (0,\pi)\times\{\pm\ell\}. \end{cases} \tag{23}$$

In a plate modeling a suspension bridge, one should also add the nonlinear restoring action due to the hangers. Then the second equation in (23) becomes

$$\Delta^2 u+\Upsilon(y)g(u)=[\Phi,u]+f-\lambda u_{xx} \quad \text{in } \Omega. \tag{24}$$

Here Υ is the characteristic function of $(-\ell,-\ell+\varepsilon)\cup(\ell-\varepsilon,\ell)$ for some small ε. This means that the restoring force due to the hangers is concentrated in two tiny parallel strips adjacent to the long edges (the free part of the boundary). The Official

Report [2, p.11] states that the region of interaction of the hangers with the plate was of approximately 2 ft on each side: this means that $\varepsilon \approx \frac{\pi}{1500}$. Augusti-Sepe [4] (see also [5]) view the restoring force at the endpoints of a cross-section of the deck as composed by two connected springs, the top one representing the action of the sustaining cable and the bottom one (connected with the deck) representing the hangers. And the action of the cables is considered by Bartoli-Spinelli [6, p.180] the main cause of the nonlinearity of the restoring force: they suggest quadratic and cubic perturbations of a linear behavior. Assuming that **the vertical axis is oriented downwards**, the restoring force acts in those parts of the deck which are below the equilibrium position (where $u > 0$) while it exerts no action where the deck is above the equilibrium position ($u < 0$). Taking into account all these facts, for the explicit action of the restoring force, we take

$$g(u) = (ku + \delta u^3)^+ \tag{25}$$

which is a compromise between the nonlinearities suggested by McKenna-Walter [31] and Plaut-Davis [33] and follows the idea of Ferrero-Gazzola [19]. Here $k > 0$ denotes the Hooke constant of elasticity of steel (hangers) while $\delta > 0$ is a small parameter reflecting the nonlinear behavior of the sustaining cables. Only the positive part is taken into account due to possible slackening, see [2, V-12]: the hangers behave as a restoring force if extended (when $u > 0$) and give no contribution when they lose tension (when $u \leq 0$).

By assuming (25), and inserting (24) into (23) leads to the problem

$$\begin{cases} \Delta^2\Phi = -[u,u] & \text{in } \Omega \\ \Delta^2 u + \Upsilon(y)(ku+\delta u^3)^+ = [\Phi,u] + f - \lambda u_{xx} & \text{in } \Omega \\ u = \Phi = u_{xx} = \Phi_{xx} = 0 & \text{on } \{0,\pi\}\times(-\ell,\ell) \\ u_{yy} + \sigma u_{xx} = u_{yyy} + (2-\sigma)u_{xxy} = 0 & \text{on } (0,\pi)\times\{\pm\ell\} \\ \Phi = \Phi_y = 0 & \text{on } (0,\pi)\times\{\pm\ell\}\,. \end{cases} \tag{26}$$

Finally, we go back to the original unknowns u, v, w. After that a solution (u, Φ) of (23) or (26) is found, (8)–(13) yield

$$w_x + \sigma v_y = \frac{1-\sigma^2}{Ed}\,\Phi_{yy} - \frac{1}{2}u_x^2 - \frac{\sigma}{2}u_y^2\,, \quad \sigma w_x + v_y = \frac{1-\sigma^2}{Ed}\,\Phi_{xx} - \frac{1}{2}u_y^2 - \frac{\sigma}{2}u_x^2$$

which immediately gives w_x and v_y. Upon integration, this gives $w = w(x, y)$ up to the addition of a function only depending on y and $v = v(x, y)$ up to the addition of a function depending only on x. These two additive functions are determined by solving the last constraint given by (8)–(13), that is,

$$w_y + v_x = -\frac{2(1+\sigma)}{Ed}\,\Phi_{xy} - u_x - u_y\,.$$

3 Main results

With no further mention, we assume (3). The first step to study (23) and (26) is to analyze the spectrum of the linear problem obtained by taking $\Phi = f = k = \delta = 0$:

$$\begin{cases} \Delta^2 u + \lambda u_{xx} = 0 & \text{in } \Omega \\ u = u_{xx} = 0 & \text{on } \{0,\pi\} \times (-\ell,\ell) \\ u_{yy} + \sigma u_{xx} = u_{yyy} + (2-\sigma)u_{xxy} = 0 & \text{on } (0,\pi) \times \{\pm\ell\}\,. \end{cases} \tag{27}$$

In Section 5 we prove the following result

Theorem 3.1. *The problem* (27) *admits a sequence of divergent eigenvalues*

$$\lambda_1 < \lambda_2 \leq \ldots \leq \lambda_k \leq \ldots$$

whose corresponding eigenfunctions $\{\overline{e}_k\}$ *form a complete orthonormal system in* $H^2_*(\Omega)$. *Moreover, the least eigenvalue* λ_1 *is simple and is the unique value of* $\lambda \in ((1-\sigma)^2, 1)$ *such that*

$$\sqrt{1-\lambda^{1/2}}\,\big(\lambda^{1/2}+1-\sigma\big)^2 \tanh(\ell\sqrt{1-\lambda^{1/2}}\,) \\ = \sqrt{1+\lambda^{1/2}}\,\big(\lambda^{1/2}-1+\sigma\big)^2 \tanh(\ell\sqrt{1+\lambda^{1/2}}\,)\,;$$

the corresponding eigenspace is generated by the positive eigenfunction

$$\overline{e}_1(x,y) = \left\{ (\lambda^{1/2}+1-\sigma)\,\frac{\cosh\left(y\sqrt{1-\lambda^{1/2}}\right)}{\cosh\left(\ell\sqrt{1-\lambda^{1/2}}\right)} \right. \\ \left. +(\lambda^{1/2}-1+\sigma)\,\frac{\cosh\left(y\sqrt{1+\lambda^{1/2}}\right)}{\cosh\left(\ell\sqrt{1+\lambda^{1/2}}\right)} \right\} \sin x\,.$$

The simplicity of the least eigenvalue was not to be expected. It is shown in [25, §3] that the eigenvalue problem (27) for a fully hinged (simply supported) rectangular plate, that is with $u = \Delta u = 0$ on the four edges, may admit a least eigenvalue of multiplicity 2.

The least eigenvalue λ_1 represents the critical buckling load and may be characterised variationally by

$$\lambda_1 := \min_{v \in H^2_*(\Omega)} \frac{\|v\|^2_{H^2_*(\Omega)}}{\|v_x\|^2_{L^2(\Omega)}}\,.$$

Ferrero-Gazzola [19] studied the eigenvalue problem $\Delta^2 u = \lambda u$ under the boundary conditions in (27): by comparing [19, Theorem 3.4] with the above Theorem 3.1 we observe that the least eigenvalues (and eigenfunctions) of the two problems coincide, that is,

$$\lambda_1 = \min_{v \in H^2_*(\Omega)} \frac{\|v\|^2_{H^2_*(\Omega)}}{\|v_x\|^2_{L^2(\Omega)}} = \min_{v \in H^2_*(\Omega)} \frac{\|v\|^2_{H^2_*(\Omega)}}{\|v\|^2_{L^2(\Omega)}} . \tag{28}$$

Therefore, the critical buckling load for a rectangular plate equals the eigenvalue relative to the first eigenmode of the plate. In turn, the first eigenmode is also the first buckling deformation of the plate. From (28) we readily infer the Poincaré-type inequalities

$$\lambda_1 \|v_x\|^2_{L^2(\Omega)} \le \|v\|^2_{H^2_*(\Omega)}, \quad \lambda_1 \|v\|^2_{L^2(\Omega)} \le \|v\|^2_{H^2_*(\Omega)} \qquad \forall v \in H^2_*(\Omega) \tag{29}$$

with strict inequality unless v minimizes the ratio in (28), that is, v is a real multiple of $\overline{e}_1$. Note also that by taking $v(x, y) = \sin x$ one finds that $\lambda_1 < 1$.

Finally, let us mention that Theorem 3.1 may be complemented with the explicit form of all the eigenfunctions: they are $\sin(mx)$ ($m \in \mathbb{N}$) multiplied by trigonometric or hyperbolic functions with respect to y: we refer again to Section 5.

Then we insert an external load f and we study the existence and multiplicity of solutions of (23).

Theorem 3.2. *For all $f \in L^2(\Omega)$ and $\lambda \ge 0$ (23) admits a solution $(u, \Phi) \in H^2_*(\Omega) \times H^2_{**}(\Omega)$. Moreover:*

(i) if $\lambda \le \lambda_1$ and $f = 0$, then (23) only admits the trivial solution $(u, \Phi) = (0, 0)$;
(ii) if $\lambda \in (\lambda_k, \lambda_{k+1}]$ for some $k \ge 1$ and $f = 0$, then (23) admits at least k pairs of nontrivial solutions;
(iii) if $\lambda < \lambda_1$ there exists $K > 0$ such that if $\|f\|_{L^2(\Omega)} < K$, then (23) admits a unique solution $(u, \Phi) \in H^2_*(\Omega) \times H^2_{**}(\Omega)$;
(iv) if $\lambda > \lambda_1$ there exists $K > 0$ such that if $\|f\|_{L^2(\Omega)} < K$, then (23) admits at least three solutions.

Theorem 3.2 gives both uniqueness and multiplicity results. Since the solutions are obtained as critical points of an action functional, they describe the stable and unstable equilibria positions of the plate. When both the buckling load λ and the external load f are small there is just one possible equilibrium position. If one of them is large, then multiple equilibrium positions may exist. The uniqueness statement (iii) has a fairly delicate proof: we will show that the corresponding action functional is "locally convex" in the region where the equilibria positions are confined.

The last step is to study the nonlinear plate modeling the suspension bridge, that is, with the action of the hangers. We first define the constants

$$\alpha := \int_\Omega \Upsilon(y)\overline{e}_1^2\,, \qquad \overline{\lambda} := (\alpha k + 1)\lambda_1 > \lambda_1\,, \tag{30}$$

where λ_1 denotes the least eigenvalue and $\overline{e}_1$ denotes here the positive least eigenfunction normalized in $H^2_*(\Omega)$, see Theorem 3.1. Then we have

Theorem 3.3. *For all $f \in L^2(\Omega)$, $\lambda \geq 0$ and $k, \delta > 0$ problem* (26) *admits a solution $(u, \Phi) \in H^2_*(\Omega) \times H^2_{**}(\Omega)$. Moreover:*

(i) if $\lambda < \lambda_1$ there exists $K > 0$ such that if $\|f\|_{L^2(\Omega)} < K$, then (26) *admits a unique solution $(u, \Phi) \in H^2_*(\Omega) \times H^2_{**}(\Omega)$;*
(ii) if $\lambda > \lambda_1$ and $f = 0$, then (26) *admits at least two solutions $(u, \Phi) \in H^2_*(\Omega) \times H^2_{**}(\Omega)$ and one of them is trivial and unstable;*
(iii) if $\overline{\lambda} < \lambda_2$ and $\overline{\lambda} < \lambda < \lambda_2$, there exists $K > 0$ such that if $\|f\|_{L^2(\Omega)} < K$, then (26) *admits at least three solutions $(u, \Phi) \in H^2_*(\Omega) \times H^2_{**}(\Omega)$, two being stable and one being unstable.*

Also Theorem 3.3 gives both uniqueness and multiplicity results. Item (ii) states that even in absence of an external load ($f = 0$), if the buckling load λ is sufficiently large, then there exists at least two equilibrium positions; we conjecture that if we further assume that $\lambda < \overline{\lambda}$ then there exist no other solutions and that the equilibrium positions look like in Figure 1. In the left picture we see the trivial equilibrium $u = 0$ which is unstable due to the buckling load. In the right picture we see the stable equilibrium for some $u < 0$ (above the horizontal position). We conjecture that it is a negative multiple of the first eigenfunction $\overline{e}_1$, see Theorem 3.1; since ℓ is very small, a rough approximation shows that this negative multiple looks like $\approx C\sin(x)$ for some $C < 0$, which is the shape represented in the right picture. The reason of this conjecture will become clear in the proof, see in particular the plots in Figure 3 in Section 7: in this pattern, a crucial role is played by the positivity of $\overline{e}_1$. Our feeling is that the action functional corresponding to this case has a qualitative shape as described in Figure 2, where O is the trivial unstable equilibrium and M is the stable equilibrium. If there were no hangers also the opposite position would be a stable equilibrium. But the presence of the restoring force requires a larger buckling term in order to generate a positive (downwards) displacement. Indeed, item (iii) states, in particular, that if $f = 0$ and the buckling load is large then there exist three equilibria: one is trivial and unstable, the second is the enlarged negative one already found in item (ii), the third should precisely be the positive one which

Fig. 1 Equilibrium positions of the buckled bridge.

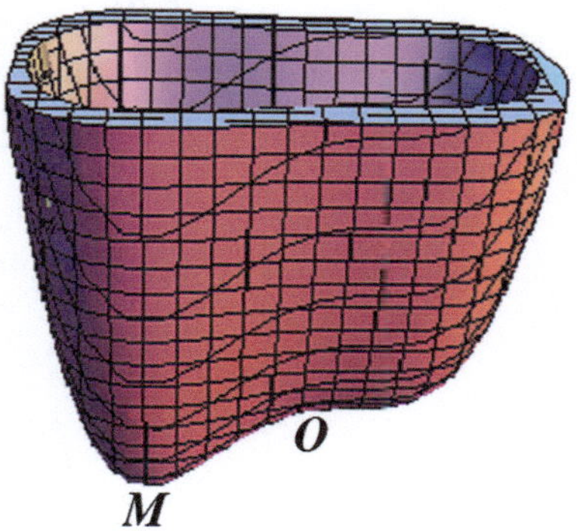

Fig. 2 Qualitative shape of the action functional for Theorem 3.3 (ii) when $\lambda < \overline{\lambda}$.

appears because the buckling load λ is stronger than the restoring force due to the hangers. All these conjectures and qualitative explanations are supported by similar results for a simplified (one dimensional) beam equation, see [12, Theorem 3.2].

Remark 3.4 (Open problem). Can the assumption $\overline{\lambda} < \lambda_2$ in Theorem 3.3 (*iii*) be weakened or removed? In our proof this assumption is needed to disconnect two open regions of negativity of the action functional. But, perhaps, other critical point theorems may be applied.

Remark 3.5 (Regularity). A weak solution satisfies $(u, \Phi) \in H^2_*(\Omega) \times H^2_{**}(\Omega)$: then the assumption $f \in L^2(\Omega)$ implies that $\Delta^2 u \in L^1(\Omega)$. By an embedding and elliptic regularity we infer that $u \in H^2_*(\Omega) \cap H^{3-\varepsilon}(\Omega)$ for all $\varepsilon > 0$ and then $D^2 u \in H^{1-\varepsilon}(\Omega)$. Therefore, $[u, u] \in L^q(\Omega)$ for all $1 \leq q < \infty$. Hence, $\Phi \in W^{4,q}(\Omega)$ and, in turn, $[u, \Phi] \in L^q(\Omega)$ for all $1 \leq q < \infty$. Moreover, $f \in L^2(\Omega)$ implies $\Delta^2 u \in L^2(\Omega)$ and then $u \in H^4(\Omega)$. This means that the generalized solution (u, Φ) is also a strong solution. For smoother f, the regularity of (u, Φ) can be increased.

4 Preliminaries: some useful operators and functionals

For all $v, w \in H^2_*(\Omega)$, consider the problem

$$\begin{cases} \Delta^2 \Phi = -[v, w] \text{ in } \Omega \\ \Phi = \Phi_{xx} = 0 \quad \text{on } \{0, \pi\} \times (-\ell, \ell) \\ \Phi = \Phi_y = 0 \quad \text{on } (0, \pi) \times \{\pm \ell\}. \end{cases} \tag{31}$$

We claim that (31) has a unique solution $\Phi = \Phi(v, w)$ and $\Phi \in H^2_{**}(\Omega)$.

Since $\Omega \subset \mathbb{R}^2$, we have $H^{1+\varepsilon}(\Omega) \Subset L^\infty(\Omega) = (L^1(\Omega))'$, for all $\varepsilon > 0$. On the other hand, $L^1(\Omega) \subset (L^\infty(\Omega))' \Subset H^{-(1+\varepsilon)}(\Omega)$. If $v, w \in H^2_*(\Omega) \subset H^2(\Omega)$, then $[v, w] \in L^1(\Omega)$. Therefore,

$$[v, w] \in H^{-(1+\varepsilon)}(\Omega) \qquad \forall \varepsilon > 0\,.$$

Then by the Lax-Milgram Theorem and the regularity theory of elliptic equations, there exists a unique solution of (31) and $\Phi \in H^{3-\varepsilon}(\Omega)$ for all $\varepsilon > 0$. An embedding and the boundary conditions show that $\Phi \in H^2_{**}(\Omega)$, which completes the proof of the claim.

This result enables us to define a bilinear form $B = B(v, w) = -\Phi$, where Φ is the unique solution of (31); this form is implicitly characterized by

$$B : (H^2_*(\Omega))^2 \to H^2_{**}(\Omega)\,,$$

$$(B(v,w),\varphi)_{H^2_{**}(\Omega)} = \int_\Omega [v,w]\varphi \quad \forall v,w \in H^2_*(\Omega)\,,\ \varphi \in H^2_{**}(\Omega)\,.$$

Similarly, one can prove that for all $v \in H^2_*(\Omega)$ and $\varphi \in H^2_{**}(\Omega)$ there exists a unique solution $\Psi \in H^2_*(\Omega)$ of the problem

$$\begin{cases} \Delta^2\Psi = -[v,\varphi] & \text{in } \Omega \\ \Psi = \Psi_{xx} = 0 & \text{on } \{0,\pi\}\times(-\ell,\ell) \\ \Psi_{yy} + \sigma\Psi_{xx} = \Psi_{yyy} + (2-\sigma)\Psi_{xxy} = 0 & \text{on } (0,\pi)\times\{\pm\ell\}\,. \end{cases}$$

This defines another bilinear form $C = C(v,\varphi) = -\Psi$ which is implicitly characterized by

$$C : H^2_*(\Omega)\times H^2_{**}(\Omega) \to H^2_*(\Omega)\,,$$

$$(C(v,\varphi),w)_{H^2_*(\Omega)} = \int_\Omega [v,\varphi]w \quad \forall v,w \in H^2_*(\Omega)\,,\ \varphi \in H^2_{**}(\Omega)\,.$$

Then we prove

Lemma 4.1. *The trilinear form*

$$(H^2_*(\Omega))^3 \ni (v,w,\varphi) \mapsto \int_\Omega [v,w]\varphi \tag{32}$$

is independent of the order of v, w, φ if at least one of them is in $H^2_{**}(\Omega)$. Moreover, if $\varphi \in H^2_{**}(\Omega)$, $v, w \in (H^2_*(\Omega))^2$, then

$$(B(v,w),\varphi)_{H^2_{**}(\Omega)} = (B(w,v),\varphi)_{H^2_{**}(\Omega)} = (C(v,\varphi),w)_{H^2_*(\Omega)} = (C(w,\varphi),v)_{H^2_*(\Omega)}. \tag{33}$$

Finally, the operators B and C are compact.

Proof. By a density argument and by continuity it suffices to prove all the identities for smooth functions v, w, φ, in such a way that third interior derivatives and second boundary derivatives are well defined and integration by parts is allowed. In the trilinear form (32) one can exchange the order of v and w by exploiting the symmetry of the Monge-Ampère operator, that is, $[v,w] = [w,v]$ for all v and w.

So, we may assume that one among w, φ is in $H^2_{**}(\Omega)$: note that this function also has vanishing x-derivative on $(0,\pi)\times\{\pm\ell\}$. Then some integration by parts enable to switch the position of w and φ.

From the just proved symmetry of the trilinear form (32) we immediately infer (33).

If $\varphi \in H^2_{**}(\Omega)$, then $\varphi_{xx} = \varphi_{xy} = 0$ on $(0,\pi)\times\{\pm\ell\}$ and an integration by parts yields

$$\begin{aligned}(B(v,w),\varphi)_{H^2_{**}(\Omega)} &= \int_\Omega [v,w]\varphi = \int_\Omega [\varphi,w]v \\ &= \int_\Omega \varphi_{xy}(w_x v_y + w_y v_x) - \int_\Omega (\varphi_{xx} w_y v_y + \varphi_{yy} w_x v_x).\end{aligned}$$

In turn, this shows that

$$|(B(v,w),\varphi)_{H^2_{**}(\Omega)}| \le c\|\varphi\|_{H^2_{**}(\Omega)}\|v\|_{W^{1,4}(\Omega)}\|w\|_{W^{1,4}(\Omega)},$$

$\forall v, w \in H^2_*(\Omega)$, $\forall \varphi \in H^2_{**}(\Omega)$. Therefore,

$$\|B(v,w)\|_{H^2_{**}(\Omega)} = \sup_{0\neq\varphi\in H^2_{**}(\Omega)} \frac{(B(v,w),\varphi)_{H^2_{**}(\Omega)}}{\|\varphi\|_{H^2_{**}(\Omega)}} \le c\|v\|_{W^{1,4}(\Omega)}\|w\|_{W^{1,4}(\Omega)}. \tag{34}$$

Assume that the sequence $\{(v_n, w_n)\} \subset H^2_*(\Omega)^2$ weakly converges to $(v,w) \in H^2_*(\Omega)^2$. Then the triangle inequality and the just proved estimate yield

$$\begin{aligned}\|B(v_n,w_n) - B(v,w)\|_{H^2_{**}(\Omega)} &\le \|B(v_n - v, w_n)\|_{H^2_{**}(\Omega)} + \|B(v, w_n - w)\|_{H^2_{**}(\Omega)} \\ &\le c\|v_n - v\|_{W^{1,4}(\Omega)}\|w_n\|_{W^{1,4}(\Omega)} + c\|v\|_{W^{1,4}(\Omega)}\|w_n - w\|_{W^{1,4}(\Omega)}.\end{aligned}$$

The compact embedding $H^2_*(\Omega) \Subset W^{1,4}(\Omega)$ then shows that

$$\|B(v_n,w_n) - B(v,w)\|_{H^2_{**}(\Omega)} \to 0$$

and hence that B is a compact operator. The proof for C is similar. □

We now define another operator $D : H^2_*(\Omega) \to H^2_*(\Omega)$ by

$$D(v) = C(v, B(v,v)) \qquad \forall v \in H^2_*(\Omega)$$

and we prove

Lemma 4.2. *The operator D is compact.*

Proof. Assume that the sequence $\{v_n\} \subset H^2_*(\Omega)$ weakly converges to $v \in H^2_*(\Omega)$. Then, by Lemma 4.1,

$$B(v_n, v_n) \to B(v, v) \text{ in } H^2_{**}(\Omega), \qquad C(v_n, B(v_n, v_n)) \to C(v, B(v, v)) \text{ in } H^2_*(\Omega).$$

This proves that $D(v_n) \to D(v)$ in $H^2_*(\Omega)$ and that D is a compact operator. □

In turn, the operator D enables us to define a functional $d : H^2_*(\Omega) \to \mathbb{R}$ by

$$d(v) = \frac{1}{4}(D(v), v)_{H^2_*(\Omega)} \qquad \forall v \in H^2_*(\Omega).$$

In the next statement we prove some of its properties.

Lemma 4.3. *The functional $d : H^2_*(\Omega) \to \mathbb{R}$ has the following properties:*

(i) d is nonnegative and $d(v) = 0$ if and only if $v = 0$ in Ω. Moreover,

$$d(v) = \frac{1}{4}\|B(v, v)\|^2_{H^2_{**}(\Omega)};$$

(ii) d is quartic, i.e.,

$$d(rv) = r^4 d(v), \quad \forall r \in \mathbb{R}, \forall v \in H^2_*(\Omega);$$

(iii) d is differentiable in $H^2_*(\Omega)$ and

$$\langle d'(v), w\rangle = (D(v), w)_{H^2_*(\Omega)}, \qquad v, w \in H^2_*(\Omega);$$

(iv) d is weakly continuous on $H^2_*(\Omega)$.

Proof. (i) By (33) we know that for any $v \in H^2_*(\Omega)$,

$$(D(v), v)_{H^2_*(\Omega)} = (C(v, B(v, v)), v)_{H^2_*(\Omega)} = (B(v, v), B(v, v))_{H^2_{**}(\Omega)} = \|B(v, v)\|^2_{H^2_{**}(\Omega)}.$$

Whence, if $d(v) = 0$, then $B(v, v) = 0$ and $[v, v] = 0$, see (31). But $[v, v]$ is proportional to the Gaussian curvature and since it vanishes identically this implies that the surface $v = v(x, y)$ is covered by straight lines. By using the boundary condition (19) we finally infer that $v \equiv 0$. This idea of the last part of this proof is taken from [11, Lemma 3.2'].

(ii) The functional d is quartic as a trivial consequence of its definition.

(iii) From (33) we infer that

$$(C(v, B(v, w)), v)_{H^2_*(\Omega)} = (B(v, v), B(v, w))_{H^2_{**}(\Omega)} = (C(v, B(v, v)), w)_{H^2_*(\Omega)} \tag{35}$$

for all $v, w \in H^2_*(\Omega)$. Then we compute

$$\langle d'(v), w\rangle = \lim_{\varepsilon\to 0} \frac{1}{4\varepsilon}\{(D(v + \varepsilon w), v + \varepsilon w)_{H^2_*(\Omega)} - (D(v), v)_{H^2_*(\Omega)}\}$$

$$= \lim_{\varepsilon\to 0} \frac{1}{4\varepsilon}\{(C(v+\varepsilon w, B(v+\varepsilon w, v+\varepsilon w)), v+\varepsilon w)_{H^2_*(\Omega)}$$
$$- (C(v,B(v,v)),v)_{H^2_*(\Omega)}\}$$
$$= \frac{1}{4}\{(C(w,B(v,v)),v)_{H^2_*(\Omega)} + (C(v,B(v,v)),w)_{H^2_*(\Omega)}$$
$$+ 2(C(v,B(v,w)),v)_{H^2_*(\Omega)}\}$$
$$\text{by (33)} \;= \frac{1}{2}\{(C(v,B(v,v)),w)_{H^2_*(\Omega)} + (C(v,B(v,w)),v)_{H^2_*(\Omega)}\}$$
$$\text{by (35)} \;= (D(v),w)_{H^2_*(\Omega)},$$

which proves (iii).

(iv) Assume that the sequence $\{v_n\} \subset H^2_*(\Omega)$ weakly converges to $v \in H^2_*(\Omega)$. Then by Lemma 4.2 we know that

$$\lim_{n\to\infty} \|D(v_n) - D(v)\|_{H^2_*(\Omega)} = 0.$$

This shows that

$$\lim_{n\to\infty} (D(v_n) - D(v), v_n)_{H^2_*(\Omega)} = 0.$$

Finally, this yields

$$d(v_n) - d(v) = \frac{1}{4}(D(v_n) - D(v), v_n)_{H^2_*(\Omega)} + \frac{1}{4}(D(v), v_n - v)_{H^2_*(\Omega)} \to 0$$

which proves (iv). □

5 Proof of Theorem 3.1

In this section we prove Theorem 3.1 and we give some more details about the eigenvalues and eigenfunctions of (27). We proceed as in [19, Theorem 3.4], see also [1, Theorem 4], with some changes due to the presence of the buckling term. We write the eigenvalue problem (27) as

$$(u_x, v_x)_{L^2(\Omega)} = \frac{1}{\lambda}(u,v)_{H^2_*(\Omega)} \qquad \forall v \in H^2_*(\Omega).$$

Define the linear operator $T : H^2_*(\Omega) \to H^2_*(\Omega)$ such that

$$(Tu, v)_{H^2_*(\Omega)} = (u_x, v_x)_{L^2(\Omega)} \qquad \forall v \in H^2_*(\Omega).$$

The operator T is self-adjoint since

$$(Tu,v)_{H^2_*(\Omega)} = (u_x,v_x)_{L^2(\Omega)} = (v_x,u_x)_{L^2(\Omega)} = (u,Tv)_{H^2_*(\Omega)} \qquad \forall u,v \in H^2_*(\Omega).$$

Moreover, by the compact embedding $H^2_*(\Omega) \Subset H^1(\Omega)$ and the definition of T, the following implications hold:

$$u_n \rightharpoonup u \text{ in } H^2_*(\Omega) \Longrightarrow (u_n)_x \to u_x \text{ in } L^2(\Omega) \Longrightarrow \sup_{\|v\|_{H^2_*(\Omega)}=1} ((u_n-u)_x, v_x)_{L^2(\Omega)} \to 0$$

$$\Longrightarrow \sup_{\|v\|_{H^2_*(\Omega)}=1} (T(u_n-u), v)_{H^2_*(\Omega)} \to 0 \Longrightarrow Tu_n \to Tu \text{ in } H^2_*(\Omega)$$

which shows that T is also compact. Then the spectral theory of linear compact self-adjoint operator yields that (27) admits an ordered increasing sequence of eigenvalues and the corresponding eigenfunctions form an Hilbertian basis of $H^2_*(\Omega)$. This proves the first part of Theorem 3.1.

According to the boundary conditions on $x = 0, \pi$, we seek eigenfunctions in the form:

$$u(x,y) = \sum_{m=1}^{+\infty} h_m(y) \sin(mx) \qquad \text{for } (x,y) \in (0,\pi) \times (-\ell,\ell)\,. \tag{36}$$

Then we are led to find nontrivial solutions of the ordinary differential equation

$$h_m''''(y) - 2m^2 h_m''(y) + (m^4 - m^2\lambda) h_m(y) = 0\,, \qquad (\lambda > 0) \tag{37}$$

with the boundary conditions

$$h_m''(\pm\ell) - \sigma m^2 h_m(\pm\ell) = 0\,, \qquad h_m'''(\pm\ell) + (\sigma - 2)m^2 h_m'(\pm\ell) = 0\,. \tag{38}$$

The characteristic equation related to (37) is $\alpha^4 - 2m^2\alpha^2 + m^4 - m^2\lambda = 0$ and then

$$\alpha^2 = m^2 \pm m\sqrt{\lambda}\,. \tag{39}$$

For a given $\lambda > 0$ three cases have to be distinguished.

- **The case $m^2 > \lambda$.** By (39) we infer

$$\alpha = \pm\beta \text{ or } \alpha = \pm\gamma \qquad \text{with} \qquad \sqrt{m^2 - m\sqrt{\lambda}} =: \gamma < \beta := \sqrt{m^2 + m\sqrt{\lambda}}\,. \tag{40}$$

Nontrivial solutions of (37) have the form

$$h_m(y) = a\cosh(\beta y) + b\sinh(\beta y) + c\cosh(\gamma y) + d\sinh(\gamma y) \qquad (a,b,c,d \in \mathbb{R})\,. \tag{41}$$

By imposing the boundary conditions (38) and arguing as in [19] we see that a nontrivial solution of (37) exists if and only if one of the two following equalities holds:

$$\frac{\gamma}{(\gamma^2-m^2\sigma)^2}\tanh(\ell\gamma)=\frac{\beta}{(\beta^2-m^2\sigma)^2}\tanh(\ell\beta)\,, \tag{42}$$

$$\frac{\beta}{(\beta^2-m^2\sigma)^2}\coth(\ell\beta)=\frac{\gamma}{(\gamma^2-m^2\sigma)^2}\coth(\ell\gamma)\,. \tag{43}$$

For any integer $m>\sqrt{\lambda}$ such that (42) holds, the function h_m in (41) with $b=d=0$ and suitable $a=a_m\neq 0$ and $c=c_m\neq 0$ yields the eigenfunction $h_m(y)\sin(mx)$ associated with the eigenvalue λ. Similarly, for any integer $m>\sqrt{\lambda}$ such that (43) holds, the function h_m in (41) with $a=c=0$ and suitable $b=b_m\neq 0$ and $d=d_m\neq 0$ yields the eigenfunction $h_m(y)\sin(mx)$ associated with the eigenvalue λ. Clearly, the number of both such integers is finite. In particular, when $m=1$ the equation (37) coincides with [19, (57)]. Therefore, the statement about the least eigenvalue and the explicit form of the corresponding eigenfunction hold.

- **The case** $m^2=\lambda$**.** This case is completely similar to the second case in [19]. By (39) we infer that possible nontrivial solutions of (37)-(38) have the form

$$h_m(y)=a\cosh(\sqrt{2}my)+b\sinh(\sqrt{2}my)+c+dy\qquad(a,b,c,d\in\mathbb{R})\,.$$

Then one sees that $a=c=0$ if (3) holds. Moreover, let $\bar{s}>0$ the unique solution of $\tanh(s)=\left(\frac{\sigma}{2-\sigma}\right)^2 s$. If $m_*:=\bar{s}/\ell\sqrt{2}$ is an integer, and only in this case, then $\lambda=m_*^2$ is an eigenvalue and the corresponding eigenfunction is

$$\left[\sigma\ell\sinh(\sqrt{2}m_*y)+(2-\sigma)\sinh(\sqrt{2}m_*\ell)\,y\right]\sin(m_*x)\,.$$

- **The case** $m^2<\lambda$**.** By (39) we infer that

$$\alpha=\pm\beta \text{ or } \alpha=\pm i\gamma\ \text{ with }\ \sqrt{m\sqrt{\lambda}-m^2}=\gamma<\beta=\sqrt{m\sqrt{\lambda}+m^2}\,.$$

Therefore, possible nontrivial solutions of (37) have the form

$$h_m(y)=a\cosh(\beta y)+b\sinh(\beta y)+c\cos(\gamma y)+d\sin(\gamma y)\qquad(a,b,c,d\in\mathbb{R})\,.$$

Differentiating h_m and imposing the boundary conditions (38) yields the two systems:

$$\begin{cases}(\beta^2-m^2\sigma)\cosh(\beta\ell)a-(\gamma^2+m^2\sigma)\cos(\gamma\ell)c=0\\(\beta^3-m^2(2-\sigma)\beta)\sinh(\beta\ell)a+(\gamma^3+m^2(2-\sigma)\gamma)\sin(\gamma\ell)c=0\,,\end{cases}$$

$$\begin{cases}(\beta^2-m^2\sigma)\sinh(\beta\ell)b-(\gamma^2+m^2\sigma)\sin(\gamma\ell)d=0\\(\beta^3-m^2(2-\sigma)\beta)\cosh(\beta\ell)b-(\gamma^3+m^2(2-\sigma)\gamma)\cos(\gamma\ell)d=0\,.\end{cases}$$

Due to the presence of trigonometric sine and cosine, for any integer m there exists a sequence $\zeta_k^m \uparrow +\infty$ such that $\zeta_k^m > m^2$ for all $k \in \mathbb{N}$ and such that if $\lambda = \zeta_k^m$ for some k then one of the above systems admits a nontrivial solution. On the other hand, for any eigenvalue λ there exists at most a finite number of integers m such that $m^2 < \lambda$; if these integers yield nontrivial solutions h_m, then the function $h_m(y)\sin(mx)$ is an eigenfunction corresponding to λ.

6 Proof of Theorem 3.2

By Lemma 4.3 we know that a functional whose critical points are solutions of the problem (23) reads

$$J(u) = \frac{1}{2}\|u\|^2_{H^2_*(\Omega)} + d(u) - \frac{\lambda}{2}\|u_x\|^2_{L^2(\Omega)} - \int_\Omega fu \qquad \forall u \in H^2_*(\Omega).$$

By combining Lemmas 4.1–4.2–4.3, we obtain a one-to-one correspondence between solutions of (23) and critical points of the functional J:

Lemma 6.1. *Let $f \in L^2(\Omega)$. The couple $(u, \Phi) \in H^2_*(\Omega) \times H^2_{**}(\Omega)$ is a weak solution of* (23) *if and only if $u \in H^2_*(\Omega)$ is a critical point of J and if $\Phi \in H^2_{**}(\Omega)$ weakly solves $\Delta^2\Phi = -[u,u]$ in Ω.*

The first step is then to prove geometrical properties (coercivity) and compactness properties (Palais-Smale condition) of J. Although the former may appear straightforward, it requires delicate arguments. The reason is that no useful lower bound for $d(u)$ is available. We prove

Lemma 6.2. *For any $f \in L^2(\Omega)$ and any $\lambda \geq 0$, the functional J is coercive in $H^2_*(\Omega)$ and it is bounded from below. Moreover, it satisfies the Palais-Smale (PS) condition.*

Proof. Assume for contradiction that there exists a sequence $\{v_n\} \subset H^2_*(\Omega)$ and $M > 0$ such that

$$\lim_{n\to\infty} \|v_n\|_{H^2_*(\Omega)} \to \infty, \qquad J(v_n) \leq M.$$

Put $w_n = \frac{v_n}{\|v_n\|_{H^2_*(\Omega)}}$ so that $v_n = \|v_n\|_{H^2_*(\Omega)} w_n$ and

$$\|w_n\|_{H^2_*(\Omega)} = 1 \qquad \forall n. \tag{44}$$

By combining the Hölder inequality with (29), we infer that

$$M \geq J(v_n) \geq \frac{1}{2}\|v_n\|^2_{H^2_*(\Omega)} + \|v_n\|^4_{H^2_*(\Omega)} d(w_n)$$
$$- \frac{\lambda}{2}\|v_n\|^2_{H^2_*(\Omega)}\|(w_n)_x\|^2_{L^2(\Omega)} - \frac{\|f\|_{L^2(\Omega)}}{\sqrt{\lambda_1}}\|v_n\|_{H^2_*(\Omega)}, \tag{45}$$

where we also used Lemma 4.3 (ii). By letting $n \to \infty$, this shows that $d(w_n) \to 0$ which, combined with Lemma 4.3 and (44), shows that $w_n \rightharpoonup 0$ in $H^2_*(\Omega)$; then, $(w_n)_x \to 0$ in $L^2(\Omega)$ by compact embedding. Hence, since $d(w_n) \geq 0$, (45) yields

$$o(1) = \frac{M}{\|v_n\|^2_{H^2_*(\Omega)}} \geq \frac{1}{2} + \|v_n\|^2_{H^2_*(\Omega)} d(w_n)$$

$$- \frac{\lambda}{2}\|(w_n)_x\|^2_{L^2(\Omega)} - \frac{\|f\|_{L^2(\Omega)}}{\|v_n\|_{H^2_*(\Omega)}\sqrt{\lambda_1}} \geq \frac{1}{2} + o(1)$$

which leads to a contradiction by letting $n \to \infty$. Therefore J is coercive. Since the lower bound for $J(v_n)$ in (45) only depends on $\|v_n\|_{H^2_*(\Omega)}$, we also know that J is bounded from below.

In order to prove that J satisfies the (PS) condition we consider a sequence $\{u_n\} \subset H^2_*(\Omega)$ such that $J(u_n)$ is bounded and $J'(u_n) \to 0$ in $\mathcal{H}_*(\Omega)$. By what we just proved, we know that $\{u_n\}$ is bounded and therefore, there exists $\overline{u} \in H^2_*(\Omega)$ such that $u_n \rightharpoonup \overline{u}$ and, by weak continuity, $J'(\overline{u}) = 0$. Moreover, by Lemma 4.3,

$$\langle J'(u_n), u_n\rangle = \|u_n\|^2_{H^2_*(\Omega)} + (D(u_n), u_n)_{H^2_*(\Omega)} - \lambda\|(u_n)_x\|^2_{L^2(\Omega)} - \int_\Omega fu_n \to$$

$$\to 0 = \langle J'(\overline{u}), \overline{u}\rangle = \|\overline{u}\|^2_{H^2_*(\Omega)} + (D(\overline{u}), \overline{u})_{H^2_*(\Omega)} - \lambda\|\overline{u}_x\|^2_{L^2(\Omega)} - \int_\Omega f\overline{u}\,.$$

Since $(D(u_n), u_n)_{H^2_*(\Omega)} \to (D(\overline{u}), \overline{u})_{H^2_*(\Omega)}$ by Lemma 4.2, $\|(u_n)_x\|^2_{L^2(\Omega)} \to \|\overline{u}_x\|^2_{L^2(\Omega)}$ and $\int_\Omega fu_n \to \int_\Omega f\overline{u}$ by compact embedding, this proves that $\|u_n\|_{H^2_*(\Omega)} \to \|\overline{u}\|_{H^2_*(\Omega)}$. This fact, together with the weak convergence $u_n \rightharpoonup \overline{u}$ proves that, in fact, $u_n \to \overline{u}$ strongly; this proves (PS). □

Lemma 6.2 shows that the (smooth) functional J admits a global minimum in $H^2_*(\Omega)$ for any f and λ. This minimum is a critical point for J and hence, by Lemma 6.1, it gives a weak solution of (23). This proves the first part of Theorem 3.2. Let us now prove the items.

(i) If $\lambda \leq \lambda_1$ and $f = 0$, we see that any critical point u of J satisfies

$$0 = \langle J'(u), u\rangle = \|u\|^2_{H^2_*(\Omega)} + 4d(u) - \lambda\|u_x\|^2_{L^2(\Omega)}$$

where we also used Lemma 4.3 (iii). By Lemma 4.3 i) and (29), this proves that $u = 0$. Then we apply again Lemma 6.1 and find $(u, \Phi) = (0, 0)$.

(ii) If $f = 0$ and $\lambda \in (\lambda_k, \lambda_{k+1}]$, then the twice differentiable functional J is even and its second derivative $J''(0)$ at 0 has Morse index k. By Lemma 6.2 we may then apply [1, Theorem 11] (which is a variant of Theorem 5.2.23 p.369 in [14]), to infer that J has at least k pairs of district nonzero critical points. Then by Lemma 6.1 there exist at least k pairs of nontrivial solutions of (23).

(iii) For any $f \in L^2(\Omega)$, if u is a critical point of the functional J it satisfies $\langle J'(u), u\rangle = 0$ and therefore, by the Hölder inequality,

$$\|u\|^2_{H^2_*(\Omega)} + 4d(u) - \lambda\|u_x\|^2_{L^2(\Omega)} \leq \|f\|_{L^2(\Omega)}\|u\|_{L^2(\Omega)} \,.$$

In turn, by using Lemma 4.3 i) and twice (29), we obtain

$$\left(1 - \frac{\lambda}{\lambda_1}\right)\|u\|^2_{H^2_*(\Omega)} \leq \frac{\|f\|_{L^2(\Omega)}}{\sqrt{\lambda_1}}\|u\|_{H^2_*(\Omega)} \,.$$

This gives the a priori bound

$$\|u\|_{H^2_*(\Omega)} \leq \frac{\sqrt{\lambda_1}}{\lambda_1 - \lambda}\|f\|_{L^2(\Omega)} \,. \tag{46}$$

Next, we prove a local convexity property of the functional J. Let

$$Q(u) := \|u\|^2_{H^2_*(\Omega)} - \lambda\|u_x\|^2_{L^2(\Omega)} \quad \forall u \in H^2_*(\Omega) \,.$$

Then, for all $u, v \in H^2_*(\Omega)$ and all $t \in [0, 1]$, we have

$$Q\Big(tu + (1-t)v\Big) - tQ(u) - (1-t)Q(v) = -t(1-t)\Big(\|u-v\|^2_{H^2_*(\Omega)} - \lambda\|u_x - v_x\|^2_{L^2(\Omega)}\Big). \tag{47}$$

Moreover, for all $u, v \in H^2_*(\Omega)$ and all $t \in [0, 1]$, some tedious computations show that

$$d\Big(tu + (1-t)v\Big) - td(u) - (1-t)d(v) =$$

$$= -\frac{t(1-t)}{4}\Big\{(t^2 - 3t + 1)(\|B(v, u-v)\|^2_{H^2_{**}(\Omega)} - \|B(u, u-v)\|^2_{H^2_{**}(\Omega)})$$

$$+2\big(B(v, v), B(v-u, v-u)\big)_{H^2_{**}(\Omega)} + 2(t^2 - t + 1)\big(B(u, u-v), B(u+v, u-v)\big)_{H^2_{**}(\Omega)}$$

$$-4t(1-t)\big(B(u-v, u), B(v-u, v)\big)_{H^2_{**}(\Omega)}\Big\}$$

$$\text{by (34)} \quad \leq C\,t(1-t)\,(\|u\|^2_{H^2_*(\Omega)} + \|v\|^2_{H^2_*(\Omega)})\,\|u-v\|^2_{H^2_*(\Omega)} \,; \tag{48}$$

here $C > 0$ is a constant independent of t, u, v. Consider the "unforced" functional

$$J_0(u) = \frac{1}{2}\|u\|^2_{H^2_*(\Omega)} + d(u) - \frac{\lambda}{2}\|u_x\|^2_{L^2(\Omega)} = \frac{Q(u)}{2} + d(u) \,; \tag{49}$$

by putting together (47) and (48) we see that

$$\begin{aligned} J_0\Big(tu+(1-t)v\Big) &- tJ_0(u)-(1-t)J_0(v) \\ &\le -\frac{t(1-t)}{2}\Big(\|u-v\|^2_{H^2_*(\Omega)}-\lambda\|u_x-v_x\|^2_{L^2(\Omega)}\Big) \\ &\quad + C\,t(1-t)\,(\|u\|^2_{H^2_*(\Omega)}+\|v\|^2_{H^2_*(\Omega)})\,\|u-v\|^2_{H^2_*(\Omega)} \\ &\le t(1-t)\Big(C\,(\|u\|^2_{H^2_*(\Omega)}+\|v\|^2_{H^2_*(\Omega)})-\frac{\lambda_1-\lambda}{2\lambda_1}\Big)\|u-v\|^2_{H^2_*(\Omega)}\,. \end{aligned} \tag{50}$$

Take f sufficiently small such that

$$\|f\|^2_{L^2(\Omega)} < K^2 := \frac{(\lambda_1-\lambda)^3}{4C\,\lambda_1^2}\,. \tag{51}$$

By (46) and (51) we know that any critical point of J satisfies

$$\|u\|^2_{H^2_*(\Omega)} \le \frac{\lambda_1}{(\lambda_1-\lambda)^2}\,K^2 = \frac{\lambda_1-\lambda}{4C\,\lambda_1} =: \rho^2\,;$$

put $B_\rho=\{u\in H^2_*(\Omega);\ \|u\|_{H^2_*(\Omega)}\le\rho\}$. Moreover, from (50) we know that

$$J_0\Big(tu+(1-t)v\Big)-tJ_0(u)-(1-t)J_0(v)\le 0 \qquad \forall u,v\in B_\rho\,,$$

with strict inequality if $u\neq v$ and $t\notin\{0,1\}$. This proves that J_0 is strictly convex in B_ρ and since $J(u)$ equals $J_0(u)$ plus a linear term (with respect to u), also J is strictly convex in B_ρ.

Summarising, if (51) holds, then we know that:

- by (46) all the critical points of J belong to B_ρ;
- by the first part of the proof we then know that there exists at least a critical point in B_ρ;
- J is strictly convex in B_ρ.

We then deduce that J admits a unique critical point in B_ρ (its absolute minimum) and no other critical points elsewhere. Together with Lemma 6.1, this completes the proof of item (iii).

(iv) If $\lambda>\lambda_1$ we know from item (ii) that the unforced functional J_0 defined in (49) has two global minima $\pm\bar u\neq 0$. Then a sufficiently small linear perturbation of J_0 has a local minimum in a neighborhood of both $\pm\bar u$. Whence, if f is sufficiently small, say $\|f\|_{L^2(\Omega)}<K$, then the functional J defined by $J(u)=J_0(u)-\int_\Omega fu$ admits two local minima in two neighborhoods of both $\pm\bar u$. These local minima, which we name u_1 and u_2, are the first two critical points of J. A minimax procedure then yields an additional (mountain-pass) solution. Indeed, consider the set of continuous paths connecting u_1 and u_2:

$$\Gamma := \Big\{p \in C^0([0,1], H^2_*(\Omega));\ p(0) = u_1,\ p(1) = u_2\Big\}.$$

Since by Lemma 6.2 the functional J satisfies the (PS) condition, the mountain-pass Theorem guarantees that the level

$$\min_{p\in\Gamma} \max_{t\in[0,1]} J\big(p(t)\big) > \max\Big\{J(u_1), J(u_2)\Big\}$$

is a critical level for J; this yields a third critical point. By Lemma 6.1 this proves the existence of (at least) three weak solutions of (23).

7 Proof of Theorem 3.3

Similar to Lemma 6.1, the functional whose critical points are solutions of problem (26) is

$$J(u) = \frac{1}{2}\|u\|^2_{H^2_*(\Omega)} + \int_\Omega \Upsilon(y)\left(\frac{k}{2}(u^+)^2 + \frac{\delta}{4}(u^+)^4\right) + d(u) - \frac{\lambda}{2}\|u_x\|^2_{L^2(\Omega)} - \int_\Omega fu\,.$$

And similar to Lemma 6.2 one can prove that for any $f \in L^2(\Omega)$ and any $\lambda \geq 0$, the functional J is coercive in $H^2_*(\Omega)$, it is bounded from below and it satisfies the (PS) condition. Then the smooth functional J admits a global minimum in $H^2_*(\Omega)$ for any f and λ. This minimum is a critical point for J and hence a weak solution of (26). This proves the first part of Theorem 3.3. Let us now prove the items.

(i) The proof of this item follows the same steps as item (iii) of Theorem 3.2: it suffices to notice that the additional term $\int_\Omega \Upsilon(y)\left(\frac{k}{2}(u^+)^2 + \frac{\delta}{4}(u^+)^4\right)$ is also convex.

(ii) If $f = 0$, then $u = 0$ is a solution for any $\lambda \geq 0$. We just need to show that it is not the global minimum which we know to exist. Let $\overline{e}_1$ and α be as in (30) and consider the function

$$g(t) := J(t\overline{e}_1) = -\frac{\lambda - \lambda_1}{2\lambda_1}t^2 + \frac{k\alpha}{2}(t^+)^2 + \frac{\delta\,(t^+)^4}{4}\int_\Omega \Upsilon(y)\overline{e}_1^4 + t^4\,d(\overline{e}_1) \qquad t \in \mathbb{R}\,. \tag{52}$$

Since $\lambda > \lambda_1$, the coefficient of $(t^-)^2$ is negative and the qualitative graph of g is as in Figure 3 (on the left the case where $\lambda < \overline{\lambda}$ so that the coefficient of $(t^+)^2$ is nonnegative, on the right the case where also the coefficient of $(t^+)^2$ is negative). It is clear that there exists $\overline{t} < 0$ such that $g(\overline{t}) < 0$. This means that $J(\overline{t}\overline{e}_1) < 0$ and that 0 is not the absolute minimum of J. This completes the proof of item (ii).

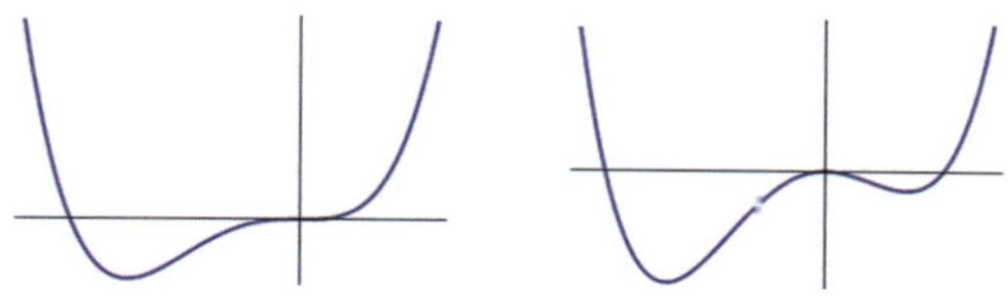

Fig. 3 Qualitative graphs of the functions g (left) and h (right).

(iii) We study first the case where $f = 0$ and we name J_0 the unforced functional, that is,

$$J_0(u) = \frac{1}{2}\|u\|^2_{H^2_*(\Omega)} + \int_\Omega \Upsilon(y)\left(\frac{k}{2}(u^+)^2 + \frac{\delta}{4}(u^+)^4\right) + d(u) - \frac{\lambda}{2}\|u_x\|^2_{L^2(\Omega)} .$$

We consider again the function g in (52) that we name here h in order to distinguish their graphs, $h(t) = g(t)$ as in (52). Since $\lambda > \overline{\lambda}$, the coefficient of $(t^+)^2$ is now also negative and the qualitative graph of h is as in the right picture of Figure 3. Then the function h has a nondegenerate local maximum at $t = 0$ which means that also the map $t \mapsto J_0(t\overline{e}_1)$ has a local maximum at $t = 0$ and it is strictly negative in a punctured interval containing $t = 0$. Let $E = \text{span}\{\overline{e}_k;\ k \geq 2\}$ denote the infinite dimensional space of codimension 1 being the orthogonal complement of $\text{span}\{\overline{e}_1\}$. By the improved Poincaré inequality

$$\lambda_2\|v_x\|^2_{L^2(\Omega)} \leq \|v\|^2_{H^2_*(\Omega)} \qquad \forall v \in E$$

and by taking into account Lemma 4.3 (i) and $\lambda \leq \lambda_2$, we see that

$$J_0(u) \geq \frac{\lambda_2 - \lambda}{2\lambda_2}\|u\|^2_{H^2_*(\Omega)} + \int_\Omega \Upsilon(y)\left(\frac{k}{2}(u^+)^2 + \frac{\delta}{4}(u^+)^4\right) \geq 0 \qquad \forall u \in E .$$

Therefore, the two open sets

$$A^+ = \{u \in H^2_*(\Omega);\ (u, \overline{e}_1)_{H^2_*(\Omega)} > 0,\ J_0(u) < 0\} ,$$

$$A^- = \{u \in H^2_*(\Omega);\ (u, \overline{e}_1)_{H^2_*(\Omega)} < 0,\ J_0(u) < 0\}$$

are disconnected. Since J_0 satisfies the (PS) condition and is bounded from below, J_0 admits a global minimum u^+ (resp. u^-) in A^+ (resp. A^-) and $J_0(u^\pm) < 0$.

A sufficiently small linear perturbation of J_0 then has a local minimum in a neighborhood of both $u^\pm$. Whence, if f is sufficiently small, say $\|f\|_{L^2(\Omega)} < K$, then the functional J defined by $J(u) = J_0(u) - \int_\Omega fu$ admits a local minimum in two neighborhoods of both $u^\pm$. A minimax procedure then yields an additional (mountain-pass) critical point, see the proof of Theorem 3.2 (iv) for the details. This yields a third solution of (26).

References

1. Al-Gwaiz, M., Benci, V., Gazzola, F.: Bending and stretching energies in a rectangular plate modeling suspension bridges. Nonlinear Anal. Theory Methods Appl. **106**, 18–34 (2014)
2. Ammann, O.H., von Kármán, T., Woodruff, G.B.: The failure of the Tacoma Narrows Bridge. Federal Works Agency, Washington (1941)
3. Arioli, G., Gazzola, F.: A new mathematical explanation of what triggered the catastrophic torsional mode of the Tacoma Narrows Bridge collapse. Appl. Math. Model. **39**, 901–912 (2015)
4. Augusti, G., Sepe, V.: A "deformable section" model for the dynamics of suspension bridges. Part I: Model and linear response. Wind Struct. **4**, 1–18 (2001)
5. Augusti, G., Diaferio, M., Sepe, V.: A "deformable section" model for the dynamics of suspension bridges. Part II: Nonlinear analysis and large amplitude oscillations. Wind Struct. **6**, 451–470 (2003)
6. Bartoli, G., Spinelli, P.: The stochastic differential calculus for the determination of structural response under wind. J. Wind Eng. Ind. Aerodyn. **48**, 175–188 (1993)
7. Berchio, E., Gazzola, F.: A qualitative explanation of the origin of torsional instability in suspension bridges. Nonlinear Anal. Theory Methods Appl. **121**, 54–72 (2015)
8. Berger, H.M.: A new approach to the analysis of large deflections of plates. J. Appl. Mech. **22**, 465–472 (1955)
9. Berger, M.S.: On von Kármán's equations and the buckling of a thin elastic plate, I. The clamped plate. Commun. Pure Appl. Math. **20**, 687–719 (1967)
10. Berger, M.S., Fife, P.C.: Von Kármán's equations and the buckling of a thin elastic plate. Bull. Am. Math. Soc. **72**, 1006–1011 (1966)
11. Berger, M.S., Fife, P.C.: Von Kármán's equations and the buckling of a thin elastic plate, II. Plate with general edge conditions. Commun. Pure Appl. Math. **21**, 227–241 (1968)
12. Bochicchio, I., Giorgi, C., Vuk, E.: Long-term dynamics of the coupled suspension bridge system. Math. Models Methods Appl. Sci. **22**, 22 pp. (2012)
13. Brownjohn, J.M.W.: Observations on non-linear dynamic characteristics of suspension bridges. Earthq. Eng. Struct. Dyn. **23**, 1351–1367 (1994)
14. Chang, K.C.: Methods in Nonlinear Analysis. Springer Monographs in Mathematics. Springer, Berlin (2005)
15. Ciarlet, P.G., A justification of the von Kármán equations. Arch. Ration. Mech. Anal. **73**, 349–389 (1980)
16. Ciarlet, P.G., Rabier, P.: Les équations de von Kármán. Studies in Mathematics and its Applications, vol. 27. Springer, Berlin (1980)
17. Davet, J.L.: Justification de modèles de plaques nonlinéaires pour des lois de comportment générales. Modélisation Math. Anal. Numér. **20**, 147–192 (1986)
18. de Figueiredo, D.G.: Semilinear elliptic systems. In: Nonlinear Functional Analysis and Applications to Differential Equations (Trieste, 1997), pp. 122–152. World Scientific, River Edge (1998)
19. Ferrero, A., Gazzola, F.: A partially hinged rectangular plate as a model for suspension bridges. Discrete. Cont. Dynam. Syst. A. **35**, 5879–5908 (2015)
20. Gazzola, F.: Nonlinearity in oscillating bridges. Electron. J. Differ. Equ. **211**, 1–47 (2013)
21. Gazzola, F., Grunau, H.-Ch., Sweers, G.: Polyharmonic Boundary Value Problems. Lecture Notes in Mathematics, vol. 1991. Springer, Heidelberg (2010)
22. Kirchhoff, G.R.: Über das gleichgewicht und die bewegung einer elastischen scheibe. J. Reine Angew. Math. **40**, 51–88 (1850)
23. Knightly, G.H.: An existence theorem for the von Kármán equations. Arch. Ration. Mech. Anal. **27**, 233–242 (1967)
24. Knightly, G.H., Sather, D.: On nonuniqueness of solutions of the von Kármán equations. Arch. Ration. Mech. Anal. **36**, 65–78 (1970)

25. Knightly, G.H., Sather, D.: Nonlinear buckled states of rectangular plates. Arch. Ration. Mech. Anal. **54**, 356–372 (1974)
26. Lacarbonara, W.: Nonlinear Structural Mechanics. Springer, New York (2013)
27. Lagnese, J.E.: Boundary Stabilization of Thin Plates. SIAM Studies in Applied Mathematics, Philadelphia (1989)
28. Lagnese, J.E., Lions, J.L.: Modelling Analysis and Control of Thin Plates. Masson - Collection RMA, Paris (1988)
29. Levy, S.: Bending of rectangular plates with large deflections. National Advisory Committee for Aeronautics, Washington, Report n.737, 139–157 (1942)
30. Levy, S., Goldenberg, D., Zibritosky, G.: Simply supported long rectangular plate under combined axial load and normal pressure. National Advisory Committee for Aeronautics, Washington, Technical Note 949, 24 pp. (1944)
31. McKenna, P.J., Walter, W.: Nonlinear oscillations in a suspension bridge. Arch. Ration. Mech. Anal. **98**, 167–177 (1987)
32. Menn, C.: Prestressed Concrete Bridges. Birkhäuser, Boston (1990)
33. Plaut, R.H., Davis, F.M.: Sudden lateral asymmetry and torsional oscillations of section models of suspension bridges. J. Sound Vib. **307**, 894–905 (2007)
34. Pugsley, A.: The Theory of Suspension Bridges. Edward Arnold, London (1968)
35. Robinson, A.R., West, H.H.: A Re-examination of the Theory of Suspension Bridges. Civil Engineering Series, Structural Research Series, vol. 322. Doctoral Dissertation, Urbana (1967)
36. Scanlan, R.H.: Developments in low-speed aeroelasticity in the civil engineering field. AIAA J. **20**, 839–844 (1982)
37. Scanlan, R.H., Tomko, J.J.: Airfoil and bridge deck flutter derivatives. J. Eng. Mech. **97**, 1717–1737 (1971)
38. Scott, R.: In the Wake of Tacoma. Suspension Bridges and the Quest for Aerodynamic Stability. ASCE, Reston (2001)
39. Tacoma Narrows Bridge Collapse: http://www.youtube.com/watch?v=3mclp9QmCGs (1940)
40. Truesdell, C.: Some challenges offered to analysis by rational thermomechanics. In: de la Penha, G.M., Medeiros, L.A. (eds.) Contemporary Developments in Continuum Mechanics and Partial Differential Equations, pp. 495–603. North-Holland, Amsterdam (1978)
41. Ventsel, E., Krauthammer, T.: Thin Plates and Shells: Theory, Analysis, and Applications. Marcel Dekker, New York (2001)
42. Villaggio, P.: Mathematical Models for Elastic Structures. Cambridge University Press, Cambridge (1997)
43. von Kármán, T.: Festigkeitsprobleme im maschinenbau. Encycl. der Mathematischen Wissenschaften, vol. IV/4 C, pp. 348–352. Leipzig (1910)
44. Wang, Y.: Finite time blow-up and global solutions for fourth order damped wave equations. J. Math. Anal. Appl. **418**, 713–733 (2014)
45. Wang, Y.: A new mathematical model for suspension bridges with energy dependent boundary conditions, preprint (2014)

Nonlinear Klein-Gordon-Maxwell systems with Neumann boundary conditions on a Riemannian manifold with boundary

Marco Ghimenti and Anna Maria Micheletti

Mathematics Subject Classification (2010): 35J57, 35J60, 58E05, 81V10

1 Introduction

Let (M, g) be a smooth compact, n dimensional Riemannian manifold, $n = 3, 4$ with boundary ∂M which is the union of a finite number of connected, smooth, boundaryless, $n - 1$ submanifolds embedded in M. Here g denotes the Riemannian metric tensor. By Nash theorem we can consider (M, g) as a regular submanifold embedded in $\mathbb{R}^N$.

We search for the positive solutions of the following Klein Gordon Maxwell Proca system with homogeneous Neumann boundary conditions

$$\begin{cases} -\varepsilon^2 \Delta_g u + au = |u|^{p-2}u + \omega^2(qv-1)^2 u & \text{in } M \\ -\Delta_g v + (1 + q^2u^2)v = qu^2 & \text{in } M \\ \frac{\partial u}{\partial \nu} = \frac{\partial v}{\partial \nu} = 0 & \text{on } \partial M \end{cases} \tag{1}$$

or Klein Gordon Maxwell system with mixed Dirichlet Neumann homogeneous boundary conditions

M. Ghimenti (✉) • A.M. Micheletti
Dipartimento di Matematica, Università di Pisa, Largo B. Pontecorvo 5, 56100 Pisa, Italy
e-mail: marco.ghimenti@dma.unipi.it; a.micheletti@dma.unipi.it

A.N. Carvalho et al. (eds.), *Contributions to Nonlinear Elliptic Equations and Systems*, Progress in Nonlinear Differential Equations and Their Applications 86,
DOI 10.1007/978-3-319-19902-3_19

$$\begin{cases} -\varepsilon^2 \Delta_g u + au = |u|^{p-2}u + \omega^2(qv-1)^2 u & \text{in } M \\ -\Delta_g v + q^2 u^2 v = qu^2 & \text{in } M \\ v = 0 & \text{on } \partial M \\ \frac{\partial u}{\partial \nu} = 0. & \text{on } \partial M \end{cases} \tag{2}$$

Here $2 < p < 2^* = \frac{2n}{n-2}$, ν is the external normal to ∂M, $a > 0$, $q > 0$, $\omega \in (-\sqrt{a}, \sqrt{a})$ and ε is a positive perturbation parameter.

We are interested in finding solutions $u, v \in H_g^1(M)$ to problem (1) and (2). Also, we show that, for ε sufficiently small, the function u has a peak near a stable critical point of the mean curvature of the boundary.

Definition 1.1. Let $f \in C^1(N, \mathbb{R})$, where (N, g) is a Riemannian manifold. We say that $K \subset N$ is a C^1-stable critical set of f if $K \subset \{x \in N : \nabla_g f(x) = 0\}$ and for any $\mu > 0$ there exists $\delta > 0$ such that, if $h \in C^1(N, \mathbb{R})$ with

$$\max_{d_g(x,K) \le \mu} |f(x) - h(x)| + |\nabla_g f(x) - \nabla_g h(x)| \le \delta,$$

then h has a critical point x_0 with $d_g(x_0, K) \le \mu$. Here d_g denotes the geodesic distance associated with the Riemannian metric g.

Now we state the main theorem.

Theorem 1.2. *Assume $K \subset \partial M$ is a C^1-stable critical set of the mean curvature of the boundary. Then there exists $\varepsilon_0 > 0$ such that, for any $\varepsilon \in (0, \varepsilon_0)$, Problem (1) has a solution $(u_\varepsilon, v_\varepsilon) \in H_g^1(M) \times H_g^1(M)$. Analogously, problem (2) has a solution $(u_\varepsilon, v_\varepsilon) \in H_g^1(M) \times H_{0,g}^1(M)$. Moreover, the function u_ε has a peak in some $\xi_\varepsilon \in \partial M$ which converges to a point $\xi_0 \in K$ as ε goes to zero.*

From the seminal paper of [2] many authors studied KGM systems on a flat domain. We cite [1, 4, 6–10, 21].

For KGM and KGMP system on Riemannian manifolds, as far as we know the first paper which deals with this problem is by Druet and Hebey [11]. In this work the authors study the case $\varepsilon = 1$ and prove the existence of a solution for KGMP systems on a closed manifold, by the mountain pass theorem. Thereafter several works are devoted to the study of KGMP system on Riemaniann closed manifold. We limit ourselves to cite [5, 16–19].

Klein Gordon Maxwell system provides a model for a particle u interacting with its own electrostatic field v. Thus, it is somewhat more natural to prescribe Neumann condition on the second equation as d'Avenia Pisani and Siciliano nicely explained in the introduction of [9].

So, recently we moved to study KGMP systems in a Riemaniann manifold M with boundary ∂M with Neumann boundary condition on the second equation. In [14] the authors proved that the topological properties of the boundary ∂M, namely the Lusternik Schnirelmann category of the boundary, affect the number of the low energy solution for the systems. Also, we notice that the natural dimension for KGM and KGMP systems is $n = 3$, since this systems arises from a physical model.

However, the case $n = 4$ is interesting from a mathematical point of view, since the second equation of systems (1) and (2) becomes energy critical by the presence of the u^2v term. For further comments on this subject, we refer to [18]

We can compare [14] and Theorem 1.2. In [15] we proved that the set of metrics for which the mean curvature has only nondegenerate critical points is an open dense set among all the C^k metrics on M, $k \geq 3$. Thus, generically with respect to the metric, the mean curvature has $P_1(\partial M)$ nondegenerate (hence stable) critical points, where $P_1(\partial M)$ is the Poincaré polynomial of ∂M, namely $P_t(\partial M)$, evaluated in $t = 1$. Hence, generically with respect to metric, Problem (1) has $P_1(\partial M)$ solution and holds $P_1(\partial M) \geq \text{cat}\partial M$. Also, in many cases the strict inequality $P_1(\partial M) > \text{cat}\partial M$ holds.

The paper is organized as follows. In Section 2 we summarize some results that are necessary to frame the problem. Namely, we recall some well-known notion of Remannian geometry, we introduce the variational setting and we study some properties of the second equation of the systems. In Section 3 we perform the finite dimensional reduction and we sketch the proof of Theorem 1.2. A collection of technical results is contained in Appendix 5.

2 Preliminary results

We recall some well-known fact about Riemannian manifold with boundary.

First of all we define the Fermi coordinate chart.

Definition 2.1. If q belongs to the boundary ∂M, let $\bar{y} = (z_1, \ldots, z_{n-1})$ be Riemannian normal coordinates on the $n-1$ manifold ∂M at the point q. For a point $\xi \in M$ close to q, there exists a unique $\bar{\xi} \in \partial M$ such that $d_g(\xi, \partial M) = d_g(\xi, \bar{\xi})$. We set $\bar{z}(\xi) \in \mathbb{R}^{n-1}$ the normal coordinates for $\bar{\xi}$ and $z_n(\xi) = d_g(\xi, \partial M)$. Then we define a chart $\Psi_q^\partial : \mathbb{R}_+^n \to M$ such that $(\bar{z}(\xi), z_n(\xi)) = \left(\Psi_q^\partial\right)^{-1}(\xi)$. These coordinates are called *Fermi coordinates* at $q \in \partial M$. The Riemannian metric $g_q(\bar{z}, z_n)$ read through the Fermi coordinates satisfies $g_q(0) = \text{Id}$.

We note by d_g^∂ and $\exp^\partial$, respectively, the geodesic distance and the exponential map on by ∂M. By compactness of ∂M, there is an R^∂ and a finite number of points $q_i \in \partial M$, $i = 1, \ldots, k$ such that

$$I_{q_i}(R^\partial, R_M) := \left\{x \in M,\ d_g(x, \partial M) = d_g(x, \bar{\xi}) < R_M,\ d_g^\partial(q_i, \bar{\xi}) < R^\partial\right\}$$

form a covering of $(\partial M)_\rho$ and on every I_{q_i} the Fermi coordinates are well defined. In the following we choose, $R = \min\left\{R^\partial, R_M\right\}$, such that we have a finite covering

$$M \subset \left\{\cup_{i=1}^k B(q_i, R)\right\} \bigcup \left\{\cup_{i=k+1}^l I_{\xi_i}(R, R)\right\}$$

where $k, l \in \mathbb{N}$, $q_i \in M \smallsetminus \partial M$, and $\xi_i \in \partial M$.

Given the Fermi coordinates in a neighborhood of p, and we denoted by the matrix $(h_{ij})_{i,j=1,\dots,n-1}$ the second fundamental form, we have the well-known formulas (see [3, 12])

$$g^{ij}(y) = \delta_{ij} + 2h_{ij}(0)y_n + O(|y|^2) \text{ for } i,j = 1,\dots n-1 \tag{3}$$

$$g^{in}(y) = \delta_{in} \tag{4}$$

$$\sqrt{g}(y) = 1 - (n-1)H(0)y_n + O(|y|^2) \tag{5}$$

where $(y_1, \dots, y_n)$ are the Fermi coordinates and the mean curvature H is

$$H = \frac{1}{n-1}\sum_i^{n-1} h_{ii} \tag{6}$$

To solve our system, using an idea of Benci and Fortunato [2], we reduce the system to a single equation. We introduce the map ψ defined by the equation

$$\begin{cases} -\Delta_g\psi + (1+q^2u^2)\psi = qu^2 & \text{in } M \\ \frac{\partial\psi}{\partial\nu} = 0 & \text{on } \partial M \end{cases} \tag{7}$$

in case of Neumann boundary condition or by

$$\begin{cases} -\Delta_g\psi + qu^2\psi = qu^2 & \text{in } M \\ \psi = 0 & \text{on } \partial M \end{cases} \tag{8}$$

in case of Dirichlet boundary condition.

In what follows we call $H = H_g^1$ for the Neumann problem and $H = H_{0,g}^1$ for the Dirichlet problem. Thus with abuse of language we will say that $\psi : H \to H$ in both (7) and (8). Moreover, from standard variational arguments, it easy to see that ψ is well defined in H and it holds

$$0 \le \psi(u) \le 1/q \tag{9}$$

for all $u \in H$. We collect now some well-known result on the map ψ. For a more extensive presentation of these properties, we refer to [11]

Lemma 2.2. *The map $\psi : H \to H$ is C^2 and its differential $\psi'(u)[h] = V_u[h]$ at u is the map defined by*

$$-\Delta_g V_u[h] + (1+q^2u^2)V_u[h] = 2qu(1-q\psi(u))h \textit{ for all } h \in H. \tag{10}$$

in case of Neumann boundary condition or

$$-\Delta_g V_u[h] + q^2u^2V_u[h] = 2qu(1-q\psi(u))h \textit{ for all } h \in H. \tag{11}$$

in case of Dirichlet boundary condition.

Also, we have

$$0 \leq \psi'(u)[u] \leq \frac{2}{q}.$$

Finally, the second derivative $(h,k) \to \psi''(u)[h,k] = T_u(h,k)$ *is the map defined by the equation*

$$-\Delta_g T_u(h,k) + (1 + q^2 u^2) T_u(h,k) = -2q^2 u(kV_u(h) + hV_u(k)) + 2q(1 - q\psi(u))hk$$

in case of Neumann boundary condition or

$$-\Delta_g T_u(h,k) + q^2 u^2 T_u(h,k) = -2q^2 u(kV_u(h) + hV_u(k)) + 2q(1 - q\psi(u))hk$$

in case of Dirichlet boundary condition.

Lemma 2.3. *The map* $\Theta : H \to \mathbb{R}$ *given by*

$$\Theta(u) = \frac{1}{2} \int_M (1 - q\psi(u))u^2 d\mu_g$$

is C^2 *and*

$$\Theta'(u)[h] = \int_M (1 - q\psi(u))^2 uh d\mu_g$$

for any $u, h \in H$

For the proofs of these results we refer to, in which the case of KGMP is treated. For KGM systems, the proof is identical.

Now, we introduce the functionals $I_\varepsilon, J_\varepsilon, G_\varepsilon : H \to \mathbb{R}$

$$I_\varepsilon(u) = J_\varepsilon(u) + \frac{\omega^2}{2} G_\varepsilon(u), \tag{12}$$

where

$$J_\varepsilon(u) := \frac{1}{2\varepsilon^n} \int\limits_M \left[\varepsilon^2 |\nabla_g u|^2 + (a - \omega^2)u^2\right] d\mu_g - \frac{1}{p\varepsilon^n} \int\limits_M \left(u^+\right)^p d\mu_g \tag{13}$$

and

$$G_\varepsilon(u) := \frac{1}{\varepsilon^n} q \int_M \psi(u) u^2 d\mu_g. \tag{14}$$

By Lemma 2.3 we deduce that

$$\frac{1}{2}G'_{\varepsilon}(u)[\varphi] = \frac{1}{\varepsilon^n}\int_M [2q\psi(u) - q^2\psi^2(u)]u\varphi d\mu_g. \tag{15}$$

If $u \in H$ is a critical point of I_ε, then the pair $(u, \psi(u))$ is the desired solution of Problem (1) or (2).

Finally, we introduce a model function for the solution u. It is well known that, in $\mathbb{R}^n$, there is a unique positive radially symmetric function $V(z) \in H^1(\mathbb{R}^n)$ satisfying

$$-\Delta V + (a - \omega^2)V = V^{p-1} \text{ on } \mathbb{R}^n. \tag{16}$$

Moreover, the function V exponentially decays at infinity as well as its derivative, that is, for some $c > 0$

$$\lim_{|z|\to\infty} V(|z|)|z|^{\frac{n-1}{2}}e^{|z|} = c \quad \lim_{|z|\to\infty} V'(|z|)|z|^{\frac{n-1}{2}}e^{|z|} = -c.$$

We can define on the half space $\mathbb{R}^n_+ = \{(z_1, \dots, z_n) \in \mathbb{R}^n\ ,\ z_n \geq 0\}$ the function

$$U(x) = V|_{x_n \geq 0}\,.$$

The function Usatisfies the following Neumann problem in $\mathbb{R}^n_+$

$$\begin{cases} -\Delta U + (a - \omega^2)U = U^{p-1} & \text{in } \mathbb{R}^n_+ \\ \dfrac{\partial U}{\partial z_n} = 0 & \text{on } \{z_n = 0\}\,. \end{cases} \tag{17}$$

and it is easy to see that the space solution of the linearized problem

$$\begin{cases} -\Delta\varphi + (a - \omega^2)\varphi = (p-1)U^{p-2}\varphi & \text{in } \mathbb{R}^n_+ \\ \dfrac{\partial \varphi}{\partial z_n} = 0 & \text{on } \{z_n = 0\}\,. \end{cases} \tag{18}$$

is generated by the linear combination of

$$\varphi^i = \frac{\partial U}{\partial z_i}(z) \text{ for } i = 1, \dots, n-1.$$

We endow $H^1_g(M)$ with the scalar product $\langle u, v\rangle_\varepsilon := \frac{1}{\varepsilon^n}\int_M \varepsilon^2\nabla_g u\nabla_g v + (a-\omega^2)$ $uvd\mu_g$ and the norm $\|u\|_\varepsilon = \langle u, u\rangle_\varepsilon^{1/2}$. We call H_ε the space H^1_g equipped with the norm $\|\cdot\|_\varepsilon$. We also define L^p_ε as the space $L^p_g(M)$ endowed with the norm $|u|_{\varepsilon,p} = \frac{1}{\varepsilon^n}\left(\int_M u^p d\mu_g\right)^{1/p}$.

For any $p \in [2, 2^*)$, the embedding $i_\varepsilon : H_\varepsilon \hookrightarrow L_{\varepsilon,p}$ is a compact, continuous map, and it holds $|u|_{\varepsilon,p} \leq c\|u\|_\varepsilon$ for some constant c not depending on ε. We define the adjoint operator $i_\varepsilon^* : L_{\varepsilon,p'} :\hookrightarrow H_\varepsilon$ as

$$u = i_\varepsilon^*(v) \Leftrightarrow \langle u, \varphi \rangle_\varepsilon = \frac{1}{\varepsilon^n} \int_M v\varphi d\mu_g.$$

Now on set

$$f(u) = |u^+|^{p-1}$$

and

$$g(u) := \left(q^2\psi^2(u) - 2q\psi(u)\right) u.$$

we can rewrite problem (1) in an equivalent formulation

$$u = i_\varepsilon^* \left[f(u) + \omega^2 g(u)\right], \ u \in H_\varepsilon.$$

Remark 2.4. We have that $\|i_\varepsilon^*(v)\|_\varepsilon \leq c|v|_{p',\varepsilon}$ with c independent by ε.

Remark 2.5. We recall the following two estimates, that can be obtained by trivial computations

$$\|u\|_{H_g^1} \leq c\varepsilon^{\frac{1}{2}} \|u\|_\varepsilon \text{ for } n = 3 \tag{19}$$

$$\|u\|_{H_g^1} \leq c\varepsilon \|u\|_\varepsilon \text{ for } n = 4 \tag{20}$$

We often will use the estimate (19) also when $n = 4$, which is still true even if weaker, to simplify the exposition.

Finally, we define an important class of functions on the manifold, modeled on the function U. For all $\xi \in \partial M$ we define

$$W_{\varepsilon,\xi} = \begin{cases} U_\varepsilon\left(\left(\Psi_\xi^\partial\right)^{-1}(x)\right) \chi_R\left(\left(\Psi_\xi^\partial\right)^{-1}(x)\right) & x \in I_\xi(R) := I_\xi(R, R); \\ 0 & \text{elsewhere.} \end{cases}$$

We recall a fundamental limit property for the function $W_{\varepsilon,\xi}$.

Remark 2.6. Since U decays exponentially, it holds, uniformly with respect to $q \in \partial M$,

$$\lim_{\varepsilon \to 0} \left|W_{\varepsilon,\xi}\right|_{t,\varepsilon}^t = \int_{\mathbb{R}_+^n} U^t(z)dz \tag{21}$$

for all $1 \le t \le 2^*$, and

$$\lim_{\varepsilon\to 0} \varepsilon^2 \left|\nabla_g W_{\varepsilon,\xi}\right|^2_{2,\varepsilon} = \int_{\mathbb{R}^n_+} |\nabla U|^2 (z) dz \tag{22}$$

We also have the following estimate for the function ψ and for its differential ψ'.

Lemma 2.7. *It holds, for any $\varphi \in H$ and for any $\xi \in \partial M$*

$$\|\psi(W_{\varepsilon,\xi} + \varphi)\|_H \le c_1 \left(\varepsilon^{\frac{n+2}{2}} + \|\varphi\|^2_H\right) \tag{23}$$

$$\|\psi(W_{\varepsilon,\xi} + \varphi)\|_H \le c_2 \varepsilon^{\frac{n+2}{2}} \left(1 + \|\varphi\|^2_\varepsilon\right) \tag{24}$$

for some positive constants c_1, c_2, when ε is sufficiently small.

Proof. We prove the claim for the Neumann boundary condition. For the Dirichlet boundary condition the proof is completely analogous taking into account the gradient norm on H.

To simplify the notations we set $v = \psi(W_{\varepsilon,\xi} + \varphi)$. By definition of ψ we have

$$\begin{aligned}\|v\|^2_H &\le \int_M |\nabla_g v|^2 + v^2 + q^2 (W_{\varepsilon,\xi} + \varphi)^2 v^2 = q \int (W_{\varepsilon,\xi} + \varphi)^2 v \\ &\le \left(\int_M v^{2^*}\right)^{\frac{1}{2^*}} \left(\int_M (W_{\varepsilon,\xi} + \varphi)^{\frac{4n}{n+2}}\right)^{\frac{n+2}{2n}} \le c\|v\|_{H^1_g} \left|W_{\varepsilon,\xi} + \varphi\right|^2_{\frac{4n}{n+2},g} \\ &\le c\|v\|_{H^1_g} \left(\left|W_{\varepsilon,\xi}\right|^2_{\frac{4n}{n+2},g} + |\varphi|^2_{\frac{4n}{n+2},g}\right)\end{aligned}$$

Thus $\|v\|_H \le c\left(\left|W_{\varepsilon,\xi}\right|_{\frac{4n}{n+2},g} + |\varphi|_{\frac{4n}{n+2},g}\right)$. Taking into account (21) of Remark 2.6 we have that, for ε small $\left|W_{\varepsilon,\xi}\right|^2_{\frac{4n}{n+2},g} \le C\varepsilon^{\frac{2n}{n+2}} |U|^2_{\frac{4n}{n+2},g}$. Thus we have

$$\|v\|_{H^1_g} \le c_1 \left(\varepsilon^{\frac{2n}{n+2}} + |\varphi|^2_{\frac{4n}{n+2},g}\right) \le c_1 (\varepsilon^{\frac{2n}{n+2}} + \|\varphi\|^2_{H^1_g}) \tag{25}$$

and

$$\|v\|_{H^1_g} \le c_2 \varepsilon^{\frac{2n}{n+2}} \left(1 + |\varphi|^2_{\frac{4n}{n+2},\varepsilon}\right) \le c_2 \varepsilon^{\frac{2n}{n+2}} \left(1 + \|\varphi\|^2_\varepsilon\right). \tag{26}$$

that prove (23) and (24).For any $\xi \in M$ and $h, k \in H^1_g$ it holds. □

Lemma 2.8. *It holds, for any $h, k \in H$ and for any $\xi \in \partial M$*

$$\|\psi'(W_{\varepsilon,\xi} + k)[h]\|_H \le c\|h\|_H \left\{\varepsilon^2 + \|k\|_H\right\}$$

for some positive constant c when ε is sufficiently small.

Proof. Again, we prove the claim for the Neumann boundary condition being the other case completely analogous. By (10) and since $0 < \psi < 1/q$,

$$\begin{aligned}\|\psi'(W_{\varepsilon,\xi}+k)[h]\|^2_{H^1_g} &= 2q\int_M (W_{\varepsilon,\xi}+k)(1-q\psi(W_{\varepsilon,\xi}+k))h\psi'(W_{\varepsilon,\xi}+k)[h] \\ &\quad -q^2\int_M (W_{\varepsilon,\xi}+k)^2(\psi'(W_{\varepsilon,\xi}+k)[h])^2 \\ &\le \int_M W_{\varepsilon,\xi}|h|\left|\psi'(W_{\varepsilon,\xi}+k)[h]\right| + \int_M |k||h|\left|\psi'(W_{\varepsilon,\xi}+k)[h]\right| \\ &:= I_1 + I_2\end{aligned}$$

We estimate the two terms I_1 and I_2 separately. We have

$$I_1 \le \left|\psi'(W_{\varepsilon,\xi}+k)[h]\right|_{2^*,g} |h|_{2^*,g} \left|W_{\varepsilon,\xi}\right|_{\frac{2}{n},g} \le \varepsilon^2\|\psi'\|_{H^1_g}\|h\|_{H^1_g}\left|W_{\varepsilon,\xi}\right|_{\frac{n}{2},\varepsilon}$$

$$I_2 \le \|k\|_{L^3_g}\|h\|_{L^3_g}\|\psi'(W_{\varepsilon,\xi}+k)[h]\|_{L^3_g} \le \|k\|_{H^1_g}\|h\|_{H^1_g}\|\psi'\|_{H^1_g}$$

and, in light of Remark 2.6, we obtain the claim. □

2.1 The Lyapunov Schmidt reduction

We want to split the space H_ε in a finite dimensional space generated by the solution of (18) and its orthogonal complement. Fixed $\xi \in \partial M$ and $R > 0$, we consider on the manifold the functions

$$Z^i_{\varepsilon,\xi} = \begin{cases} \varphi^i_\varepsilon\left(\left(\psi^\partial_\xi\right)^{-1}(x)\right)\chi_R\left(\left(\psi^\partial_\xi\right)^{-1}(x)\right) & x \in I_\xi(R) := I_\xi(R,R); \\ 0 & \text{elsewhere.}\end{cases} \tag{27}$$

where $\varphi^i_\varepsilon(z) = \varphi^i\left(\frac{z}{\varepsilon}\right)$ and $\chi_R : B^{n-1}(0,R)\times[0,R) \to \mathbb{R}^+$ is a smooth cut-off function such that $\chi_R \equiv 1$ on $B^{n-1}(0,R/2)\times[0,R/2)$ and $|\nabla\chi| \le 2$.

In the following, for the sake of simplicity, we denote

$$D^+(R) = B^{n-1}(0,R)\times[0,R) \subset \mathbb{R}^n_+ \tag{28}$$

Let

$$K_{\varepsilon,\xi} := \mathrm{Span}\left\{Z^1_{\varepsilon,\xi},\cdots,Z^{n-1}_{\varepsilon,\xi}\right\}.$$

We can split H_ε in the sum of the $(n-1)$-dimensional space and its orthogonal complement with respect of $\langle\cdot,\cdot\rangle_\varepsilon$, i.e.

$$K_{\varepsilon,\xi}^{\perp} := \left\{ u \in H_\varepsilon \,,\ \left\langle u, Z_{\varepsilon,\xi}^{i} \right\rangle_\varepsilon = 0. \right\}.$$

We solve problem (1) by a Lyapunov Schmidt reduction: we look for a function of the form $W_{\varepsilon,\xi} + \phi$ with $\phi \in K_{\varepsilon,\xi}^{\perp}$ such that

$$\Pi_{\varepsilon,\xi}^{\perp} \left\{ W_{\varepsilon,\xi} + \phi - i_\varepsilon^* \left[f\left(W_{\varepsilon,\xi} + \phi\right) + \omega^2 g\left(W_{\varepsilon,\xi} + \phi\right)\right]\right\} = 0 \tag{29}$$

$$\Pi_{\varepsilon,\xi} \left\{ W_{\varepsilon,\xi} + \phi - i_\varepsilon^* \left[f\left(W_{\varepsilon,\xi} + \phi\right) + \omega^2 g\left(W_{\varepsilon,\xi} + \phi\right)\right]\right\} = 0 \tag{30}$$

where $\Pi_{\varepsilon,\xi} : H_\varepsilon \to K_{\varepsilon,\xi}$ and $\Pi_{\varepsilon,\xi}^{\perp} : H_\varepsilon \to K_{\varepsilon,\xi}^{\perp}$ are, respectively, the projection on $K_{\varepsilon,\xi}$ and $K_{\varepsilon,\xi}^{\perp}$. We see that $W_{\varepsilon,\xi} + \phi$ is a solution of (1) if and only if $W_{\varepsilon,\xi} + \phi$ solves (29–30).

3 Reduction to finite dimensional space

In this section we find a solution for equation (29). In particular, we prove that for all $\varepsilon > 0$ and for all $\xi \in \partial M$ there exists $\phi_{\varepsilon,\xi} \in K_{\varepsilon,\xi}^{\perp}$ solving (29). The main part of the reduction is performed in [13] and in [20]. Here we explicitly estimate only the term appearing in this specific contest.

We can rewrite equation (29) as

$$L_{\varepsilon,\xi}(\phi) = N_{\varepsilon,\xi}(\phi) + R_{\varepsilon,\xi} + S_{\varepsilon,\xi}(\phi)$$

where $L_{\varepsilon,\xi}$ is the linear operator

$$L_{\varepsilon,\xi} \ :\ K_{\varepsilon,\xi}^{\perp} \to K_{\varepsilon,\xi}^{\perp}$$
$$L_{\varepsilon,\xi}(\phi) := \Pi_{\varepsilon,\xi}^{\perp} \left\{ \phi - i_\varepsilon^* \left[f'(W_{\varepsilon,\xi})\phi \right] \right\},$$

$N_{\varepsilon,\xi}(\phi)$ is the nonlinear term

$$N_{\varepsilon,\xi} := \Pi_{\varepsilon,\xi}^{\perp} \left\{ i_\varepsilon^* \left[f(W_{\varepsilon,\xi} + \phi) - f(W_{\varepsilon,\xi}) - f'(W_{\varepsilon,\xi})\phi \right] \right\}$$

$R_{\varepsilon,\xi}$ is a remainder term

$$R_{\varepsilon,\xi} := \Pi_{\varepsilon,\xi}^{\perp} \left\{ i_\varepsilon^* \left[f(W_{\varepsilon,\xi}) \right] - W_{\varepsilon,\xi} \right\}$$

and $S_{\varepsilon,\xi}$ is the coupling term

$$S_{\varepsilon,\xi} = \Pi^{\perp}_{\varepsilon,\xi} \left\{ i^*_{\varepsilon} \left[\omega^2 g \left(W_{\varepsilon,\xi} + \phi \right) \right] \right\}.$$

Proposition 3.1. *There exists $\varepsilon_0 > 0$ and $C > 0$ such that for any $\xi \in \partial M$ and for all $\varepsilon \in (0, \varepsilon_0)$ there exists a unique $\phi_{\varepsilon,\xi} = \phi(\varepsilon, \xi) \in K^{\perp}_{\varepsilon,\xi}$ which solves (29). Moreover*

$$\|\phi_{\varepsilon,\xi}\|_{\varepsilon} < C\varepsilon^2.$$

Finally, $\xi \mapsto \phi_{\varepsilon,\xi}$ is a C^1 map.

In order to proof result, we premise some technical lemma.

Remark 3.2. We summarize here the results on $L_{\varepsilon,\xi}$, $N_{\varepsilon,\xi}$ and $R_{\varepsilon,\xi}$ contained in [13].
There exist ε_0 and $c > 0$ such that, for any $\xi \in \partial M$ and $\varepsilon \in (0, \varepsilon_0)$

$$\|L_{\varepsilon,\xi}\|_{\varepsilon} \geq c\|\phi\|_{\varepsilon} \text{ for any } \phi \in K^{\perp}_{\varepsilon,\xi}.$$

Also it holds

$$\|R_{\varepsilon,\xi}\|_{\varepsilon} \leq c\varepsilon^{1+\frac{n}{p'}}$$

and

$$\|N_{\varepsilon,\xi}(\phi)\|_{\varepsilon} \leq c \left(\|\phi\|^2_{\varepsilon} + \|\phi\|^{p-1}_{\varepsilon} \right)$$

We further remark that $\frac{n}{p'} > 1$ since $2 \leq p < 2^*$

We have now to estimate the coupling term $S_{\varepsilon,\xi}$.

Lemma 3.3. *If $\|\phi\|_{\varepsilon}, \|\phi_1\|_{\varepsilon}, \|\phi_2\|_{\varepsilon} = O(\varepsilon^2)$, it holds*

$$\|S_{\varepsilon,\xi}(\phi)\|_{\varepsilon} \leq c\varepsilon^2 \tag{31}$$

$$\|S_{\varepsilon,\xi}(\phi_1) - S_{\varepsilon,\xi}(\phi_2)\|_{\varepsilon} \leq l_{\varepsilon}\|\phi_1 - \phi_2\|_{\varepsilon} \tag{32}$$

where $l_{\varepsilon} \to 0$ as $\varepsilon \to 0$.

Proof. We have, by the properties of the map i^*_{ε}, that

$$\begin{aligned}
\|S_{\varepsilon,\xi}(\phi)\|_{\varepsilon} &\leq c \left| \psi^2 (W_{\varepsilon,\xi} + \phi)(W_{\varepsilon,\xi} + \phi) \right|_{\varepsilon,p'} + c \left| \psi (W_{\varepsilon,\xi} + \phi)(W_{\varepsilon,\xi} + \phi) \right|_{\varepsilon,p'} \\
&\leq c \left| \psi (W_{\varepsilon,\xi} + \phi)(W_{\varepsilon,\xi} + \phi) \right|_{\varepsilon,p'} \\
&\leq \frac{c}{\varepsilon^{\frac{n}{p'}}} \left(\int \psi (W_{\varepsilon,\xi} + \phi)^{2^*} \right)^{\frac{1}{2^*}} \left(\int |W_{\varepsilon,\xi} + \phi|^{p' \left(\frac{2^*}{p'} \right)'} \right)^{\frac{1}{p' \left(\frac{2^*}{p'} \right)'}}
\end{aligned}$$

$$\leq c\,\varepsilon^{-\frac{n}{p'}+\frac{n}{p'\left(\frac{2^*}{p'}\right)'}}\|\psi(W_{\varepsilon,\xi}+\phi)\|_H\left|W_{\varepsilon,\xi}+\phi\right|_{\varepsilon,p'\left(\frac{2^*}{p'}\right)'}$$

$$\leq\; c\varepsilon^{-\frac{n}{2^*}}\|\psi(W_{\varepsilon,\xi}+\phi)\|_H\leq c\varepsilon^{-\frac{n}{2^*}}\varepsilon^{\frac{n+2}{2}}=c\varepsilon^2$$

by (24) and taking into account that $\|\phi\|_\varepsilon=o(1)$ by Remark 2.5, and the first step is proved.

For the second claim, we have, since $0\leq\psi\leq 1/q$

$$\begin{aligned}\|S_{\varepsilon,\xi}(\phi_1)-S_{\varepsilon,\xi}(\phi_2)\|_\varepsilon \leq& c\left|\psi^2(W_{\varepsilon,\xi}+\phi_1)(W_{\varepsilon,\xi}+\phi_1)-\psi^2(W_{\varepsilon,\xi}+\phi_2)(W_{\varepsilon,\xi}+\phi_2)\right|_{\varepsilon,p'}\\ &+c\left|\psi(W_{\varepsilon,\xi}+\phi_1)(W_{\varepsilon,\xi}+\phi_1)-\psi(W_{\varepsilon,\xi}+\phi_2)(W_{\varepsilon,\xi}+\phi_2)\right|_{\varepsilon,p'}\\ \leq& c\left|\psi(W_{\varepsilon,\xi}+\phi_1)(W_{\varepsilon,\xi}+\phi_1)-\psi(W_{\varepsilon,\xi}+\phi_2)(W_{\varepsilon,\xi}+\phi_2)\right|_{\varepsilon,p'}\\ \leq& c\left|\left[\psi(W_{\varepsilon,\xi}+\phi_1)-\psi(W_{\varepsilon,\xi}+\phi_2)\right](W_{\varepsilon,\xi}+\phi_1)\right|_{\varepsilon,p'}\\ &+\left|\psi(W_{\varepsilon,\xi}+\phi_2)\left[\phi_1-\phi_2\right]\right|_{\varepsilon,p'}\\ \leq& c\left|\left[\psi'(W_{\varepsilon,\xi}+(1-\theta)\phi_1+\theta\phi_2)[\phi_1-\phi_2]\right](W_{\varepsilon,\xi}+\phi_1)\right|_{\varepsilon,p'}\\ &+\left|\psi(W_{\varepsilon,\xi}+\phi_2)\left[\phi_1-\phi_2\right]\right|_{\varepsilon,p'}:=D_1+D_2\end{aligned}$$

for some $\theta\in(0,1)$. Arguing as in the first part of the proof we get, in light of (24), that

$$D_2\leq c\varepsilon^{-\frac{n}{p^*}}\|\psi(W_{\varepsilon,\xi}+\phi)\|_H\left|\phi_1-\phi_2\right|_{\varepsilon,p'\left(\frac{2^*}{p'}\right)'}\leq c\varepsilon^{-\frac{n}{2^*}}\varepsilon^{\frac{n+2}{2}}\|\phi_1-\phi_2\|_\varepsilon$$

and, using Lemma 2.8, that

$$\begin{aligned}D_1&\leq c\varepsilon^{-\frac{n}{p'}}\left\|\left[\psi'(W_{\varepsilon,\xi}+(1-\theta)\phi_1+\theta\phi_2)[\phi_1-\phi_2]\right]\right\|_H\left|W_{\varepsilon,\xi}+\phi_1\right|_{\varepsilon,p'\left(\frac{2^*}{p'}\right)'}\\ &\leq c\varepsilon^{-\frac{n}{p'}}\left\{\varepsilon^2+(1-\theta)\|\phi_1\|_H+\theta\|\phi_2\|_H\right\}\|\phi_1-\phi_2\|_H.\end{aligned}$$

If $n=3$, by (19) and since $\|\phi_1\|_\varepsilon,\|\phi_2\|_\varepsilon=o(\varepsilon)$ by hypothesis we have

$$\begin{aligned}D_1&\leq c\varepsilon^{-\frac{3}{p'}}\left\{\varepsilon^2+\varepsilon^{1/2}(1-\theta)\|\phi_1\|_\varepsilon+\varepsilon^{1/2}\theta\|\phi_2\|_\varepsilon\right\}\varepsilon^{1/2}\|\phi_1-\phi_2\|_\varepsilon\\ &\leq c\varepsilon^{\frac{5}{2}-\frac{3}{p'}}\|\phi_1-\phi_2\|_\varepsilon\end{aligned}$$

and the claim is proved since $\frac{5}{2}-\frac{3}{p'}>0$ if $p'>\frac{6}{5}$ that is true since $p<6$. For $n=4$, analogously we have, by (20)

$$\begin{aligned}D_1&\leq c\varepsilon^{-\frac{4}{p'}}\left\{\varepsilon^2+\varepsilon(1-\theta)\|\phi_1\|_\varepsilon+\varepsilon\theta\|\phi_2\|_\varepsilon\right\}\varepsilon\|\phi_1-\phi_2\|_\varepsilon\\ &\leq c\varepsilon^{3-\frac{4}{p'}}\|\phi_1-\phi_2\|_\varepsilon\end{aligned}$$

and $3-\frac{4}{p'}>0$ iff $p'>\frac{4}{3}$ that is $p<4$. □

We can now prove the main result of this section.

Proof (Proof of Proposition 3.1). The proof is similar to Proposition 3.5 of [20], which we refer to for all details. We want to solve (29) by a fixed point argument. We define the operator

$$T_{\varepsilon,\xi} \ : \ K^{\perp}_{\varepsilon,\xi} \to K^{\perp}_{\varepsilon,\xi}$$
$$T_{\varepsilon,\xi}(\phi) = L^{-1}_{\varepsilon,\xi}\left(N_{\varepsilon,\xi}(\phi) + R_{\varepsilon,\xi}S_{\varepsilon,\xi}(\phi)\right)$$

By Remark 3.2 $T_{\varepsilon,\xi}$ is well defined and it holds

$$\|T_{\varepsilon,\xi}(\phi)\|_{\varepsilon} \le c\left(\|N_{\varepsilon,\xi}(\phi)\|_{\varepsilon} + \|R_{\varepsilon,\xi}\|_{\varepsilon} + \|S_{\varepsilon,\xi}(\phi)\|_{\varepsilon}\right)$$
$$\|T_{\varepsilon,\xi}(\phi_1) - T_{\varepsilon,\xi}(\phi_2)\|_{\varepsilon} \le c\left(\|N_{\varepsilon,\xi}(\phi_1) - N_{\varepsilon,\xi}(\phi_2)\|_{\varepsilon} + \|S_{\varepsilon,\xi}(\phi_1) - S_{\varepsilon,\xi}(\phi_2)\|_{\varepsilon}\right)$$

for some suitable constant $c > 0$. By the mean value theorem (and by the properties of i^*) we get

$$\|N_{\varepsilon,\xi}(\phi_1) - N_{\varepsilon,\xi}(\phi_2)\|_{\varepsilon} \le c\left|f'(W_{\varepsilon,\xi} + \phi_2 + t(\phi_1 - \phi_2)) - f'(W_{\varepsilon,\xi})\right|_{\frac{p}{p-2},\varepsilon} \|\phi_1 - \phi_2\|_{\varepsilon}.$$

By [20], Remark 3.4 we have that $\left|f'(W_{\varepsilon,\xi} + \phi_2 + t(\phi_1 - \phi_2)) - f'(W_{\varepsilon,\xi})\right|_{\frac{p}{p-2},\varepsilon} << 1$ provided $\|\phi_1\|_{\varepsilon}$ and $\|\phi_2\|_{\varepsilon}$ small enough. This, combined with (32) proves that there exists $0 < L < 1$ such that $\|T_{\varepsilon,\xi}(\phi_1) - T_{\varepsilon,\xi}(\phi_2)\|_{\varepsilon} \le L\|\phi_1 - \phi_2\|_{\varepsilon}$.

We recall that by Lemma 3.2 we have

$$\|N_{\varepsilon,\xi}(\phi)\|_{\varepsilon} \le c\left(\|\phi\|^2_{\varepsilon} + \|\phi\|^{p-1}_{\varepsilon}\right)$$
$$\|R_{\varepsilon,\xi}\|_{\varepsilon} \le \varepsilon^{1+\frac{n}{p'}} = o(\varepsilon^2)$$

This, combined with (31) gives us

$$\|T_{\varepsilon,\xi}(\phi)\|_{\varepsilon} \le c\left(\|N_{\varepsilon,\xi}(\phi)\|_{\varepsilon} + \|R_{\varepsilon,\xi}\|_{\varepsilon} + \|S_{\varepsilon,\xi}(\phi)\|_{\varepsilon}\right)$$
$$\le c\left(\|\phi\|^2_{\varepsilon} + \|\phi\|^{p-1}_{\varepsilon} + \varepsilon^{1+\frac{n}{p'}} + c\varepsilon^2\right)$$

So, there exists a positive constant C such that $T_{\varepsilon,\xi}$ maps a ball of center 0 and radius $C\varepsilon^2$ in $K^{\perp}_{\varepsilon,\xi}$ into itself and it is a contraction. So there exists a fixed point $\phi_{\varepsilon,\xi}$ with norm $\|\phi_{\varepsilon,\xi}\|_{\varepsilon} \le C\varepsilon^2$.

The continuity of $\phi_{\varepsilon,\xi}$ with respect to ξ is standard. □

4 The reduced functional

In this section we define the reduced functional in a finite dimensional space and we solve equation (30). This leads us to the proof of main theorem.

We have introduced $I_\varepsilon(u)$ in the introduction. We now define the reduced functional

$$\tilde{I}_\varepsilon : \partial M \to \mathbb{R}$$
$$\tilde{I}_\varepsilon(\xi) = I_\varepsilon(W_{\varepsilon,\xi} + \phi_{\varepsilon,\xi})$$

where $\phi_{\varepsilon,\xi}$ is uniquely determined by Proposition 3.1.

Lemma 4.1. *Let ξ_0 be a critical point of $\tilde{I}_\varepsilon$, that is, if $\xi = \xi(y) = \exp^\partial_{\xi_0}(y)$, $y \in B^{n-1}(0, r)$, then*

$$\left(\frac{\partial}{\partial y_h}\tilde{I}_\varepsilon(\xi(y))\right)_{|y=0} = 0, \quad h = 1, \dots, n-1.$$

Thus the function $\phi_{\varepsilon,\xi} + W_{\varepsilon,\xi}$ solves equation (30).

Proof. The proof of this lemma is just a computation. □

Lemma 4.2. *It holds*

$$\tilde{I}_\varepsilon(\xi) = C - \varepsilon H(\xi) + o(\varepsilon)$$

C^1 uniformly with respect to $\xi \in \partial M$ as ε goes to zero. Here $H(\xi)$ is the mean curvature of the boundary ∂M at ξ.

To prove Lemma we study the asymptotic expansion of $\tilde{I}_\varepsilon(\xi)$ with respect to ε. We recall the result contained in [13].

Remark 4.3. It holds

$$\begin{aligned}\tilde{J}_\varepsilon(\xi) :&= J_\varepsilon(W_{\varepsilon,\xi} + \phi_{\varepsilon,\xi}) = J_\varepsilon(W_{\varepsilon,\xi}) + o(\varepsilon) \\ &= C - \varepsilon\alpha H(\xi) + o(\varepsilon)\end{aligned} \tag{33}$$

C^1 uniformly with respect to $\xi \in \partial M$ as ε goes to zero, where

$$C := \int_{\mathbb{R}^n_+} \frac{1}{2}|\nabla U(z)|^2 + \frac{1}{2}U^2(z) - \frac{1}{p}U^p(z)dz$$

$$\alpha := \frac{(n-1)}{2}\int_{\mathbb{R}^n_+}\left(\frac{U'(|z|)}{|z|}\right)^2 z_n^3 dz$$

In light of this result, it remains to estimate the coupling functional $\mathcal{G}_\varepsilon$ to prove Lemma 4.2. We split this proof into several lemmas.

Lemma 4.4. *It holds*

$$G_\varepsilon\left(W_{\varepsilon,\xi}+\phi_{\varepsilon,\xi}\right)-G_\varepsilon\left(W_{\varepsilon,\xi}\right)=o(\varepsilon) \tag{34}$$

$$\left[G'_\varepsilon\left(W_{\varepsilon,\xi_0}+\phi_{\varepsilon,\xi_0}\right)-G'_\varepsilon\left(W_{\varepsilon,\xi_0}\right)\right]\left[\left(\frac{\partial}{\partial y_h}W_{\varepsilon,\xi(y)}\right)_{|y=0}\right]=o(\varepsilon) \tag{35}$$

$$G'_\varepsilon\left(W_{\varepsilon,\xi(y)}+\phi_{\varepsilon,\xi(y)}\right)\left[\frac{\partial}{\partial y_h}\phi_{\varepsilon,\xi(y)}\right]=o(\varepsilon) \tag{36}$$

Proof. Let us prove (34). We have (for some $\theta\in[0,1]$)

$$\begin{aligned}
&G_\varepsilon\left(W_{\varepsilon,\xi}+\phi_{\varepsilon,\xi}\right)-G_\varepsilon\left(W_{\varepsilon,\xi}\right)\\
&\quad=\frac{1}{\varepsilon^n}\int_M\left[\psi\left(W_{\varepsilon,\xi}+\phi_{\varepsilon,\xi}\right)\left(W_{\varepsilon,\xi}+\phi_{\varepsilon,\xi}\right)^2-\psi\left(W_{\varepsilon,\xi}\right)\left(W_{\varepsilon,\xi}\right)^2\right]\\
&\qquad=\frac{1}{\varepsilon^n}\int_M\psi'\left(W_{\varepsilon,\xi}+\theta\phi_{\varepsilon,\xi}\right)[\phi_{\varepsilon,\xi}]\left(W_{\varepsilon,\xi}\right)^2\\
&\qquad\quad+\frac{1}{\varepsilon^n}\int_M\psi\left(W_{\varepsilon,\xi}+\phi_{\varepsilon,\xi}\right)\left(2\phi_{\varepsilon,\xi}W_{\varepsilon,\xi}+\phi^2_{\varepsilon,\xi}\right):=I_1+I_2.
\end{aligned}$$

By Lemma 2.8 and Remark 2.6 we have

$$\begin{aligned}
I_1\le&\frac{1}{\varepsilon^n}\left(\int_M\left(\psi'\left(W_{\varepsilon,\xi}+\theta\phi_{\varepsilon,\xi}\right)[\phi_{\varepsilon,\xi}]\right)^2d\mu_g\right)^{\frac12}\left(\int_M W^4_{\varepsilon,\xi}d\mu_g\right)^{\frac12}\\
\le&\frac{\varepsilon^{\frac n2}}{\varepsilon^n}\|\psi'\left(W_{\varepsilon,\xi}+\theta\phi_{\varepsilon,\xi}\right)[\phi_{\varepsilon,\xi}]\|_H\left|W_{\varepsilon,\xi}\right|^2_{\varepsilon,2}\\
\le&\varepsilon^{-\frac n2}\left(\varepsilon^2\|\phi_{\varepsilon,\xi}\|_H+\|\phi_{\varepsilon,\xi}\|^2_H\right)\le\varepsilon^{\frac{9-n}{2}}=o(\varepsilon).
\end{aligned}$$

since $\|\phi_{\varepsilon,\xi}\|_H\le\varepsilon^{1/2}\|\phi_{\varepsilon,\xi}\|_\varepsilon\le\varepsilon^{5/2}$ by Proposition 3.1.

For I_2 we have, by (24) and Remark 2.6 in a similar way we get

$$I_2\le\frac{1}{\varepsilon^n}\left(\int_M\psi^2\left(W_{\varepsilon,\xi}+\phi_{\varepsilon,\xi}\right)d\mu_g\right)^{\frac12}\left(\int_M\phi^4_{\varepsilon,\xi}d\mu_g\right)^{\frac12}$$

$$+\frac{1}{\varepsilon^n}\left(\int\limits_M \psi^3\left(W_{\varepsilon,\xi}+\phi_{\varepsilon,\xi}\right)d\mu_g\right)^{\frac{1}{3}}\left(\int\limits_M \phi^3_{\varepsilon,\xi}d\mu_g\right)^{\frac{1}{3}}\left(\int\limits_M W^3_{\varepsilon,\xi}d\mu_g\right)^{\frac{1}{3}}$$

$$\leq\frac{1}{\varepsilon^n}\left\|\psi\left(W_{\varepsilon,\xi}+\phi_{\varepsilon,\xi}\right)\right\|_H\|\phi_{\varepsilon,\xi}\|^2_H+$$

$$+\frac{\varepsilon^{\frac{n}{3}}}{\varepsilon^n}\left\|\psi\left(W_{\varepsilon,\xi}+\phi_{\varepsilon,\xi}\right)\right\|_H\|\phi_{\varepsilon,\xi}\|_H\left|W_{\varepsilon,\xi}\right|_{\varepsilon,3}$$

$$\leq\varepsilon^{-n+\frac{n+2}{2}+5}+\varepsilon^{-\frac{2}{3}n+\frac{n+2}{2}+\frac{5}{2}}=\varepsilon^{\frac{12-n}{2}}+\varepsilon^{\frac{21-n}{6}}=o(\varepsilon)$$

since $n=3,4$. Then (34) follows.

Let us prove (35). Since $0\leq\psi\leq 1/q$ we have

$$\left[G'_\varepsilon\left(W_{\varepsilon,\xi_0}+\phi_{\varepsilon,\xi_0}\right)-G'_\varepsilon\left(W_{\varepsilon,\xi_0}\right)\right]\left[\left(\frac{\partial}{\partial y_h}W_{\varepsilon,\xi(y)}\right)_{|y=0}\right]$$

$$\leq\left|\frac{c}{\varepsilon^n}\int_M\left\{\psi\left(W_{\varepsilon,\xi}+\phi_{\varepsilon,\xi}\right)-\psi\left(W_{\varepsilon,\xi}\right)\right\}W_{\varepsilon,\xi_0}\left(\frac{\partial}{\partial y_h}W_{\varepsilon,\xi(y)}\right)_{|y=0}\right|$$

$$+\left|\frac{c}{\varepsilon^n}\int_M\left\{\psi^2\left(W_{\varepsilon,\xi}+\phi_{\varepsilon,\xi}\right)-\psi^2\left(W_{\varepsilon,\xi}\right)\right\}W_{\varepsilon,\xi_0}\left(\frac{\partial}{\partial y_h}W_{\varepsilon,\xi(y)}\right)_{|y=0}\right|$$

$$+\left|\frac{c}{\varepsilon^n}\int_M\psi\left(W_{\varepsilon,\xi}+\phi_{\varepsilon,\xi}\right)\phi_{\varepsilon,\xi_0}\left(\frac{\partial}{\partial y_h}W_{\varepsilon,\xi(y)}\right)_{|y=0}\right|$$

$$+\left|\frac{c}{\varepsilon^n}\int_M\psi^2\left(W_{\varepsilon,\xi}+\phi_{\varepsilon,\xi}\right)\phi_{\varepsilon,\xi_0}\left(\frac{\partial}{\partial y_h}W_{\varepsilon,\xi(y)}\right)_{|y=0}\right|$$

$$\leq\left|\frac{c}{\varepsilon^n}\int_M\left\{\psi\left(W_{\varepsilon,\xi}+\phi_{\varepsilon,\xi}\right)-\psi\left(W_{\varepsilon,\xi}\right)\right\}W_{\varepsilon,\xi_0}\left(\frac{\partial}{\partial y_h}W_{\varepsilon,\xi(y)}\right)_{|y=0}\right|$$

$$+\left|\frac{c}{\varepsilon^n}\int_M\psi\left(W_{\varepsilon,\xi}+\phi_{\varepsilon,\xi}\right)\phi_{\varepsilon,\xi_0}\left(\frac{\partial}{\partial y_h}W_{\varepsilon,\xi(y)}\right)_{|y=0}\right|$$

$$\leq\left|\frac{c}{\varepsilon^n}\int_M\left\{\psi'\left(W_{\varepsilon,\xi}+\theta\phi_{\varepsilon,\xi}\right)\left[\phi_{\varepsilon,\xi}\right]\right\}W_{\varepsilon,\xi_0}\left(\frac{\partial}{\partial y_h}W_{\varepsilon,\xi(y)}\right)_{|y=0}\right|$$

$$+\left|\frac{c}{\varepsilon^n}\int_M\psi\left(W_{\varepsilon,\xi}+\phi_{\varepsilon,\xi}\right)\phi_{\varepsilon,\xi_0}\left(\frac{\partial}{\partial y_h}W_{\varepsilon,\xi(y)}\right)_{|y=0}\right|:=D_1+D_2$$

for some $0<\theta<1$.

By Lemma 2.8, Remark 2.6, recalling that $\|\phi_{\varepsilon,\xi}\|_H \leq \varepsilon^{1/2}\|\phi_{\varepsilon,\xi}\|_H \leq \varepsilon^{5/2}$ and that $\left\|\frac{\partial}{\partial y_h}W_{\varepsilon,\xi(y)}\right\|_\varepsilon = O\left(\frac{1}{\varepsilon}\right)$ (cfr. eq (38)) we have

$$\begin{aligned}
D_1 \leq & \frac{c}{\varepsilon^n}\left(\int_M \{\psi'(W_{\varepsilon,\xi}+\theta\phi_{\varepsilon,\xi})[\phi_{\varepsilon,\xi}]\}^3\right)^{\frac{1}{3}}\left(\int_M W^3_{\varepsilon,\xi(y)}\right)^{\frac{1}{3}}\left(\int_M\left(\frac{\partial}{\partial y_h}W_{\varepsilon,\xi(y)}\right)^3\right)^{\frac{1}{3}} \\
\leq & c\frac{\varepsilon^{\frac{2}{3}n}}{\varepsilon^n}\left\|\psi'(W_{\varepsilon,\xi}+\theta\phi_{\varepsilon,\xi})[\phi_{\varepsilon,\xi}]\right\|_H \|W_{\varepsilon,\xi(y)}\|_\varepsilon\left\|\frac{\partial}{\partial y_h}W_{\varepsilon,\xi(y)}\right\|_\varepsilon \\
\leq & c\varepsilon^{-1-\frac{n}{3}}\left\|\psi'(W_{\varepsilon,\xi}+\theta\phi_{\varepsilon,\xi})[\phi_{\varepsilon,\xi}]\right\|_H \leq c\varepsilon^{-1-\frac{n}{3}}\|\phi_{\varepsilon,\xi}\|_H\left\{\varepsilon^2+\|\phi_{\varepsilon,\xi}\|_H\right\} \\
\leq & c\varepsilon^{-1-\frac{n}{3}}\varepsilon^{\frac{5}{2}}\varepsilon^2 = c\varepsilon^{\frac{7}{2}-\frac{n}{3}} = o(\varepsilon).
\end{aligned}$$

In a similar way, using (24) and the above estimates we get

$$\begin{aligned}
D_2 \leq & \frac{c}{\varepsilon^n}\left(\int_M \psi^3(W_{\varepsilon,\xi}+\phi_{\varepsilon,\xi})\right)^{\frac{1}{3}}\left(\int_M \phi^3_{\varepsilon,\xi_0}\right)^{\frac{1}{3}}\left(\int_M\left(\frac{\partial}{\partial y_h}W_{\varepsilon,\xi(y)}\right)^3\right)^{\frac{1}{3}} \\
\leq & c\frac{\varepsilon^{\frac{n}{3}}}{\varepsilon^n}\left\|\psi(W_{\varepsilon,\xi}+\phi_{\varepsilon,\xi})\right\|_H \left\|\phi_{\varepsilon,\xi}\right\|_H\left\|\frac{\partial}{\partial y_h}W_{\varepsilon,\xi(y)}\right\|_\varepsilon \\
\leq & c\varepsilon^{-\frac{2}{3}n-1}\varepsilon^{\frac{5}{2}}\left\|\psi(W_{\varepsilon,\xi}+\phi_{\varepsilon,\xi})\right\|_H \leq c\varepsilon^{-\frac{2}{3}n+\frac{3}{2}}\varepsilon^{\frac{n+2}{n}}\left(1+\|\phi_{\varepsilon,\xi}\|_\varepsilon\right) \\
\leq & c\varepsilon^{\frac{15-n}{6}} = o(\varepsilon)
\end{aligned}$$

and (35) is proved.

The proof of (36) requires to estimate that

$$I := \left|\frac{1}{\varepsilon^n}\int_M \left[q^2\psi^2\left(W_{\varepsilon,\xi(y)}+\phi_{\varepsilon,\xi(y)}\right)-2q\psi\left(W_{\varepsilon,\xi(y)}+\phi_{\varepsilon,\xi(y)}\right)\right]\times \times\left(W_{\varepsilon,\xi(y)}+\phi_{\varepsilon,\xi(y)}\right)Z^l_{\varepsilon,\xi(y)}\right| = o(\varepsilon), \tag{37}$$

where the functions $Z^l_{\varepsilon,\xi(y)}$ are defined in (27). By (37) it is possible to complete the proof the lemma, with the same arguments in the proof of (5.10) in [20], which we refer to for the missing details. To prove (37), since $0<\psi<1/q$, we get, as before,

$$I \leq \left|\frac{c}{\varepsilon^n}\int_M \psi\left(W_{\varepsilon,\xi(y)}+\phi_{\varepsilon,\xi(y)}\right)\left(W_{\varepsilon,\xi(y)}+\phi_{\varepsilon,\xi(y)}\right)Z^l_{\varepsilon,\xi(y)}\right| \leq$$

$$\leq c\frac{\varepsilon^{\frac{n+2}{2}}}{\varepsilon^n}\left(\int\limits_M \psi^{2^*}\left(W_{\varepsilon,\xi(y)}+\phi_{\varepsilon,\xi(y)}\right)\right)^{\frac{1}{2^*}}\left(\frac{1}{\varepsilon^n}\int\limits_M\left(W_{\varepsilon,\xi(y)}+\phi_{\varepsilon,\xi(y)}\right)^{\frac{4n}{n+2}}\right)^{\frac{n+2}{4n}}$$

$$\times\left(\frac{1}{\varepsilon^n}\int\limits_M\left(Z^l_{\varepsilon,\xi(y)}\right)^{\frac{4n}{n+2}}\right)^{\frac{n+2}{4n}}$$

$$\leq c\varepsilon^{\frac{2-n}{2}}\left\|\psi\left(W_{\varepsilon,\xi(y)}+\phi_{\varepsilon,\xi(y)}\right)\right\|_H\left|W_{\varepsilon,\xi(y)}+\phi_{\varepsilon,\xi(y)}\right|_{\varepsilon,\frac{4n}{n+2}}\left|Z^l_{\varepsilon,\xi(y)}\right|_{\varepsilon,\frac{4n}{n+2}}.$$

Arguing as in Remark 2.6, we have that $\left|Z^l_{\varepsilon,\xi(y)}\right|_{\varepsilon,\frac{4n}{n+2}} \to |\varphi^l|_{\frac{4n}{n+2}}$, so, by (24) we obtain

$$I \leq c\varepsilon^{\frac{2-n}{2}}\varepsilon^{\frac{n+2}{2}} = c\varepsilon^2.$$

This concludes the proof. □

Lemma 4.5. *It holds that*

$$G_\varepsilon(W_{\varepsilon,\xi}) := \frac{1}{\varepsilon^n}\int\limits_M \psi(W_{\varepsilon,\xi})W^2_{\varepsilon,\xi}d\mu_g = o(\varepsilon)$$

C^1–uniformly with respect to $\xi \in M$ as ε goes to zero.

Proof. At first we have, by Remark 2.6 and by (23)

$$G_\varepsilon(W_{\varepsilon,\xi}) \leq c\frac{1}{\varepsilon^n}\left(\int\limits_M \psi^3(W_{\varepsilon,\xi})\right)^{\frac{1}{3}}\left(\int\limits_M W^3_{\varepsilon,\xi}\right)^{\frac{2}{3}} \leq c\frac{1}{\varepsilon^n}\varepsilon^{\frac{n+2}{2}}\varepsilon^{\frac{2}{3}n} = c\varepsilon^{\frac{n}{6}+1} = o(\varepsilon).$$

We want now to prove the C^1 convergence, id est, if $\xi(y) = \exp_\xi(y)$ for $y \in B(0,r)$, we will prove that

$$\left.\frac{\partial}{\partial y_h}G_\varepsilon(W_{\varepsilon,\xi})\right|_{y=0} = \frac{2}{\varepsilon^n}\int_M\left(2q\psi(W_{\varepsilon,\xi})-q^2\psi^2(W_{\varepsilon,\xi})\right)W_{\varepsilon,\xi}\left[\left.\frac{\partial}{\partial y_h}W_{\varepsilon,\xi(h)}\right|_{y=0}\right]d\mu_g$$

for $h = 1,\dots,n-1$. Since $0 < \psi < 1/q$, immediately we have

$$\left|\left.\frac{\partial}{\partial y_h}G_\varepsilon(W_{\varepsilon,\xi})\right|_{y=0}\right| \leq c\left|\frac{1}{\varepsilon^n}\int\limits_M \psi(W_{\varepsilon,\xi(y)})W_{\varepsilon,\xi(h)}\left.\frac{\partial}{\partial y_h}W_{\varepsilon,\xi(h)}\right|_{y=0}d\mu_g\right|$$

Set I_1 the quantity inside the absolute value at the r.h.s. of the above equation. Using the Fermi coordinates and the previous estimates we get

$$\frac{1}{\varepsilon^2} I_1(\varepsilon,\xi) = \int_{\mathbb{R}^n_+} \frac{\tilde{v}_{\varepsilon,\xi}(z)}{\varepsilon^2} 2U(z)\chi_R(\varepsilon z)|g_\xi(\varepsilon z)|^{1/2} \times$$

$$\times \left\{ \sum_{k=1}^{3} \left[\frac{1}{\varepsilon}\frac{\partial U(z)}{\partial z_k}\chi_R(\varepsilon z) + U(z)\frac{\partial \chi_r(\varepsilon z)}{\partial z_k} \right] \frac{\partial}{\partial y_h}\mathscr{H}_k(0, \exp_\xi(\varepsilon z)) \right\} dz.$$

where $\mathscr{H}_k(x,y)$ is introduced in Definition (5.4). Since $|g_\xi(\varepsilon z)|^{1/2} = 1 + O(\varepsilon|z|)$ and by Lemma 5.6 we have

$$\begin{aligned} I_1(\varepsilon,\xi) =& 2\varepsilon \int_{\mathbb{R}^n_+} \tilde{v}_{\varepsilon,\xi}(z)U(z)\frac{\partial U(z)}{\partial z_h}\chi_R^2(\varepsilon z)dz + o(\varepsilon) \\ =& 2\varepsilon \int_{\mathbb{R}^n_+} \tilde{v}_{\varepsilon,\xi}(z)U(z)\frac{\partial U(z)}{\partial z_h}dz + o(\varepsilon) \end{aligned}$$

By Lemma 5.2 we have that $\left\{ \frac{1}{\varepsilon_n^2}\tilde{v}_{\varepsilon_n,\xi} \right\}_n$ converges to γ weakly in $L^{2^*}(\mathbb{R}^n_+)$, so we have

$$I_1(\varepsilon,\xi) = 2\varepsilon \int_{\mathbb{R}^n} \gamma U(z)\frac{\partial U(z)}{\partial z_h}dz + o(\varepsilon)$$

where $h = 1,\dots,n-1$. Finally, we have that $\int_{\mathbb{R}^n} \gamma(z)U(z)\frac{\partial U(z)}{\partial z_h}dz = 0$ because both γ (see Remark 5.3) and U are symmetric with respect to $z_1,\dots,z_{n-1}$ while $\frac{\partial U(z)}{\partial z_h}$ is antisymmetric. This concludes the proof. □

We can now prove Lemma 4.2.

Proof (Proof of Lemma 4.2). We want to estimate

$$I_\varepsilon(W_{\varepsilon,\xi} + \phi_{\varepsilon,\xi}) = J_\varepsilon(W_{\varepsilon,\xi} + \phi_{\varepsilon,\xi}) + \frac{\omega^2}{2}G_\varepsilon(W_{\varepsilon,\xi} + \phi_{\varepsilon,\xi}),$$

By Remark 4.3 we have that

$$J_\varepsilon(W_{\varepsilon,\xi} + \phi_{\varepsilon,\xi}) = J_\varepsilon(W_{\varepsilon,\xi}) + o(\varepsilon) = C - \varepsilon\alpha H(\xi) + o(\varepsilon)$$

C^1 uniformly with respect to $\xi \in \partial M$ as ε goes to zero. Moreover by Lemma 4.4 and by Lemma 4.5 we have that

$$G_\varepsilon(W_{\varepsilon,\xi} + \phi_{\varepsilon,\xi}) = o(\varepsilon)$$

C^1 uniformly with respect to $\xi \in \partial M$ and this concludes the proof. □

4.1 Sketch of the proof of Theorem 1.2

In section 3, Proposition 3.1 we found a function $\phi_{\varepsilon,\xi}$ solving (29). By Lemma 4.1 we can solve (30) once we have a critical point of functional $\tilde{I}_\varepsilon$. At this point by Lemma 4.2 and by definition of C^1 stable critical point (Def. 1.1) we can complete the proof.

5 Appendix: Technical lemmas

Lemma 5.1. *There exists $\varepsilon_0 > 0$ and $c > 0$ such that, for any $\xi_0 \in \partial M$ and for any $\varepsilon \in (0, \varepsilon_0)$ it holds*

$$\left\| \frac{\partial}{\partial y_h} Z^l_{\varepsilon,\xi(y)} \right\|_\varepsilon = O\left(\frac{1}{\varepsilon}\right), \quad \left\| \frac{\partial}{\partial y_h} W_{\varepsilon,\xi(y)} \right\|_\varepsilon = O\left(\frac{1}{\varepsilon}\right), \tag{38}$$

for $h = 1, \dots, n-1$, $l = 1, \dots, n$

Lemma 5.2. *Let us consider the functions*

$$\tilde{v}_{\varepsilon,\xi}(z) = \begin{cases} \psi(W_{\varepsilon,\xi}) \left(\Psi^\partial_\xi(\varepsilon z)\right) & \text{for } z \in D^+(R/\varepsilon) \\ 0 & \text{for } z \in \mathbb{R}^3 \smallsetminus D^+(R/\varepsilon) \end{cases}$$

where $D^+(r/\varepsilon) = \{z = (\bar{z}, z_n),\ \bar{z} \in \mathbb{R}^{n-1}, |\bar{z}| < r/\varepsilon,\ 0 \le z_n < R/\varepsilon)\}$. Then there exists a constant $c > 0$ such that

$$\|\tilde{v}_{\varepsilon,\xi}(z)\|_{L^{2^*}(\mathbb{R}^n_+)} \le c\varepsilon^2.$$

Furthermore, take a sequence $\varepsilon_n \to 0$, up to subsequences, $\left\{\frac{1}{\varepsilon_n^2}\tilde{v}_{\varepsilon_n,\xi}\right\}_n$ converges weakly in $L^{2^*}(\mathbb{R}^n_+)$ as ε goes to 0 to a function $\gamma \in D^{1,2}(\mathbb{R}^3)$. The function γ solves, in a weak sense, the equation

$$-\Delta\gamma = qU^2 \text{ in } \mathbb{R}^n_+ \tag{39}$$

Proof. We prove the Lemma for Problem (1), being the Problem (2) completely analogous. By definition of $\tilde{v}_{\varepsilon,\xi}(z)$ and by (1) we have, for all $z \in D^+(r/\varepsilon)$,

$$-\sum_{ij} \partial_j \left(|g_\xi(\varepsilon z)|^{1/2} g_\xi^{ij}(\varepsilon z) \partial_i \tilde{v}_{\varepsilon,\xi}(z) \right) =$$

$$= \varepsilon^2 |g_\xi(\varepsilon z)|^{1/2} \left\{ qU^2(z)\chi_r^2(\varepsilon z) - \left[1 + q^2U^2(z)\chi_R^2(\varepsilon z)\right] \tilde{v}_{\varepsilon,\xi}(z) \right\} \tag{40}$$

By (40), and remarking that $\tilde{v}_{\varepsilon,\xi}(z) \geq 0$ we have

$$\|\tilde{v}_{\varepsilon,\xi}(z)\|^2_{D^{1,2}(D^+(r/\varepsilon))} \leq C \int_{D^+(R/\varepsilon)} |g_\xi(\varepsilon z)|^{1/2} g_\xi^{ij}(\varepsilon z) \partial_i \tilde{v}_{\varepsilon,\xi}(z) \partial_j \tilde{v}_{\varepsilon,\xi}(z) dz$$

$$= C\varepsilon^2 \int_{D^+(R/\varepsilon)} |g_\xi(\varepsilon z)|^{1/2} \left\{ qU^2(z)\chi_R^2(\varepsilon z)\tilde{v}_{\varepsilon,\xi}(z) - \left[1 + q^2U^2(z)\chi_R^2(\varepsilon z)\right] \tilde{v}^2_{\varepsilon,\xi}(z) \right\} dz$$

$$\leq C\varepsilon^2 \int_{D^+(R/\varepsilon)} |g_\xi(\varepsilon z)|^{1/2} qU^2(z)\chi_R^2(\varepsilon|z|)\tilde{v}_{\varepsilon,\xi}(z) dz$$

$$\leq C\varepsilon^2 \|\tilde{v}_{\varepsilon,\xi}(z)\|_{L^{2^*}(D^+(R/\varepsilon))} \|U\|^2_{L^{\frac{4n}{n+2}}} \leq C\varepsilon^2 \|\tilde{v}_{\varepsilon,\xi}(z)\|_{D^{1,2}(D^+(R/\varepsilon))}$$

Thus we have

$$\|\tilde{v}_{\varepsilon,\xi}(z)\|_{D^{1,2}(D^+(R/\varepsilon))} \leq C\varepsilon^2 \text{ and } |\tilde{v}_{\varepsilon,\xi}(z)|_{L^{2^*}(\mathbb{R}^n_+)} \leq C\varepsilon^2. \tag{41}$$

By (41), if ε_n is a sequence which goes to zero, the sequence $\left\{ \frac{1}{\varepsilon_n^2} \tilde{v}_{\varepsilon_n,\xi} \right\}_n$ is bounded in $L^{2^*}(\mathbb{R}^n_+)$. Then, up to subsequence, $\left\{ \frac{1}{\varepsilon_n^2} \tilde{v}_{\varepsilon_n,\xi} \right\}_n$ converges to some $\tilde{\gamma} \in L^{2^*}(\mathbb{R}^n_+)$ weakly in $L^{2^*}(\mathbb{R}^n_+)$.

Moreover, by (40), for any $\varphi \in C_0^\infty(\mathbb{R}^n_+)$, it holds

$$\int_{\operatorname{supp}\varphi} \sum_{ij} |g_\xi(\varepsilon z)|^{1/2} g_\xi^{ij}(\varepsilon z) \partial_i \frac{\tilde{v}_{\varepsilon,\xi}(z)}{\varepsilon_n^2} \partial_j \varphi(z) dz =$$

$$\int_{\operatorname{supp}\varphi} \left\{ qU^2(z)\chi_r^2(\varepsilon|z|) - \left[1 + q^2U^2(z)\chi_R^2(\varepsilon z)\right] \tilde{v}_{\varepsilon,\xi}(z) \right\} |g_\xi(\varepsilon z)|^{1/2} \varphi(z) dz. \tag{42}$$

Consider now the functions

$$v_{\varepsilon,\xi}(z) := \psi(W_{\varepsilon,\xi}) \left(\Psi_\xi^\partial(\varepsilon z) \right) \chi_R(\varepsilon z) = \tilde{v}_{\varepsilon,\xi}(z) \chi_r(\varepsilon z) \text{ for } z \in \mathbb{R}^n_+.$$

We have immediately that $v_{\varepsilon,\xi}(z)$ is bounded in $D^{1,2}(\mathbb{R}^n_+)$, thus the sequence $\left\{\frac{1}{\varepsilon_n^2}v_{\varepsilon_n,\xi}\right\}_n$ converges to some $\gamma \in D^{1,2}(\mathbb{R}^3)$ weakly in $D^{1,2}(\mathbb{R}^n_+)$ and in $L^{2^*}(\mathbb{R}^n_+)$. Finally, for any compact set $K \subset \mathbb{R}^n_+$ eventually $v_{\varepsilon_n,\xi} \equiv \tilde{v}_{\varepsilon_n,\xi}$ on K. So it is easy to see that $\tilde{\gamma} = \gamma$.

We recall that $|g_\xi(\varepsilon z)|^{1/2} = 1 + O(\varepsilon|z|)$ and $g_\xi^{ij}(\varepsilon z) = \delta_{ij} + O(\varepsilon|z|)$ so, by the weak convergence of $\left\{\frac{1}{\varepsilon_n^2}v_{\varepsilon_n,\xi}\right\}_n$ in $D^{1,2}(\mathbb{R}^n_+)$, for any $\varphi \in C_0^\infty(\mathbb{R}^n_+)$ we get

$$\int_{\operatorname{supp}\varphi} \sum_{ij} |g_\xi(\varepsilon_n z)|^{1/2} g_\xi^{ij}(\varepsilon_n z)\partial_i \frac{\tilde{v}_{\varepsilon_n,\xi}(z)}{\varepsilon_n^2}\partial_j\varphi(z)dz$$

$$= \int_{\operatorname{supp}\varphi} \sum_{ij} |g_\xi(\varepsilon_n z)|^{1/2} g_\xi^{ij}(\varepsilon_n z)\partial_i \frac{v_{\varepsilon_n,\xi}(z)}{\varepsilon_n^2}\partial_j\varphi(z)dz$$

$$\to \int_{\mathbb{R}^3} \sum_i \partial_i\gamma(z)\partial_i\varphi(z)dz \text{ as } n \to \infty. \tag{43}$$

Thus by (42) and by (43) and because $\left\{\frac{1}{\varepsilon_n^2}\tilde{v}_{\varepsilon_n,\xi}\right\}_n$ converges to γ weakly in $L^{2^*}(\mathbb{R}^n_+)$ we get

$$\int_{\mathbb{R}^n_+} \sum_i \partial_i\gamma(z)\partial_i\varphi(z)dz = q\int_{\mathbb{R}^n_+} U^2(z)\varphi(z)dz \text{ for all } \varphi \in C_0^\infty(\mathbb{R}^n_+).$$

So, finally, up to subsequences, $\left\{\frac{1}{\varepsilon_n^2}\tilde{v}_{\varepsilon_n,\xi}\right\}_n$ converges to γ, weakly in $L^{2^*}(\mathbb{R}^n_+)$ and the function $\gamma \in D^{1,2}(\mathbb{R}^n_+)$ is a weak solution of $-\Delta\gamma = qU^2$ in $\mathbb{R}^n_+$. □

Remark 5.3. We remark that γ is positive and decays exponentially at infinity with its first derivative because it solves $-\Delta\gamma = qU^2$ in $\mathbb{R}^n_+$. Moreover it is symmetric with respect to the first $n-1$ variables.

Definition 5.4. Let $\xi_0 \in \partial M$. We introduce the functions $\mathscr{E}$ and $\tilde{\mathscr{E}}$ as follows.

$$\mathscr{E}(y,x) = \left(\exp^\partial_{\xi(y)}\right)^{-1}(x) = \left(\exp^\partial_{\exp^\partial_{\xi_0} y}\right)^{-1}(\exp^\partial_{\xi_0}\bar{\eta}) = \tilde{\mathscr{E}}(y,\bar{\eta})$$

where $x, \xi(y) \in \partial M$, $y, \bar{\eta} \in B(0,R) \subset \mathbb{R}^{n-1}$ and $\xi(y) = \exp^\partial_{\xi_0} y$, $x = \exp^\partial_{\xi_0}\bar{\eta}$. Using Fermi coordinates, in a similar way we define

$$\mathscr{H}(y,x) = \left(\psi^\partial_{\xi(y)}\right)^{-1}(x) = \left(\psi^\partial_{\exp^\partial_{\xi_0} y}\right)^{-1}\left(\psi^\partial_{\xi_0}(\bar{\eta},\eta_n)\right) = \tilde{\mathscr{H}}(y,\bar{\eta},\eta_n) = (\tilde{\mathscr{E}}(y,\bar{\eta}),\eta_n)$$

where $x \in M$, $\eta = (\bar{\eta}, \eta_n)$, with $\bar{\eta} \in B(0,R) \subset \mathbb{R}^{n-1}$ and $0 \leq \eta_n < R$, $\xi(y) = \exp^{\partial}_{\xi_0} y \in \partial M$ and $x = \psi^{\partial}_{\xi_0}(\eta)$.

Lemma 5.5. *It holds*

$$\frac{\partial \tilde{\mathscr{E}}_k}{\partial y_j}(0,0) = -\delta_{jk} \text{ for } j,k = 1,\ldots,n-1$$

Proof. We recall that $\tilde{\mathscr{E}}(y,\bar{\eta}) = \left(\exp^{\partial}_{\xi(y)}\right)^{-1}(\exp^{\partial}_{\xi_0}\bar{\eta})$. Let us introduce, for $y, \bar{\eta} \in B(0,R) \subset \mathbb{R}^{n-1}$

$$F(y,\bar{\eta}) = \left(\exp^{\partial}_{\xi_0}\right)^{-1}\left(\exp^{\partial}_{\xi(y)}(\bar{\eta})\right)$$
$$\Gamma(y,\bar{\eta}) = (y, F(y,\bar{\eta}))\,.$$

We notice that $\Gamma^{-1} = (y, \tilde{\mathscr{E}}(y,\bar{\eta}))$. We can easily compute the derivative of Γ. We have

$$\Gamma'(\hat{y},\hat{\eta})[\tilde{y},\tilde{\eta}] = \begin{pmatrix} \mathrm{Id}_{\mathbb{R}^{n-1}} & 0 \\ F'_y(\hat{y},\hat{\eta}) & F'_\eta(\hat{y},\hat{\eta}) \end{pmatrix}\begin{pmatrix} \tilde{y} \\ \tilde{\eta} \end{pmatrix},$$

thus

$$\left(\Gamma^{-1}\right)'(\hat{y},\hat{\eta})[\tilde{y},\tilde{\eta}] = \begin{pmatrix} \mathrm{Id}_{\mathbb{R}^{n-1}} & 0 \\ -\left(F'_\eta(\hat{y},\hat{\eta})\right)^{-1} F'_y(\hat{y},\hat{\eta}) & \left(F'_\eta(\hat{y},\hat{\eta})\right)^{-1} \end{pmatrix}\begin{pmatrix} \tilde{y} \\ \tilde{\eta} \end{pmatrix}$$

Now, by direct computation we have that

$$F'_\eta(0,\hat{\eta}) = \mathrm{Id}_{\mathbb{R}^{n-1}} \text{ and } F'_y(\hat{y},0) = \mathrm{Id}_{\mathbb{R}^{n-1}},$$

so $\frac{\partial \tilde{\mathscr{E}}_k}{\partial y_j}(0,0) = \left(-\left(F'_\eta(0,0)\right)^{-1} F'_y(0,0)\right)_{jk} = -\delta_{jk}$. □

Lemma 5.6. *We have that*

$$\tilde{\mathscr{H}}(0,\bar{\eta},\eta_n) = (\bar{\eta},\eta_n) \text{ for } \bar{\eta} \in \mathbb{R}^{n-1}, \eta_n \in \mathbb{R}_+$$
$$\frac{\partial \tilde{\mathscr{H}}_k}{\partial y_j}(0,0,\eta_n) = -\delta_{jk} \text{ for } j,k = 1,\ldots,n-1, \eta_n \in \mathbb{R}_+$$
$$\frac{\partial \tilde{\mathscr{H}}_n}{\partial y_j}(y,\bar{\eta},\eta_n) = 0 \text{ for } j = 1,\ldots,n-1, y,\bar{\eta} \in \mathbb{R}^{n-1}, \eta_n \in \mathbb{R}_+$$

Proof. The first two claims follow immediately by Definition 5.4 and Lemma 5.5. For the last claim, observe that $\tilde{\mathscr{H}}_k(y, \bar{\eta}, \eta_n) = \tilde{\mathscr{E}}_k(y, \bar{\eta})$ which does not depend on η_n as well as its derivatives. □

Acknowledgements M. Ghimenti was supported by Gruppo Nazionale per l'Analisi Matematica, la Probabilità e le loro Applicazioni (GNAMPA) of Istituto Nazionale di Alta Matematica (INdAM).

References

1. Azzollini, A., Pomponio, A.: Ground state solutions for the nonlinear Klein-Gordon-Maxwell equations. Topol. Methods Nonlinear Anal. **35**, 33–42 (2010)
2. Benci, V., Fortunato, D.: Solitary waves of the nonlinear Klein-Gordon equation coupled with the Maxwell equations. Rev. Math. Phys. **14**(4), 409–420 (2002)
3. Byeon, J., Park, J.: Singularly perturbed nonlinear elliptic problems on manifolds. Calc. Var. Partial Differ. Equ. **24**(4), 459–477 (2005)
4. Cassani, D.: Existence and non-existence of solitary waves for the critical Klein-Gordon equation coupled with Maxwell's equations. Nonlinear Anal. **58**(7–8), 733–747 (2004)
5. Clapp, M., Ghimenti, M., Micheletti, A.M.: Semiclassical states for a static supercritical Klein-Gordon-Maxwell-Proca system on a closed riemannian manifold. http://arxiv.org/abs/1401.5406 (2013)
6. D'Aprile, T., Mugnai, D.: Non-existence results for the coupled Klein-Gordon-Maxwell equations. Adv. Nonlinear Stud. **4**, 307–322 (2004)
7. D'Aprile, T., Wei, J.: Clustered solutions around harmonic centers to a coupled elliptic system. Ann. Inst. H. Poincaré Anal. Non Linéaire **226**, 605–628 (2007)
8. d'Avenia, P., Pisani, L.: Nonlinear Klein-Gordon equations coupled with Born-Infeld type equations. Electron. J. Differ. Equ. **26**, 13 (2002)
9. d'Avenia, P., Pisani, L., Siciliano, G.: Dirichlet and Neumann problems for Klein-Gordon-Maxwell systems. Nonlinear Anal. **71**(12), e1985–e1995 (2009)
10. d'Avenia, P., Pisani, L., Siciliano, G.: Klein-Gordon-Maxwell systems in a bounded domain. Discret. Cont. Dyn. Syst. **26**(1), 135–149 (2010)
11. Druet, O., Hebey, E.: Existence and a priori bounds for electrostatic Klein-Gordon-Maxwell systems in fully inhomogeneous spaces. Commun. Contemp. Math. **12**(5), 831–869 (2010)
12. Escobar, J.F.: Conformal deformation of a Riemannian metric to a scalar flat metric with constant mean curvature on the boundary. Ann. Math. (2) **136**(1), 1–50 (1992)
13. Ghimenti, M., Micheletti, A.M.: The role of the mean curvature of the boundary in a nonlinear elliptic problem on Riemannian manifolds. (2014) http://arxiv.org/abs/1410.8841
14. Ghimenti, M., Micheletti, A.M.: Low energy solutions for singularly perturbed coupled nonlinear systems on a Riemannian manifold with boundary. Nonlinear Anal. **119**, 315–329 (2015) http://arxiv.org/abs/1407.1182
15. Ghimenti, M., Micheletti, A.M.: Nondegeneracy of critical points of the mean curvature of the boundary for Riemannian manifolds. J. Fixed Point Theory Appl. **14**(1), 71–78 (2013)
16. Ghimenti, M., Micheletti, A.M.: Number and profile of low energy solutions for singularly perturbed Klein-Gordon-Maxwell systems on a Riemannian manifold. J. Differ. Equ. **256**(7), 2502–2525 (2014)
17. Ghimenti, M., Micheletti, A.M., Pistoia, A.: The role of the scalar curvature in some singularly perturbed coupled elliptic systems on Riemannian manifolds. Discret. Contin. Dyn. Syst. **34**(6), 2535–2560 (2014)
18. Hebey, E., Truong, T.T.: Static Klein-Gordon-Maxwell-Proca systems in 4-dimensional closed manifolds. J. Reine Angew. Math. **667**, 221–248 (2012)

19. Hebey, E., Wei, J.: Resonant states for the static Klein-Gordon-Maxwell-Proca system. Math. Res. Lett. **19**(4), 953–967 (2012)
20. Micheletti, A.M., Pistoia, A.: The role of the scalar curvature in a nonlinear elliptic problem on Riemannian manifolds. Calc. Var. Partial Differ. Equ. **34**(2), 233–265 (2009)
21. Mugnai, D.: Coupled Klein-Gordon and Born-Infeld type equations: Looking for solitary waves. Proc. R. Soc. Lond. Ser. A Math. Phys. Eng. Sci. **460**, 1519–1527 (2004)

Asymptotic analysis for radial sign-changing solutions of the Brezis-Nirenberg problem in low dimensions

Alessandro Iacopetti and Filomena Pacella

Mathematics Subject Classification (2010): 35J91, 35J61, 35B33, 35B40, 35J20

1 Introduction

We consider the Brezis-Nirenberg problem

$$\begin{cases} -\Delta u = \lambda u + |u|^{2^*-2}u & \text{in } B_1 \\ u = 0 & \text{on } \partial B_1, \end{cases} \tag{1}$$

where $\lambda > 0$, $2^* = \frac{2N}{N-2}$ and B_1 is the unit ball of $\mathbb{R}^N$, $N \geq 3$.

The aim of the paper is to get asymptotic results for radial sign-changing solutions u_λ of (1) in dimensions $N = 3, 4, 5, 6$. This will give the asymptotic profile of the positive and negative part of u_λ as λ tends to some limit value.

To motivate our analysis and to explain our results we need to recall a few known results.

A. Iacopetti (✉)
Dipartimento di Matematica e Fisica, Università degli Studi di Roma Tre,
L.go S. Leonardo Murialdo 1, 00146 Roma, Italy
e-mail: iacopetti@mat.uniroma3.it

F. Pacella
Dipartimento di Matematica, Università di Roma "La Sapienza,"
P.le Aldo Moro 5, 00185 Roma, Italy
e-mail: pacella@mat.uniroma1.it

A.N. Carvalho et al. (eds.), *Contributions to Nonlinear Elliptic Equations and Systems*, Progress in Nonlinear Differential Equations and Their Applications 86,
DOI 10.1007/978-3-319-19902-3_20

The first fundamental results about the existence of positive solutions were obtained by H. Brezis and L. Nirenberg in 1983 in the celebrated paper [9]. From their results it came out that the dimension was going to play a crucial role in the study of (1) in a general bounded domain Ω. Indeed they proved that if $N \geq 4$ there exists a positive solution of (1) for every $\lambda \in (0, \lambda_1(\Omega))$, $\lambda_1(\Omega)$ being the first eigenvalue of $-\Delta$ in Ω with Dirichlet boundary conditions, while if $N = 3$ positive solutions exist only for λ away from zero.

Since then several other interesting results were obtained for positive solutions, in particular about the asymptotic behavior of solutions, mainly for $N \geq 5$, because also the case $N = 4$ presents more difficulties compared to the higher dimensional ones.

Concerning the case of sign-changing solutions, existence results hold if $N \geq 4$ both for $\lambda \in (0, \lambda_1(\Omega))$ and $\lambda > \lambda_1(\Omega)$ as shown in [7, 10, 12].

The case $N = 3$ presents even more difficulties than in the study of positive solutions. In particular in the case of the ball is not yet known what is the least value $\bar{\lambda}$ of the parameter λ for which sign-changing solutions exist, neither whether $\bar{\lambda}$ is larger or smaller than $\lambda_1(B_1)/4$. This question, posed by H. Brezis, has been given a partial answer in [8].

However it is interesting to observe that in the study of sign-changing solutions even the "low dimensions" $N = 4, 5, 6$ exhibit some peculiarities. Indeed it was first proved by Atkinson, Brezis, and Peletier in [6] and [7] that if Ω is the ball B_1 there exists $\lambda^* = \lambda^*(N)$ such that there are no radial sign-changing solutions of (1) for $\lambda \in (0, \lambda^*)$. Later this result was proved in [1] in a different way.

Moreover, for $N \geq 7$ a recent result of Schechter and Zou [17] shows that in any bounded smooth domain there exist infinitely many sign-changing solutions for any $\lambda > 0$. Instead if $N = 4, 5, 6$ only $N + 1$ pairs of solutions, for all $\lambda > 0$, have been proved to exist in [12] but it is not clear that they change sign.

Coming back to radial sign-changing solutions and to the question of existence or nonexistence of them, according to the dimension, as shown by Atkinson, Brezis, and Peletier, it is interesting to understand in which way these results can be extended to other bounded domains and to which kind of solutions.

In order to analyze this question let us divide the discussion into two cases: the first one when the dimension N is greater or equal than 7 and the second one when $N < 7$.

In the first case ($N \geq 7$) radial sign-changing solutions u_λ exist for all $\lambda > 0$, if the domain is a ball, and analyzing the asymptotic behavior of those of least energy, as $\lambda \to 0$, it is proved in [14] that their limit profile is that of a "tower of two bubbles." This terminology means that the positive part and the negative part of the solutions u_λ concentrate at the same point (which is obviously the center of the ball) as $\lambda \to 0$ and each one has the limit profile, after suitable rescaling, of a "standard bubble" in $\mathbb{R}^N$, i.e. of a positive solution of the critical exponent problem in $\mathbb{R}^N$. More precisely the solutions can be written in the following way:

$$u_\lambda = PU_{\delta_1,\xi} - PU_{\delta_2,\xi} + w_\lambda, \tag{2}$$

where $PU_{\delta_i,\xi}$, $i = 1, 2$ is the projection on $H_0^1(\Omega)$ of the regular positive solution of the critical problem in $\mathbb{R}^N$, centered at $\xi = 0$, with rescaling parameter δ_i and w_λ is a remainder term which converges to zero in $H_0^1(\Omega)$ as $\lambda \to 0$.

Inspired by this result one could then search for solutions of type (2) in general bounded domains since this kind of solutions can be viewed as the ones which play the same role of the radial solutions in the case of the ball. This has been done recently in [16], where solutions of the type (2) have been constructed for λ close to zero in some symmetric bounded domains (the symmetry makes their construction a bit easier, but the same result should be true in any bounded domain).

On the contrary, coming to the case $N < 7$, in view of the nonexistence result of nodal radial solutions of [7] it is natural to conjecture that, in general bounded domains, there should not be solutions of the form (2) for λ close to zero. Indeed this has been recently proved in [15] if $N = 4, 5, 6$, the case $N = 3$ being obvious.

On the other side, if $N < 7$, radial nodal solutions exist for λ bigger than a certain value $\bar{\lambda}_2$ which can be studied by analyzing the associated ordinary differential equation (see [3, 7, 13]).

Therefore, to the aim of getting analogous existence results in other bounded domains, the first step would be to analyze the asymptotic behavior of nodal radial solutions in the ball, for $\lambda \to \bar{\lambda}_2$, in order to understand their limit profile and guess what kind of solutions one can construct in other domains, and for which values of the parameter λ.

This is the subject of our paper.

Denoting by u_λ a nodal radial solutions of (1) having two nodal regions and such that $u_\lambda(0) > 0$ we get the following results:

- **(i)**: if $N = 6$, then $\bar{\lambda}_2 \in (0, \lambda_1(B_1))$, $\lambda_1(B_1)$ being the first eigenvalue of $-\Delta$ in $H_0^1(B_1)$, and we have that, as $\lambda \to \bar{\lambda}_2$, u_λ^+ concentrate at the center of the ball, $\|u_\lambda^+\|_\infty \to +\infty$, and a suitable rescaling of u_λ^+ converges to the standard positive solution of the critical problem in $\mathbb{R}^N$. Instead u_λ^- converges to the unique positive solution of (1) in B_1, as $\lambda \to \bar{\lambda}_2$;
- **(ii)**: if $N = 4, 5$ then $\bar{\lambda}_2 = \lambda_1(B_1)$ and u_λ^+ behaves as for the case $N = 6$, while u_λ^- converges to zero uniformly in B_1;
- **(iii)**: if $N = 3$, then $\bar{\lambda}_2 = \frac{9}{4}\lambda_1(B_1)$ and u_λ^+ behaves as for the case $N = 6$, while u_λ^- converges to zero uniformly in B_1.

In view of these results we conjecture that, in general bounded domains Ω, for some "limit value" $\bar{\lambda}_2 = \bar{\lambda}_2(N, \Omega)$ there should exist solutions with similar asymptotic profile as $\lambda \to \bar{\lambda}_2$. The number $\bar{\lambda}_2$ should be $\lambda_1(\Omega)$ in dimension $N = 4, 5$. Some work in this direction is in progress.

The paper is divided into three sections. In Section 2 we mainly recall some preliminary results. In Section 3 we analyze the asymptotic behavior of the positive part of the solutions, for all dimensions $N = 3, 4, 5, 6$. In Section 4 we analyze the negative part in the case $N = 6$ and in Section 5 we complete the cases $N = 3, 4, 5$.

2 Some preliminary results

If u_λ is a radial sign-changing solution of (1), then we can write $u_\lambda = u_\lambda(r)$, where $r = |x|$ and $u_\lambda(r)$ is a solution of the problem

$$\begin{cases} u_\lambda'' + \frac{n-1}{r} u_\lambda' + \lambda u_\lambda + |u_\lambda|^{2^*-2} u_\lambda = 0, & \text{in } (0,1), \\ u_\lambda'(0) = 0, \ \ u_\lambda(1) = 0. \end{cases} \tag{3}$$

We consider the following transformation

$$r \mapsto \left(\frac{N-2}{\sqrt{\lambda}\, r} \right)^{N-2}, \quad u_\lambda \mapsto y(t) := \lambda^{-1/(2^*-2)} u_\lambda \left(\frac{N-2}{\sqrt{\lambda}\, t^{\frac{1}{N-2}}} \right). \tag{4}$$

It is elementary to see that since u_λ is a solution of the differential equation in (3) then $y = y(t)$ solves

$$y'' + t^{-k}(y + |y|^{2^*-2} y) = 0, \tag{5}$$

in the interval $\left(\left(\frac{N-2}{\sqrt{\lambda}} \right)^{N-2}, +\infty \right)$, where $k := 2\frac{N-1}{N-2}$. It is clear that the transformation (4) generates a one-to-one correspondence between solutions of the differential equation in (3) and solutions of (5). Equation (5) is an Emden-Fowler type equation and since $k > 2$ it is well known that, for any $\gamma \in \mathbb{R}$ the problem

$$\begin{cases} y'' + t^{-k} f(y) = 0, & \text{in } (0, +\infty), \\ y(t) \to \gamma, & \text{as } t \to +\infty, \end{cases} \tag{6}$$

where $f(y) := y + |y|^{2^*-2} y$, has a unique solution defined in the whole $\mathbb{R}^+$ which we denote by $y(t;\gamma)$. Let us recall some results on the functions $y(t;\gamma)$ which are proved in [7].

Lemma 2.1. *Let $y = y(t,\gamma)$ be a solution of Problem (6), then:*

(a) y is oscillatory near $t = 0$;
(b) the set $\{|y(\bar{t})|;\ \bar{t}$ extremum point of $y\}$ is an increasing sequence with respect to t;
(c) the set $\{|y'(t_0)|;\ t_0$ zero of $y\}$ is a decreasing sequence with respect to t.

Proof. See Lemma 1 in [7]. □

Lemma 2.2. *Let $y = y(t,\gamma)$ be a solution of Problem (6) and let $T > 0$ be one of its zeros, then*

$$|y(t)| < |y'(T)|(T - t),$$

for all $0 < t < T$.

Proof. See Lemma 2 in [7]. □

We shall denote the sequence of zeros of $y(t;\gamma)$ by $T_n(\gamma)$, ordered backwards, precisely:

$$\cdots < T_3(\gamma) < T_2(\gamma) < T_1(\gamma) < +\infty.$$

We recall some results on the asymptotic behavior of the largest zero $T_1(\gamma)$ and on the slope $y'(T_1(\gamma);\gamma)$ as $\gamma \to +\infty$.

Lemma 2.3. *Let y be a solution of Problem (6) and $T_1(\gamma)$ its largest zero, then:*

(a) if $2 < k < 3$ (which corresponds to $N > 4$), then

$$T_1(\gamma) = A(k)\gamma^{6-2k}(1 + o(1)) \ \ as\ \gamma \to +\infty,$$

where $A(k) := (k-1)^{\frac{k-3}{k-2}} \frac{\Gamma(3-k)/(k-2)\Gamma((k-1)/(k-2))}{\Gamma(2/(k-2))}$, Γ is the Gamma function.

(b) if $k = 3$ (which corresponds to $N = 4$), then

$$T_1(\gamma) = 2\log\gamma(1 + o(1)) \ \ as\ \gamma \to +\infty;$$

(c) if $k = 4$ (which corresponds to $N = 3$), then there exists $\gamma_0 \in \mathbb{R}^+$ and two positive constants A, B such that

$$A < T_1(\gamma) < B \ \ for\ all\ \gamma \geq \gamma_0.$$

Proof. The proof of (a), (b) is contained in [7], Lemma 3 and the proof of (c) is contained in [4], Theorem 3. □

Lemma 2.4. *For any $k > 2$, let y be a solution of Problem (6) and $T_1(\gamma)$ its largest zero, then*

$$y'(T_1(\gamma)) = (k-1)^{\frac{1}{k-2}}\gamma^{-1}(1 + o(1)), \ as\ \gamma \to +\infty.$$

Proof. See [7], Lemma 4. □

To prove the existence of radial sign-changing solutions of (1), with exactly two nodal regions, we consider the second zero $T_2(\gamma)$ of $y(t;\gamma)$. If we choose $\lambda = \lambda(\gamma)$ so that $T_2(\gamma) = \left(\frac{N-2}{\sqrt{\lambda}}\right)^{N-2}$, then the inverse transformation of (4) maps $t = T_2$ in $r = 1$ and $y \mapsto u_\lambda$. Hence, for $\lambda = (N-2)^2 T_2(\gamma)^{-\frac{2}{N-2}}$, we obtain a function u_λ which is a radial solution of (1) having exactly two nodal regions; moreover, $u_\lambda(0) = \lambda^{1/(2^*-2)}\gamma$. We observe also that thanks to the invertibility of (4) every radial sign-changing solution u_λ of (1) with two nodal regions corresponds to a solution $y = y(t;\gamma)$ of (6) with $\gamma = \lambda^{-1/(2^*-2)}u_\lambda(0)$, $T_2(\gamma) = \left(\frac{N-2}{\sqrt{\lambda}}\right)^{N-2}$.

We are interested in the study of the behavior of the map $\lambda_2 : \mathbb{R}^+ \to \mathbb{R}^+$, defined by $\lambda_2(\gamma) := (N-2)^2 T_2(\gamma)^{-\frac{2}{N-2}}$. Clearly this map is continuous. In [3] (see Proposition 2 and Remark 4), it is proved that for $N = 4$ it holds that $\lim_{\gamma\to 0}\lambda_2(\gamma) = \lambda_2(B_1)$, where $\lambda_2(B_1)$ is the second radial eigenvalue of $-\Delta$ in $H_0^1(B_1)$. Moreover the authors observe that this result holds for all dimensions $N \geq 3$. For the sake of completeness we give a complete proof of this fact. We begin with a preliminary lemma.

Lemma 2.5. *Let u_λ be a radial solution of (1), then we have $|u_\lambda(0)| = \|u_\lambda\|_\infty$.*

Proof. See [14], Proposition 2 or [3], Lemma 8. □

Proposition 2.6. *Let $N \geq 3$ and $\lambda_2 : \mathbb{R}^+ \to \mathbb{R}^+$ the function defined by $\lambda_2(\gamma) := (N-2)^2 T_2(\gamma)^{-\frac{2}{N-2}}$, where $T_2(\gamma)$ is the second zero of the function $y(t,\gamma)$, $y(t,\gamma)$ is the unique solution of (6). We have:*

(a) $\lambda_2(\gamma) < \lambda_2(B_1)$, for all $\gamma \in \mathbb{R}^+$;
(b) $\lim_{\gamma\to 0}\lambda_2(\gamma) = \lambda_2(B_1)$,

where $\lambda_2(B_1)$ is the second radial eigenvalue of $-\Delta$ in $H_0^1(B_1)$.

Proof. To prove (a) we observe that (a) is equivalent to show that $T_2(\gamma) > \tau_2$ for all $\gamma \in \mathbb{R}^+$, where τ_2 is the second zero of the function $\alpha : \mathbb{R}^+ \to \mathbb{R}$ defined by $\alpha(t) := A_\nu \sqrt{t} J_\nu(2\nu t^{-\frac{1}{2\nu}})$, where $A_\nu := \nu^{-\nu}\Gamma(\nu+1)$, $\nu := \frac{1}{k-2} = \frac{N-2}{2}$, J_ν is the first kind (regular) Bessel function of order ν, namely $J_\nu(s) := \sum_{j=0}^{\infty} \frac{(-1)^j}{\Gamma(j+1)\Gamma(j+\nu+1)}\left(\frac{s}{2}\right)^{\nu+2j}$. In fact, by a tedious computation, we see that α solves

$$\begin{cases} \alpha'' + t^{-k}\alpha = 0, & \text{in } (0,+\infty), \\ \alpha(t) \to 1, & \text{as } t \to +\infty. \end{cases} \tag{7}$$

Furthermore, let τ_2 be the second zero of α, then by elementary computations we see that the function $\varphi_2(x) := \alpha(\tau_2 |x|^{-(N-2)})$ solves

$$\begin{cases} -\Delta\varphi_2 = \mu_2\varphi_2 & \text{in } B_1 \\ \varphi_2 = 0 & \text{on } \partial B_1, \end{cases} \tag{8}$$

with $\mu_2 = (N-2)^2 \tau_2^{-\frac{2}{N-2}}$. Clearly $\mu_2 = \lambda_2(B_1)$. Hence $\lambda_2(\gamma) < \lambda_2(B_1)$ if and only if $T_2(\gamma) > \tau_2$.

To show that $T_2(\gamma) > \tau_2$ for all $\gamma \in \mathbb{R}^+$ first observe that for all $\gamma \in \mathbb{R}^+$ we have $T_1(\gamma) > \tau_1$. In fact, setting $\lambda_1(\gamma) := (N-2)^2 T_1(\gamma)^{-\frac{2}{N-2}}$ as before we have that $\lambda_1(\gamma) < \lambda_1(B_1)$ if and only if $T_1(\gamma) > \tau_1$. Since we know from [9] that equation (3) has positive solutions only for $\lambda \in (0, \lambda_1(B_1))$ if $N \geq 4$, and only for $\lambda \in (\frac{\lambda_1(B_1)}{4}, \lambda_1(B_1))$ if $N = 3$, we deduce $T_1(\gamma) > \tau_1$ for all $\gamma \in \mathbb{R}^+$. Now we apply the Sturm's comparison theorem to the functions $y(t;\gamma)$, $\alpha(t)$, which are, respectively, solutions of the equations in (6), (7). To this end we write $y'' + t^{-k} q_2(t) y = 0$ with $q_2(t) := 1 + |y|^{2^*-2}$ and since $\alpha'' + t^{-k}\alpha = 0$ we set $q_1(t) :\equiv 1$. Clearly $q_2(t) \geq q_1(t)$ for all $t > 0$ (for all $\gamma \in \mathbb{R}^+$), thus y is a Sturm majorant for α, and applying the Sturm's comparison theorem in the interval $[\tau_2, \tau_1]$, since $T_1(\gamma) > \tau_1$ we deduce that $T_2(\gamma) \in (\tau_2, \tau_1)$. This concludes the proof of (a).

Let us prove (b). We consider $u_{\lambda_2(\gamma)} = u_{\lambda_2(\gamma)}(r)$ which is a solution of (3) with exactly one zero in $(0, 1)$, and $u_{\lambda_2(\gamma)}(0) = [\lambda_2(\gamma)]^{1/(2^*-2)}\gamma$. Setting $\varphi(x) := u_{\lambda_2}(|x|)$ it is clear that φ is the second radial eigenfunction of

$$\begin{cases} -\Delta\varphi = \lambda\varphi + |u_{\lambda_2(\gamma)}|^{2^*-2}\varphi & \text{in } B_1 \\ \varphi = 0 & \text{on } \partial B_1, \end{cases} \tag{9}$$

with eigenvalue $\lambda = \lambda_2(\gamma)$. Let us denote by $H^1_{0,rad}(B_1)$ the subspace of radially symmetric functions in $H^1_0(B_1)$. Thanks to the variational characterization of eigenvalues and Lemma 2.5 we have

$$\begin{aligned} \lambda_2(\gamma) &= \min_{\substack{V \subset H^1_{0,rad}(B_1) \\ \dim V = 2}} \max_{\substack{\varphi \in V \\ |\varphi|_2 = 1}} \left(\int_{B_1} |\nabla\varphi|^2\, dx - \int_{B_1} |u_{\lambda_2(\gamma)}|^{2^*-2}\varphi^2\, dx \right) \\ &> \min_{\substack{V \subset H^1_{0,rad}(B_1) \\ \dim V = 2}} \max_{\substack{\varphi \in V \\ |\varphi|_2 = 1}} \left(\int_{B_1} |\nabla\varphi|^2\, dx - [\lambda_2(\gamma)]^{2/(2^*-2)}\gamma^2 \right) \\ &= \lambda_2(B_1) - [\lambda_2(\gamma)]^{2/(2^*-2)}\gamma^2. \end{aligned} \tag{10}$$

Since $\lambda_2(\gamma)$ is bounded (because by (a) we have $\lambda_2(\gamma) < \lambda_2(B_1)$ and by definition $\lambda_2(\gamma) > 0$), from (10), we deduce that $\liminf_{\gamma\to 0} \lambda_2(\gamma) \geq \lambda_2(B_1)$. On the other hand, by the first step we get that $\limsup_{\gamma\to 0} \lambda_2(\gamma) \leq \lambda_2(B_1)$. Hence we deduce that $\lim_{\gamma\to 0} \lambda_2(\gamma) = \lambda_2(B_1)$ and the proof is concluded. □

More interesting is the behavior of $\lambda_2(\gamma)$ as $\gamma \to +\infty$. The next result that we recall shows how it strongly depends on the dimension N.

Theorem 2.7. *Let $\lambda_2 : \mathbb{R}^+ \to \mathbb{R}^+$ be the function defined by $\lambda_2(\gamma) := (N-2)^2 T_2(\gamma)^{-\frac{2}{N-2}}$, where $T_2(\gamma)$ is the second zero of the function $y(t, \gamma)$, being $y(t, \gamma)$ is the unique solution (6), and let $\lambda_1(B_1)$ be the first eigenvalue of $-\Delta$ in $H^1_0(B_1)$, then:*

(a) if $N \geq 7$, we have $\lim_{\gamma\to+\infty} \lambda_2(\gamma) = 0$;
(b) if $N = 6$, we have $\lim_{\gamma\to+\infty} \lambda_2(\gamma) = \lambda_0$, for some $\lambda_0 \in (0, \lambda_1(B_1))$;
(c) if $N = 4$ or $N = 5$, we have $\lim_{\gamma\to+\infty} \lambda_2(\gamma) = \lambda_1(B_1)$;
(d) if $N = 3$, we have $\lim_{\gamma\to+\infty} \lambda_2(\gamma) = \frac{9}{4}\lambda_1(B_1) = \frac{9}{4}\pi^2$.

Proof. Statement (a) is a consequence of Theorem B in [11]. Statements (b), (c) are proved in [7], Theorem B. In Section 4 we give an alternative proof of (b). Statement (d) is proved in [5]. □

Let us define $\lambda_2^\star := \inf\{\lambda_2(\gamma),\ \gamma \in \mathbb{R}^+\}$. Gazzola and Grunau proved in [13] that for $N = 5$ it holds $\lim_{\gamma\to+\infty} \lambda_2(\gamma) = \lambda_1(B_1)^-$, in particular we deduce that for $N = 5$ we have $\lambda_2^\star < \lambda_1(B_1)$ and hence $\lambda_2^\star = \lambda_2(\gamma_0)$ for some $\gamma_0 \in \mathbb{R}^+$. In the same paper it is also proved that for $N = 4$ $\lim_{\gamma\to+\infty} \lambda_2(\gamma) = \lambda_1(B_1)^+$. Recently

Arioli, Gazzola, Grunau, Sassone proved in [3] a stronger result: for $N = 4$ we have $\lambda_2(\gamma) > \lambda_1(B_1)$ for all $\gamma \in \mathbb{R}^+$. Thus for $N = 4$, we have $\lambda_2^\star = \lambda_1(B_1)$ and $\lambda_2^\star$ is not achieved.

The asymptotic behavior of $\lambda_2(\gamma)$ as $\gamma \to +\infty$ for $N = 6$ is still unknown. Nevertheless in Section 3 we give a characterization of the number λ_0 appearing in (b) of Theorem 2.7.

3 Energy and asymptotic analysis of the positive part

Let $u_{\lambda_2(\gamma)}$ be the radial solution with exactly two nodal regions of (1), for $\lambda = \lambda_2(\gamma)$, obtained in the previous section. To simplify the notation we omit the dependence on γ and write u_{λ_2}. We recall that, by definition, for $\gamma \in \mathbb{R}^+$ we have $u_{\lambda_2}(0) > 0$ and we denote by $r_{\lambda_2} \in (0,1)$ its node.

The aim of this section is to compute the limit energy of the positive part $u_{\lambda_2}^+$, as $\gamma \to +\infty$, as well as, to study the asymptotic behavior of a suitable rescaling of $u_{\lambda_2}^+$. We begin with recalling an elementary but crucial fact:

Lemma 3.1. *Let $u \in H^1_{0,rad}(B)$, where B is a ball or an annulus centered at the origin of $\mathbb{R}^N$ and consider the rescaling $\tilde{u}(y) := M^{1/\beta}u(My)$, where $M > 0$ is a constant, $\beta := \frac{2}{N-2}$. We have:*

(i): $\|u\|_B^2 = \|\tilde{u}\|^2_{M^{-1}B}$,
(ii): $|u|^{2^*}_{2^*,B} = |\tilde{u}|^{2^*}_{2^*,M^{-1}B}$,
(iii): $|u|^2_{2,B} = M^2|\tilde{u}|^2_{2,M^{-1}B}$.

Proof. It suffices to apply the formula of change of variable for the integrals in (i), (ii), (iii). For the details see Lemma 2 in [14]. □

In order to state the main result of this section we introduce some notation. We define the rescaled functions

$$\tilde{u}^+_{\lambda_2}(y) := \frac{1}{M_{\lambda_2,+}} u^+_{\lambda_2}\left(\frac{y}{M^\beta_{\lambda_2,+}}\right), \quad y \in B_{\sigma_{\lambda_2}},$$

where $\beta := \frac{2}{N-2}$, $\sigma_{\lambda_2} = M^\beta_{\lambda_2,+} r_{\lambda_2}$, $M_{\lambda_2,+} := \|u^+_{\lambda_2}\|_{\infty,B_1}$. We observe that thanks to Lemma 2.5 and since $u_{\lambda_2}(0) > 0$ we have $M_{\lambda_2,+} = \|u_{\lambda_2}\|_{\infty,B_1} = u_{\lambda_2}(0)$. The following theorem holds for all dimensions $N \geq 3$, here we discuss the case $3 \leq N \leq 6$ (the case $N \geq 7$ is studied in [14]).

Theorem 3.2. *Let $N = 3,4,5,6$ and let u_{λ_2} be the radial solution with exactly two nodal regions of (1) with $\lambda = \lambda_2(\gamma)$ obtained in the previous section. Then* *(i):*

$$J_{\lambda_2}(u^+_{\lambda_2}) \to \frac{1}{N} S^{N/2},$$

as $\gamma \to +\infty$, *where* $J_\lambda(u) := \frac{1}{2}\left(\int_{B_1} |\nabla u|^2 - \lambda |u|^2\, dx\right) - \frac{1}{2^*}\int_{B_1} |u|^{2^*}\, dx$ *is the energy functional related to (1), S is the best Sobolev constant for the embedding of* $\mathscr{D}^{1,2}(\mathbb{R}^N)$ *into* $L^{2^*}(\mathbb{R}^N)$.

(ii): Up to a subsequence, the rescaled function $\tilde{u}^+_\lambda$ *converges in* $C^2_{loc}(\mathbb{R}^N)$ *to* $U_{0,\mu}$, *as* $\gamma \to +\infty$, *where* $U_{0,\mu}$ *is the solution of the critical exponent problem in* $\mathbb{R}^N$ *centered at* $x_0 = 0$ *and with concentration parameter* $\mu = \sqrt{N(N-2)}$. *We recall that such functions are defined by*

$$U_{x_0,\mu}(x) := \frac{[N(N-2)\mu^2]^{(N-2)/4}}{[\mu^2 + |x - x_0|^2]^{(N-2)/2}}.$$

Proof. We start by proving (i). Let (u_{λ_2}) be this family of solutions. Since $u^+_{\lambda_2}$ solves $-\Delta u = \lambda_2 u + u^{2^*-1}$ in $B_{r_{\lambda_2}}$ then, considering the rescaling $\hat{u}^+_{\lambda_2}(y) := r^{1/\beta}_{\lambda_2} u^+_{\lambda_2}(r_{\lambda_2} y)$, where $\beta := \frac{2}{N-2}$, we see that $\tilde{u}^+_{\lambda_2}$ solves

$$\begin{cases} -\Delta u = \lambda_2 r^2_{\lambda_2} u + u^{2^*-1} & \text{in } B_1, \\ u > 0 & \text{in } B_1, \\ u = 0 & \text{on } \partial B_1. \end{cases} \tag{11}$$

Now we distinguish between two cases: $N = 4, 5, 6$ and $N = 3$.

If $N = 4, 5, 6$, then, from Lemma 2.3 we deduce that $r_{\lambda_2} \to 0$ as $\gamma \to +\infty$, in particular this is true for $\lambda_2 r^2_{\lambda_2}$. From [2] we know that $\hat{u}^+_{\lambda_2}$ is unique and it coincides with the solution found in [9], which minimizes the energy $J_{\lambda_2 r^2_{\lambda_2}}$; thus, since $\lambda_2 r^2_{\lambda_2} \to 0$ as $\gamma \to +\infty$ we get that $J_{\lambda_2 r^2_{\lambda_2}}(\hat{u}^+_{\lambda_2}) \to \frac{1}{N} S^{N/2}$. Thanks to Lemma 3.1 we get that $J_{\lambda_2}(u^+_{\lambda_2}) = J_{\lambda_2 r^2_{\lambda_2}}(\hat{u}^+_{\lambda_2}) \to \frac{1}{N} S^{N/2}$ as $\gamma \to +\infty$.

Assume now that $N = 3$. As stated in Lemma 2.3 we have that r_{λ_2} is bounded away from zero. From a well-known result of Brezis and Nirenberg (see [9], Theorem 1) we have that (11) has a positive solution if and only if $\lambda_2 r^2_{\lambda_2} \in (\frac{\pi^2}{4}, \pi^2)$. As $\gamma \to +\infty$ we must have $\lambda_2 r^2_{\lambda_2} \to \frac{\pi^2}{4}$. Hence, the only possibility is that $J_{\lambda_2 r^2_{\lambda_2}}(\hat{u}^+_{\lambda_2}) \to \frac{1}{3} S^{3/2}$ as $\gamma \to +\infty$. As before thanks to Lemma 3.1 we have $J_{\lambda_2}(u^+_{\lambda_2}) = J_{\lambda_2 r^{1/\beta}_{\lambda_2}}(\hat{u}^+_{\lambda_2})$ and hence $J_{\lambda_2}(u^+_{\lambda_2}) \to \frac{1}{3} S^{3/2}$ as $\gamma \to +\infty$. The proof of (i) is complete.

We now prove (ii). By definition the rescaled function $\tilde{u}_{\lambda_2}^{+}$ solves the following problem

$$\begin{cases} -\Delta u = \frac{\lambda_2}{M_{\lambda_2,+}^{2\beta}} u + u^{2^*-1} & \text{in } B_{\sigma_{\lambda_2}}, \\ u > 0 & \text{in } B_{\sigma_{\lambda_2}}, \\ u = 0 & \text{on } \partial B_{\sigma_{\lambda_2}}, \end{cases} \tag{12}$$

where $\sigma_{\lambda_2} := M_{\lambda_2}^{\beta} r_{\lambda_2}$.

Since the family $(\tilde{u}_{\lambda_2}^{+})$ is uniformly bounded, then by standard elliptic theory we get that $\tilde{u}_{\lambda_2}^{+} \to \tilde{u}$ in $C^2_{loc}(B_l)$, where l is the limit of σ_{λ_2} as $\gamma \to +\infty$. We want to show that

$$\lim_{\gamma \to +\infty} \sigma_{\lambda_2} = +\infty,$$

so that the limit domain is the whole $\mathbb{R}^N$. We can proceed in two different ways: one is to apply directly the estimates contained in Section 1, the other one is to apply the methods of [14]. We choose the second approach: arguing as in the proof of Proposition 9 in [14], taking into account that by (i) of Theorem 3.2, $J_{\lambda_2}(u_{\lambda_2}^{+}) \to \frac{1}{N} S^{N/2}$, as $\gamma \to +\infty$, we see that up to a subsequence it cannot happen that $\lim_{\gamma \to +\infty} \sigma_{\lambda_2}$ is finite.

Since $\frac{\lambda_2}{M_{\lambda_2,+}^{2\beta}} \to 0$, as $\gamma \to +\infty$, $\tilde{u}_{\lambda_2}^{+}$ converges in $C^2_{loc}(\mathbb{R}^N)$ to a positive solution $\tilde{u}$ of

$$\begin{cases} -\Delta u = |u|^{2^*-2} u & \text{in } \mathbb{R}^N \\ u \to 0 & \text{as } |y| \to +\infty. \end{cases}$$

Observe that this holds even in the case $N = 3$, in fact by definition and Remark 3.3 we have

$$\frac{\lambda_1(B_{\sigma_{\lambda_2}})}{4} = \frac{\pi^2}{4 M_{\lambda_2,+}^4 r_{\lambda_2}^2} = \frac{9}{4}\pi^2(1 + o(1)) \frac{1}{M_{\lambda_2,+}^4} = \frac{\lambda_2}{M_{\lambda_2,+}^4}(1 + o(1)) \to 0,$$

as $\gamma \to +\infty$.

Since $\tilde{u}$ is radial and $\tilde{u}(0) = 1$ then $\tilde{u} = U_{0,\mu}$ where $\mu = \sqrt{N(N-2)}$ (see Proposition 2.2 in [11]). The proof is complete. □

Remark 3.3. We observe that for $N = 3$, since $\lambda_2 r_{\lambda_2}^2 \to \frac{\pi^2}{4}$ and (d) of Theorem 2.7 holds, then, we deduce that $r_{\lambda_2} \to \frac{1}{3}$. On the contrary, if $N = 4, 5, 6$, as seen in the proof of Theorem 3.2, we have $r_{\lambda_2} \to 0$ as $\gamma \to +\infty$ (this also holds for $N \geq 7$, see [14], Proposition 4).

4 Asymptotic analysis of the negative part in dimension $N = 6$

In this section we focus on the case $N = 6$ which means to take $k = 5/2$ in (6). As in [7] we define

$$\begin{aligned} t_0(\gamma) &:= \inf\{t \in (0, +\infty);\ y' > 0 \text{ on } (t, +\infty)\}, \\ y_0(\gamma) &:= y(t_0(\gamma); \gamma). \end{aligned} \tag{13}$$

We have the following:

Proposition 4.1. *Assume $k = 5/2$. Then*

(a) $y_0(\gamma) = -\frac{1}{2}(1 + o(1))$, *as* $\gamma \to +\infty$;
(b) $t_0(\gamma) = (\frac{2}{9}\gamma)^{2/3}(1 + o(1))$, *as* $\gamma \to +\infty$.

Proof. See [7], Theorem 2. □

Let u_λ be any radial solution of (1) with exactly two nodal regions and without loss of generality assume that $u_\lambda(0) > 0$. We denote by s_λ the global minimum point of u_λ. As in the previous section we set $M_{\lambda,+} := \|u_\lambda^+\|_\infty$, $M_{\lambda,-} := \|u_\lambda^-\|_\infty$, where u_λ^+, u_λ^- are, respectively, the positive and the negative part of u_λ Clearly, by definition, we have $u_\lambda^-(s_\lambda) = M_{\lambda,-}$. In order to estimate the energy of such solutions we need the following preliminary result.

Proposition 4.2. *Let $N = 6$ and let (u_λ) be any family of radial sign-changing solutions of (1) with exactly two nodal regions and such that $u_\lambda(0) > 0$ for all λ. Assume that there exists $\lambda_0 \in \mathbb{R}^+$ such that $M_{\lambda,+} \to \infty$ as $\lambda \to \lambda_0$. Then*

$$M_{\lambda,-} \le \frac{\lambda}{2}(1 + o(1)),$$

for all λ sufficiently close to λ_0.

Proof. Let (u_λ) be such a family of solutions. Since $N = 6$, we have $2^* - 2 = \frac{4}{N-2} = 1$ and thanks to the transformation (4) we have

$$u_\lambda(r(t)) = \lambda\, y(t; \gamma), \tag{14}$$

for $t \in \left(\left(\frac{N-2}{\sqrt{\lambda}}\right)^{N-2}, +\infty\right)$, where $\gamma = \lambda^{-1}M_{\lambda,+}$. We observe that the global minimum point s_λ corresponds, through the transformation (4), to the number $t_0(\gamma)$ defined in (13). In fact by definition we have $u_\lambda'(s_\lambda) = 0$ so it suffices to show that $u_\lambda'(r) < 0$ for all $r \in (0, s_\lambda)$. By Corollary 1 in [14] we know that $u_\lambda'(r) < 0$ for all $r \in (0, r_\lambda)$, and for all $r \in (r_\lambda, s_\lambda)$. Moreover since u_λ^+ solves (1) in B_{r_λ}, then, by Hopf lemma it follows that $u_\lambda'(r_\lambda) < 0$. Now, thanks to the assumptions, as $\lambda \to \lambda_0$ we have $\gamma = \lambda^{-1}M_{\lambda,+} \to +\infty$ and the result follows immediately from (14) and Proposition 4.1. □

Remark 4.3. A straight important consequence of Proposition 4.2 is that $M_{\lambda,-}$ is uniformly bounded for all λ sufficiently close to λ_0. In particular there cannot exist radial sign-changing solutions of (1) with the shape of a tower of two bubbles in dimension $N = 6$ (this fact also holds for the dimensions $N = 3, 4, 5$, as we will see later). This is in deep contrast with the case of higher dimensions $N \geq 7$ as showed in [14].

Remark 4.4. In the case of the solutions obtained in the previous section, thanks to Theorem 2.7 we deduce that $M_{\lambda_2(\gamma),-} \leq \frac{\lambda_0}{2}(1 + o(1)) \leq \frac{\lambda_1(B_1)}{2}$ for all sufficiently large $\gamma \in \mathbb{R}^+$.

In the previous section we have studied the limit energy (see Theorem 3.2) of the positive part of the solutions u_{λ_2}. Here we consider the negative part $u^-_{\lambda_2}$ and prove that its energy J_{λ_2} is uniformly bounded as $\gamma \to +\infty$. This is the content of the next proposition.

Proposition 4.5. *Let $N = 6$. Let $\lambda_2 = \lambda_2(\gamma)$ and u_{λ_2} be the radial solution with exactly two nodal regions of (1) described in Section 2. Let $J_\lambda(u) := \frac{1}{2}\left(\int_{B_1} |\nabla u|^2 - \lambda |u|^2 \, dx\right) - \frac{1}{2^*}\int_{B_1} |u|^{2^*} \, dx$ be the energy functional related to (1). Then*

$$J_{\lambda_2}(u^-_{\lambda_2}) \leq \frac{\pi^3}{36}\left(\frac{\lambda_1(B_1)}{2}\right)^3,$$

for all sufficiently large γ.

Proof. Since $u^-_{\lambda_2}$ solves $-\Delta u = \lambda_2 u + u^{2^*-1}$ in the annulus $A_{r_{\lambda_2}}$, in particular it belongs to the Nehari manifold $\mathcal{N}_{\lambda_2}$ associated with that equation, which is defined by

$$\mathcal{N}_{\lambda_2} := \{u \in H^1_0(A_{r_{\lambda_2}});\ \|u\|^2_{A_{r_{\lambda_2}}} - \lambda_2 |u|^2_{2,A_{r_{\lambda_2}}} = |u|^{2^*}_{2^*,A_{r_{\lambda_2}}}\}. \tag{15}$$

Hence we deduce that

$$J_{\lambda_2}(u^-_{\lambda_2}) = \frac{1}{6}|u^-_{\lambda_2}|^{2^*}_{2^*,A_{r_{\lambda_2}}}. \tag{16}$$

Now, thanks to Proposition 4.2, (b) of Theorem 2.7 and Remark 4.4 we have

$$|u^-_{\lambda_2}|^{2^*}_{2^*,A_{r_{\lambda_2}}} = \int_{A_{r_{\lambda_2}}} |u^-_{\lambda_2}|^3 \, dx \leq |B_1| \|u^-_{\lambda_2}\|^3_\infty \leq \frac{\pi^3}{6}\left(\frac{\lambda_1(B_1)}{2}\right)^3, \tag{17}$$

for all sufficiently large γ. From (16) and (17) we deduce the desired relation and the proof is complete. □

Remark 4.6. Since λ_2 is a bounded function, by the same proof of Proposition 4.5, but without using (b) of Theorem, 2.7 we deduce anyway that $J_{\lambda_2}(u^-_{\lambda_2})$ is uniformly bounded for all sufficiently large γ.

We are interested now in studying the asymptotic behavior of the family $(u^-_{\lambda_2})$. More precisely we show that, as $\gamma \to \infty$, the family $(u^-_{\lambda_2})$ converges in $C^2_{loc}(B_1 - \{0\})$ to the unique positive solution u_0 of (1) with $\lambda = \lambda_0$, for some $\lambda_0 \in (0, \lambda_1(B_1))$. We point out that these results will improve the energy estimate of $u^-_{\lambda_2}$ obtained before.

The pointwise convergence of $(u^-_{\lambda_2})$ to u_0 is contained in Theorem 3 of [7], but here we use a different approach which is based on the arguments of [14]. Our result is the following:

Theorem 4.7. *Let $N = 6$, up to a subsequence, we have $\lambda_2(\gamma) \to \lambda_0$, as $\gamma \to +\infty$, for some $\lambda_0 \in (0, \lambda_1(B_1))$, and $(u^-_{\lambda_2})$ converges in $C^2_{loc}(B_1 - \{0\})$ to the unique positive solution u_0 of (1) with $\lambda = \lambda_0$.*

Proof. Let us consider the family $(u^-_{\lambda_2})$. These functions solve

$$\begin{cases} -\Delta u = \lambda_2 u + u^2 & \text{in } A_{r_{\lambda_2}}, \\ u > 0 & \text{in } A_{r_{\lambda_2}}, \\ u = 0 & \text{on } \partial A_{r_{\lambda_2}}. \end{cases} \tag{18}$$

Since λ_2 is bounded, up to a subsequence we have $\lim_{\gamma\to+\infty} \lambda_2 = \lambda_0$. Thanks to Proposition 4.2 we have that $u^-_{\lambda_2}$ is uniformly bounded for all sufficiently large γ and by Lemma 2.3 and the inverse transformation of (4) we have $r_{\lambda_2} \to 0$. Hence by standard elliptic theory, up to a subsequence, for any $0 < \delta < 1$, $u^-_{\lambda_2}$ converges in $C^2(\overline{B_1} - B_\delta)$ as $\gamma \to +\infty$ to a solution u_0 of $-\Delta u = \lambda_0 u + u^2$ in $B_1 - \{0\}$, where B_δ is the ball centered at the origin having radius δ. We now proceed in three steps.

Step 1: we have

$$\lim_{r\to 0} u_0(r) = \frac{\lambda_0}{2}. \tag{19}$$

Since $u^-_{\lambda_2}$ is a radial solution of (1) and thanks to Proposition 4.2, for all sufficiently large γ, we have

$$u^-_{\lambda_2} \leq \frac{\lambda_0}{2}(1 + o(1)), \tag{20}$$

and then we deduce that

$$[(u^-_{\lambda_2})' r^5]' = -\lambda_2 u^-_{\lambda_2}(r) r^5 - [u^-_{\lambda}(r)]^2 r^5 \geq -\lambda_2 \frac{\lambda_0}{2}(1 + o(1)) r^5 - \left[\frac{\lambda_0}{2}(1 + o(1))\right]^2 r^5$$

$$= -\frac{\lambda_0^2}{2}(1 + o(1))^2 r^5 - \frac{\lambda_0^2}{4}(1 + o(1))^2 r^5 \geq -\lambda_0^2 \, r^5.$$

Integrating between s_{λ_2} and r (with $s_{\lambda_2} < r < 1$) we get that

$$(u_{\lambda_2}^-)'(r)r^5 \geq -\lambda_0^2 \int_{s_{\lambda_2}}^r t^5 dt \geq -\frac{\lambda_0^2}{6} r^6.$$

Hence $(u_{\lambda_2}^-)'(r) \geq -\frac{\lambda_0^2}{6} r$ for all $r \in (s_{\lambda_2}, 1)$. Integrating again between s_{λ_2} and r we have

$$u_{\lambda_2}^-(r) - \frac{\lambda_0}{2}(1 + o(1)) \geq -\frac{\lambda_0^2}{12}(r^2 - s_{\lambda_2}^2) \geq -\frac{\lambda_0^2}{12} r^2.$$

Hence $u_{\lambda_2}^-(r) \geq \frac{\lambda_0}{2}(1+o(1)) - \frac{\lambda_0^2}{12} r^2$ for all sufficiently large γ, for all $r \in (s_{\lambda_2}, 1)$. Since $s_{\lambda_2} \to 0$, then, passing to the limit as $\gamma \to \infty$, we get that $u_0(r) \geq \frac{\lambda_0}{2} - \frac{\lambda_0^2}{12} r^2$, for all $0 < r < 1$. From this inequality and (20) we deduce that $\lim_{r\to 0} u_0(r) = \frac{\lambda_0}{2}$. The proof of Step 1 is complete.

Step 2: we have

$$\lim_{r\to 0} u_0'(r) = 0. \tag{21}$$

As in the previous step, integrating the equation between s_{λ_2} and r, with $s_{\lambda_2} < r < 1$, we get that

$$-(\tilde{u}_{\lambda_2}^-)'(r)r^5 = \lambda_2 \int_{s_{\lambda_2}}^r u_{\lambda_2}^- t^5 dt + \int_{s_{\lambda_2}}^r (u_{\lambda_2}^-)^2 t^5 dt.$$

Thanks to (20), for all sufficiently large γ we have

$$|(u_{\lambda_2}^-)'(r)r^5| \leq \lambda_2 \frac{\lambda_0}{2}(1 + o(1)) \int_{s_{\lambda_2}}^r t^5 dt + \frac{\lambda_0^2}{4}(1 + o(1))^2 \int_{s_{\lambda_2}}^r t^5 dt \leq \lambda_0^2 \frac{r^6}{6}.$$

Dividing by r^5 the previous inequality and passing to the limit, as $\gamma \to +\infty$, we get that

$$|u_0'(r)| \leq \frac{\lambda_0^2}{6} r,$$

for all $0 < r < 1$. Hence $\lim_{r\to 0} u_0'(r) = 0$ and the proof of Step 2 is complete. From Step 1 and Step 2 it follows that the radial function $u_0(x) = u_0(|x|)$ can be extended to a $C^1(B_1)$ function. We still denote by u_0 this extension.

Step 3: The function u_0 is a weak solution in B_1 of

$$-\Delta u = \lambda_0 u + u^2. \tag{22}$$

Let us fix a test function $\phi \in C_0^\infty(B_1)$. If $0 \notin supp(\phi)$ the proof is trivial so from now on we assume $0 \in supp(\phi)$. Applying Green's formula to $\Omega(\delta) := B_1 - B_\delta$, since u_0 is a $C^2(B_1 - \{0\})$-solution of (22) and $\phi \equiv 0$ on ∂B_1, we have

$$\int_{\Omega(\delta)} \nabla u_0 \cdot \nabla \phi \, dx = \lambda_0 \int_{\Omega(\delta)} \phi \, u_0 \, dx + \int_{\Omega(\delta)} \phi \, u_0^2 \, dx + \int_{\partial B_\delta} \phi \left(\frac{\partial u_0}{\partial \nu} \right) d\sigma. \tag{23}$$

We show now that $\int_{\partial B_\delta} \phi \left(\frac{\partial u_0}{\partial \nu} \right) d\sigma \to 0$ as $\delta \to 0$. In fact since u_0 is a radial function we have $\frac{\partial u_0}{\partial \nu}(x) = u_0'(\delta)$ for all $x \in \partial B_\delta$, and hence we get that

$$\left| \int_{\partial B_\delta} \phi \left(\frac{\partial u_0}{\partial \nu} \right) d\sigma \right| \leq |u_0'(\delta)| \int_{\partial B_\delta} |\phi| \, d\sigma \leq \omega_6 |u_0'(\delta)| \delta^5 ||\phi||_\infty.$$

Thanks to (21) we have $|u_0'(\delta)|\delta^5 \to 0$ as $\delta \to 0$. To complete the proof we pass to the limit in (23) as $\delta \to 0$. We observe that

$$\begin{aligned} |\nabla u_0 \cdot \nabla \phi| \, \chi_{\Omega(\delta)} &\leq |\nabla u_0|^2 \, \chi_{\{|\nabla u_0|>1\}} |\nabla \phi| + |\nabla u_0| \, \chi_{\{|\nabla u_0| \leq 1\}} |\nabla \varphi| \\ &\leq |\nabla u_0|^2 \, \chi_{\{|\nabla u_0|>1\}} |\nabla \phi| + \, \chi_{\{|\nabla u_0| \leq 1\}} |\nabla \phi|. \end{aligned} \tag{24}$$

We point out that $\int_{B_1} |\nabla u_0|^2 dx$ is finite: this is an easy consequence of the fact that $u_{\lambda_2} \to u_0$ in $C^2_{loc}(B_1 - \{0\})$, the family (u_{λ_2}) is uniformly bounded, (15) and Lebesgue's theorem.

Thus, since $\int_{B_1} |\nabla u_0|^2 dx$ is finite and ϕ has compact support, the right-hand side of (24) belongs to $L^1(B_1)$. Hence from Lebesgue's theorem we have

$$\lim_{\delta \to 0} \int_{\Omega(\delta)} \nabla u_0 \cdot \nabla \phi \, dx = \int_{B_1} \nabla u_0 \cdot \nabla \phi \, dx. \tag{25}$$

Since ϕ has compact support by Lebesgue's theorem we have

$$\begin{aligned} \lim_{\delta \to 0} \int_{\Omega(\delta)} \phi \, u_0 \, dx &= \int_{B_1} \phi \, u_0 \, dx, \\ \lim_{\delta \to 0} \int_{\Omega(\delta)} \phi \, u_0^2 \, dx &= \int_{B_1} \phi \, u_0^2 \, dx. \end{aligned} \tag{26}$$

From (23), (25), (26) and since we have proved $\int_{\partial B(\delta)} \phi \left(\frac{\partial \tilde{u}}{\partial \nu} \right) d\sigma \to 0$ as $\delta \to 0$ it follows that

$$\int_{B_1} \nabla u_0 \cdot \nabla \phi \, dx = \lambda_0 \int_{B_1} \phi \, u_0 \, dx + \int_{B_1} \phi \, u_0^2 \, dx,$$

which completes the proof of Step 3.

Thanks to Step 1 - Step 3 we get that $u_0 \in H^1_{0,rad}(B_1)$ is a weak solution of

$$\begin{cases} -\Delta u = \lambda_0 u + u^2 & \text{in } B_1, \\ u > 0 & \text{in } B_1, \\ u = 0 & \text{on } \partial B_1. \end{cases} \tag{27}$$

In particular, as a consequence of a well known result of Brezis and Kato (for instance see Lemma 1.5 in [9]) it is possible to show that u_0 is a classical solution of (27) (see Appendix B of [18]). Thanks to [2] we know that u_0 is the unique positive radial solution of (27), which is the one found by Brezis and Nirenberg in [9]. Hence we must have $\lambda_0 < \lambda_1(B_1)$ and $J_{\lambda_0}(u_{\lambda_0}) < \frac{1}{6}S^3$. □

Next result gives a characterization of the value $\lambda_0 \in (0, \lambda_1(B_1))$ appearing in Theorem 4.7.

Theorem 4.8. *Let $N = 6$. Let $\lambda_0 := \lim_{\gamma \to +\infty} \lambda_2(\gamma)$. We have that λ_0 is the unique $\lambda \in (0, \lambda_1(B_1))$ such that $u_\lambda(0) = \frac{\lambda}{2}$, where u_λ is the unique positive solution of*

$$\begin{cases} -\Delta u = \lambda u + u^2 & \text{in } B_1, \\ u > 0 & \text{in } B_1, \\ u = 0 & \text{on } \partial B_1. \end{cases} \tag{28}$$

Proof. Thanks to Theorem 4.7 and (19) we have that the set

$$\Gamma := \left\{ \lambda \in (0, \lambda_1(B_1));\ u_\lambda(0) = \frac{\lambda}{2}, \text{ where } u_\lambda \text{ is the unique solution of (28)} \right\},$$

is not empty since $\lambda_0 \in \Gamma$. We want to prove that $\Gamma = \{\lambda_0\}$. To this end assume that $\bar{\lambda} \in \Gamma$ and $\bar{\lambda} \neq \lambda_0$. In particular the functions u_{λ_0} and $u_{\bar{\lambda}}$ are different. Thanks to the definition of Γ and applying (4) (with $2^* - 2 = 1$ because $N = 6$) we get that u_{λ_0} and $u_{\bar{\lambda}}$ are, respectively, transformed to a solution of (6) with $\gamma = \frac{1}{2}$, but, for a given γ, the solution of (6) is unique and this gives a contradiction. □

Now we have all the tools to estimate the energy of the solutions u_{λ_2}. This is the content of the next result.

Corollary 4.9. *Let $N = 6$ and let u_{λ_2} be the radial solution with exactly two nodal regions of (1) with $\lambda = \lambda_2(\gamma)$ obtained in Section 2. Then*

$$J_{\lambda_2}(u_{\lambda_2}) < \frac{1}{3}S^3,$$

for all sufficiently large $\gamma \in \mathbb{R}^+$, where

$$J_\lambda(u) := \frac{1}{2}\left(\int_{B_1} |\nabla u|^2 - \lambda |u|^2 \, dx\right) - \frac{1}{2^*}\int_{B_1} |u|^{2^*} \, dx$$

is the energy functional related to (1), S is the best Sobolev constant for the embedding of $\mathscr{D}^{1,2}(\mathbb{R}^6)$ *into* $L^{2^*}(\mathbb{R}^6)$.

Proof. Let (u_{λ_2}) be this family of solutions. Observe that $J_{\lambda_2}(u_{\lambda_2}) = J_{\lambda_2}(u^+_{\lambda_2}) + J_{\lambda_2}(u^-_{\lambda_2})$ hence it suffices to estimate separately the energy of the positive and negative part of u_{λ_2}. The energy of $u^+_{\lambda_2}$ has been determinated in Theorem 3.2, and in particular we have $J_{\lambda_2}(u^+_{\lambda_2}) \to \frac{1}{6}S^3$, as $\gamma \to +\infty$.

Now we estimate $J_{\lambda_2}(u^-_{\lambda_2})$. Since $u^-_{\lambda_2}$ solves $-\Delta u = \lambda_2 u + u^{2^*-1}$ in the annulus $A_{r_{\lambda_2}}$, in particular it belongs to the Nehari manifold $\mathscr{N}_{\lambda_2}$ associated with this equation, (see (15)). Hence we deduce that $J_{\lambda_2}(u^-_{\lambda_2}) = \frac{1}{6}|u^-_{\lambda_2}|^{2^*}_{2^*,A_{r_{\lambda_2}}}$. To complete the proof it will suffice to show that

$$|u^-_{\lambda_2}|^{2^*}_{2^*,A_{r_{\lambda_2}}} \to |u^-_{\lambda_0}|^{2^*}_{2^*,B_1},$$

where u_0 is the unique solution of (27). In fact, thanks to Theorem 4.7 we know that, up to a subsequence, $(u^-_{\lambda_2})$ converges in $C^2_{loc}(B_1 - \{0\})$ to the unique solution u_0 of (27). Hence to prove our assertion it suffices to apply Lebesgue's theorem, which clearly holds since $(u^-_{\lambda_2})$ is uniformly bounded as $\gamma \to +\infty$.

Now since $J_{\lambda_2}(u^-_{\lambda_2}) \to J_{\lambda_0}(u_{\lambda_0})$ and $J_{\lambda_0}(u_{\lambda_0}) < \frac{1}{6}S^3$ we deduce the desired relation. □

5 Asymptotic analysis of the negative part in dimension $N = 3, 4, 5$

Here we prove:

Theorem 5.1. *Let* $N = 3, 4, 5$ *and let* (u_λ) *be any family of radial sign-changing solutions of (1) with exactly two nodal regions and such that* $u_\lambda(0) > 0$ *for all* λ. *Assume that there exists* $\bar{\lambda} \in \mathbb{R}^+$ *such that* $M_{\lambda,+} \to \infty$, *as* $\lambda \to \bar{\lambda}$. *Then*

(i): $M_{\lambda,-} \to 0$, *as* $\lambda \to \bar{\lambda}$;
(ii): (u^-_λ) *converges to zero uniformly in* B_1, *as* $\lambda \to \bar{\lambda}$.

Proof. We start by proving (i). Let (u_λ) be such a family of solutions. Thanks to the transformation (4) we have

$$u_\lambda(r(t)) = \lambda^{\frac{1}{2^*-2}}\, y(t;\gamma), \tag{29}$$

for $t \in \left(\left(\frac{N-2}{\sqrt{\lambda}}\right)^{N-2}, +\infty\right)$, where $\gamma = \lambda^{-\frac{1}{2^*-2}} M_{\lambda,+}$ and $y = y(t;\gamma)$ solves (6). Clearly, as $\lambda \to \bar{\lambda}$, we have $\gamma \to +\infty$. As in the proof of Proposition 4.2 we have that the global minimum point s_λ corresponds, through the transformation (4), to the number $t_0(\gamma)$ defined in (13).

Hence, thanks to Lemma 2.2, it holds

$$|y(t_0(\gamma);\gamma)| < |y'(T_1(\gamma))|\,(T_1(\gamma) - t_0(\gamma)). \tag{30}$$

For $N = 3$, which corresponds to $k = 4$, by Lemma 2.3 we have that $T_1(\gamma)$ is uniformly bounded for all sufficiently large γ, while, by Lemma 2.4 it holds $y'(T_1(\gamma)) = (k-1)^{\frac{1}{k-2}}\gamma^{-1}(1+o(1))$. Thus, since $0 < t_0(\gamma) < T_1(\gamma)$, from (30), (29) we get that $M_{\lambda,-} = \lambda^{\frac{1}{2^*-2}} y(t_0;\gamma) \to 0$ as $\lambda \to \bar{\lambda}$.

For $N = 4$, which corresponds to $k = 3$, by Lemma 2.3 we have that $T_1(\gamma) = 2\log(\gamma)(1 + o(1))$ for all sufficiently large γ, and hence as in the previous case, we get that $M_{\lambda,-} = \lambda^{\frac{1}{2^*-2}} y(t_0;\gamma) \to 0$ as $\lambda \to \bar{\lambda}$. The same happens for $N = 5$ ($k = 8/3$); in fact by Lemma 2.3 we have that $T_1(\gamma) = A\gamma^{2/3}(1 + o(1))$ for all sufficiently large γ, where $A = A(k)$ is a positive constant depending only on k (see Lemma 2.3 for its definition). The proof of (i) is complete.

Now we prove (ii). We recall that u_λ^- is nonzero in the annulus $A_{r_\lambda}(0) = \{x \in \mathbb{R}^N;\ r_\lambda < |x| < 1\}$ and vanishes outside. Thanks to (i), we have $\|u_\lambda\|_{\infty,B_1} = M_{\lambda,-} \to 0$ as $\lambda \to \bar{\lambda}$ and we are done. □

Acknowledgements Research partially supported by MIUR-PRIN project-201274FYK7_005 and GNAMPA-INDAM.

References

1. Adimurthi, Yadava, S.L.: Elementary proof of the nonexistence of nodal solutions for the semilinear elliptic equations with critical Sobolev exponent. Nonlinear Anal. **14**(9), 785–787 (1990)
2. Adimurthi, Yadava, S.L.: An Elementary proof of the uniqueness of positive radial solutions of a quasilinear Dirichlet problem. Arch. Ration. Mech. Anal. **127**, 219–229 (1994)
3. Arioli, G., Gazzola, F., Grunau, H.-C., Sassone, E.: The second bifurcation branch for radial solutions of the Brezis-Nirenberg problem in dimension four. Nonlinear Differ. Equat. Appl. **15**(1–2), 69–90 (2008)
4. Atkinson, F.V., Peletier, L.A.: Emden-Fowler equations involving critical exponents. Nonlinear Anal. Theory Meth. Appl. **10**(8), 755–776 (1986)
5. Atkinson, F.V., Peletier, L.A.: Large solutions of elliptic equations involving critical exponents. Asymptot. Anal. **1**, 139–160 (1988)
6. Atkinson, F.V., Brezis, H., Peletier, L.A.: Solutions d'equations elliptiques avec exposant de Sobolev critique qui changent de signe. C. R. Acad. Sci. Paris Sér. I Math. **306**(16), 711–714 (1988)
7. Atkinson, F.V., Brezis, H., Peletier, L.A.: Nodal solutions of elliptic equations with critical Sobolev exponents. J. Differ. Equat. **85**(1), 151–170 (1990)
8. Ben Ayed, M., El Mehdi, K., Pacella, F.: Blow-up and nonexistence of sign-changing solutions to the Brezis-Nirenberg problem in dimension three. Ann. Inst. H. Poincaré Anal. Non linéaire **23**(4), 567–589 (2006)
9. Brezis, H., Nirenberg, L.: Positive solutions of nonlinear elliptic equations involving critical Sobolev exponents. Commun. Pure Appl. Math. **36**, 437–477 (1983)

10. Capozzi, A., Fortunato, D., Palmieri, G.: An existence result for nonlinear elliptic problems involving critical Sobolev exponent. Ann. Inst. H. Poincaré Anal. Non linéaire 2(6), 463–470 (1985)
11. Cerami, G., Solimini, S., Struwe, M.: Some existence results for superlinear elliptic boundary value problems involving critical exponents. J. Funct. Anal. **69**, 289–306 (1986)
12. Clapp, M., Weth, T.: Multiple solutions for the Brezis-Nirenberg problem. Adv. Differ. Equat. **10**(4), 463–480 (2005)
13. Gazzola, F., Grunau, H.C.: On the role of space dimension $n = 2 + 2\sqrt{2}$ in the semilinear Brezis-Nirenberg eigenvalue problem. Analysis **20**, 395–399 (2000)
14. Iacopetti, A.: Asymptotic analysis for radial sign-changing solutions of the Brezis-Nirenberg problem. Ann. Mat. Pura Appl. (2014). doi:10.1007/s10231-014-0438-y
15. Iacopetti, A., Pacella, F.: A nonexistence result for sign-changing solutions of the Brezis-Nirenberg problem in low dimensions. Journal of Differential Equations **258**(12), 4180–4208 (2015)
16. Iacopetti, A., Vaira, G.: Sign-changing tower of bubbles for the Brezis-Nirenberg problem. Commun. Contemp. Math. DOI: 10.1142/S0219199715500364
17. Schechter, M., Zou, W.: On the Brezis Nirenberg problem. Arch. Ration. Mech. Anal. **197**, 337–356 (2010)
18. Struwe, M.: Variational Methods - Applications to Nonlinear Partial Differential Equations and Hamiltonian Systems, 4th edn. Springer, New York (2008)

Critical and noncritical regions on the critical hyperbola

Jéssyca Lange Ferreira Melo and Ederson Moreira dos Santos

To Djairo with affection and esteem

Mathematics Subject Classification (2010): 35J30, 35B33, 58E05

1 Introduction

We start referring to the celebrated results of Brezis and Nirenberg [8] on the second order equation

$$\begin{cases} -\Delta u = \lambda u + u^{2^*-1}, \ \ u > 0 \ \text{ in } \ \Omega, \\ u = 0 \ \text{ on } \ \partial\Omega, \end{cases} \tag{BN}$$

where $2^* = 2N/(N-2)$, $N \geq 3$, and $\Omega \subset \mathbb{R}^N$ is a bounded smooth domain. Let $\lambda_1(\Omega)$ be the first eigenvalue of $(-\Delta, H_0^1(\Omega))$. We recall that if $N \geq 4$, then (BN) has a solution if, and only if, $0 < \lambda < \lambda_1(\Omega)$. However, the problem becomes much more delicate in the case of $N = 3$ and a sharp result on existence of solution is known only if Ω is a ball and, in this case, (BN) has a solution if, and only if, $\frac{1}{4}\lambda_1(\Omega) < \lambda < \lambda_1(\Omega)$. For this reason $N = 3$ is referred to as the critical dimension associated with the critical growth problem (BN), whereas $N \geq 4$ are called noncritical dimensions.

J.L.F. Melo (✉)
Departamento de Matemática, Universidade Federal de Viçosa,
CEP 36570-000 Viçosa, MG, Brazil
e-mail: jessycalange@gmail.com

E. Moreira dos Santos
Instituto de Ciências Matemáticas e de Computação, Universidade de São Paulo,
Caixa Postal 668, CEP 13560-970 São Carlos, SP, Brazil
e-mail: ederson@icmc.usp.br

A.N. Carvalho et al. (eds.), *Contributions to Nonlinear Elliptic Equations and Systems*, Progress in Nonlinear Differential Equations and Their Applications 86,
DOI 10.1007/978-3-319-19902-3_21

On the other hand, it is well known that the functions that realize the Sobolev constant

$$\inf\left\{\int_{\mathbb{R}^N} |\nabla u|^2 dx;\ u \in \mathscr{D}^{1,2}(\mathbb{R}^N),\ \int_{\mathbb{R}^N} |u|^{2^*}\,dx = 1\right\},$$

behave like $|x|^{-(N-2)}$, as $|x| \to \infty$. According to the arguments in [8, Lemmas 1.1 and 1.2], the definition of critical and noncritical dimensions associated with (BN) is also linked to the L^2-integrability of $|x|^{-(N-2)}$ in $\mathbb{R}^N \backslash B_1(0)$, where $B_1(0)$ stands for the open unit ball in $\mathbb{R}^N$, and $2(N-2) \geq N$ corresponds to the noncritical dimensions of (BN), whereas $2(N-2) < N$ corresponds to the critical dimension. This turns out to be an alternative, although less precise, manner to characterize the critical and noncritical dimensions associated with (BN).

Similar asymptotic analysis was introduced in [4, §7 and §8] to classify the critical and noncritical dimensions associated with the corresponding critical growth problem involving the p-Laplacian

$$\begin{cases} -\Delta_p u = \lambda u^{p-1} + u^{p^*-1}, & u > 0 \text{ in } \Omega, \\ u = 0 \text{ on } \partial\Omega, \end{cases} \tag{P_p}$$

where $p^* = pN/(N-p)$ for $N > p$. The functions that realize

$$\inf\left\{\int_{\mathbb{R}^N} |\nabla u|^p dx;\ u \in \mathscr{D}^{1,p}(\mathbb{R}^N),\ \int_{\mathbb{R}^N} |u|^{p^*}\,dx = 1\right\},$$

behave like $|x|^{-\frac{N-p}{p-1}}$ as $|x| \to \infty$ and, in this framework, the definition of critical and noncritical dimensions associated with the problem (P_p) is related to the L^p-integrability of $|x|^{-\frac{N-p}{p-1}}$ in $\mathbb{R}^N \backslash B_1(0)$, and $N \geq p^2$ are the noncritical dimensions of (P_p), whereas $p < N < p^2$ are the critical dimensions of (P_p).

In this paper we consider the Hamiltonian elliptic system

$$\begin{cases} -\Delta u = |v|^{p-1}v \text{ in } \Omega, \\ -\Delta v = \mu |u|^{s-1}u + |u|^{q-1}u \text{ in } \Omega, \\ u, v = 0 \text{ on } \partial\Omega, \end{cases} \tag{S}$$

where $\Omega \subset \mathbb{R}^N$ is a bounded smooth domain, $N \geq 3$, $\mu > 0$, and we are interested in existence of positive classical solutions, i.e., solutions such that $u, v > 0$ in Ω and $u, v \in C^2(\Omega) \cap C_0(\overline{\Omega})$.

Throughout in this paper, even if not explicitly stated, we assume that the point (p, q) lies on the critical hyperbola

$$\frac{1}{p+1}+\frac{1}{q+1}=\frac{N-2}{N}, \tag{1}$$

which was introduced by de Figueiredo et al. [9] and van der Vorst [32], see also [11, 25–27], and that s satisfies

$$\frac{p+1}{p}\leq s+1<q+1. \tag{2}$$

One can use various variational settings to deal with the system (S), see, for instance, the surveys [7, 10, 31]. Here we rewrite (S) as the fourth order equation under Navier boundary conditions

$$\begin{cases} \Delta(|\Delta u|^{\frac{1}{p}-1}\Delta u)=\mu|u|^{s-1}u+|u|^{q-1}u \text{ in } \Omega, \\ u, \Delta u=0 \text{ on } \partial\Omega, \end{cases} \tag{E}$$

and the positivity of u and v corresponds to $u, -\Delta u>0$ in Ω. Then note that condition $\frac{p+1}{p}<s+1$ says that the perturbation $\mu|u|^{s-1}u$ is "superlinear" with respect to $\Delta(|\Delta u|^{\frac{1}{p}-1}\Delta u)$, whereas $s=1/p$ implies that $|u|^{s-1}u$ and $\Delta(|\Delta u|^{\frac{1}{p}-1}\Delta u)$ have the same homogeneity. Moreover, the condition $s+1<q+1$ means that the term $\mu|u|^{s-1}u$ is a lower order perturbation in the Lane-Emden system

$$-\Delta u=|v|^{p-1}v \text{ in } \Omega, \qquad -\Delta v=|u|^{q-1}u \text{ in } \Omega, \qquad u, v=0 \text{ on } \partial\Omega. \tag{3}$$

The contribution of this paper is twofold: to indicate that the location, *critical* or *noncritical*, of the point (p, q) on the critical hyperbola (1), cf. Definition 1.1, can interfere on the existence of positive solutions for the critical growth system (S); inspired by the works of Rey [30] and Lazzo [20] on the second order equation (BN), we prove that if Ω has a rich topology, described by its Lusternik-Schnirelmann category, then the system (S) has multiple positive solutions, at least as many as $\text{cat}_\Omega(\Omega)$, in case the parameter $\mu>0$ is sufficiently small and if s satisfies some suitable and natural conditions, namely condition (H) ahead, which depend on the *critical* or *noncritical* location of (p, q).

The suitable environment to work with (E), and so with (S), is the space $E(\Omega):=W^{2,\frac{p+1}{p}}(\Omega)\cap W_0^{1,\frac{p+1}{p}}(\Omega)$. Consider the Sobolev constant associated with the critical embedding $E(\Omega)\hookrightarrow L^{q+1}(\Omega)$, namely

$$S(\Omega):=\inf\left\{\int_\Omega|\Delta u|^{\frac{p+1}{p}}dx;\ u\in E(\Omega), \int_\Omega|u|^{q+1}dx=1\right\}. \tag{4}$$

It is known that $S(\Omega)$ does not depend on Ω, $S(\Omega)$ is not achieved and that $S(\Omega)=S$ with

$$S := \inf\left\{\int_{\mathbb{R}^N} |\Delta u|^{\frac{p+1}{p}}dx;\ u \in \mathscr{D}^{2,\frac{p+1}{p}}(\mathbb{R}^N),\ \int_{\mathbb{R}^N} |u|^{q+1}dx = 1\right\}. \tag{5}$$

Lions [21] proved that S is achieved and that all of the solutions of S have definite sign and are radially symmetric with respect to some point and are radially monotone with respect to the point of symmetry. Let $\varphi \in \mathscr{D}^{2,\frac{p+1}{p}}(\mathbb{R}^N)$ be an extremal function that realizes S, positive and radially symmetric with respect to the origin. Lions [21, Corollary I.1] proved that all of the positive solutions that realize S are given by

$$\varphi_{\delta,a}(x) = \delta^{-\frac{N}{q+1}}\varphi\left(\frac{x-a}{\delta}\right),\quad x \in \mathbb{R}^N,\ \ a \in \mathbb{R}^N \text{ and } \delta > 0. \tag{6}$$

Due to the scaling invariance (6), we can choose φ with $\varphi(0) = 1$. We recall that the functions $S^\kappa\varphi_{\delta,a}$, with $\varphi_{\delta,a}$ as in (6) and $\kappa = \frac{p(p(N-2)-2)}{2(p+1)^2}$, are precisely the regular positive solutions of

$$\Delta(|\Delta u|^{\frac{1}{p}-1}\Delta u) = u^q \text{ in } \mathbb{R}^N.$$

According to [18, (1.14), (1.15), and (1.16)], the function φ has the following asymptotic behavior:

$$\lim_{r\to\infty} r^{p(N-2)-2}\varphi(r) = b \text{ if } p < \frac{N}{N-2},\qquad \lim_{r\to\infty}\frac{r^{N-2}}{\log r}\varphi(r) = b \text{ if } p = \frac{N}{N-2}, \tag{7}$$

and

$$\lim_{r\to\infty} r^{N-2}\varphi(r) = b \text{ if } p > \frac{N}{N-2}, \tag{8}$$

where $b > 0$ is a constant and $r = |x|$.

Observe that, in the case of $p < \frac{N}{N-2}$, the function φ behaves like $|x|^{-[p(N-2)-2]}$ as $|x| \to \infty$. Hence, as in [8, Lemma 1.2], to study the existence of a solution to (E), equivalently to (S), in the case of $s = 1/p$, it is natural to require

$$[p(N-2)-2]\left(\frac{p+1}{p}\right) \geq N,\ \text{ or equivalently, }\ \frac{2+\sqrt{2N}}{N-2} \leq p < \frac{N}{N-2}.$$

Note that $\frac{2+\sqrt{2N}}{N-2} < \frac{N}{N-2}$ if, and only if, $N \geq 6$. Moreover, observe that

$$[p(N-2)-2]\left(\frac{p+1}{p}\right) \geq N \text{ is equivalent to } N \geq 2\left(\frac{p+1}{p}\right)^2.$$

Then, we point out that the last inequality seems very natural when compared with the noncritical dimensions associated with the biharmonic operator, i.e., the case of $p = 1$ and $N \geq 8$; cf. [14, 28, 33]. In addition, since $\Delta(|\Delta|^{\frac{1}{p}-1}\Delta)$ is a perturbation of the biharmonic operator as well as the p-Laplacian is a perturbation of the Laplacian operator, it is reasonable to compare the condition $N \geq 2\left(\frac{p+1}{p}\right)^2$ for the operator $\Delta(|\Delta|^{\frac{1}{p}-1}\Delta)$ with the condition $N \geq p^2$ for the p-Laplacian operator.

In the case of $p \geq \frac{N}{N-2}$, φ behaves like $|x|^{-(N-2)}$ as $|x| \to \infty$ and in this case we are induced to consider

$$(N-2)\left(\frac{p+1}{p}\right) \geq N, \quad \text{or equivalently,} \quad \frac{N}{N-2} \leq p \leq \frac{N-2}{2}$$

and again, the inequality $\frac{N}{N-2} \leq \frac{N-2}{2}$ is valid if, and only if, $N \geq 6$.

Then we introduce the notion of *critical* and *noncritical* regions (on the critical hyperbola) associated with the system (S).

Definition 1.1. Consider the system (S). The point (p, q) of the critical hyperbola (1) is on a:

(a) Noncritical region, if $N \geq 6$ and p satisfies

$$\frac{2+\sqrt{2N}}{N-2} \leq p \leq \frac{N-2}{2}. \tag{9}$$

(b) Critical region, if $N \geq 6$ and p does not satisfy (9); or $N = 3, 4, 5$ and any p.

Note that the noncritical region associated with (S) is not necessarily symmetric with respect to the line $p = q$ because (S) is not a symmetric perturbation of the system (3).

Throughout in this paper the hypothesis (H), with respect to s, indicates the following situation

$$\left.\begin{array}{l}
\textbf{(a)} \text{ In the noncritical region, i.e., } N \geq 6 \text{ and } \frac{2+\sqrt{2N}}{N-2} \leq p \leq \frac{N-2}{2}, \text{ suppose} \\
\qquad \frac{p+1}{p} \leq s+1 < q+1. \\
\textbf{(b)} \text{ In the critical regions:} \\
\quad \text{(b1) } N \geq 6 \text{ and } \frac{2}{N-2} < p < \frac{2+\sqrt{2N}}{N-2}, \text{ suppose } q-p < s+1 < q+1; \\
\quad \text{(b2) } N \geq 6 \text{ and } p > \frac{N-2}{2}, \text{ suppose } \frac{(p-1)(q+1)}{p} < s+1 < q+1; \\
\quad \text{(b3) } N = 3,4,5 \text{ and } \frac{2}{N-2} < p < \frac{N}{N-2}, \text{ suppose } q-p < s+1 < q+1; \\
\quad \text{(b4) } N = 3,4,5 \text{ and } p \geq \frac{N}{N-2}, \text{ suppose } \frac{(p-1)(q+1)}{p} < s+1 < q+1,
\end{array}\right\} \tag{H}$$

which is quite natural to study (S); cf. the proof of Lemma 2.6. Indeed, consider the equivalent formulation of (S) as the fourth order equation (E). First, observe that in the cases of (b1) and (b3) the inequality $\frac{p+1}{p} < q - p$ holds, and that in the cases

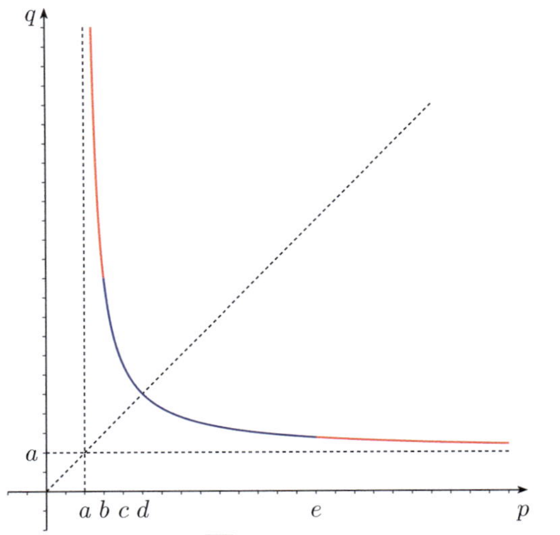

(b2) and (b4) then $\frac{p+1}{p} < \frac{(p-1)(q+1)}{q}$; cf. (41) and (43) ahead for more details. Then we mention that a hypothesis similar to (H) was introduced in [8, eq. (0.6)] to study the problem

$$-\Delta u = \lambda u^s + u^{2^*-1}, \quad u > 0 \quad \text{in} \quad \Omega, \ u = 0 \quad \text{on} \quad \partial\Omega$$

and in [4, §7 and §8] and [5, Section 3] to study

$$-\Delta_p u = \lambda u^s + u^{p^*-1}, \ u > 0 \text{ in } \Omega, \ u = 0 \text{ on } \partial\Omega.$$

Moreover, in the biharmonic case, that is, in the case of $p = 1$, the hypothesis (H) turns out to be the hypothesis on $s + 1$ assumed in [23, Theorem 1.1], in such a way that the hypothesis in the noncritical region is equivalent to the hypothesis in the noncritical dimensions $N \geq 8$, whereas the hypotheses in the critical regions correspond to the hypothesis in the critical dimensions $N = 5, 6, 7$.

We will denote by $\mu_1(\Omega)$ the first eigenvalue of the problem

$$\begin{cases} \Delta(|\Delta u|^{\frac{1}{p}-1}\Delta u) = \mu|u|^{\frac{1}{p}-1}u & \text{in} \quad \Omega, \\ u, \Delta u = 0 \quad \text{on} \quad \partial\Omega, \end{cases} \tag{10}$$

and we state the first result in this paper.

Theorem 1.2. *Let $\mu > 0$ and in the case of $s = 1/p$ also assume that $\mu < \mu_1(\Omega)$. If* (H) *is satisfied, then* (S) *has a classical positive solution.*

We point out that Theorem 1.2 includes all of the results in [19, Theorem 2] regarding the system (S), since [19, Theorem 2] treats the particular case of $s = 1$ and, in this case, our hypothesis (H) is equivalent to the hypothesis in [19, Theorem 2]. Moreover, we believe that it is more natural to study (S) with $s = 1/p$ instead of $s = 1$.

Next, based on some topological arguments of Lusternik and Schnirelmann [22], we prove existence of multiple positive solutions to the system (S).

Theorem 1.3. *If* (H) *is satisfied, then there exists* $\overline{\mu} > 0$ *such that, for each* $0 < \mu < \overline{\mu}$, *the system* (S) *has at least* $cat_{\Omega}(\Omega)$ *classical positive solutions.*

To prove the results in this paper we consider the equivalent formulation of (S) as the fourth order equation (E). We follow some of the arguments in [23, 24], which consider (E) in the particular case of $p = 1$, i.e., the corresponding problem involving the biharmonic operator. However, in the nonlinear regime of $\Delta(|\Delta|^{\frac{1}{p}-1}\Delta)$ some extra difficulties have be overcame. In particular, in [24] and [23], the comparison principle for the biharmonic operator under Navier boundary conditions is the key argument to get the positivity of the solutions at the Lusternik-Schnirelmann critical levels. However, the same procedure seems not suitable in the nonlinear setting and then we use, instead, an energy argument, namely Lemma 5.1, along with the regularity result of Lemma 2.2. Finally, we mention that we also present some results on qualitative properties on the ground state solutions of the system (S), which are included in Section 3.

2 Compactness and proof of Theorem 1.2

First we fix some notations. We recall that a strong solution of (S) is a pair (u, v) with

$$u \in W^{2,\frac{p+1}{p}}(\Omega) \cap W_0^{1,\frac{p+1}{p}}(\Omega), \qquad v \in W^{2,\frac{q+1}{q}}(\Omega) \cap W_0^{1,\frac{q+1}{q}}(\Omega)$$

satisfying the system in (S) for a.e. $x \in \Omega$. We consider the space $E(\Omega) := W^{2,\frac{p+1}{p}}(\Omega) \cap W_0^{1,\frac{p+1}{p}}(\Omega)$ endowed with the norm $\|u\| := |\Delta u|_{\frac{p+1}{p}}$. If $0 < \mu < \mu_1(\Omega)$, then we set

$$\|u\|_\mu := \left(|\Delta u|_{\frac{p+1}{p}}^{\frac{p+1}{p}} - \mu |u|_{\frac{p+1}{p}}^{\frac{p+1}{p}} \right)^{\frac{p}{p+1}}, \quad \forall\, u \in E(\Omega). \tag{11}$$

So $\|\cdot\|_\mu$ satisfies

$$\|u\|_\mu \le \|u\| \le c(\Omega,\mu)\|u\|_\mu \quad \text{and} \quad \|tu\|_\mu = |t|\|u\|_\mu \;\; \forall\, u \in E(\Omega),\ \forall\, t \in \mathbb{R}, \tag{12}$$

where $c(\Omega,\mu) = \left(1 - \frac{\mu}{\mu_1(\Omega)}\right)^{-\frac{p}{p+1}} > 0$.

To prove existence of classical solutions (u, v) to the system (S), we rewrite (S) as the fourth order equation (E). We recall that $\Omega \subset \mathbb{R}^N$, with $N \geq 3$, stands for a bounded smooth domain and that (1) and (2) are our basic assumptions. Associated with (E), we consider the $C^1(E(\Omega), \mathbb{R})$ functional

$$I_\mu(u) := \frac{p}{p+1}\int_\Omega |\Delta u|^{\frac{p+1}{p}} dx - \frac{\mu}{s+1}\int_\Omega |u|^{s+1} dx - \frac{1}{q+1}\int_\Omega |u|^{q+1} dx. \qquad (13)$$

Definition 2.1. We say that $u \in E(\Omega)$ is a weak solution of (E) if u is a critical point of I_μ, that is, if u satisfies

$$\int_\Omega |\Delta u|^{\frac{1}{p}-1}\Delta u \Delta v dx = \mu \int_\Omega |u|^{s-1} uv dx + \int_\Omega |u|^{q-1} uv dx, \ \forall\, v \in E(\Omega).$$

We say that u is a classical solution of (E) if $u \in \mathscr{C}^2(\overline{\Omega})$, $|\Delta u|^{\frac{1}{p}-1}\Delta u \in \mathscr{C}^2(\Omega)$ and u satisfies (E) pointwise.

Hence, if u is a classical solution of (E) and if we set $v = |\Delta u|^{\frac{1}{p}-1}(-\Delta u)$, then (u, v) is a classical solution of (S).

Lemma 2.2. *If u is a weak solutions of* (E)*, then it is classical solution of* (E) *and vice versa.*

Proof. Let $u \in E(\Omega)$ be a weak solution of (E) and set $v = |\Delta u|^{\frac{1}{p}-1}(-\Delta u)$. As in [12, Section 4], we can show that $v \in W^{2,\frac{q+1}{q}}(\Omega) \cap W_0^{1,\frac{q+1}{q}}(\Omega)$ and that (u, v) is a strong solution of (S). Then we argue as in [17, Section 3] to show that $u, v \in L^r(\Omega)$ for all $1 \leq r < \infty$. So, we apply the classical regularity results for second order elliptic equations to each of the equations of the system (S) to get that $(u, v) \in C^{2,\alpha}(\overline{\Omega}) \times C^{2,\alpha}(\overline{\Omega})$ for some α which depends on p and q. On the other hand, it is clear that any classical solution of (E) is a weak solution of (E). □

Lemma 2.3. *Let $\mu > 0$ and in the case of $s = 1/p$ also assume that $\mu < \mu_1(\Omega)$. Then the functional I_μ has a mountain pass geometry around its local minimum at zero, with associated mountain pass level*

$$c_\mu = \inf_{\gamma \in \Gamma} \max_{t \in [0,1]} I_\mu(\gamma(t)), \qquad (14)$$

where $\Gamma = \{\gamma \in \mathscr{C}([0, 1], E(\Omega)); \gamma(0) = 0, I_\mu(\gamma(1)) < 0\}$. Moreover,

$$c_\mu = \tilde{c}_\mu = \hat{c}_\mu, \qquad (15)$$

where

$$\tilde{c}_\mu = \inf_{u\in E(\Omega)\setminus\{0\}} \max_{t\geq 0} I_\mu(tu), \qquad \hat{c}_\mu = \inf_{u\in \mathcal{N}_\mu} I_\mu(u), \tag{16}$$

and $\mathcal{N}_\mu$ is the Nehari manifold

$$\mathcal{N}_\mu := \{u \in E(\Omega)\setminus\{0\}; I'_\mu(u)u = 0\}. \tag{17}$$

Proof. The proof that I_μ has a mountain pass geometry around its local minimum at zero is quite standard and will be omitted. The proof of the identities at (15) follows as in [29, Proposition 3.11]. □

Lemma 2.4. *Let $\mu > 0$ and in the case of $s = 1/p$ also assume that $\mu < \mu_1(\Omega)$. Then any (PS)-sequence of I_μ is bounded.*

Proof. Suppose that $I_\mu(u_n) \to d$ and $I'_\mu(u_n) \to 0$. Then there exist $C > 0$ and $(\sigma_n) \subset [0, +\infty)$, $\sigma_n \to 0$, such that, for all $n \in \mathbb{N}$ and $w \in E(\Omega)$

$$\begin{aligned}|I_\mu(u_n)| = \bigg|&\frac{p}{p+1}\int_\Omega |\Delta u_n|^{\frac{p+1}{p}}dx - \mu\frac{p}{p+1}\int_\Omega |u_n|^{\frac{p+1}{p}}\\ &\times dx - \frac{1}{q+1}\int_\Omega |u_n|^{q+1}dx\bigg| \leq C\end{aligned}$$

and

$$\begin{aligned}|I'_\mu(u_n)w| &= \left|\int_\Omega |\Delta u_n|^{\frac{1}{p}-1}\Delta u_n \Delta w dx - \mu\int_\Omega |u_n|^{\frac{1}{p}-1}u_n w dx - \int_\Omega |u_n|^{q-1}u_n w dx\right|\\ &\leq \sigma_n\|w\|.\end{aligned}$$

Case 1: $s = \frac{1}{p}$ and $0 < \mu < \mu_1(\Omega)$. From (11) and (12) we infer that

$$\begin{aligned}(q+1)I_\mu(u_n) - I'_\mu(u_n)u_n &= \left[\frac{p}{p+1}(q+1) - 1\right]\int_\Omega [|\Delta u_n|^{\frac{p+1}{p}} - \mu|u_n|^{\frac{p+1}{p}}]dx\\ &\geq \frac{1}{c(\Omega)}\frac{pq-1}{p+1}\|u_n\|^{\frac{p+1}{p}},\end{aligned}$$

whence

$$\frac{1}{c(\Omega)}\frac{pq-1}{p+1}\|u_n\|^{\frac{p+1}{p}} \leq (q+1)I_\mu(u_n) - I'_\mu(u_n)u_n \leq (q+1)C + \sigma_n\|u_n\|,$$

which implies that (u_n) is bounded in $E(\Omega)$.

Case 2: $\frac{p+1}{p} < s+1 < q+1$ and $\mu > 0$. In this case $ps > 1$ and

$$I_\mu(u_n) - \frac{1}{s+1} I'_\mu(u_n)u_n = \left(\frac{p}{p+1} - \frac{1}{s+1}\right)\|u_n\|^{\frac{p+1}{p}} + \left(\frac{1}{s+1} - \frac{1}{q+1}\right)|u_n|_{q+1}^{q+1}$$
$$\geq \frac{ps-1}{(p+1)(s+1)}\|u_n\|^{\frac{p+1}{p}},$$

whence

$$\frac{ps-1}{(p+1)(s-1)}\|u_n\|^{\frac{p+1}{p}} \leq I_\mu(u_n) - \frac{1}{s+1} I'_\mu(u_n)u_n \leq C + \frac{1}{s+1}\sigma_n\|u_n\|$$

and so that (u_n) is bounded in $E(\Omega)$.

□

In order to apply the classical mountain pass theorem [3], it is necessary to know at which levels the functional I_μ satisfies the (PS)-condition. This is done at Lemma 2.5, whose proof is very similar to the proof of [13, Proposition 3.2].

Lemma 2.5. *Let $\mu > 0$ and in the case of $s = 1/p$ also assume that $\mu < \mu_1(\Omega)$. Then I_μ satisfies the $(PS)_c$-condition for all $c < \frac{2}{N}S^{\frac{pN}{2(p+1)}}$.*

Lemma 2.6. *Let $\mu > 0$ and in the case of $s = 1/p$ also assume that $\mu < \mu_1(\Omega)$. If* (H) *is satisfied, then the mountain pass level c_μ of the functional I_μ, given by* (14)*, is such that $c_\mu \in (0, \frac{2}{N}S^{\frac{pN}{2(p+1)}})$.*

Proof. See Appendix 6 ahead. □

The next result is a consequence of Lemmas 2.5 and 2.6, and the classical mountain pass theorem [3].

Proposition 2.7. *Let $\mu > 0$ and in the case of $s = 1/p$ also assume that $\mu < \mu_1(\Omega)$. If* (H) *is satisfied, then the mountain pass level c_μ is a critical value of I_μ, that is, there exists $u_\mu \in E(\Omega)$ such that $I_\mu(u_\mu) = c_\mu$ and $I'_\mu(u_\mu) = 0$.*

Proof (Proof of Theorem 1.2). The existence of a nontrivial classical solution for the problem (E) follows from Lemma 2.2 and Proposition 2.7. Moreover, we have observed that if u is a classical solution of (E) and if we set $v = |\Delta u|^{\frac{1}{p}-1}(-\Delta u)$, then (u, v) is a classical solution of (S). In addition, any solution u of (E) associated with the mountain pass level c_μ satisfies $u, -\Delta u > 0$ in Ω, up to replace u by $-u$; cf. Theorem 3.2 (iii) ahead. □

3 On the ground state solutions of (S)

Various (equivalent) variational settings are available to deal with the system (S); cf. [7, 10, 31]. Each of these settings has an energy functional associated with (S). We recall that, given a solution (u, v) of (S), then all of these energy functionals have the same energy level at (u, v). Therefore, the definition of ground state solution of (S) does not depend on the variational setting under consideration. Here we rewrite (S) as the fourth order equation (E) and hence the notions of ground state solution of (S) and (E) coincide.

Definition 3.1. Let $\mu > 0$. We say that $u \in E(\Omega)$ is a ground state solution of the problem (E) if u is a nontrivial weak solution of (E) and minimizes the energy I_μ among the nontrivial weak solutions of (E).

We point out that the conditions (1), (2) and $\mu > 0$ guarantee that $I_\mu(u) > 0$ for every nontrivial weak solutions u of (E), cf. (21) ahead.

Theorem 3.2 (Ground state solutions).
Let $\mu > 0$ and in the case of $s = 1/p$ also assume that $\mu < \mu_1(\Omega)$. If (H) *is satisfied, then:*

(i) The problem (E) *has a ground state solution.*
(ii) The ground state solutions of (E) *are critical points of I_μ of mountain pass type at the level c_μ.*
(iii) If u is a ground state solution of (E)*, then $u, -\Delta u > 0$ in Ω, up to replace u by $-u$.*
(iv) If Ω is a ball centered in the origin of $\mathbb{R}^N$, then all of the ground state solutions of (E) *are such that u and $-\Delta u$ are radially symmetric.*

Proof. (i) and (ii) We proved, at Proposition 2.7, the existence of a mountain pass solution of (E) associated with the critical level c_μ. From the identities given by (15), it follows that any mountain pass solution of (E) at the level c_μ is a ground state solution and vice versa.

(iii) Let u be a ground state solution of (E). According to Lemma 2.2, we know that $u \in \mathscr{C}^2(\overline{\Omega})$. So, by the strong maximum principle, it is enough to prove that Δu has definite sign in Ω. By contradiction, suppose that Δu changes sign in Ω. Let w be the solution of

$$-\Delta w = |\Delta u| \text{ in } \Omega, \qquad w = 0 \text{ on } \partial\Omega.$$

So, by the strong maximum principle, $w > |u|$ in Ω. Then we infer that

$$
\begin{aligned}
c_\mu &\le \max_{t\ge 0} I_\mu(tw)\\
&= \max_{t\ge 0}\left\{\frac{p}{p+1}t^{\frac{p+1}{p}}\int_\Omega |\Delta w|^{\frac{p+1}{p}}dx - \frac{\mu}{s+1}t^{s+1}\int_\Omega |w|^{s+1}dx - \frac{t^{q+1}}{q+1}\int_\Omega |w|^{q+1}dx\right\}\\
&< \max_{t\ge 0}\left\{\frac{p}{p+1}t^{\frac{p+1}{p}}\int_\Omega |\Delta u|^{\frac{p+1}{p}}dx - \frac{\mu}{s+1}t^{s+1}\int_\Omega |u|^{s+1}dx - \frac{t^{q+1}}{q+1}\int_\Omega |u|^{q+1}dx\right\}\\
&\qquad\qquad\qquad\qquad\qquad\qquad\qquad\qquad\qquad\qquad\qquad\qquad = \max_{t\ge 0} I_\mu(tu) = c_\mu,
\end{aligned}
$$

which is a contradiction.

(iv) Suppose $\Omega = B_r = B_r(0)$. Let u be a ground state solution of (E). According to item (iii) of this theorem, we can assume that $u, -\Delta u > 0$ in B_r. Denote by u^* and $(-\Delta u)^*$ Schwarz symmetrization of u and $-\Delta u$, respectively. Let w be the solution of

$$
-\Delta w = (-\Delta u)^* \ \text{ in } B_r, \qquad w = 0 \ \text{ on } \partial B_r.
$$

Then $w = w^*$ and it is enough to show that $u = w$. By [2], see also [21, p. 165] and [6, Lemma 2.8], we have $w \ge u^*$ and

$$
|w > u^*| = 0 \Leftrightarrow -\Delta u = (-\Delta u)^*.
$$

If $|w > u^*| > 0$, then

$$
\begin{aligned}
c_\mu &\le \max_{t\ge 0} I_\mu(tv)\\
&= \max_{t\ge 0}\left\{\frac{p}{p+1}t^{\frac{p+1}{p}}\int_{B_r} |\Delta w|^{\frac{p+1}{p}}dx - \frac{\mu}{s+1}t^{s+1}\int_{B_r} |w|^{s+1}dx - \frac{t^{q+1}}{q+1}\int_{B_r} |w|^{q+1}dx\right\}\\
&< \max_{t\ge 0}\left\{\frac{pt^{\frac{p+1}{p}}}{p+1}\int_{B_r} |(-\Delta u)^*|^{\frac{p+1}{p}}dx - \frac{\mu t^{s+1}}{s+1}\int_{B_r} |u^*|^{s+1}dx - \frac{t^{q+1}}{q+1}\int_{B_r} |u^*|^{q+1}dx\right\}\\
&= \max_{t\ge 0}\left\{\frac{p}{p+1}t^{\frac{p+1}{p}}\int_{B_r} |-\Delta u|^{\frac{p+1}{p}}dx - \frac{\mu}{s+1}t^{s+1}\int_{B_r} |u|^{s+1}dx - \frac{t^{q+1}}{q+1}\int_{B_r} |u|^{q+1}dx\right\}\\
&\qquad\qquad\qquad\qquad\qquad\qquad\qquad\qquad\qquad\qquad\qquad\qquad = \max_{t\ge 0} I_\mu(tu) = c_\mu
\end{aligned}
$$

which cannot happen. Thus, $-\Delta u = (-\Delta u)^*$ and since u and w are both solutions to the problem

$$
-\Delta z = (-\Delta u)^* \ \text{ in } B_r, \qquad z = 0 \ \text{ on } \partial B_r,
$$

it follows that $u = w$. □

4 More on compactness

In this section we prove a compactness result, namely Lemma 4.2, which is essential for the proof of Theorem 1.3. With this lemma we will be able to compare the category of Ω with the category of an appropriated level set of the functional I_μ to which we can conveniently apply the Lusternik-Schnirelmann theory. We stress that the main ingredient in the proof of Lemma 4.2 is a compactness result due to Lions [21, Corollary I.2].

Lemma 4.1 ([21], Corollary I.2).

Let $(u_n) \subset \mathscr{D}^{2,\frac{p+1}{p}}(\mathbb{R}^N)$ *such that* $|u_n|_{q+1} = 1$ *and* $\|u_n\|^{\frac{p+1}{p}} = |\Delta u_n|^{\frac{p+1}{p}}_{\frac{p+1}{p}} \to S.$

Then there exist $(y_n, \alpha_n) \subset \mathbb{R}^N \times (0, +\infty)$ *and* $v \in \mathscr{D}^{2,\frac{p+1}{p}}(\mathbb{R}^N)$ *such that, up to a subsequence,*

$$v_n \to v \ \text{in} \ \mathscr{D}^{2,\frac{p+1}{p}}(\mathbb{R}^N) \ \text{with} \ v_n(x) := \alpha_n^{\frac{N}{q+1}} u_n(\alpha_n x + y_n).$$

Lemma 4.2. *Let* $(u_n) \subset E(\Omega)$ *be a sequence such that*

$$|u_n|_{q+1} = 1 \ \text{ and } \quad \|u_n\|^{\frac{p+1}{p}} = |\Delta u_n|^{\frac{p+1}{p}}_{\frac{p+1}{p}} = S + o_n(1).$$

For each $n \in \mathbb{N}$, *let* w_n *be the Newtonian potential of* $\widetilde{|-\Delta u_n|}$, *where* $\sim$ *denotes the zero extension outside* Ω. *Then, there exist a sequence* $(y_n, \alpha_n) \subset \mathbb{R}^N \times (0, +\infty)$ *and* $v \in \mathscr{D}^{2,\frac{p+1}{p}}(\mathbb{R}^N)$ *such that, up to a subsequence,*

$$v_n \to v \ \text{in} \ \mathscr{D}^{2,\frac{p+1}{p}}(\mathbb{R}^N) \ \text{where} \ v_n(x) := \alpha_n^{\frac{N}{q+1}} w_n(\alpha_n x + y_n).$$

Moreover, $y_n \to y \in \overline{\Omega}$ *and* $\alpha_n \to 0$.

Proof. We recall that, if w_n is the Newtonian potential of $\widetilde{|-\Delta u_n|}$, then by [16, Theorem 9.9] we know that $w_n \in \mathscr{D}^{2,\frac{p+1}{p}}(\mathbb{R}^N)$ and

$$-\Delta w_n = \widetilde{|-\Delta u_n|} \ \text{a.e. in } \mathbb{R}^N, \tag{18}$$

whence we infer that

$$|w_n|_{q+1} \geq |\widetilde{u_n}|_{q+1} = |u_n|_{q+1} = 1 \ \text{ and } \ |\Delta w_n|^{\frac{p+1}{p}}_{\frac{p+1}{p}} = |\Delta u_n|^{\frac{p+1}{p}}_{\frac{p+1}{p}} = S + o_n(1).$$

Define $z_n = \frac{w_n}{|w_n|_{q+1}}$. Then $z_n \in \mathscr{D}^{2,\frac{p+1}{p}}(\mathbb{R}^N)$, $|z_n|_{q+1} = 1$ and $S \leq |\Delta z_n|_{\frac{p+1}{p}}^{\frac{p+1}{p}}$. Since (w_n) is bounded in $\mathscr{D}^{2,\frac{p+1}{p}}(\mathbb{R}^N)$, then (w_n) is bounded in $L^{q+1}(\mathbb{R}^N)$ and, up to a subsequence, $\lim |w_n|_{q+1} = a \geq 1$. Suppose $a > 1$. Then, for n large,

$$S \leq |\Delta z_n|_{\frac{p+1}{p}}^{\frac{p+1}{p}} = \frac{|\Delta w_n|_{\frac{p+1}{p}}^{\frac{p+1}{p}}}{|w_n|_{q+1}^{\frac{p+1}{p}}} = \frac{S + o_n(1)}{|w_n|_{q+1}^{\frac{p+1}{p}}} < S,$$

a contradiction. So, $\lim |w_n|_{q+1} = 1$, whence we infer that

$$|z_n|_{q+1} = 1 \quad \text{and} \quad |\Delta z_n|_{\frac{p+1}{p}}^{\frac{p+1}{p}} = S + o_n(1).$$

It follows from Lemma 4.1 that, there exist $(y_n, \alpha_n) \subset \mathbb{R}^N \times (0, +\infty)$ and $v \in \mathscr{D}^{2,\frac{p+1}{p}}(\mathbb{R}^N)$ such that, up to a subsequence, and since $\lim |w_n|_{q+1} = 1$,

$$v_n \to v \quad \text{in} \quad \mathscr{D}^{2,\frac{p+1}{p}}(\mathbb{R}^N) \quad \text{with} \quad v_n(x) := \alpha_n^{\frac{N}{q+1}} w_n(\alpha_n x + y_n).$$

Then we argue as in Step 4 in the proof of [23, Lemma 3.2], to show that $y_n \to y \in \overline{\Omega}$ and $\alpha_n \to 0$. □

5 Multiplicity of solutions: proof of Theorem 1.3

The multiple positive solution of (S) will be obtained as critical points of the functional I_μ. As we shall see, all of these solutions have critical levels below $\frac{2}{N} S^{\frac{pN}{2(p+1)}}$ and their positivity is a consequence of the next lemma.

Lemma 5.1. *There exists $0 < \tilde{\mu} < \mu_1(\Omega)$ such that if $0 < \mu < \tilde{\mu}$, then any nontrivial critical point $u \in E(\Omega)$ of I_μ with $I_\mu(u) < \frac{2}{N} S^{\frac{pN}{2(p+1)}}$ satisfies $u, -\Delta u > 0$ in Ω, up to replace u by $-u$.*

Proof. If $u \in E(\Omega)$ is a critical point of I_μ, then

$$\int_\Omega |\Delta u|^{\frac{1}{p}-1} \Delta u \Delta v dx = \mu \int_\Omega |u|^{s-1} u v dx + \int_\Omega |u|^{q-1} u v dx, \ \forall\, v \in E(\Omega), \tag{19}$$

$$\int_\Omega |\Delta u|^{\frac{p+1}{p}} dx = \mu \int_\Omega |u|^{s+1} dx + \int_\Omega |u|^{q+1} dx, \tag{20}$$

and since $\frac{p}{p+1} - \frac{1}{q+1} = \frac{2}{N}$,

$$I_\mu(u) = \left(\frac{p}{p+1} - \frac{1}{s+1}\right)\mu \int_\Omega |u|^{s+1}dx + \frac{2}{N}\int_\Omega |u|^{q+1}dx. \tag{21}$$

Moreover, by (2), $\frac{p}{p+1} - \frac{1}{s+1} \geq 0$. Therefore, from (20) and (21), we infer that all of the solutions u that satisfy $I_\mu(u) < \frac{2}{N}S^{\frac{pN}{2(p+1)}}$ are a priori bounded, that is, there exists $C > 0$ such that

$$\|u\| \leq C. \tag{22}$$

Suppose that $-\Delta u$ changes sign in Ω and let u_1 and u_2 be the solutions of

$$\begin{cases} -\Delta u_1 = (-\Delta u)^+ & \text{in } \Omega, \\ u_1 = 0 & \text{on } \partial\Omega, \end{cases} \quad \text{and} \quad \begin{cases} -\Delta u_2 = -(-\Delta u)^- & \text{in } \Omega, \\ u_2 = 0 & \text{on } \partial\Omega, \end{cases}$$

where $(-\Delta u)^+ = \max\{-\Delta u, 0\}$ and $(-\Delta u)^- = \max\{-(-\Delta u), 0\}$. It follows that $u_1, u_2 \in E(\Omega)$, $u = u_1 + u_2$, $u_1 \geq 0$ in Ω and $u_2 \leq 0$ in Ω. In addition

$$\int_\Omega |\Delta u|^{\frac{p+1}{p}}dx = \int_\Omega |\Delta u_1|^{\frac{p+1}{p}}dx + \int_\Omega |\Delta u_2|^{\frac{p+1}{p}}dx, \tag{23}$$

and, for $i = 1, 2$, the following inequalities hold

$$|u(x)|^{s-1}u(x)u_i(x) \leq |u_i(x)|^{s+1}, \quad |u(x)|^{q-1}u(x)u_i(x) \leq |u_i(x)|^{q+1} \quad \text{a.e. } x \in \Omega; \tag{24}$$

cf. [15, eq. (16)] for a similar argument. Thus, from (4), (19), (24), and from embedding $L^{q+1}(\Omega) \hookrightarrow L^{s+1}(\Omega)$ we infer that

$$\begin{aligned} S|u_i|_{q+1}^{\frac{p+1}{p}} \leq |\Delta u_i|_{\frac{p+1}{p}}^{\frac{p+1}{p}} &= \int_\Omega |\Delta u|^{\frac{1}{p}-1}\Delta u \Delta u_i dx \\ &= \mu \int_\Omega |u|^{s-1}uu_i dx + \int_\Omega |u|^{q-1}uu_i dx \\ &\leq \mu \int_\Omega |u_i|^{s+1}dx + \int_\Omega |u_i|^{q+1}dx \leq \mu C|u_i|_{q+1}^{s+1} + |u_i|_{q+1}^{q+1}, \end{aligned}$$

whence, since $u_i \neq 0$,

$$|u_i|_{q+1}^{\frac{pq-1}{p}} \geq S - \mu C|u_i|_{q+1}^{\frac{ps-1}{p}}, \quad i = 1, 2. \tag{25}$$

In view of (22) and (23), we consider $\mu > 0$ small enough such that $S - \mu C|u_i|_{q+1}^{\frac{ps-1}{p}} > 0$. Then, from (4), (20), (23), (25), and the a priori bound (22), we obtain

$$\begin{aligned}
I_\mu(u) &= \frac{2}{N}|\Delta u|_{\frac{p+1}{p}}^{\frac{p+1}{p}} - \mu\left(\frac{1}{s+1} - \frac{1}{q+1}\right)|u|_{s+1}^{s+1} \\
&= \frac{2}{N}\left(|\Delta u_1|_{\frac{p+1}{p}}^{\frac{p+1}{p}} + |\Delta u_2|_{\frac{p+1}{p}}^{\frac{p+1}{p}}\right) - \mu\left(\frac{1}{s+1} - \frac{1}{q+1}\right)|u|_{s+1}^{s+1} \\
&\geq \frac{2}{N}S\left(|u_1|_{q+1}^{\frac{p+1}{p}} + |u_2|_{q+1}^{\frac{p+1}{p}}\right) - \mu\left(\frac{1}{s+1} - \frac{1}{q+1}\right)|u|_{s+1}^{s+1} \\
&\geq \frac{2S}{N}\left[\sum_{i=1,2}(S - \mu C|u_i|_{q+1}^{\frac{ps-1}{p}})^{\frac{p+1}{pq-1}}\right] - \mu\left(\frac{1}{s+1} - \frac{1}{q+1}\right)|u|_{s+1}^{s+1} \\
&\qquad\qquad\qquad\qquad \to \frac{4}{N}S^{\frac{p(q+1)}{pq-1}} = \frac{4}{N}S^{\frac{pN}{2(p+1)}},
\end{aligned}$$

as $\mu \to 0^+$ and so, for $\mu > 0$ small enough, $I_\mu(u) > \frac{2}{N}S^{\frac{pN}{2(p+1)}}$.

Therefore, there exists $0 < \tilde{\mu} < \mu_1(\Omega)$ such that if $0 < \mu < \tilde{\mu}$, then any nontrivial critical point $u \in E(\Omega)$ of I_μ with $I_\mu(u) < \frac{2}{N}S^{\frac{pN}{2(p+1)}}$ is such that Δu does not change sign in Ω. Hence, by the strong maximum principle, $u, -\Delta u > 0$ in Ω, up to replace u by $-u$. □

Hereafter in this section we assume all of the hypotheses from Theorem 1.2 and we borrow some arguments from [23]. Without loss of generality, suppose that $0 \in \Omega$. We fix $r > 0$ small enough such that

$$\Omega_r^+ := \{x \in \mathbb{R}^N; \mathrm{dist}(x, \Omega) \leq r\} \text{ and } \Omega_r^- := \{x \in \Omega; \mathrm{dist}(x, \partial\Omega) \geq r\}$$

are homotopically equivalent to Ω and such that $B_r = B_r(0) \subset\subset \Omega$. Consider the functional $I_{\mu,r} : E(B_r) \to \mathbb{R}$ defined by

$$I_{\mu,r}(u) := \frac{p}{p+1}\int_{B_r}|\Delta u|^{\frac{p+1}{p}}dx - \frac{\mu}{s+1}\int_{B_r}|u|^{s+1}dx - \frac{1}{q+1}\int_{B_r}|u|^{q+1}dx,$$

where $E(B_r) := W^{2,\frac{p+1}{p}}(B_r) \cap W_0^{1,\frac{p+1}{p}}(B_r)$. Set

$$\mathcal{N}_{\mu,r} := \{u \in E(B_r)\backslash\{0\}; I'_{\mu,r}(u)u = 0\}$$

and

$$m(\mu) := \inf_{u \in \mathcal{N}_{\mu,r}} I_{\mu,r}(u) = \inf_{u \in E(B_r)\backslash\{0\}} \max_{t \geq 0} I_{\mu,r}(tu).$$

We set the barycenter map $\beta : \mathcal{N}_\mu \to \mathbb{R}^N$, with $\mathcal{N}_\mu$ as in Lemma 2.:, by

$$\beta(u) := \frac{\int_\Omega |\Delta u|^{\frac{p+1}{p}} x dx}{\int_\Omega |\Delta u|^{\frac{p+1}{p}} dx} \quad \text{and} \quad I_\mu^{m(\mu)} := \{u \in E(\Omega); I_\mu(u) \leq m(\mu)\}. \tag{26}$$

Since $B_r = B_r(0) \subset\subset \Omega$, arguing as in [23, Lemma 2.6], we can show that $c_\mu \leq m(\mu)$ and then, by Proposition 2.7, we infer that $I_\mu^{m(\mu)} \cap \mathcal{N}_\mu \neq \emptyset$.

We denote by c_0 and $\mathcal{N}_0$, respectively, the mountain pass level and the Nehari manifold associated with the functional

$$I_0(u) := \frac{p}{p+1} \int_\Omega |\Delta u|^{\frac{p+1}{p}} dx - \frac{1}{q+1} \int_\Omega |u|^{q+1} dx, \; u \in E(\Omega).$$

Lemma 5.2. *(i)* $c_0 = \frac{2}{N} S^{\frac{pN}{2(p+1)}}$.
(ii) Under hypothesis (H), *if* $\mu_n \to 0^+$, *then* $c_{\mu_n} \to c_0$.

Proof. See [1, Lemmas 2.4 and 2.5], respectively. □

Lemma 5.3. *Assume* (H). *There exists* $\hat{\mu} > 0$ *small such that, if* $\mu \in (0, \hat{\mu})$ *and* $u \in \mathcal{N}_\mu$ *with* $I_\mu(u) \leq m(\mu)$, *then* $\beta(u) \in \Omega_r^+$.

Proof. The proof is similar to the proof of [23, Lemma 3.6], using now the Lemmas 4.2 and 5.2. □

Let $\hat{\mu}$ be as in Lemma 5.3. For each $0 < \mu < \hat{\mu}$ we define $\gamma_\mu : \Omega_r^- \to I_\mu^{m(\mu)}$ by

$$\begin{aligned} \gamma_\mu(y) : \Omega &\longrightarrow \mathbb{R} \\ x &\longmapsto \gamma_\mu(y)(x) = w_y(x), \end{aligned} \tag{27}$$

where w_y is the solution for the problem

$$\begin{cases} -\Delta w_y = z_y \text{ in } \Omega, \\ w_y = 0 \text{ on } \partial\Omega, \end{cases} \quad \text{with} \quad z_y(x) = \begin{cases} -\Delta v_\mu(x-y), \text{ if } x \in B_r(y), \\ 0, \text{ if } x \in \Omega \backslash B_r(y), \end{cases}$$

and, cf. Lemma 3.2 (iv), v_μ is radially symmetric with respect to zero, $v_\mu, -\Delta v_\mu > 0$ in B_r,

$$I_{\mu,r}(v_\mu) = m(\mu) \text{ and } I'_{\mu,r}(v_\mu) = 0.$$

By the strong maximum principle, we can show that γ_μ is well defined, that is, $I_\mu(\gamma_\mu(y)) \leq m(\mu)$ for all $y \in \Omega_r^-$. The continuity of γ_μ is a consequence of a regularity of v_μ. Indeed, if $y_n \to y$ in Ω_r^-, then

$$\|\gamma_\mu(y_n)-\gamma_\mu(y)\|^{\frac{p+1}{p}} = |\Delta(\gamma_\mu(y_n)-\gamma_\mu(y))|_{\frac{p+1}{p}}^{\frac{p+1}{p}} = |\Delta w_{y_n}-\Delta w_y|_{\frac{p+1}{p}}^{\frac{p+1}{p}} = |z_{y_n}-z_y|_{\frac{p+1}{p}}^{\frac{p+1}{p}}$$
$$= \int_{B_r(y_n)\cap B_r(y)} |\Delta v_\mu(x-y_n)-\Delta v_\mu(x-y)|^{\frac{p+1}{p}}\,dx + \int_{B_r(y_n)\setminus B_r(y)} |\Delta v_\mu(x-y_n)|^{\frac{p+1}{p}}\,dx$$
$$+ \int_{B_r(y)\setminus B_r(y_n)} |\Delta v_\mu(x-y)|^{\frac{p+1}{p}}\,dx \to 0,$$

because $\Delta v_\mu : \overline{B_r(0)} \to \mathbb{R}$ is continuous.

Again by the strong maximum principle, we infer that $w_y(x) > v_\mu(x-y) > 0$, for all $x \in B_r(y)$ and $y \in \Omega_r^-$. Then for all $y \in \Omega_r^-$,

$$I'_\mu(\gamma_\mu(y))\gamma_\mu(y) = \int_\Omega |\Delta w_y|^{\frac{p+1}{p}}dx - \mu\int_\Omega |w_y|^{s+1}dx - \int_\Omega |w_y|^{q+1}dx$$
$$< \int_{B_r(0)} |\Delta v_\mu|^{\frac{p+1}{p}}dx - \mu\int_{B_r(0)} |v_\mu|^{s+1}dx - \int_{B_r(0)} |v_\mu|^{q+1}dx = I'_{\mu,r}(v_\mu)v_\mu = 0$$

and so $\gamma_\mu(y) \notin \mathcal{N}_\mu$. Nevertheless, for each $y \in \Omega_r^-$ there exists a unique $t_y > 0$ such that $t_y\gamma_\mu(y) \in \mathcal{N}_\mu$. In addition, for all $t > 0$,

$$I_\mu(t\gamma_\mu(y)) = \frac{pt^{\frac{p+1}{p}}}{p+1}\int_\Omega |\Delta w_y|^{\frac{p+1}{p}}dx - \frac{\mu t^{s+1}}{s+1}\int_\Omega |w_y|^{s+1}dx - \frac{t^{q+1}}{q+1}\int_\Omega |w_y|^{q+1}dx$$
$$< \frac{p}{p+1}t^{\frac{p+1}{p}}\int_{B_r(0)} |\Delta v_\mu|^2dx - \frac{\mu}{s+1}t^{s+1}\int_{B_r(0)} |v_\mu|^{s+1}dx - \frac{t^{q+1}}{q+1}\int_{B_r(0)} |v_\mu|^{q+1}dx$$
$$= I_{\mu,r}(tv_\mu) \le I_{\mu,r}(v_\mu) = m(\mu), \tag{28}$$

since $v_\mu \in \mathcal{N}_{\mu,r}$ implies that $I_{\mu,r}(v_\mu) = \max_{t\ge 0} I_{\mu,r}(tv_\mu)$. Then we define

$$\begin{aligned} \overline{\gamma}_\mu : \Omega_r^- &\longrightarrow I_\mu^{m(\mu)} \cap \mathcal{N}_\mu \\ y &\longmapsto \overline{\gamma}_\mu(y) = t_y\gamma_\mu(y). \end{aligned} \tag{29}$$

To complete the proof of Theorem 1.3 we need some auxiliary results, whose proofs we refer to [23].

Lemma 5.4 ([23], Lemma 2.4 (ii) and (iv)).

(i) If u_μ is a critical point of I_μ constrained to $\mathcal{N}_\mu$, then u is a nontrivial critical point of I_μ in $E(\Omega)$.

(ii) The functional I_μ constrained to the manifold $\mathcal{N}_\mu$ satisfies the $(PS)_c$-condition for every $c < \frac{2}{N}S^{\frac{pN}{2(p+1)}}$.

Lemma 5.5 ([23], Lemma 3.8). *Assume* (H) *and take $\hat{\mu}$ as in Lemma 5.3. Then, for every $0 < \mu < \hat{\mu}$, the function $\overline{\gamma}_\mu$, given by* (29)*, is well defined, continuous and satisfies*

$$(\beta \circ \overline{\gamma}_\mu)(y) = y, \quad \forall\, y \in \Omega_r^-. \tag{30}$$

Then we point out that Lemma 5.5 is the key argument in the proof of the next lemma.

Lemma 5.6 ([23], Lemma 4.1). *Assume* (H) *and take* $\hat{\mu}$ *as in Lemma 5.3. Then, for every* $0 < \mu < \hat{\mu}$,

$$cat_{I_{\mathcal{N}_\mu}^{m(\mu)}}(I_{\mathcal{N}_\mu}^{m(\mu)}) \geq cat_\Omega(\Omega) \;\; where \;\; I_{\mathcal{N}_\mu}^{m(\mu)} := \{u \in \mathcal{N}_\mu; I_\mu(u) \leq m(\mu)\}.$$

Proof (Proof of Theorem 1.3 completed). Denote by $I_{\mathcal{N}_\mu}$ the restriction of I_μ to $\mathcal{N}_\mu$. By Lemma 2.6, we know that

$$c_\mu < \frac{2}{N} S^{\frac{pN}{2(p+1)}} \text{ and } m(\mu) < \frac{2}{N} S^{\frac{pN}{2(p+1)}}.$$

Moreover, by Lemma 5.4 (ii), $I_{\mathcal{N}_\mu}$ satisfies the (PS)$_c$-condition for all $c < \frac{2}{N} S^{\frac{pN}{2(p+1)}}$. Let $\overline{\mu} := \min\{\tilde{\mu}, \hat{\mu}\}$, with $\tilde{\mu}$ from Lemma 5.1, $\hat{\mu}$ from Lemma 5.3 and consider $0 < \mu < \overline{\mu}$. Applying the standard Lusternik-Schnirelmann theory (cf. [34, Theorem 5.20]) and Lemma 5.6, we conclude that $I_{\mathcal{N}_\mu}^{m(\mu)}$ has at least $\text{cat}_\Omega(\Omega)$ critical points of $I_{\mathcal{N}_\mu}$. Finally, by Lemma 5.4 (i), we conclude that I_μ has at least $\text{cat}_\Omega(\Omega)$ critical points. Then the result follows from Lemmas 2.2 and 5.1 and the fact that if u is a classical solution of (E) and if we set $v = |\Delta u|^{\frac{1}{p}-1}(-\Delta u)$, then (u, v) is a classical solution of (S). □

6 Appendix: Proof of Lemma 2.6

Let $\mu > 0$ and in the case of $s = 1/p$ also assume that $\mu < \mu_1(\Omega)$. Let c_μ be the mountain pass level as defined in (14). Here we prove that if (H) is satisfied, then $c_\mu \in (0, \frac{2}{N} S^{\frac{pN}{2(p+1)}})$.

Without loss of generality, suppose $0 \in \Omega$. Let $\xi \in \mathcal{C}_c^\infty(\mathbb{R}^N)$ be a function such that $0 \leq \xi(x) \leq 1$ for all $x \in \mathbb{R}^N$, $\xi \equiv 1$ in $B(0, \rho/2)$, $\xi \equiv 0$ in $B(0, \rho)^c$ and $B(0, \rho) \subset\subset \Omega$, $\rho > 0$. Set

$$U_\delta(x) := \xi(x)\psi_\delta(x),\; x \in \mathbb{R}^N,\; 0 < \delta < \rho/2,$$

where $\psi_\delta = S^{\frac{p[p(2-N)+2]}{2(p+1)^2}} \varphi_\delta$ and $\varphi_\delta(x) = \varphi_{\delta,0}(x)$ is given by (6). So $\int_{\mathbb{R}^N} |\Delta \psi_\delta|^{\frac{p+1}{p}} dx = S$ and $\int_{\mathbb{R}^N} |\psi_\delta|^{q+1} dx = 1$. Then, cf. [13] eq. (6.4) and eq. (6.3), respectively,

$$|\Delta U_\delta|_{\frac{p+1}{p},\Omega}^{\frac{p+1}{p}} = \begin{cases} S + O(\delta^{p(N-2)-2}), & \text{if } p < \frac{N}{N-2}, \\ S + |\log\delta|^{\frac{p+1}{p}} O(\delta^{\frac{N}{p}}), & \text{if } p = \frac{N}{N-2}, \\ S + O(\delta^{\frac{N}{p}}), & \text{if } p > \frac{N}{N-2}, \end{cases} \tag{31}$$

and

$$|U_\delta|_{q+1,\Omega}^{q+1} = \begin{cases} 1 + O(\delta^{pN}), & \text{if } p < \frac{N}{N-2}, \\ 1 + |\log\delta|^{q+1} O(\delta^{q(N-2)-2}), & \text{if } p = \frac{N}{N-2}, \\ 1 + O(\delta^{q(N-2)-2}), & \text{if } p > \frac{N}{N-2}. \end{cases} \tag{32}$$

Set

$$V_\delta(x) = \frac{U_\delta(x)}{|U_\delta|_{q+1}}, \ x \in \mathbb{R}^N. \tag{33}$$

Then

$$|\Delta V_\delta|_{\frac{p+1}{p},\Omega}^{\frac{p+1}{p}} = \begin{cases} S + O(\delta^{p(N-2)-2}), & \text{if } p < \frac{N}{N-2}, \\ S + |\log\delta|^{\frac{p+1}{p}} O(\delta^{\frac{N}{p}}), & \text{if } p = \frac{N}{N-2}, \\ S + O(\delta^{\frac{N}{p}}), & \text{if } p > \frac{N}{N-2}, \end{cases} \tag{34}$$

Using the asymptotic behavior of φ, (6), and since $\frac{p+1}{p} \leq s+1 < q+1$, we infer that:

(i) if $p < \frac{N}{N-2}$, then

$$|V_\delta|_{s+1}^{s+1} = \begin{cases} C\delta^{(s+1)\frac{p}{p+1}[p(N-2)-2]} + o(\delta^{(s+1)\frac{p}{p+1}[p(N-2)-2]}), & \text{if } s+1 < \frac{N}{p(N-2)-2}, \\ C\delta^{\frac{Np}{p+1}}|\log\delta| + O(\delta^{\frac{Np}{p+1}}), & \text{if } s+1 = \frac{N}{p(N-2)-2}, \\ C\delta^{N-\frac{N(s+1)}{q+1}} + o(\delta^{N-\frac{N(s+1)}{q+1}}), & \text{if } s+1 > \frac{N}{p(N-2)-2}; \end{cases} \tag{35}$$

(ii) if $p > \frac{N}{N-2}$, then

$$|V_\delta|_{s+1}^{s+1} = \begin{cases} C\delta^{\frac{N(s+1)}{p+1}} + o(\delta^{\frac{N(s+1)}{p+1}}), & \text{if } s+1 < \frac{N}{N-2}, \\ C\delta^{\frac{N^2}{(p+1)(N-2)}}|\log\delta| + O(\delta^{\frac{N^2}{(p+1)(N-2)}}), & \text{if } s+1 = \frac{N}{N-2}, \\ C\delta^{N-\frac{N(s+1)}{q+1}} + o(\delta^{N-\frac{N(s+1)}{q+1}}), & \text{if } s+1 > \frac{N}{N-2}; \end{cases} \tag{36}$$

(iii) if $p = \frac{N}{N-2}$, then

$$|V_\delta|_{s+1}^{s+1} = \begin{cases} C\delta^{\frac{N(s+1)}{p+1}}|\log\delta|^{s+1} + |\log\delta|^{s+1} o(\delta^{\frac{N(s+1)}{p+1}}), & \text{if } s+1 < \frac{N}{N-2}, \\ C\delta^{\frac{N^2}{(p+1)(N-2)}}|\log\delta| + |\log\delta|^{s+1} O(\delta^{\frac{N^2}{(p+1)(N-2)}}), & \text{if } s+1 = \frac{N}{N-2}, \\ C\delta^{N-\frac{N(s+1)}{q+1}} + o(\delta^{N-\frac{N(s+1)}{q+1}}), & \text{if } s+1 > \frac{N}{N-2}; \end{cases} \tag{37}$$

where C is a positive constant. Our goal is to show that, for δ small enough,

$$\max_{t\geq 0} I_\mu(tV_\delta) < \frac{2}{N} S^{\frac{pN}{2(p+1)}}. \tag{38}$$

Note that

$$I_\mu(tV_\delta) = \frac{p}{p+1} t^{\frac{p+1}{p}} |\Delta V_\delta|_{\frac{p+1}{p}}^{\frac{p+1}{p}} - \frac{\mu}{s+1} t^{s+1} |V_\delta|_{s+1}^{s+1} - \frac{t^{q+1}}{q+1},$$

$$\lim_{t\to\infty} I_\mu(tV_\delta) = -\infty,$$

and then $\max_{t\geq 0} I_\mu(tV_\delta)$ is achieved at some $t_\delta > 0$. Thus, $|\Delta V_\delta|_{\frac{p+1}{p}}^{\frac{p+1}{p}} = \mu t_\delta^{s-\frac{1}{p}} |V_\delta|_{s+1}^{s+1} + t_\delta^{q-\frac{1}{p}} \geq t_\delta^{q-\frac{1}{p}}$ and so

$$t_\delta \leq |\Delta V_\delta|_{\frac{p+1}{p}}^{\frac{p+1}{pq-1}}. \tag{39}$$

The last inequality implies that

$$|\Delta V_\delta|_{\frac{p+1}{p}}^{\frac{p+1}{p}} = t_\delta^{q-\frac{1}{p}} + \mu t_\delta^{s-\frac{1}{p}} |V_\delta|_{s+1}^{s+1} \leq t_\delta^{q-\frac{1}{p}} + \mu |\Delta V_\delta|_{\frac{p+1}{p}}^{\frac{p+1}{pq-1}\frac{ps-1}{p}} |V_\varepsilon|_{s+1}^{s+1}$$

and then

$$\begin{aligned} t_\delta^{q-\frac{1}{p}} &\geq |\Delta V_\delta|_{\frac{p+1}{p}}^{\frac{p+1}{p}} - \mu |\Delta V_\delta|_{\frac{p+1}{p}}^{\frac{p+1}{pq-1}\frac{ps-1}{p}} |V_\delta|_{s+1}^{s+1} \\ &= |\Delta V_\delta|_{\frac{p+1}{p}}^{\frac{p+1}{p}} \left[1 - \mu |\Delta V_\delta|_{\frac{p+1}{p}}^{\frac{p+1}{pq-1}\frac{ps-1}{p} - \frac{p+1}{p}} |V_\delta|_{s+1}^{s+1}\right]. \end{aligned}$$

Using the estimates in (34), the estimates in (35), (36), and (37), and the inequality (39), we infer that

$$t_\delta \to S^{\frac{p}{pq-1}} \text{ as } \delta \to 0. \tag{40}$$

Consider the function $g(t) = \frac{p}{p+1} t^{\frac{p+1}{p}} |\Delta V_\delta|_{\frac{p+1}{p}}^{\frac{p+1}{p}} - \frac{t^{q+1}}{q+1}$ for $t \geq 0$. Observe that g achieves its maximum at $t_0 = |\Delta V_\delta|_{\frac{p+1}{p}}^{\frac{p+1}{pq-1}}$ and that g is increasing on the interval $[0, |\Delta V_\delta|_{\frac{p+1}{p}}^{\frac{p+1}{pq-1}}]$. Thus, from (39),

$$\max_{t\geq 0} I_\mu(tV_\delta) = I_\mu(t_\delta V_\delta) = g(t_\delta) - \frac{\mu}{s+1} t_\delta^{s+1} |V_\delta|_{s+1}^{s+1}$$

$$\leq \frac{p}{p+1} (|\Delta V_\delta|_{\frac{p+1}{p}}^{\frac{p+1}{pq-1}})^{\frac{p+1}{p}} |\Delta V_\delta|_{\frac{p+1}{p}}^{\frac{p+1}{p}} - \frac{(|\Delta V_\delta|_{\frac{p+1}{p}}^{\frac{p+1}{pq-1}})^{q+1}}{q+1} - \frac{\mu}{s+1} t_\delta^{s+1} |V_\delta|_{s+1}^{s+1}$$

$$= \frac{2}{N} (|\Delta V_\delta|_{\frac{p+1}{p}}^{\frac{p+1}{p}})^{\frac{pN}{2(p+1)}} - \frac{\mu}{s+1} t_\delta^{s+1} |V_\delta|_{s+1}^{s+1}.$$

Case 1: $p < \frac{N}{N-2}$.

It follows from (34) that

$$\max_{t\geq 0} I_\mu(tV_\delta) \leq \frac{2}{N} S^{\frac{pN}{2(p+1)}} + O(\delta^{p(N-2)-2}) - \frac{\mu}{s+1} t_\delta^{s+1} |V_\delta|_{s+1}^{s+1},$$

and so, for δ small enough, it follows from (40) that

$$\max_{t\geq 0} I_\mu(tV_\delta) \leq \frac{2}{N} S^{\frac{pN}{2(p+1)}} + O(\delta^{p(N-2)-2}) - \frac{\mu}{s+1} \left(\frac{S^{\frac{p}{pq-1}}}{2} \right)^{s+1} |V_\delta|_{s+1}^{s+1}$$

$$= \frac{2}{N} S^{\frac{pN}{2(p+1)}} + O(\delta^{p(N-2)-2}) - \hat{C} |V_\delta|_{s+1}^{s+1}.$$

Note that the first line in (35) is not useful for our purpose because

$$s+1 \geq \frac{p+1}{p} \Rightarrow (s+1)\frac{p}{p+1} \geq 1 \Rightarrow (s+1)\frac{p}{p+1}[p(N-2)-2] \geq [p(N-2)-2].$$

Now, the third line (35) is useful if $N - \frac{N(s+1)}{q+1} < p(N-2) - 2$, and since $q+1 = \frac{N(p+1)}{p(N-2)-2}$, the last inequality is equivalent to

$$N - \frac{N(s+1)}{q+1} < \frac{N(p+1)}{q+1}, \quad \text{that is,} \quad q - p < s+1.$$

Now, since $\frac{p+1}{p} \leq s+1$, we infer that

$$q - p > \frac{p+1}{p} \Leftrightarrow p < \frac{2+\sqrt{2N}}{N-2}. \tag{41}$$

Moreover,

$$\min\left\{ \frac{N}{N-2}, \frac{2+\sqrt{2N}}{N-2} \right\} = \begin{cases} \frac{N}{N-2}, & \text{if } N = 3,4,5, \\ \frac{2+\sqrt{2N}}{N-2}, & \text{if } N \geq 6. \end{cases} \tag{42}$$

Hence,

(i) For $N \geq 6$

(i.1) $\frac{2+\sqrt{2N}}{N-2} \leq p < \frac{N}{N-2}$

Considering the second line of (35), we infer that (38) is true for all $\frac{p+1}{p} \leq s+1 < q+1$;

(i.2) $\frac{2}{N-2} < p < \frac{2+\sqrt{2N}}{N-2}$

It follows from (41) and (42) that (38) is true if $q-p < s+1 < q+1$;

(ii) For $N = 3, 4, 5$ and $\frac{2}{N-2} < p < \frac{N}{N-2}$

It follows from (41) and (42) that (38) is true if $q-p < s+1 < q+1$.

Case 2: $p > \frac{N}{N-2}$**.**

It follows from (34) that

$$\max_{t\geq 0} I_\mu(tV_\delta) \leq \frac{2}{N} S^{\frac{pN}{2(p+1)}} + O(\delta^{\frac{N}{p}}) - \frac{\mu}{s+1} t_\delta^{s+1} |V_\delta|_{s+1}^{s+1},$$

and so, for δ small enough, it follows from (40) that

$$\begin{aligned}\max_{t\geq 0} I_\mu(tV_\delta) &\leq \frac{2}{N} S^{\frac{pN}{2(p+1)}} + O(\delta^{\frac{N}{p}}) - \frac{\mu}{s+1} \left(\frac{S^{\frac{p}{pq-1}}}{2}\right)^{s+1} |V_\delta|_{s+1}^{s+1} \\ &= \frac{2}{N} S^{\frac{pN}{2(p+1)}} + O(\delta^{\frac{N}{p}}) - \hat{C} |V_\delta|_{s+1}^{s+1}.\end{aligned}$$

Note that the first line in (36) is not useful for our purpose because

$$s+1 \geq \frac{p+1}{p} \Rightarrow \frac{s+1}{p+1} \geq \frac{1}{p} \Rightarrow \frac{N(s+1)}{p+1} \geq \frac{N}{p}.$$

Now, the third line (36) is useful if $N - \frac{N(s+1)}{q+1} < \frac{N}{p}$, that is, $\frac{(p-1)(q+1)}{p} < s+1$. Now, since $\frac{p+1}{p} \leq s+1$,

$$\frac{(p-1)(q+1)}{p} > \frac{p+1}{p} \Leftrightarrow p > \frac{N-2}{2}. \tag{43}$$

Moreover

$$\max\left\{\frac{N}{N-2}, \frac{N-2}{2}\right\} = \begin{cases} \frac{N}{N-2}, & \text{if } N = 3, 4, 5, \\ \frac{N-2}{2}, & \text{if } N \geq 6. \end{cases} \tag{44}$$

Hence,

(i) For $N \geq 6$

(i.1) $\frac{N}{N-2} < p \le \frac{N-2}{2}$
Considering the second line of (36), we have (38) is true for all $\frac{p+1}{p} \le s+1 < q+1$;

(i.2) $p > \frac{N-2}{2}$
It follows from (43) and (44) that (38) is true if $\frac{(p-1)(q+1)}{p} < s+1 < q+1$;

(ii) For $N = 3, 4, 5$ and $p > \frac{N}{N-2}$
It follows from (43) and (44) that (38) is true if $\frac{(p-1)(q+1)}{p} < s+1 < q+1$.

Case 3: $p = \frac{N}{N-2}$.
If follows from (34) that

$$\max_{t\ge 0} I_\mu(tV_\delta) \le \frac{2}{N} S^{\frac{pN}{2(p+1)}} + |\log\delta|^{\frac{p+1}{p}} O(\delta^{\frac{N}{p}}) - \frac{\mu}{s+1} t_\delta^{s+1} |V_\delta|_{s+1}^{s+1},$$

and so, for δ small enough, it follows from (40) that

$$\begin{aligned}\max_{t\ge 0} I_\mu(tV_\delta) &\le \frac{2}{N} S^{\frac{pN}{2(p+1)}} + |\log\delta|^{\frac{p+1}{p}} O(\delta^{\frac{N}{p}}) - \frac{\mu}{s+1}\left(\frac{S^{\frac{p}{pq-1}}}{2}\right)^{s+1} |V_\delta|_{s+1}^{s+1} \\ &= \frac{2}{N} S^{\frac{pN}{2(p+1)}} + |\log\delta|^{\frac{p+1}{p}} O(\delta^{\frac{N}{p}}) - \hat{C} |V_\delta|_{s+1}^{s+1}.\end{aligned}$$

The analysis of this case is analogous to the Case 2, using now the estimates in (37). Hence,

(i) For $N \ge 6$

(i.1) $\frac{N}{N-2} = p \le \frac{N-2}{2}$
Considering the second line of (37), we have (38) is true for all $\frac{p+1}{p} \le s+1 < q+1$;

(i.2) $p > \frac{N-2}{2}$
It follows from (43) and (44) that (38) is true if $\frac{(p-1)(q+1)}{p} < s+1 < q+1$;

(ii) For $N = 3, 4, 5$ and $p \ge \frac{N}{N-2}$
It follows from (43) and (44) that (38) is true if $\frac{(p-1)(q+1)}{p} < s+1 < q+1$.

Now we summarize the above conditions on s. Combining the items (i.1) of the 3 cases above, we obtain (a) of hypothesis (H). The items (i.2) and (ii) in Case 1 are equivalent to (b1) and (b3) of (H), while the items (i.2) and (ii) in Cases 2 and 3 are equivalent to (b2) and (b4) of (H). Therefore, if (H) is satisfied, then $c_\mu \in (0, \frac{2}{N} S^{\frac{pN}{2(p+1)}})$. □

Acknowledgements J.L.F. Melo was supported by FAPESP #2010/00603-9 grant; E. Moreira dos Santos was partially supported by CNPq #309291/2012-7 grant and FAPESP #2014/03805-2 grant. Part of this work was done while the second author was visiting the Dipartimento di Matematica of the Université di Roma *Sapienza*, whose hospitality he gratefully acknowledges.

References

1. Alves, C.O., Figueiredo, G.M.: Multiplicity of nontrivial solutions to a biharmonic equation via Lusternik-Schnirelman theory. Math. Methods Appl. Sci. **36**(6), 683–694 (2013)
2. Alvino, A., Lions, P.-L., Trombetti, G.: A remark on comparison results via symmetrization. Proc. R. Soc. Edinb. Sect. A **102**(1–2), 37–48 (1986)
3. Ambrosetti, A., Rabinowitz, P.H.: Dual variational methods in critical point theory and applications. J. Funct. Anal. **14**, 349–381 (1973)
4. Azorero, J.P.G., Alonso, I.P.: Existence and nonuniqueness for the p-Laplacian: nonlinear eigenvalues. Commun. Partial Differ. Equat. **12**(12), 1389–1430 (1987)
5. Azorero, J.G., Alonso, I.P.: Multiplicity of solutions for elliptic problems with critical exponent or with a nonsymmetric term. Trans. Am. Math. Soc. **323**(2), 877–895 (1991)
6. Bonheure, D., dos Santos, E.M., Ramos, M.: Ground state and non-ground state solutions of some strongly coupled elliptic systems. Trans. Am. Math. Soc. **364**(1), 447–491 (2012)
7. Bonheure, D., dos Santos, E.M., Tavares, H.: Hamiltonian elliptic systems: a guide to variational methods. Port. Math. **71**(3), 301–395 (2014)
8. Brézis, H., Nirenberg, L.: Positive solutions of nonlinear elliptic equations involving critical Sobolev exponents. Commun. Pure Appl. Math. **36**(4), 437–477 (1983)
9. Clément, P., de Figueiredo, D.G., Mitidieri, E.: Positive solutions of semilinear elliptic systems. Commun. Partial Differ. Equat. **17**(5–6), 923–940 (1992)
10. de Figueiredo, D.G.: Semilinear elliptic systems: existence, multiplicity, symmetry of solutions. In: Chipot, M. (ed.) Handbook of Differential Equations: Stationary Partial Differential Equations, vol. V, pp. 1–48. Elsevier/North-Holland, Amsterdam (2008)
11. de Figueiredo, D.G., Felmer, P.L.: On superquadratic elliptic systems. Trans. Am. Math. Soc. **343**(1), 99–116 (1994)
12. dos Santos, E.M.. Multiplicity of solutions for a fourth-order quasilinear nonhomogeneous equation. J. Math. Anal. Appl. **342**(1), 277–297 (2008)
13. dos Santos, E.M.: Positive solutions for a fourth-order quasilinear equation with critical Sobolev exponent. Commun. Contemp. Math. **12**(1), 1–33 (2010)
14. Edmunds, D.E., Fortunato, D., Jannelli, E.: Critical exponents, critical dimensions and the biharmonic operator. Arch. Ration. Mech. Anal. **112**(3), 269–289 (1990)
15. Gazzola, F., Grunau, H.-C., Squassina, M.: Existence and nonexistence results for critical growth biharmonic elliptic equations. Calc. Var. Partial Differ. Equat. **18**(2), 117–143 (2003)
16. Gilbarg, D., Trudinger, N.S.: Elliptic partial differential equations of second order. Classics in Mathematics. Springer, Berlin (2001). Reprint of the 1998 edition
17. Hulshof, J., van der Vorst, R.: Differential systems with strongly indefinite variational structure. J. Funct. Anal. **114**(1), 32–58 (1993)
18. Hulshof, J., van der Vorst, R.C.A.M.: Asymptotic behaviour of ground states. Proc. Am. Math. Soc. **124**(8), 2423–2431 (1996)
19. Hulshof, J., Mitidieri, E., van der Vorst, R.: Strongly indefinite systems with critical Sobolev exponents. Trans. Am. Math. Soc. **350**(6), 2349–2365 (1998)
20. Lazzo, M.: Solutions positives multiples pour une équation elliptique non linéaire avec l'exposant critique de Sobolev. C. R. Acad. Sci. Paris Sér. I Math. **314**(1), 61–64 (1992)
21. Lions, P.-L.: The concentration-compactness principle in the calculus of variations. The limit case. I. Rev. Mat. Iberoam. **1**(1), 145–201 (1985)
22. Lusternik, L., Schnirelmann, L.: Méthodes topologiques dans les problèmes variationnels. I. Pt. Espaces à un nombre fini de dimensions. Traduit du russe par J. Kravtchenko. Hermann et Cie., Paris (1934)
23. Melo, J.L.F., dos Santos, E.M.: Positive solutions to a fourth-order elliptic problem by the Lusternik-Schnirelmann category. J. Math. Anal. Appl. **420**(1), 532–550 (2014)
24. Melo, J.L.F., dos Santos, E.M.: A fourth-order equation with critical growth: the effect of the domain topology. Topological Methods in Nonlinear Analysis **45**(2), 551–574 (2015)

25. Mitidieri, E.: A Rellich type identity and applications. Commun. Partial Differ. Equat. **18**(1–2), 125–151 (1993)
26. Peletier, L.A., Van der Vorst, R.C.A.M.: Existence and nonexistence of positive solutions of nonlinear elliptic systems and the biharmonic equation. Differ. Integr. Equat. **5**(4), 747–767 (1992)
27. Pucci, P., Serrin, J.: A general variational identity. Indiana Univ. Math. J. **35**(3), 681–703 (1986)
28. Pucci, P., Serrin, J.: Critical exponents and critical dimensions for polyharmonic operators. J. Math. Pures Appl. (9) **69**(1), 55–83 (1990)
29. Rabinowitz, P.H.: On a class of nonlinear Schrödinger equations. Z. Angew. Math. Phys. **43**(2), 270–291 (1992)
30. Rey, O.: A multiplicity result for a variational problem with lack of compactness. Nonlinear Anal. **13**(10), 1241–1249 (1989)
31. Ruf, B.: Superlinear elliptic equations and systems. In: Chipot, M. (ed.) Handbook of Differential Equations: Stationary Partial Differential Equations, vol. V, pp. 211–276. Elsevier/North-Holland, Amsterdam (2008)
32. Van der Vorst, R.C.A.M.: Variational identities and applications to differential systems. Arch. Ration. Mech. Anal. **116**(4), 375–398 (1992)
33. Van der Vorst, R.C.A.M.: Fourth-order elliptic equations with critical growth. C. R. Acad. Sci. Paris Sér. I Math. **320**(3), 295–299 (1995)
34. Willem, M.: Minimax Theorems. Progress in Nonlinear Differential Equations and Their Applications, vol. 24. Birkhäuser, Boston, MA (1996)

A nonlinear Steklov problem arising in corrosion modeling

C.D. Pagani, D. Pierotti, G. Verzini, and A. Zilio

1 Introduction

In a simple mathematical model of electrochemical corrosion, i.e. a deterioration of a metal by electrochemical reaction with its environment, a (suitably defined) galvanic potential is represented by a function u harmonic in a domain $\Omega \subset \mathbb{R}^N$ whose boundary is partly electrochemically active and partly inert. In the inactive boundary region the current density flow $J \cdot \nu$ (ν is the outward unit normal to $\partial\Omega$) is of course zero, but in the active part it is modeled (by interpolating experimental data) by a difference of two exponentials according to the so-called Butler-Volmer formula:

$$\left(J \cdot \nu\right)(x) = \lambda\mu(x)\left(e^{\beta u(x)} - e^{-(1-\beta)u(x)}\right) + g(x), \qquad x \in \partial\Omega \tag{1}$$

Here $\beta \in (0, 1)$ is a constant depending on the constituents of the electrochemical system, the function $\mu(x)$ distinguishes between the active and the inert boundary regions (typically μ is the characteristic function of some subset $\subseteq \partial\Omega$), λ is a real parameter which may take negative as well as positive values, and g is an externally imposed current (see [1] and the references therein for a detailed discussion).

C.D. Pagani (✉) • D. Pierotti • G. Verzini
Dipartimento di Matematica, Politecnico di Milano, p.za Leonardo da Vinci 32, 20133 Milano, Italy
e-mail: carlo.pagani@polimi.it; dario.pierotti@polimi.it; gianmaria.verzini@polimi.it

A. Zilio
Centre d'analyse et de mathématique sociales, École des Hautes Études en Sciences Sociales, 190-198 Avenue de France, 75244 Paris Cedex 13, France
e-mail: azilio@ehess.fr

A.N. Carvalho et al. (eds.), *Contributions to Nonlinear Elliptic Equations and Systems*, Progress in Nonlinear Differential Equations and Their Applications 86, DOI 10.1007/978-3-319-19902-3_22

Assuming $\mu(x) \geq 0$ and not identically vanishing, the resulting mathematical problem is quite different in the two cases, λ negative or positive; for, the corresponding *linearized problem*

$$\begin{aligned} \Delta u(x) &= 0 \quad \text{in } \Omega \\ \partial_\nu u(x) &= \lambda\, \mu(x) u(x) + g(x) \quad \text{on } \partial\Omega \end{aligned} \tag{2}$$

is a classical elliptic problem with a Robin (or mixed Neumann-Robin) boundary condition if $\lambda < 0$, while for $\lambda > 0$ it is a Steklov problem. We stress that in the latter case, there are nontrivial solutions of the problem with $g = 0$ (Steklov eigenvalue problem).

Another quite sensible parameter of the problem is the dimension N of the space. In fact, if $N = 2$, the nonlinear problem is subcritical in the energy space $H^1(\Omega)$ (thanks to the Moser-Trudinger inequality); on the other hand, if $N \geq 3$ (and therefore in the physically relevant case $N = 3$), the problem is supercritical (see the discussion in [2]). The two-dimensional case has been considered by various authors [1, 3–7].

The literature concerning the supercritical case is much more lacking and seems to take into consideration mainly the case $\lambda < 0$ (that is, with the Robin boundary condition; see, e.g., [8]). A first attempt to investigate the three-dimensional problem with $\lambda > 0$ (and vanishing external current g) is in [2]. For the reader's convenience, let us summarize with few details the main results obtained in [2].

The authors discuss the following problem: find a (nonidentically vanishing) function u in a bounded domain $\Omega \subset \mathbb{R}^3$ with Lipschitz boundary, satisfying the system

$$\begin{aligned} \Delta u(x) &= 0 \quad \text{in } \Omega \\ \partial_\nu u(x) &= \lambda\, \mu(x) \sinh[u(x)] \quad \text{on } \partial\Omega \end{aligned} \tag{3}$$

where $\lambda > 0$ and μ is a nonnegative function in $L^\infty(\partial\Omega)$.

By observing that the above problem has the line of trivial solutions $\{(\lambda, 0) \mid \lambda \in \mathbb{R}\}$, they look for *bifurcation solutions*. By applying classical results of Bifurcation Theory [9, 10], the authors prove that, for every eigenvalue κ of the linearized problem

$$\begin{aligned} \Delta u(x) &= 0 \quad \text{in } \Omega \\ \partial_\nu u(x) &= \lambda\, \mu(x) u(x) \quad \text{on } \partial\Omega \end{aligned} \tag{4}$$

(which is a classical *Steklov eigenvalue problem* [11]) the pair $(\kappa, 0)$ is a bifurcation point for (3). Further results on global existence are proved by assuming specific symmetries of the domain. By restricting the study of the problem (3) in the unit ball of $\mathbb{R}^3$ and taking $\mu(x) \equiv 1$, they prove the existence of a branch of global solutions bifurcating from the first eigenvalue $\lambda = 1$ of the linearized problem.

In the present paper, after recalling some general results about existence of global solutions (section 2) the analysis of the branch bifurcating from the first eigenvalue is expanded (section 3) and some new properties (local analiticy, blow up of the solutions, etc.) as well as open problems are presented. In section 4 we describe some numerical results illustrating the properties of the previously investigated bifurcation branch.

2 Global existence of the bifurcation solutions

Hereafter, we consider the problem (3) with μ nonnegative and bounded. For more details and some proofs of the results of this section, see [2]. As we will see below, it is convenient to search three dimensional solutions in the *Hilbert* space $H^{3/2}(\Omega)$.

Let $f \in L^2(\partial\Omega)$ satisfy $\int_{\partial\Omega} f = 0$; define the *Neumann to Dirichlet map*

$$\mathscr{G}f = v_0|_{\partial\Omega} \tag{5}$$

where v_0 is the *unique* harmonic function in Ω with Neumann datum f and such that $\int_{\partial\Omega} \mu v_0 = 0$.

By known regularity results [12] we have $v_0 \in H^{3/2}(\Omega)$ and therefore $\mathscr{G}f \in H^1(\partial\Omega)$.

Let us define the subspace

$$\dot{H}^1(\partial\Omega) = \left\{\phi \in H^1(\partial\Omega), \quad \int_{\partial\Omega} \mu\phi = 0\right\} \tag{6}$$

and the operator

$$G(\lambda, \phi) = \lambda \mathscr{G}\Big(\mu \sinh[\phi + s(\phi)]\Big) \tag{7}$$

where

$$s(\phi) = -\tanh^{-1}\left(\frac{\int_{\partial\Omega} \mu \sinh(\phi)}{\int_{\partial\Omega} \mu \cosh(\phi)}\right) = \frac{1}{2}\log\left(\frac{\int_{\partial\Omega} \mu e^{-\phi}}{\int_{\partial\Omega} \mu e^{\phi}}\right) \tag{8}$$

By known estimates on *two dimensional* manifolds, the exponentials $e^{\pm\phi}$ lie in $L^p(\partial\Omega)$ for every $p \geq 1$; moreover, by the definition (8) the argument of $\mathscr{G}$ at the right-hand side of (7) has vanishing integral on $\partial\Omega$. Then, by standard calculations one can show that the operator $G(\lambda, \cdot)$ is a $\mathscr{C}^1$ map from $\dot{H}^1(\partial\Omega)$ in itself. Assume now that ϕ solves the *functional equation*

$$\phi = G(\lambda, \phi) = \lambda \mathscr{G}\Big(\mu \sinh[\phi + s(\phi)]\Big) \tag{9}$$

Then, the unique harmonic function $u_0 \in H^1(\Omega)$ such that $u_0|_{\partial\Omega} = \phi$ satisfies the variational equation

$$\int_\Omega \nabla u_0 \nabla v = \lambda \int_{\partial\Omega} \mu \sinh[u_0|_{\partial\Omega} + s(u_0|_{\partial\Omega})]v \tag{10}$$

for every v such that $\int_{\partial\Omega} \mu v = 0$.

Finally, by standard regularity results the function

$$u(x) = u_0(x) + s(u_0|_{\partial\Omega})$$

satisfies the boundary value problem (3). Then, the following result holds [2]:

Theorem 2.1. *Let $\Omega \subset \mathbb{R}^3$ be a bounded domain with Lipschitz boundary and let μ be a bounded nonnegative function on $\partial\Omega$. Moreover, let κ be an eigenvalue of multiplicity n_κ of the linear problem* (4)*. Then, there is an $r_0 > 0$ such that for each $r \in (0, r_0)$ the bifurcation equation* (9) *has at least n_κ distinct pairs of nontrivial solutions $(\lambda_m(r), \pm\phi_m(r)) \subset \mathbb{R} \times \dot{H}^1(\partial\Omega)$, $m = 1, 2, \ldots, n_\kappa$; moreover, as $r \to 0$, $\lambda_m(r) \to \kappa$ and $\|\phi_m(r)\|_{H^1(\partial\Omega)} = O(r)$.*

Thus, by the previous discussion, the nonlinear boundary value problem (3) has at least n_κ distinct pairs of nontrivial solutions $(\lambda_m(r), \pm u_m(r)) \subset \mathbb{R} \times H^{3/2}(\Omega)$, $m = 1, 2, \ldots n_\kappa$ for $r \in (0, r_0)$.

In the case of bifurcation from eigenvalues of *odd multiplicity*, a global result holds (see [10], Theorem 1.10). By denoting with $\mathscr{S} \subset \mathbb{R} \times \dot{H}^1(\partial\Omega)$ the closure of the set of the nontrivial solutions (λ, ϕ) to (9), we have

Proposition 2.2. *Let κ be an eigenvalue of odd multiplicity of the linear problem* (4) *and let $\mathscr{C}$ be the component (i.e., a closed connected subset maximal with respect to inclusion) of $\mathscr{S}$ to which $(\kappa, 0)$ belongs. Then, either $\mathscr{C}$ is unbounded or contains $(\bar{\kappa}, 0)$, where $\bar{\kappa} \neq \kappa$.*

From now on, *we consider the problem* (3) *with $\Omega \equiv B$ the unit ball of $\mathbb{R}^3$ and $\mu \equiv 1$.* It is well known that the eigenfunctions of the corresponding linear Steklov problem are the *homogenous harmonic polynomials* of degree n and that the Steklov eigenvalues are precisely n, $n = 0, 1, 2, \ldots$ Moreover, the dimension of each eigenspace is $2n + 1$. Hence, Proposition 2.2 applies to the component of $\mathscr{S}$ containing $(n, 0)$ for every $n = 1, 2, \ldots$.

In a spherical domain it is natural to look for solutions with an *axial symmetry* with respect to a diameter (note that there are no nontrivial *radially* symmetric solutions to (3) in the ball). By suitably choosing the coordinate system, we may consider solutions symmetric with respect to the z axis, i.e. solutions which are constant along the parallel lines of the sphere; in spherical coordinates, they will only depend on the distance $r = \sqrt{x^2 + y^2 + z^2}$ from the origin, and on the polar angle θ.

Let us denote by $H^{3/2}_{ax}(B)$ the subspace of the functions $v \in H^{3/2}(B)$ with the above axial symmetry; the boundary traces $v|_{\partial B}$ with vanishing integral on the sphere will belong to a subspace of (6) denoted by $\dot{H}^1_{ax}(\partial B)$. Now, by rotational invariance of the Laplacian, by the symmetry of the Neumann condition on the sphere and by uniqueness of the solution to the Neumann problem, one can check that *the operator* $G(\lambda, \cdot)$ *defined by* (7) *maps* $\dot{H}^1_{ax}(\partial B)$ *in itself.* Moreover, the (nonconstant) axially symmetric eigenfunctions of the Steklov problem in the ball are those harmonic polynomials which (in polar coordinates) are independent of the azimuthal angle, that is $r^n P_n(\cos\theta)$, $n = 1, 2, \ldots$ where the P_n are the *Legendre polynomials.* The restrictions of these eigenfunctions to the spherical surface span the subspace of axially symmetric, zero mean functions of $L^2(\partial B)$.

We now define axially symmetric $\phi \in \dot{H}^1_{ax}(\partial B)$ and $u_0 \in H^{3/2}_{ax}(B)$ as in (6) and (10), respectively; then (see [2], section 4) we find nontrivial solutions (λ, u) of (3) bifurcating from $(n, 0)$, $n = 1, 2, \ldots$ and such that $u \in H^{3/2}_{ax}(B)$. We stress that there is a *unique* (normalized) axially symmetric eigenfunction for every eigenvalue n, so that all the eigenvalues of the linear problem in $H^{3/2}_{ax}(B)$ are *simple.* Thus, we get

Proposition 2.3. *Let B be the unit ball and let $\mu \equiv 1$. Then, for any $n = 1, 2, \ldots$ there is a component $\mathscr{C}_n \subset \mathbb{R} \times \dot{H}^1_{ax}(\partial B)$ of $\mathscr{S}$ which meets the point $(n, 0)$; each $\mathscr{C}_n$ is either unbounded or meets $(m, 0)$, with $m \neq n$.*

Remark 2.4. It is worthwhile to recall the following properties of the solutions bifurcating from a simple eigenvalue λ_0 (see [10, 13, 14]): the set of *nontrivial solutions* near to $(\lambda_0, 0)$ consists precisely of a smooth (even analytic in our case, see below) curve $(\lambda(s), \Phi(s))$, where $s \in \mathscr{I}$, an open neighborhood of the origin. Moreover, $\Phi(s) = sv_0 + o(s)$, where v_0 is an eigenfunction corresponding to λ_0.

Hence, by Theorem 2.1, it follows that near to $(n, 0)$ each component $\mathscr{C}_n$ defined in the above proposition is represented by a curve $(\lambda(s), \Phi(s))$ such that $\Phi(-s) = -\Phi(s)$ for s small.

Since $G : \mathbb{R} \times \dot{H}^1_{ax}(\partial B) \to \dot{H}^1_{ax}(\partial B)$ is real analytic, further properties of $\mathscr{S}$ can be deduced in the framework of the analytic bifurcation theory due to Dancer (see [15–17]).

Proposition 2.5. *Let B be the unit ball, $\mu \equiv 1$, and, for any $n = 1, 2, \ldots$, let $\mathscr{C}_n$ denote the component of $\mathscr{S}$ which meets the point $(n, 0)$, according to Proposition 2.3. Then there exists a curve $\mathfrak{C}_n$ with the following properties:*

1. *$\mathfrak{C}_n = \{(\Lambda(s), \Phi(s)) : s \in [0, \infty)\}$, where $(\Lambda, \Phi) : [0, \infty) \to \mathbb{R} \times \dot{H}^1_{ax}(\partial B)$ is continuous;*
2. *$(\Lambda(0), \Phi(0)) = (n, 0)$, $\mathfrak{C}_n \subset \mathscr{C}_n$;*
3. *the set $\Sigma_n = \{s \geq 0 : \ker(\mathrm{Id} - \partial_\phi G(\Lambda(s), \Phi(s))) \neq \{0\}\}$ has no accumulation point;*
4. *at each point, $\mathfrak{C}_n$ has a local analytic re-parameterization (this holds, in particular, at each point of Σ_n);*
5. *one of the following occurs:*

a. $\|(\Lambda(s), \Phi(s))\| \to \infty$ *as* $s \to \infty$ *(which is much stronger than the claim that* $\mathfrak{C}_n$ *is unbounded in* $\mathbb{R} \times \dot{H}^1_{ax}(\partial B)$*);*
b. $\mathfrak{C}_n$ *is a closed loop.*

In particular, we can assume without loss of generality that (Λ, Φ) *is* C^∞*; furthermore, outside the singular set* Σ_n*, which is discrete,* ϕ *(and hence its harmonic extension* u*) can be smoothly parameterized with respect to* λ *along* $\mathfrak{C}_n$*.*

The previous result is simply [18, Theorem 9.1.1] written in our context.

Remark 2.6. Since for every solution $(\Lambda(s), \Phi(s))$ there is another solution $(\Lambda(s), -\Phi(s))$, we can define the curves

$$\tilde{\mathfrak{C}}_n = \{(\Lambda(-s), -\Phi(-s)) : s \in (-\infty), 0]\}$$

By Remark 2.4 above, the union of $\tilde{\mathfrak{C}}_n$ with $\mathfrak{C}_n$ form a continuous, locally analytic curve, which in a neighborhood of the origin takes the form

$$\big(\Lambda(s), \Phi(s)\big) = \big(n + o(1), sv_n + o(s)\big)$$

where v_n is an eigenfunction corresponding to the eigenvalue n.

It would be interesting to establish which of the alternatives of the previous propositions actually holds. For the analogous two-dimensional problem in a disk, the results obtained by variational methods seem to indicate that, in the $(\lambda, \|\phi\|)$ plane, the branches of solutions outgoing from $(n, 0)$ become asymptotic to the $\lambda = 0$ axis. Actually in [3] an explicit family of solutions of problem (3) in the case of the unit disk and for $\mu = 1$ is constructed. These solutions bifurcate from the Steklov eigenfunctions of the disk and become asymptotic to the $\lambda = 0$ axis, blowing up at equidistant points on the boundary (for any smooth two-dimensional domain, it has been proved in [7] that there are at least two distinct families of solutions which for $\lambda \to 0$ exhibit the same qualitative behaviour of the explicit solutions in the disk).

The analysis of the three dimensional problem, even in the case of axially symmetric solutions in the unit ball (with $\mu \equiv 1$) is much more complicated; hence, we will study in detail the component of the set of nontrivial solutions bifurcating from $(1, 0)$.

3 Analysis of the first branch

We first prove that we can further restrict our problem to the subspace of the axially symmetric functions u (in the unit ball) which are *odd* with respect to z; such subspace only contains the components of $\mathscr{S}$ which meet the points $(2k + 1, 0)$, $k = 0, 1, 2, \dots$

In spherical coordinates, we may represent an axially symmetric function u by $u = \hat{u}(r, \cos\theta)$; by putting $\cos\theta = t$, $-1 \le t \le 1$, we get $u = \hat{u}(r,t)$. Then, if u is odd with respect to z, we have $\hat{u}(r,-t) = -\hat{u}(r,t)$. We still denote by ϕ the traces $\phi = \hat{u}(1,\cdot)$.

Now, let V be the subspace of the functions $\phi \in \dot{H}^1_{ax}(\partial B)$ such that $\phi(-t) = -\phi(t)$; by the invariance of the Laplace operator with respect to the reflection $z \mapsto -z$ and by the symmetry of the Neumann condition on the sphere, it follows that *any solution of the Neumann problem in the ball with boundary data in V is axially symmetric and odd with respect to z.*

Hence, we can further restrict the functional formulation of the nonlinear equation (9) to the subspace V. Note that $s(\phi) = 0$ for every $\phi \in V$ (see equation (8)) so that $u = u_0$ for every solution of (3) defined below (10). Then, we can rephrase Propositions 2.3 and 2.5 in this context.

Proposition 3.1. *Let B be the unit ball and let $\mu \equiv 1$. Then, for any $k = 0, 1, 2, \ldots$ there exist a curve $\mathfrak{D}_k$, enjoying the properties of the curve $\mathfrak{C}_n$ described in Proposition 2.5, and a connected set $\mathscr{D}_k$, enjoying the properties of the set $\mathscr{C}_n$ described in Proposition 2.3, such that*

$$(2k+1, 0) \in \mathfrak{D}_k \subset \mathscr{D}_k \subset \mathscr{S} \subset \mathbb{R} \times V.$$

The main advantage of this restriction is that now we can describe some finer properties of the first branch. In fact, we can state

Proposition 3.2. *Let $\lambda > 0$, $u \in V$ be such that $(\lambda, u|_{\partial B}) \in \mathscr{D}_0$ and $u \not\equiv 0$; we may assume that $u > 0$ at some point of the upper half-sphere $\partial B \cap \{z > 0\}$ (otherwise, take $-u$). Then, $u|_{B\cap\{z>0\}} > 0$ and (by writing as before $u = \hat{u}(r, \cos\theta) = \hat{u}(r,t)$ with r, θ, spherical coordinates) the map $t \mapsto \hat{u}(r,t)$ is strictly increasing for every $r > 0$. Furthermore, $\lambda < 1$.*

Proof. By Theorem 4.1 of [2] we can assume that any solution to problem (3) in a ball (and with smooth μ) is smooth up to the boundary. In the following, we will denote by u an axially symmetric solution as a function of the *cylindrical coordinates* that is

$$u = u(\rho, z)$$

where $\rho = \sqrt{x^2 + y^2}$. We have

$$\hat{u}(r,t) = u\big(r\sqrt{1-t^2}, rt\big) \tag{11}$$

Let us now define

$$\hat{v}(r,t) = \frac{1}{r}\hat{u}_t(r,t) \tag{12}$$

By (11) we have

$$\hat{v} = u_z - \frac{t}{\sqrt{1-t^2}} u_\rho = u_z - \frac{z}{\rho} u_\rho \equiv v(\rho, z) \tag{13}$$

Then, by applying to v the Laplace operator in cylindrical coordinates

$$\Delta v = v_{\rho\rho} + \frac{1}{\rho} v_\rho + v_{zz}$$

we find after some calculations

$$\Delta v = -\Delta\Big(\frac{z}{\rho} u_\rho\Big) = -\frac{2}{\rho}\partial_\rho\Big(u_z - \frac{z}{\rho} u_\rho\Big) = -\frac{2}{\rho} v_\rho$$

Then, the function v solves the equation

$$v_{\rho\rho} + \frac{3}{\rho} v_\rho + v_{zz} = 0 \tag{14}$$

for $r = \sqrt{\rho^2 + z^2} < 1$. But the left-hand side is the expression of the Laplace operator in cylindrical coordinates in 5 dimensions applied to an axially symmetric function. Hence, v is harmonic (and axially symmetric) in the unit ball $\tilde{B} \subset \mathbb{R}^5$.

Moreover, by definition (12),

$$\hat{v}_r = -\frac{1}{r^2}\hat{u}_t + \frac{1}{r}\hat{u}_{tr} = -\frac{1}{r}\hat{v} + \frac{1}{r}\partial_t \hat{u}_r$$

and by recalling (3) we find on the unit sphere

$$\hat{v}_r(1,t) = -\hat{v}(1,t) + \partial_t\big(\lambda \sinh \hat{u}(1,t)\big) = -\hat{v}(1,t) + \lambda \cosh \hat{u}(1,t)\, \hat{v}(1,t)$$

that is

$$\hat{v}_r(1,t) = \big(\lambda \cosh \hat{u}(1,t) - 1\big)\, \hat{v}(1,t) \tag{15}$$

Hence, v is an axially symmetric solution of the *linear* eigenvalue problem (4) in a ball $\tilde{\Omega} \subset \mathbb{R}^5$, with weight $\mu(x) = \lambda \cosh u(x) - 1$ (and eigenvalue 1).

By our assumptions on u, we can write in a neighborhood of $(1, 0)$ (see Remark 2.4)

$$\big(\lambda, u\big) = \big(1 + \rho(\epsilon), \epsilon(z + w(\epsilon))\big)$$

where ϵ lies in some interval $[-\bar{\epsilon}, \bar{\epsilon}]$ and ρ, w are such that:

1. $\rho : [-\bar{\epsilon}, \bar{\epsilon}] \to \mathbb{R}$ is continuous and $\rho(0) = 0$
2. the map $\epsilon \mapsto w(\epsilon) \equiv w(\epsilon; x, y, z)$, $(x, y, z) \in \bar{B}$, is continuous from $[-\bar{\epsilon}, \bar{\epsilon}]$ to $\mathscr{C}^1(\bar{B})$ and $w(0) = 0$

Then we can write

$$v = \frac{1}{r} u_t = \epsilon(1 + \hat{w}_t(\epsilon)/r)$$

where as before we set $\hat{w}(\epsilon) \equiv \hat{w}(\epsilon; r, t) = w(\epsilon; \rho, z)$. The function $\hat{w}_t(\epsilon)/r$ is harmonic in the unit ball $\tilde{B} \subset \mathbb{R}^5$ and has the same normal derivative as v on the boundary ∂B; it follows by (15) that such normal derivative is uniformly vanishing for $\epsilon \to 0$. Then, $\lim_{\epsilon\to 0} \hat{w}_t(\epsilon)/r = c$; now, by choosing $r = 1$ and recalling that $\hat{w}(\epsilon) \to 0$ in $\mathscr{C}^1(\bar{B})$ for $\epsilon \to 0$, we conclude $c = 0$.

From the above result it follows that $v > 0$ for ϵ small enough. We claim that $v > 0$ all along $\mathscr{D}_0$; if not, by continuity there is a pair $(\lambda, u) \in \mathscr{D}_0$ such that $v \geq 0$ and $v(x) = 0$ for some $x \in \partial\tilde{B}$ (the boundary the unit ball of $\mathbb{R}^5$). Then, by (15) we get $v_r(x) = 0$, contradicting the Hopf principle. Since $\hat{u}_t = rv$, we find that $t \mapsto \hat{u}(r, t)$ is strictly increasing for every $r > 0$. But $\hat{u}(r, 0) = 0$, so that $\hat{u} > 0$ for $t > 0$, i.e. $u > 0$ on the upper half ball. Finally, by integration of both sides of (15) we get

$$\int_{\partial B} (\lambda \cosh u - 1)\, v = 0$$

which is possible for a positive v only if $\lambda < 1$. □

Theorem 3.3. *The set $\mathfrak{D}_0$ is unbounded; more precisely,* $0 < \Lambda(s) \leq 1$ *and* $\|\Phi(s)\|_{L^\infty(\partial B)} \to \infty$ *as* $s \to \infty$, *where* $\Phi(s)$, $s \in [0, +\infty)$ *are the solutions defined in Proposition 2.5.*

Proof. By Proposition 2.5 either $\mathfrak{D}_0$ is unbounded, or it is a closed loop: in the latter case, by Remark 2.4 there exist two solutions of opposite sign at the beginning and at the end of the loop near to $(1, 0)$; since the nontrivial solutions in $\mathfrak{D}_0$ only vanish at $z = 0$ (by Proposition 3.2) it is readily checked that $\mathfrak{D}_0$ must intersect the λ axis at some other point, which is necessarily $(2j + 1, 0)$ for some $j > 0$.

By recalling that $\Phi(s) = u|_{\partial B}$ with u harmonic function (axially symmetric and odd with respect to the reflection $z \mapsto -z$) it now follows by continuity (see [2]) that there exists a $(\Lambda, \Phi) \in \mathfrak{D}_0 \subset \mathscr{D}_0$, $\Phi \neq 0$ such that $\Phi = \hat{u}(1, t)$ (see (11)) and $\hat{u}_t(1, t) = 0$ at some point $t \in (-1, 1)$ contradicting the positivity of (12) on $\mathscr{D}_0$.

Thus, we conclude that $\mathfrak{D}_0$ is unbounded; but we know from Proposition 3.2 that λ is bounded along $\mathscr{D}_0 \supset \mathfrak{D}_0$; then, again by Proposition 2.5, we have $\|\Phi\|_{H^1(\partial\Omega)} \to \infty$ along $\mathfrak{D}_0$. Hence, as remarked in [2], we also have that the uniform norm $\|\Phi\|_{L^\infty(\partial\Omega)}$ becomes arbitrarily large; we stress that, due to Proposition 3.2, the sup norm of Φ is given by the value $u(0, 0, 1)$ where u is defined as above. □

Before investigating the limiting behaviour of the solutions with increasing supremum norm along $\mathfrak{D}_0$, we point out some further properties of such solutions. Let us introduce the energy

$$\mathscr{E}_\lambda(u) = \frac{1}{2} \int_B |\nabla u|^2\, dx - \lambda \int_{\partial B} (\cosh u - 1)\, d\sigma \tag{16}$$

and assume that $(\Lambda(s), u(s))$ is the solution to (3) corresponding to the point $(\Lambda(s), \Phi(s)) \in \mathfrak{D}_0$; by denoting with u', Λ' the derivatives with respect to s, we can compute

$$\begin{aligned}\frac{d\mathscr{E}_\lambda(u)}{ds} &= \int_B \nabla u \nabla u' \, dx - \Lambda \int_{\partial B} (\sinh u)\, u' \, d\sigma - \Lambda' \int_{\partial B} (\cosh u - 1)\, d\sigma \\ &= -\Lambda' \int_{\partial B} (\cosh u - 1)\, d\sigma\end{aligned} \tag{17}$$

the last equality following by the weak form of (3). Since the last integral is nonnegative, it follows by (17) that the energy is decreasing for $\Lambda' \geq 0$; in particular, if $s \notin \Sigma_0$ (see Proposition 2.5), we can take Φ, and consequently u, smoothly dependent on λ in certain intervals contained in $(0, 1)$. Then, in every such interval *$\mathscr{E}_\lambda(u)$ is strictly decreasing with respect to λ.*

As it is suggested by numerical experiments (see below) in the 3 dimensional problem we have $\Sigma_0 \neq \emptyset$, and $\Lambda'(s)$ changes its sign along $\mathfrak{D}_0$. It is an open problem to find whether Σ_0 is a finite or infinite discrete set.

Blow-up analysis

The final part of this section is devoted to the asymptotic analysis of solutions with increasing supremum norm along $\mathfrak{D}_0$. Taking into account Proposition 3.2 and Theorem 3.3, we have that any unbounded subset of $\mathfrak{D}_0$ contains a subsequence $(\lambda_j, u_j|_{\partial B})$, with u_j harmonic in B, such that $u_j > 0$ on $B \cap \{x_3 > 0\}$ and

$$0 < \lambda_j < 1, \qquad \max_{\overline{B}} u_j(x) = u_j(k) = M_j \to \infty \tag{18}$$

as $j \to \infty$ (here $x = (x_1, x_2, x_3)$, and $k = (0, 0, 1)$ denotes the north pole of B).

We choose a sequence r_j such that $r_j \to 0$ for $j \to +\infty$ and define the transformation

$$y = \frac{k - x}{r_j} \tag{19}$$

which maps B onto a sphere B_j (of radius r_j^{-1}, center at $y = r_j^{-1}k$ and outer normal $\nu_j = r_j y - k$) in the upper half plane $y_3 \geq 0$. Note that the point k is mapped to the origin $y = 0$ and that the sequence B_j exhausts $\mathbb{R}^3_+$.

Let us now define

$$v_j(y) = \frac{e^{-M_j}}{r_j}\Big[M_j - u_j(-r_j y + k)\Big] \tag{20}$$

The functions v_j are harmonic in B_j, positive and symmetric with respect to the y_3 axis, with minimum $v_j(0) = 0$. Moreover, they satisfy the following boundary conditions

$$\partial_{\nu_j} v_j(y) = (r_j y - k)\cdot \nabla_y v_j(y) = \frac{e^{-M_j}}{r_j}(k - r_j y)\cdot \nabla_y u_j(-r_j y + k)$$

$$= -e^{-M_j}\partial_\nu u_j(-r_j y + k) = -\lambda_j e^{-M_j}\sinh(u_j(-r_j y + k)), \qquad y \in \partial B_j \tag{21}$$

By the assumptions on u_j we have

$$-\frac{\lambda_j}{2}\left(1 - e^{-2M_j}\right) \le \partial_{\nu_j} v_j(y) < 0$$

for every $y \in \partial B_j \cap \{y_3 < r_j^{-1}\}$ (the lower half of the spherical surface ∂B_j).

By the above estimate, one can infer that the sequence v_j converges uniformly in every bounded set of $\mathbb{R}^3_+$. Of course, the form of the limit problem depends on the choice of r_j. By taking $r_j = e^{-M_j}$ we get from (20) and (21),

$$\partial_{\nu_j} v_j(y) = -\lambda_j e^{-M_j}\sinh(M_j - v_j(y)) = -\frac{\lambda_j}{2}\left[e^{-v_j(y)} - e^{-2M_j + v_j(y)}\right]. \tag{22}$$

Then (via suitable projections of the lower half spheres $B_j \cap \{y_3 < r_j^{-1}\}$ to the upper half space $y_3 > 0$ and letting $j \to \infty$) we obtain the following limit problem

$$\begin{aligned} &\Delta v = 0 \quad \text{in } \mathbb{R}^3_+ \\ &\partial_\nu v = -\frac{\lambda^*}{2} e^{-v} \quad \text{on } \mathbb{R}^2 \times \{y_3 = 0\} \\ &v = v(\rho, z), \quad v \ge 0 \text{ in } \mathbb{R}^3_+, \quad v(0) = 0, \end{aligned} \tag{23}$$

where $\lambda^* \in [0, 1]$ is an accumulation point for the sequence (λ_j). In this way, the study of the asymptotic behaviour of $\mathfrak{D}_0$ is related to the classification of the entire solutions of problem (23).

Remark 3.4. The function

$$v^*(y) = \frac{\lambda^*}{2} y_3$$

solves Problem (23). Moreover, if $\lambda^* > 0$ it has infinite Morse index.

Indeed, by writing the energy functional associated with the problem:

$$E(v) = \frac{1}{2}\int_{\mathbb{R}^3_+} |\nabla v|^2 - \frac{\lambda^*}{2}\int_{\mathbb{R}^2} e^{-v} \tag{24}$$

we have that v solves (23) if and only if

$$E'(v)[\phi] = \int_{\mathbb{R}^3_+} \nabla v \cdot \nabla \phi + \frac{\lambda^*}{2} \int_{\mathbb{R}^2} e^{-v} \phi = 0 \tag{25}$$

for every $\phi \in C_0^1(\mathbb{R}^3)$ (not necessarily vanishing on $\mathbb{R}^2 \times \{y_3 = 0\}$); therefore

$$E''(v)[\phi, \phi] = \int_{\mathbb{R}^3_+} |\nabla \phi|^2 - \frac{\lambda^*}{2} \int_{\mathbb{R}^2} e^{-v} \phi^2 \tag{26}$$

and

$$E''(v^*)[\phi, \phi] = \int_{\mathbb{R}^3_+} |\nabla \phi|^2 - \frac{\lambda^*}{2} \int_{\mathbb{R}^2} \phi^2. \tag{27}$$

Now, for any fixed $\eta \in C_0^1(\mathbb{R}^3)$, $r > 0$ and $\xi \in \mathbb{R}^3$, we can choose $\phi_\xi(y) = r\eta\,(ry + \xi)$, so that

$$E''(v^*)[\phi_\xi, \phi_\xi] = r^2 \int_{\mathbb{R}^3_+} |\nabla \eta|^2 - \frac{\lambda^*}{2} \int_{\mathbb{R}^2} \eta^2 < 0 \qquad \text{for sufficiently small } r.$$

As a consequence, for any $m \in \mathbb{N}$, one can easily find $\xi_1, \ldots, \xi_m \in \mathbb{R}^3$ in such a way that $W_m = \text{span}\,\{\phi_{\xi_1}, \ldots, \phi_{\xi_m}\}$ has dimension m and $E''(v^*)$ is negative defined on W_m.

It remains an open question whether Problem (23) admits other nontrivial solutions, apart from v^*, and in such a case whether solutions with finite Morse index may exist. In case $\lambda^* > 0$, the absence of finite Morse index solutions to (23) would indicate the presence of infinitely many secondary bifurcation points (turning points) along $\mathfrak{D}_0$. Such kind of behaviour is also suggested by the numerical simulations we discuss in the next section.

4 Numerical simulations

In this section we present and discuss the numerical scheme that was used in order to approximate the bifurcation branches of problem (3). We then give some comments on the numerical bifurcation diagram that we have obtained.

The mathematical literature concerning continuation methods is vast and it is not our aim to give here a complete list of the possible solutions already available, we just refer the interested reader to [19, 20] and to the references therein. As a matter of fact, the method we implemented can be traced back to the large class of predictor-corrector methods, where in our case the predictor step is obtained through a projection on a suitable space of solutions, while the corrector step consists in a time step of a suitable parabolic flow.

From a previous numerical investigation it is clear that λ cannot be used to parametrize the curves of the bifurcation diagram, due to the nonmonotonicity of this quantity along these curves. Moreover the fact that for any $\lambda > 0$ the set of trivial solutions $(\lambda, 0)$ is also the stable branch discourages the direct use of more classical methods, such as the standard parabolic flow. As a concluding remark, we recall that even from the theoretical point of view, the branches of nontrivial solutions are obtained in the functional space $\dot{H}^1(\partial B)$, which is somehow unnatural from a numerical point of view. For all these reasons, after preliminary numerical investigations we consider worthwhile to assume the $L^\infty(\partial B)$ norm of the solution to be a possible parameter to describe the curve, as it has already been shown that such norm is unbounded along any bifurcation branch.

Algorithm 1 (Continuation method)

1: initialize s as a small number and let α be a large positive constant
2: initialize $(\lambda, u) \leftarrow (1, sz)$
3: **repeat**
4: set $s \leftarrow s + \varepsilon$
5: $(\gamma, v) \leftarrow (\lambda, u)$
6: **repeat**
7: $\tilde{v} \leftarrow \dfrac{v}{\|v\|_{L^\infty}} \cdot s$ ▷ Predictor step
8: $\gamma \leftarrow \dfrac{\int_{B_1} |\nabla \tilde{v}|^2}{\int_{\partial B_1} \sinh(\tilde{v})\tilde{v}}$
9: Solve $\begin{cases} \Delta v = 0 & \text{in } B_1(0) \\ \partial_\nu v + \alpha(v - \tilde{v}) = \gamma \sinh(\tilde{v}) & \text{on } \partial B_1(0) \end{cases}$ ▷ Corrector step
10: **until** convergence with a prescribed tolerance
11: $(\lambda, u) \leftarrow (\gamma, v)$
12: **until** blow-up

Remark 4.1. Similar results can be obtained also in the case that we assume as parameter the $\dot{H}^1(\partial B)$ norm, even though the resulting method seems less efficient in terms of convergence rate.

Let us point out that both the $L^\infty(\partial B)$ and $\dot{H}^1(\partial B)$ norms constitute an unnatural choice as parameters from a point of view of the numerical method used in the corrector step (step 9 in the algorithm), which is discretized using its weak formulation in $H^1(B)$. As a particular consequence, this makes the predictor step a priori unfeasible, as the set $\{u \in H^1(B) : \|u\|_{L^\infty(\partial B)} = s\}$ (and $\{u \in H^1(B)$: $\|u\|_{\dot{H}^1(\partial B)} = s\}$) is not closed in the topology of $H^1(B)$: this complicates the convergence analysis, which is not carried out in the following.

Remark 4.2. One may also try to use a more sophisticated algorithm such as the Newton's method. As it turns out from numerical investigation, the Morse index of the solutions increases by one unit at each turning point, but the negative eigenvalues of the linearized operator diverge rather rapidly (see Figure 2). Any attempt we tried

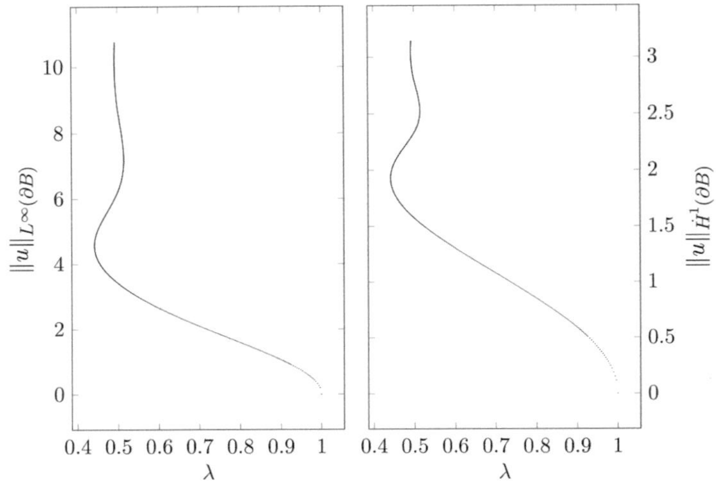

Fig. 1 Plot of the numerical simulations for the bifurcation branch $\mathfrak{D}_0$

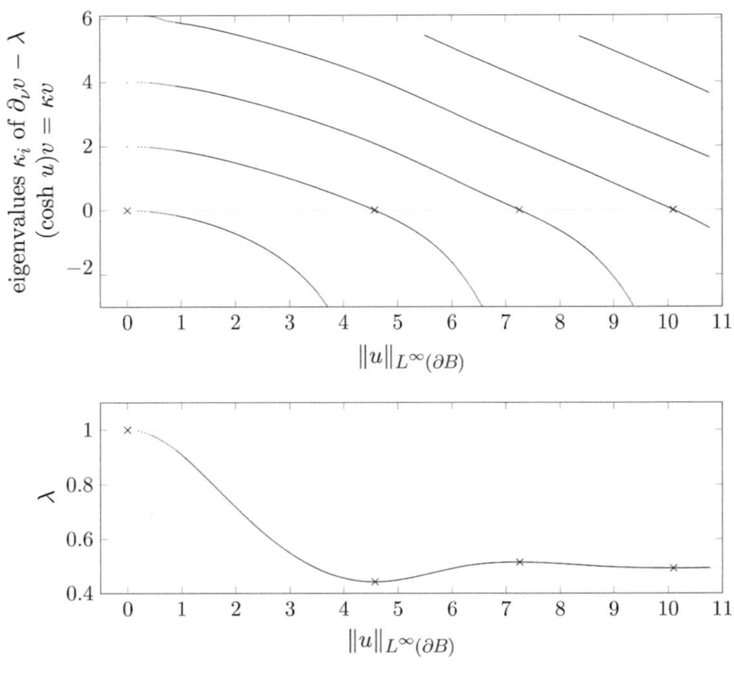

Fig. 2 Plot of the first 4 eigenvalues of the Steklov eigenvalue problem $\partial_\nu v - \lambda(\cosh u)v = \kappa v$ (above), and of the bifurcation parameter λ (below), as functions of $\|u\|_{L^\infty(\partial B)}$ along $\mathfrak{D}_0$. The turning points on the branch correspond to the increasing of the Morse index of the solution

in stabilizing such algorithm led us to lose the convergence of the original method, and for this reason we chose to focus our attention on a more stable, even if less efficient, fixed point method.

Now we proceed with some comments on the numerical bifurcation diagrams. The plots are obtained from the simulation data using the LaTeX-graphics packages Ti*k*Z and PGFPLOTS. To start with, as we already mentioned, the $L^\infty(\partial B)$ norm of the solution is increasing along the branch essentially by construction. Also the $\dot{H}^1(\partial B)$ norm appears to increase (Figure 1). On the other hand, other norms are not monotone, and the simulations suggest that $\mathfrak{D}_0$ may be bounded in $H^1(B)$ (Figure 3). The energy

$$\mathscr{E}_\lambda(u) = \frac{1}{2}\int_B |\nabla u|^2\, dx - \lambda \int_{\partial B} (\cosh u - 1)\, d\sigma$$

exhibits an analogous behaviour along the branch, but it loses smoothness: the turning points of $\mathfrak{D}_0$ become corner points for $\mathscr{E}_\lambda$. Moreover, according to (17),

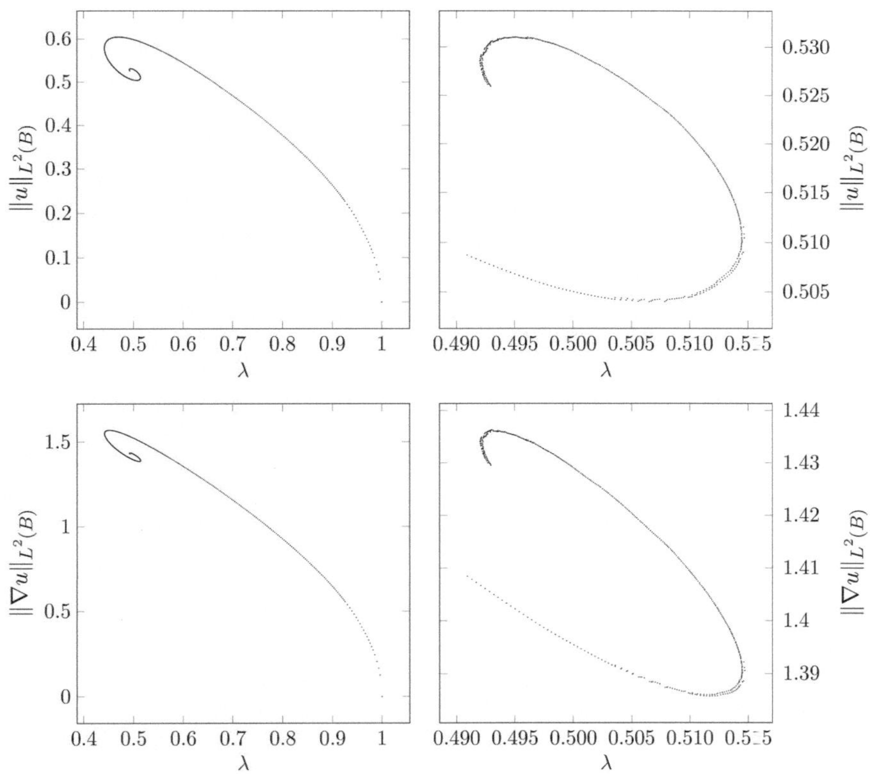

Fig. 3 The numerical simulations suggest that, on the branch $\mathscr{D}_0$, the L^2 norm and the H^1 seminorm of u are not monotone, and that the H^1 norm is bounded

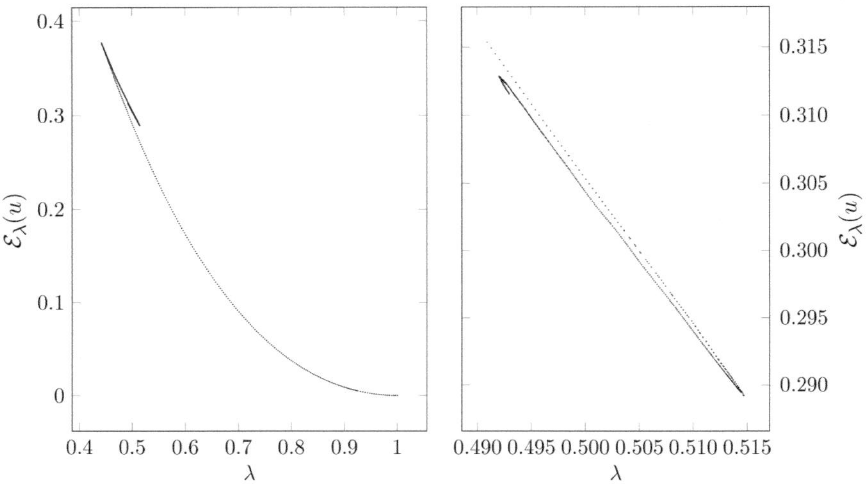

Fig. 4 The apparent non monotonicity of the energy E_λ with respect to λ suggests the presence of corner points where the energy is not a smooth function of the bifurcation parameter

$\mathcal{E}_\lambda(u)$ is decreasing with respect to λ in every interval of smooth dependence (Figure 4).

Acknowledgements G. Verzini is partially supported by the PRIN-2012-74FYK7 Grant: "Variational and perturbative aspects of nonlinear differential problems" and by the project ERC Advanced Grant 2013 n. 339958: "Complex Patterns for Strongly Interacting Dynamical Systems–COMPAT". A. Zilio is partially supported by the ERC Advanced Grant 2013 n. 321186: "ReaDi – Reaction-Diffusion Equations, Propagation and Modelling" and by the project ERC Advanced Grant 2013 n. 339958: "Complex Patterns for Strongly Interacting Dynamical Systems–COMPAT".

References

1. Vogelius, M., Xu, J.-M.: A nonlinear elliptic boundary value problem related to corrosion modeling. Q. Appl. Math. **56**, 479–505 (1998)
2. Pagani, C.D., Pierotti, D.: A three dimensional Steklov eigenvalue problem with exponential nonlinearity on the boundary. Nonlinear Anal. **79**, 28–40 (2013)
3. Bryan, K., Vogelius, M.: Singular solutions to a nonlinear elliptic boundary value problem originating from corrosion modeling. Q. Appl. Math. **60**(4), 675–694 (2002)
4. Kavian, O., Vogelius, M.: On the existence and 'blow-up' of solutions to a two-dimensional nonlinear boundary-value problem arising in corrosion modeling. Proc. Roy. Soc. Edinb. Sect A **133**, 119–149 (2003). [Corrigendum to the same, Proc. R. Soc. Edinb. Sect A **133**, 729–730 (2003)]
5. Pagani, C.D., Pierotti, D.: Variational methods for nonlinear Steklov eigenvalues problems with an indefinite weight function. Calc. Var. **39**, 35–58 (2010)
6. Pagani, C.D., Pierotti, D.: Multiple variational solutions to non linear Steklov problems. Nonlinear Differ. Equ. Appl. **19**, 417–436 (2012)

7. Dávila, J., del Pino, M., Musso, M., Wei, J.: Singular limits of a two-dimensional boundary value problem arising in corrosion modelling. Arch. Ration. Mech. Anal. **182**(2), 181–221 (2006)
8. Bhat, Y.S., Moskow, S.: Linearization of a nonlinear periodic boundary condition related to corrosion modeling. J. Comput. Math. **25**(6), 645–660 (2007)
9. Rabinowitz, P.H.: Minimax Methods in Critical Point Theory with Applications to Differential Equations. CBMS Regional Conference Series in Mathematics, vol. 65. AMS, Providence (1986)
10. Rabinowitz, P.H.: Some aspects of nonlinear eigenvalue problems. Rocky Mt. J. Math. **3**(2), 161–202 (1973)
11. Bandle, C.: Isoperimetric inequalities and applications, Monographs and Studies in Mathematics, vol. 7. Pitman (Advanced Publishing Program), Boston (1980)
12. Jerison, D.S., Kenig, C.E.: The Neumann problem on Lipschitz domains. Bull. A.M.S. **4**(2), 203–207 (1981)
13. Crandall, M., Rabinowitz, P.H.: Bifurcation from simple eigenvalues. J. Funct. Anal. **8**, 321–340 (1971)
14. Ambrosetti, A., Prodi, G.: A primer of nonlinear analysis. Cambridge Studies in Advanced Mathematics, vol. 34. Cambridge University Press, Cambridge (1995)
15. Dancer, E.N.: Global structure of the solutions of non-linear real analytic eigenvalue problems. Proc. Lond. Math. Soc. **3**(27), 747–765 (1973)
16. Dancer, E.N.: Global solution branches for positive mappings. Arch. Rational Mech. Anal. **52**, 181–192 (1973)
17. Dancer, E.N.: Bifurcation theory for analytic operators. Proc. Lond. Math. Soc. **3**(26), 359–384 (1973)
18. Buffoni, B., Toland, J.F.: Analytic Theory of Global Bifurcation, Princeton Series in Applied Mathematics. Princeton University Press, Princeton (2003)
19. Ortega, J.M., Rheinboldt, W.C.: Iterative Solution of Nonlinear Equations in Several Variables. Academic Press, New York (1970)
20. Allgower, E.L., Georg, K.: Introduction to Numerical Continuation Methods. Classics in Applied Mathematics. vol. 45. Society for Industrial and Applied Mathematics (SIAM), Philadelphia (2003). [Reprint of the 1990 edn., Springer, Berlin, MR1059455 (92a:65165)]

Remarks on the p-Laplacian on thin domains

Marcone C. Pereira and Ricardo P. Silva

To Djairo G. de Figueiredo, on the occasion of his 80th birthday

Mathematics Subject Classification (2010): 35B25, 35J92, 80A20

1 Introduction

Over the last years partial differential equations on thin domains have received considerable attention in the literature of pure and applied mathematics. They occur in applications as in mechanics of nano structures (thin rods, plates, or shells), fluids in thin channels (lubrication models, blood circulation), chemical diffusion process on membranes or narrow strips (catalytic process), homogenization of reticulated structures, as in the study of the stability (or instability) of the asymptotic dynamics of singularly perturbed parabolic equations, see, for instance, [1–4, 8, 12, 14, 15, 17–23, 26].

In all of these problems the aim is to describe the effective behavior of perturbed elements. Generally this is made establishing a formal limit in order to compare them, since, in many cases, the limit is simpler to study than the perturbed one. Indeed, this is our goal here, establish the limiting regime for the family of solutions of the elliptic equation

$$-\Delta_p u + |u|^{p-2}u = f(u), \tag{1}$$

M.C. Pereira (✉)
Instituto de Matemática e Estatística, Universidade de São Paulo, São Paulo SP, Brazil
e-mail: marcone@ime.usp.br

R.P. Silva
Instituto de Geociências e Ciências Exatas, Universidade Estadual Paulista, Rio Claro SP, Brazil
e-mail: rpsilva@rc.unesp.br

A.N. Carvalho et al. (eds.), *Contributions to Nonlinear Elliptic Equations and Systems*, Progress in Nonlinear Differential Equations and Their Applications 86,
DOI 10.1007/978-3-319-19902-3_23

posed on thin domains of the form

$$\Omega^\epsilon = \{(x, \epsilon y) \in \mathbb{R}^{m+n} : (x, y) \in \Omega\},$$

where $\Omega \subset \mathbb{R}^{m+n}$ is a smooth bounded domain and $\epsilon > 0$ is a small parameter. $\Delta_p u := \mathrm{div}(\|\nabla u\|^{p-2}\nabla u)$ denotes the p-Laplacian operator, $p > 1$ and $f : \mathbb{R} \to \mathbb{R}$ is a suitable nonlinearity which will be specified later. We consider (1) coupled by Dirichlet

$$u = 0, \qquad \text{on } \partial\Omega^\epsilon, \tag{2}$$

or Neumann boundary conditions

$$\frac{\partial u}{\partial \eta^\epsilon} = 0, \qquad \text{on } \partial\Omega^\epsilon, \tag{3}$$

where η^ϵ denotes the outward unitary normal vector field to $\partial\Omega^\epsilon$.

If Dirichlet boundary condition is considered, we show (Theorem 2.5) that solutions of the perturbed problem (1)–(2) converge to the null function as $\epsilon \to 0$. On the other hand, assuming Neumann boundary conditions we obtain a nontrivial limiting equation (see (25)) posed in a lower dimensional domain which determines the behavior of the solutions of the problem (1) and (3) as $\epsilon \to 0$.

We notice that if one considers quasilinear parabolic equations of the form

$$u_t - \Delta_p u + |u|^{p-2}u = f(u), \quad [0, \infty) \times \Omega^\epsilon, \tag{4}$$

which are relevant in a variety of physical phenomena (see [7, 10, 13, 27]), the solutions of (1) are the steady states of (4). If $\mathscr{E}_\epsilon$ is the set of solutions of (1) and (3) and $\mathscr{E}_0$ the solutions of (25), the results discussed here say with respect to the upper semicontinuity of $\mathscr{E}_\epsilon$ at $\epsilon = 0$, i.e., with some abuse of notation,

$$\sup_{u_\epsilon \in \mathscr{E}_\epsilon} \inf_{u_0 \in \mathscr{E}_0} \|u_\epsilon - u_0\|_{W^{1,p}(\Omega_\epsilon)} \overset{\epsilon \to 0}{\longrightarrow} 0.$$

Such result is the first step in order to prove the stability of the asymptotic dynamics (attractors) of (4). For recent works on this topic, we indicate [5, 6, 25].

Finally, we would like to express our gratitude to Prof. Djairo G. de Figueiredo who opened, and still opens paths through the field of PDEs for generations of mathematicians in Brazil. Besides, the way he works as professor and scientist is an inspiration for us. Indeed Djairo's work [11] gave us the idea of write down this paper which generalizes to the p-Laplacian operator some results obtained by Hale and Raugel in the seminal work [14] in the case $p = 2$.

The paper is organized as follows: in Section 2 we study the problem (1) with Dirichlet boundary conditions and in Section 3, the case of Neumann boundary conditions. In Section 4 we present as example a particular class of thin domains defined as graphic of smooth functions.

2 Dirichlet boundary condition

Let Ω be a smooth and bounded domain in $\mathbb{R}^m \times \mathbb{R}^n$ not necessary a product domain. As usual, we identify $\mathbb{R}^m \times \mathbb{R}^n$ with $\mathbb{R}^{m+n}$ writing (x, y) for a generic point of $\mathbb{R}^{m+n}$ where $x = (x_1, \cdots, x_m) \in \mathbb{R}^m$ and $y = (y_1, \cdots, y_n) \in \mathbb{R}^n$. We also write $\|(x, y)\| = \left(\|x\|^2 + \|y\|^2\right)^{1/2}$ for the euclidian norm in the space $\mathbb{R}^{m+n}$ where $\|\cdot\|$ denotes indistinctly the euclidian norm in $\mathbb{R}^m$ or $\mathbb{R}^n$, and we adopt standard notation for the inner product $(x, y) \cdot (w, z) = x \cdot w + y \cdot z$ for all $(x, y), (w, z) \in \mathbb{R}^{m+n}$. For a function $u \in W^{1,p}(\Omega)$, we denote by $\nabla u = (\nabla_x u, \nabla_y u) \in L^p(\Omega)^{m+n}$ the (distributional) gradient of u. We will endow $W^{1,p}(\Omega)$ with the equivalent norm

$$\|u\|_{W^{1,p}(\Omega)} = \left[\int_\Omega \left(\|\nabla_x u\|^p + \|\nabla_y u\|^p + |u|^p\right) dxdy\right]^{1/p}$$

which still preserves the uniformly convexity of $W^{1,p}(\Omega)$ [9]. Along the paper ϵ will represent a small positive parameter which will converge to zero.

2.1 *The thin domain problem*

Considering the squeezing operator $\Phi_\epsilon : \mathbb{R}^{m+n} \to \mathbb{R}^{m+n}$ defined by $\Phi_\epsilon(x, y) = (x, \epsilon y)$, we set up the *thin domain* $\Omega^\epsilon := \Phi_\epsilon(\Omega)$ and we analyze convergence of the solutions of the family of elliptic equations

$$\begin{aligned} -\Delta_p u + |u|^{p-2} u &= h^\epsilon, \quad &&\text{in } \Omega^\epsilon, \\ u &= 0, &&\text{on } \partial\Omega^\epsilon, \end{aligned} \tag{5}$$

where

$$\begin{aligned} \Delta_p u := \operatorname{div}(\|\nabla u\|^{p-2} \nabla u) &= \sum_{i=1}^m \frac{\partial}{\partial x_i}\left[\left(\|\nabla_x u\|^2 + \|\nabla_y u\|^2\right)^{(p-2)/2} \frac{\partial u}{\partial x_i}\right] \\ &+ \sum_{i=1}^n \frac{\partial}{\partial y_i}\left[\left(\|\nabla_x u\|^2 + \|\nabla_y u\|^2\right)^{(p-2)/2} \frac{\partial u}{\partial y_i}\right] \end{aligned}$$

denotes the p-Laplacian operator, $p > 1$, and $h^\epsilon \in L^{p'}(\Omega^\epsilon)$, $1/p + 1/p' = 1$.

Definition 2.1. Given $h^\epsilon \in L^{p'}(\Omega^\epsilon)$, we say that $u \in W_0^{1,p}(\Omega^\epsilon)$ is solution of the problem (5) if

$$\int_{\Omega^\epsilon} \Big(\|\nabla u\|^{p-2} \nabla u \cdot \nabla \varphi + |u|^{p-2} u\, \varphi \Big) dxdy = \int_{\Omega^\epsilon} h^\epsilon\, \varphi\, dxdy, \tag{6}$$

for all $\varphi \in W_0^{1,p}(\Omega^\epsilon)$.

It is well known (see, e.g., [16]) that for each value of $\epsilon > 0$, the p-Laplacian operator can be seen as

$$-\Delta_p : W_0^{1,p}(\Omega^\epsilon) \to W^{-1,p'}(\Omega^\epsilon)$$

$$\langle -\Delta_p u, v \rangle_{W^{-1,p'}, W_0^{1,p}} = \int_{\Omega^\epsilon} \|\nabla u\|^{p-2} \nabla u \cdot \nabla v\, dxdy, \qquad \text{for } u, v \in W_0^{1,p}(\Omega^\epsilon),$$

where $\langle \cdot, \cdot \rangle_{W_0^{-1,p'}, W_0^{1,p}}$ denotes the pair of duality between $W^{-1,p'}(\Omega^\epsilon)$ and $W_0^{1,p}(\Omega^\epsilon)$. With $-\Delta_p$ defined above, $-\Delta_p u + |u|^{p-2} u = J_\psi u$, where J_ψ is the duality mapping[1] corresponding to the normalization function $\psi(t) = t^{p-1}$ on the space $W_0^{1,p}(\Omega^\epsilon)$. We know from [16] that J_ψ is single valued and therefore, for each value of the parameter $\epsilon > 0$, the problem (5) has a unique solution $\bar{u}^\epsilon \in W_0^{1,p}(\Omega^\epsilon)$. Moreover, since the functional $\phi : W_0^{1,p}(\Omega^\epsilon) \to \mathbb{R}$ defined by

$$\phi(u) = \int_0^{\|u\|_{W_0^{1,p}(\Omega^\epsilon)}} \psi(t)\, dt - \int_{\Omega^\epsilon} h^\epsilon u\, dxdy = \frac{1}{p} \|u\|^p_{W_0^{1,p}(\Omega^\epsilon)} - \int_{\Omega^\epsilon} h^\epsilon u\, dxdy,$$

is Gatêaux differentiable and

$$\langle \phi'(u), v \rangle_{W^{-1,p'}, W_0^{1,p}} = \langle -J_\psi u, v \rangle_{W^{-1,p'}, W_0^{1,p}} - \int_{\Omega^\epsilon} h^\epsilon v\, dxdy,$$

we have that $\bar{u}^\epsilon$ is solution of the problem (5) if and only if $\phi'(\bar{u}^\epsilon) = 0$. Since ϕ is convex and $W_0^{1,p}(\Omega^\epsilon)$ is reflexive, we have the following characterization: $\bar{u}^\epsilon \in W_0^{1,p}(\Omega^\epsilon)$ is solution of (5) if and only if

$$\frac{1}{p} \|\bar{u}^\epsilon\|^p_{W_0^{1,p}(\Omega^\epsilon)} - \int_{\Omega^\epsilon} h^\epsilon \bar{u}^\epsilon\, dxdy = \min_{\varphi \in W_0^{1,p}(\Omega^\epsilon)} \left\{ \frac{1}{p} \|\varphi\|^p_{W_0^{1,p}(\Omega^\epsilon)} - \int_{\Omega^\epsilon} h^\epsilon \varphi\, dxdy \right\}. \tag{7}$$

Now, in order to obtain the limiting regime of the family of solutions $\{\bar{u}^\epsilon\}_{\epsilon > 0}$, we perform a dilatation of the domain Ω^ϵ by a factor ϵ^{-1} in the y-direction. Introducing

[1] $J_\psi : W_0^{1,p}(\Omega^\epsilon) \to 2^{W^{-1,p'}(\Omega^\epsilon)}$ is defined by $J_\psi u := \{u^* \in W^{-1,p'}(\Omega^\epsilon) : \|u^*\|_{W^{-1,p'}(\Omega^\epsilon)} = \psi(\|u\|_{W_0^{1,p}(\Omega^\epsilon)}) \|u\|_{W_0^{1,p}(\Omega^\epsilon)}, \langle u^*, u \rangle_{W_0^{-1,p'}, W_0^{1,p}} = \psi(\|u\|_{W_0^{1,p}(\Omega^\epsilon)})\}$

the operator

$$\Delta_p^\epsilon u := \operatorname{div}\Big(\|(\nabla_x u, \frac{1}{\epsilon}\nabla_y u)\|^{p-2}(\nabla_x u, \frac{1}{\epsilon^2}\nabla_y u)\Big), \tag{8}$$

we obtain in the fixed domain Ω the equivalent equation

$$\begin{aligned} -\Delta_p^\epsilon u + |u|^{p-2}u = f^\epsilon, &\quad \text{in } \Omega, \\ u = 0, &\quad \text{on } \partial\Omega, \end{aligned} \tag{9}$$

where $f^\epsilon(x,y) := h^\epsilon(x, \epsilon y)$.

The relation between the functions spaces set in Ω^ϵ and those set in Ω is given by the isomorphism

$$\begin{aligned} \Phi_\epsilon^* : L^p(\Omega^\epsilon) &\to L^p(\Omega) \\ u &\mapsto u \circ \Phi_\epsilon, \end{aligned} \tag{10}$$

which also define an isomorphism from $W^{1,p}(\Omega^\epsilon)$ onto $W^{1,p}(\Omega)$, as well as from $W_0^{1,p}(\Omega^\epsilon)$ onto $W_0^{1,p}(\Omega)$. It is not difficult to see that $\bar{u}^\epsilon \in W_0^{1,p}(\Omega^\epsilon)$ is solution of (5) if and only if $u^\epsilon := \Phi_\epsilon^*(\bar{u}^\epsilon) \in W_0^{1,p}(\Omega)$ is solution of (9), i e., it for all $\varphi \in W_0^{1,p}(\Omega)$, u^ϵ must satisfy

$$\int_\Omega \Big(\|(\nabla_x u^\epsilon, \frac{1}{\epsilon}\nabla_y u^\epsilon)\|^{p-2}(\nabla_x u^\epsilon \cdot \nabla_x \varphi + \frac{1}{\epsilon^2}\nabla_y u^\epsilon \cdot \nabla_y \varphi) + |u^\epsilon|^{p-2}u^\epsilon \varphi \Big)\, dxdy = \int_\Omega f^\epsilon \varphi \, dxdy. \tag{11}$$

As we can see from the last identity, when we consider the problem (1) in the fixed domain Ω (which is not thin anymore), it appears a factor ϵ^{-1} on the gradient of u^ϵ in the y-direction. Physically this means that there is a very strong diffusion mechanism acting on the y-direction, and therefore one expects solutions become homogeneous in this direction, i.e., is expected that in the limit $\epsilon \to 0$ the limiting solution does not depend on the variable y. We formalize this in the next Proposition.

Proposition 2.2. *Let $f^\epsilon \in L^{p'}(\Omega)$ be a uniformly bounded (with respect to ϵ) family of functions. If $u^\epsilon \in W_0^{1,p}(\Omega)$ is the solution of (9), there exists $u^0 \in W_0^{1,p}(\Omega)$ such that $\nabla_y u^0 = 0$ a.e. in Ω and, up to subsequence,*

$$u^\epsilon \xrightarrow{\epsilon \to 0} u^0, \text{ weakly in } W_0^{1,p}(\Omega) \text{ and strongly in } L^p(\Omega). \tag{12}$$

Proof. Taking u^ϵ as test function in (11) we obtain

$$\int_\Omega \Big(\|\nabla_x u^\epsilon\|^p + \|\epsilon^{-1}\nabla_y u^\epsilon\|^p + |u^\epsilon|^p\Big) dxdy = \int_\Omega f^\epsilon \, u^\epsilon \, dxdy.$$

Therefore

$$\|u^\epsilon\|^p_{W_0^{1,p}(\Omega)} \le \int_\Omega \Big(\|\nabla_x u^\epsilon\|^p + \|\epsilon^{-1}\nabla_y u^\epsilon\|^p + |u^\epsilon|^p\Big) dxdy$$
$$\le \|f^\epsilon\|_{L^{p'}(\Omega)} \|u^\epsilon\|_{L^p(\Omega)}, \qquad (13)$$

which implies that $\|u^\epsilon\|_{W_0^{1,p}(\Omega)} = O(1)$. Since $W_0^{1,p}(\Omega)$ is reflexive and $W_0^{1,p}(\Omega) \hookrightarrow L^p(\Omega)$ compactly, up to subsequence there exists $u^0 \in W_0^{1,p}(\Omega)$ satisfying (12). Also, due to (13), $\|\nabla_y u^\epsilon\|_{L^p(\Omega)} = O(\epsilon)$ and then, $\nabla_y u^\epsilon \overset{\epsilon\to 0}{\longrightarrow} 0$ in $L^p(\Omega)$ implying $\nabla_y u^0 = 0$ in $L^p(\Omega)$. □

Corollary 2.3. *Let f^ϵ, u^ϵ and u^0 as before. Then $u^0 = 0$.*

Proof. Since Ω is bounded, by Poincare's inequality [24, Lemma 5.1] there exists $C = C(\Omega, p)$ such that

$$\|u^0\|_{L^p(\Omega)} \le C \|\nabla_y u^0\|_{L^p(\Omega)}.$$

The result follows from Proposition 2.2. □

Corollary 2.4. $\|u^\epsilon\|_{W_0^{1,p}(\Omega)} \overset{\epsilon\to 0}{\longrightarrow} 0.$

Proof. It follows from Proposition 2.2, Corollary 2.3, and (12) that $u^\epsilon \to 0$ in $L^p(\Omega)$. Hence, we get the result by estimate (13). □

2.2 Nonlinearities

Now we consider the problem (11) with a nonlinearity f at the right side, i.e., we consider the problem: find $u^\epsilon \in W_0^{1,p}(\Omega)$ such that

$$\int_\Omega \Big(\|(\nabla_x u, \frac{1}{\varepsilon}\nabla_y u)\|^{p-2}(\nabla_x u \cdot \nabla_x \varphi + \frac{1}{\varepsilon^2}\nabla_y u \cdot \nabla_y \varphi) + |u|^{p-2} u \, \varphi\Big) dxdy$$
$$= \int_\Omega f(u) \, \varphi \, dxdy, \qquad (14)$$

for all $\varphi \in W_0^{1,p}(\Omega)$.

This class of equations were deeply studied by Djairo's et al in [11]. In order to simplify our discussion, let us assume that the nonlinearity f satisfies

$$|f(s)| \leq c + d\,|s|^{p-1-\alpha}, \quad \forall\, s \in \mathbb{R}, \tag{15}$$

for some constants $c, d \in \mathbb{R}$, $\alpha > 0$ such that $p > 1 + \alpha$. The Nemytskii operator $L^p(\Omega) \ni u \mapsto f \circ u \in L^{p/(p-1-\alpha)}(\Omega)$ maps bounded sets into bounded sets and is not difficult to see that the family of solutions u^ϵ of (14) is uniformly bounded in $W_0^{1,p}(\Omega)$. Indeed, for any $\epsilon \in (0,1)$, we have similarly to (13) that

$$\begin{aligned}\|u^\epsilon\|^p_{W_0^{1,p}(\Omega)} &\leq \int_\Omega \Big(\|\nabla_x u^\epsilon\|^p + \|\epsilon^{-1}\nabla_y u^\epsilon\|^p + |u^\epsilon|^p\Big)dxdy \\ &\leq \|f(u^\epsilon)\|_{L^{p/(p-1-\alpha)}(\Omega)}\|u^\epsilon\|_{L^{p_\alpha}(\Omega)} \leq C\|u^\epsilon\|^{p-\alpha}_{L^p(\Omega)} \end{aligned} \tag{16}$$

where $(p-1-\alpha)/p + 1/p_\alpha = 1$. Note that $L^p(\Omega) \hookrightarrow L^{p_\alpha}(\Omega)$.

Hence, we can argue as in the proofs of Proposition 2.2 and Corollary 2.3 to conclude that the family of solutions of the nonlinear problem (14) converge to the null function obtaining the following result.

Theorem 2.5. *Let u^ϵ be a family satisfying* (14) *with $f : \mathbb{R} \to \mathbb{R}$ continuous and satisfying* (15). *Then* $\|u^\epsilon\|_{W_0^{1,p}(\Omega)} \xrightarrow{\epsilon \to 0} 0$.

3 Neumann boundary condition

Now we consider a similar problem but with Neumann boundary condition rather Dirichlet boundary condition. Given a family $h^\epsilon \in L^{p'}(\Omega^\epsilon)$, we analyze convergence properties of the solutions of the family of elliptic equations

$$\begin{cases} -\Delta_p u + |u|^{p-2}u = h^\epsilon, & \text{in } \Omega^\epsilon, \\ \dfrac{\partial u}{\partial \eta^\epsilon} = 0, & \text{on } \partial\Omega^\epsilon, \end{cases} \tag{17}$$

where η^ϵ denotes the outward unitary normal vector field to the boundary of the thin domain Ω^ϵ.

Definition 3.1. We say that $u \in W^{1,p}(\Omega^\epsilon)$ is solution of the problem (17) if

$$\int_{\Omega^\epsilon} \Big(\|\nabla u\|^{p-2}\nabla u \cdot \nabla\varphi + |u|^{p-2}u\,\varphi\Big)\,dxdy = \int_{\Omega^\epsilon} h^\epsilon\,\varphi\,dxdy, \tag{18}$$

for all $\varphi \in W^{1,p}(\Omega^\epsilon)$.

As before, in order to study the limiting behavior of the family of solutions of equation (17), we stretch the domain Ω^ϵ by the factor ϵ^{-1} in the y-direction. In this case the equivalent equation in Ω has the form

$$\begin{aligned} -\Delta_p^\epsilon u + |u|^{p-2}u &= f^\epsilon, && \text{in } \Omega, \\ \nabla_x u \cdot \eta_x + \frac{1}{\epsilon^2} \nabla_y u \cdot \eta_y &= 0, && \text{on } \partial\Omega, \end{aligned} \tag{19}$$

where Δ_p^ϵ is the operator introduced in (8), $f^\epsilon(x, y) := h^\epsilon(x, \epsilon y)$ and (η_x, η_y) denotes the outward unitary normal vector field to $\partial\Omega$.

Hence, $\bar{u}^\epsilon \in W^{1,p}(\Omega^\epsilon)$ is solution of (17) if and only if $u^\epsilon := \Phi_\epsilon^*(\bar{u}^\epsilon) \in W^{1,p}(\Omega)$ is solution of (19), i.e., u^ϵ satisfies

$$\int_\Omega \Big(\|(\nabla_x u^\epsilon, \frac{1}{\epsilon} \nabla_y u^\epsilon)\|^{p-2} (\nabla_x u^\epsilon \cdot \nabla_x \varphi + \frac{1}{\epsilon^2} \nabla_y u^\epsilon \cdot \nabla_y \varphi) + |u^\epsilon|^{p-2} u^\epsilon \, \varphi \Big) \, dxdy$$
$$= \int_\Omega f^\epsilon \, \varphi \, dxdy, \tag{20}$$

for all $\varphi \in W^{1,p}(\Omega)$.

Here, we will see that the family of solutions $\{u^\epsilon\}_{\epsilon>0}$ of the problem (19) will converge to a limit u^0 which is solution of a suitable equation defined on a lower dimensional domain. In order to obtain this limiting regime, we perform a dilatation of the $(m+n)$-dimensional Lebesgue measure by a factor $1/\epsilon^n$. With this measure, namely $\rho^\epsilon := 1/\epsilon^n \times$ *Lebesgue measure*, we set the Lebesgue and Sobolev spaces $L^p(\Omega^\epsilon; \rho^\epsilon)$ and $W^{1,p}(\Omega^\epsilon; \rho^\epsilon)$. Now it is natural to consider in $W^{1,p}(\Omega)$ the equivalent norm

$$\|u\|_\epsilon := \left[\int_\Omega \Big(\|\nabla_x u\|^p + \frac{1}{\epsilon^p} \|\nabla_y u\|^p + |u|^p \Big) dxdy \right]^{1/p}.$$

It is easy to see that the isomorphism defined in (10)

$$\Phi_\epsilon^* : W^{1,p}(\Omega^\epsilon; \rho^\epsilon) \to W_\epsilon^{1,p}(\Omega)$$

is indeed an isometry where $W_\epsilon^{1,p}(\Omega) := (W^{1,p}(\Omega), \|\cdot\|_\epsilon)$. Similar observation can be done considering $\Phi_\epsilon^* : L^p(\Omega^\epsilon; \rho^\epsilon) \to L^p(\Omega)$.

We also denote by $\Omega_1 := \pi_1(\Omega)$ where $\pi_1 : \mathbb{R}^{m+n} \to \mathbb{R}^m : (x, y) \mapsto x$, is the projection onto the first m components. For each $x \in \Omega_1$, let $\Gamma_x = \{y \in \mathbb{R}^n : (x, y) \in \pi_1^{-1}\{x\}\}$ be the x-vertical section of Ω. We define $\omega : \Omega_1 \to \mathbb{R}_+$ by

$$\omega(x) = |\Gamma_x|,$$

the m-dimensional Lebesgue measure of Γ_x.

Now we consider the spaces $L^p(\Omega_1;\omega)$ and $W^{1,p}(\Omega_1;\omega)$. The norm in $W^{1,p}(\Omega_1;\omega)$ will be denoted by

$$\|u\|_0 = \left[\int_{\Omega_1} \omega\big(\|\nabla u\|^p + |u|^p\big)\, dxdy\right]^{1/p}.$$

Due to the nature of this specific kind of perturbation, we also introduce the following operators

(*Average projector*)

$$\begin{aligned} M_\epsilon : L^p(\Omega) &\to L^p(\Omega_1) \\ (M_\epsilon u)(x) &= \frac{1}{\omega(x)}\int_{\Gamma_x} u(x,y)\, dy \end{aligned} \tag{21}$$

(*Extension operator*)

$$\begin{aligned} E_\epsilon : L^p(\mathbb{R}^n) &\to L^p(\Omega^\epsilon) \\ (E_\epsilon u)(x,y) &= u(x) \end{aligned} \tag{22}$$

Notice that $M_\epsilon \circ E_\epsilon = I$, the identity operator in $L^p(\Omega_1)$. Furthermore the extension operator E_ϵ maps the family of spaces $W^{1,p}(\Omega_1)$ into $W^{1,p}(\Omega^\epsilon)$ and satisfies $\frac{\partial}{\partial y}(E_\epsilon u) = 0$.

It is easy to see from Fubini-Tonelli Theorem and Hölder inequality that the operators $M_\epsilon : L^p(\Omega^\epsilon) \to L^p(\Omega_1)$ satisfy $\|M_\epsilon\|_{\mathscr{L}(L^p(\Omega^\epsilon;\rho_\epsilon),L^p(\Omega_1))} = 1$. In fact, let $u \in L^p(\Omega^\epsilon)$,

$$\begin{aligned} \|M_\epsilon u\|_{L^p(\Omega_1;\omega)} &= \left[\int_{\Omega_1} \omega(x)|M_\epsilon u(x)|^p dx\right]^{1/p} \\ &= \left[\int_{\Omega_1} \frac{1}{\omega(x)^{p-1}}\left|\int_{\Gamma_x} u(x,y)dy\right|^p dx\right]^{1/p} \\ &\le \left[\int_{\Omega_1} \frac{1}{\omega(x)^{p-1}}\omega(x)^{p-1}\int_{\Gamma_x} |u(x,y)|^p dydx\right]^{1/p} \\ &= \left[\int_{\Omega} |u(x,y)|^p dxdy\right]^{1/p} = \|u\|_{L^p(\Omega)}. \end{aligned}$$

The equality holds taking u independent of y in Ω.

3.1 Convergence

Similarly to the previous section, we notice that $u^\epsilon \in W^{1,p}(\Omega)$ is solution of (19) if and only if u^ϵ satisfies

$$\frac{1}{p}\|u^\epsilon\|_\epsilon^p - \int_\Omega f^\epsilon u^\epsilon \, dxdy = \min_{\varphi \in W^{1,p}(\Omega)} \left\{ \frac{1}{p}\|\varphi\|_\epsilon^p - \int_\Omega f^\epsilon \varphi \, dxdy \right\}.$$

Proposition 3.2. *Assume that for all $x \in \pi_1(\Omega)$, Γ_x is connected. Let $f^\epsilon \in L^{p'}(\Omega)$ be a family of functions uniformly bounded with respect to $\epsilon > 0$ such that*

$$M_\epsilon f^\epsilon \overset{\epsilon \to 0}{\rightharpoonup} f^0, \quad w - L^{p'}(\Omega_1).$$

If we define

$$\lambda^\epsilon = \min_{\varphi \in W^{1,p}(\Omega)} \left\{ \frac{1}{p}\|\varphi\|_\epsilon^p - \int_\Omega f^\epsilon \varphi \, dxdy \right\}$$

and

$$\lambda^0 = \min_{\varphi \in W^{1,p}(\Omega_1)} \left\{ \frac{1}{p}\|\varphi\|_0^p - \int_{\Omega_1} \omega f^0 \varphi \, dx \right\},$$

then $\lambda^\epsilon \to \lambda^0$.

Proof. Taking $\varphi = \varphi(x)$ we get

$$\begin{aligned}
\lambda^\epsilon &\le \frac{1}{p}\int_\Omega \Big(\|\nabla_x \varphi\|^p + \|\epsilon^{-1}\nabla_y \varphi\|^p + |\varphi|^p \Big) dxdy - \int_\Omega f^\epsilon \varphi \, dxdy \\
&= \frac{1}{p}\int_{\Omega_1} \omega \Big(\|\nabla_x \varphi\|^p + |\varphi|^p \Big) dx - \int_{\Omega_1} \omega M_\epsilon f^\epsilon \varphi \, dx \\
&= \frac{1}{p}\int_{\Omega_1} \omega \Big(\|\nabla_x \varphi\|^p + |\varphi|^p \Big) dx - \int_{\Omega_1} \omega f^0 \varphi \, dx + \int_{\Omega_1} \omega (f^0 - M_\epsilon f^\epsilon) \varphi \, dx.
\end{aligned}$$

Taking the infimum over all $\varphi \in W^{1,p}(\Omega_1)$ we obtain that $\limsup_{\epsilon \to 0} \lambda^\epsilon \le \lambda^0$.

Recalling the previous section, since $f^\epsilon \in L^{p'}(\Omega)$ is uniformly bounded, if u^ϵ is the solution of (19), we can obtain a function $u^0 \in W^{1,p}(\Omega)$ such that $\nabla_y u = 0$ a.e. in Ω and

$$u^\epsilon \overset{\epsilon \to 0}{\longrightarrow} u^0, \text{ weakly in } W^{1,p}(\Omega) \text{ and strongly in } L^p(\Omega). \tag{23}$$

From the weak convergence in $W^{1,p}(\Omega)$ and strong convergence in $L^p(\Omega)$ we obtain that

$$\begin{aligned}\liminf_{\epsilon\to0}\lambda^\epsilon &= \liminf_{\epsilon\to0}\left\{\frac{1}{p}\|u^\epsilon\|_\epsilon^p - \int_\Omega f^\epsilon u^\epsilon\,dxdy\right\}\\ &\geq \liminf_{\epsilon\to0}\left\{\frac{1}{p}\|u^\epsilon\|_{W^{1,p}(\Omega)}^p - \int_\Omega f^\epsilon u^\epsilon\,dxdy\right\}\\ &\geq \frac{1}{p}\|u^0\|_0^p - \int_{\Omega_1}\omega f^0u^0\,dx \geq \lambda^0,\end{aligned} \tag{24}$$

which proves the statement. □

Remark 3.3. Notice that we use u^0 as a test function in the conclusion of the proof of Proposition 3.2. This was possible thanks to the hypothesis Γ_x connected.

Remark 3.4. By (24)

$$\lambda^\epsilon \geq \frac{1}{p}\|u^0\|_0^p - \int_{\Omega_1}\omega f^0u^0\,dx \geq \lambda^0,$$

consequently, we derive

$$\lambda^0 = \frac{1}{p}\|u^0\|_0^p - \int_{\Omega_1}\omega f^0u^0\,dx.$$

Therefore u^0 is characterized as solution of the equation

$$\begin{cases} -\frac{1}{\omega}\mathrm{div}(\omega\|\nabla u\|^{p-2}\nabla u) + |u|^{p-2}u = f^0, & \text{in } \Omega_1,\\ \frac{\partial u}{\partial\nu} = 0, & \text{on } \partial\Omega_1,\end{cases} \tag{25}$$

where $\Omega_1 = \pi_1(\Omega) \subset \mathbb{R}^m$ and ν denotes the outward unitary normal vector field to $\partial\Omega_1$.

As a consequence of Proposition 3.2 we obtain:

Theorem 3.5. *Let f^ϵ, u^ϵ, u^0 and f^0 be as before. Then*

$$u^\epsilon \stackrel{\epsilon\to0}{\longrightarrow} u^0,\ \textit{strongly in } W^{1,p}(\Omega).$$

Proof. From the weak convergence $u^\epsilon \rightharpoonup u^0$ in $W^{1,p}(\Omega)$ and Proposition 3.2, we have that

$$\begin{aligned}\int_\Omega\left(\|\nabla_x u^0\|^p + |u^0|^p\right)dxdy &\leq \liminf_{\epsilon\to0}\int_\Omega\left(\|\nabla_x u^\epsilon\|^p + \|\nabla_y u^\epsilon\|^p + |u^\epsilon|^p\right)dxdy\\ &\leq \limsup_{\epsilon\to0}\int_\Omega\left(\|\nabla_x u^\epsilon\|^p + \|\nabla_y u^\epsilon\|^p + |u^\epsilon|^p\right)dxdy\end{aligned}$$

$$\leq \lim_{\epsilon\to 0}\int_{\Omega}\left(\|\nabla_x u^\epsilon\|^p + \|\epsilon^{-1}\nabla_y u^\epsilon\|^p + |u^\epsilon|^p\right) dxdy$$
$$= \int_{\Omega_1}\omega f^0 u^0\, dydx = \int_{\Omega_1}\omega\,\left(\|\nabla_x u^0\|^p + |u^0|^p\right) dydx$$
$$= \int_{\Omega}\left(\|\nabla_x u^0\|^p + |u^0|^p\right) dxdy.$$

Now the result follows from the uniform convexity of $W^{1,p}(\Omega)$. □

Corollary 3.6. *Let u^ϵ and u^0 be as in Theorem 3.5. Then*

$$\|u^\epsilon - E_\epsilon u^0\|_\epsilon \xrightarrow{\epsilon\to 0} 0.$$

Proof. Since we have strong convergence of u^ϵ to u^0 in $W^{1,p}(\Omega)$, the identity

$$\int_{\Omega}(\|\nabla_x u^0\|^p + |u^0|^p)dxdy = \lim_{\epsilon\to 0}\int_{\Omega}\left(\|\nabla_x u^\epsilon\|^p + \frac{1}{\epsilon^p}\|\nabla_y u^\epsilon\|^p + |u^\epsilon|^p\right) dxdy,$$

obtained in the proof of Theorem 3.5, implies that $\epsilon^{-1}\|\nabla_y u^\epsilon\|_{L^p(\Omega)} \xrightarrow{\epsilon\to 0} 0$. □

3.2 Nonlinearities

In the case of Neumann boundary conditions if we consider a continuous nonlinearity $f:\mathbb{R}\to\mathbb{R}$, the correspondent problem assumes the form

$$\int_{\Omega}\Big(\|(\nabla_x u, \frac{1}{\varepsilon}\nabla_y u)\|^{p-2}(\nabla_x u\cdot\nabla_x\varphi + \frac{1}{\varepsilon^2}\nabla_y u\cdot\nabla_y\varphi) + |u|^{p-2}u\,\varphi\Big)dxdy$$
$$= \int_{\Omega} f(u)\,\varphi\, dxdy, \qquad (26)$$

for all $\varphi \in W^{1,p}(\Omega)$.

Assuming that the nonlinearity f also satisfies (15) we have similarly to (16) that the family of solutions u^ϵ of (26) is uniformly bounded in $W^{1,p}(\Omega)$. Therefore we still have the existence of a suitable limit $u^0 \in W^{1,p}(\Omega_1)$ (see (23)) which satisfies the nonlinear equation

$$\int_{\Omega_1}\omega\Big(\|(\nabla_x u)\|^{p-2}\nabla_x u\cdot\nabla_x\varphi + |u|^{p-2}u\,\varphi\Big)dx = \int_{\Omega_1}\omega f(u)\,\varphi\, dx,$$

for all $\varphi \in W^{1,p}(\Omega_1)$. As $f(u^\epsilon) \to f(u^0)$ in $L^{p'}(\Omega)$, we obtain, mutatis mutandis the proof of Proposition 3.5, strong convergence in $W^{1,p}(\Omega)$, and therefore a similar convergence as stated in Corollary 3.6.

4 A specific example

In this section we present the case of a thin domain considered by Hale and Raugel in the seminal paper [14] in the case $p = 2$. Let Ω_1 be a smooth bounded domain in $\mathbb{R}^m$, $n \geq 1$, and $g \in C^2(\overline{\Omega}_1; \mathbb{R})$ a positive function. We define the family of thin domains $\Omega^\epsilon \subset \mathbb{R}^{m+1}$ as

$$\Omega^\epsilon := \{(x, y) \in \mathbb{R}^{m+1} : x \in \Omega_1,\ 0 < y < \epsilon g(x)\}.$$

In Ω^ϵ we consider the family of elliptic equations

$$\begin{aligned} -\Delta_p u + |u|^{p-2}u = h^\epsilon, &\quad \text{in } \Omega^\epsilon, \\ \frac{\partial u}{\partial \eta^\epsilon} = 0, &\quad \text{on } \partial\Omega^\epsilon, \end{aligned} \tag{27}$$

where $h^\epsilon \in L^{p'}(\Omega^\epsilon)$ and η^ϵ denotes the outward unitary normal vector field to $\partial\Omega^\epsilon$.

Considering the change of coordinates $\Phi^\epsilon : \Omega \to \Omega^\epsilon$, $(x, y) \mapsto (x, \epsilon y)$, where $\Omega = \{(x, y) \in \mathbb{R}^{m+1} : x \in \Omega_1,\ 0 < y < g(x)\}$, equation (27) becomes

$$\begin{aligned} -\Delta_p^\epsilon u + |u|^{p-2}u = f^\epsilon, &\quad \text{in } \Omega, \\ \nabla_x u \cdot \eta_x + \frac{1}{\epsilon^2} \nabla_y u \cdot \eta_y = 0, &\quad \text{on } \partial\Omega, \end{aligned} \tag{28}$$

where Δ_p^ϵ is the operator introduced in (8), $f^\epsilon(x, y) := h^\epsilon(x, \epsilon y)$ and (η_x, η_y) is the outward unitary normal vector field to $\partial\Omega$.

As proved in Section 3, the limiting problem is

$$\begin{aligned} -\frac{1}{g}\mathrm{div}(g\|\nabla u\|^{p-2}\nabla u) + |u|^{p-2}u = f^0, &\quad \text{in } \Omega_1, \\ \frac{\partial u}{\partial \nu} = 0, &\quad \text{on } \partial\Omega_1, \end{aligned} \tag{29}$$

where $f^0 \in L^{p'}(\Omega_1)$ is the weak limit (in $L^{p'}(\Omega_1)$) of the family $M_\epsilon f^\epsilon$. This agrees with the limiting problem founded in [14] with $p = 2$,

$$\begin{aligned} -\frac{1}{g}\mathrm{div}(g\nabla u) + u = f^0, &\quad \text{in } \Omega_1, \\ \frac{\partial u}{\partial \nu} = 0, &\quad \text{on } \partial\Omega_1. \end{aligned} \tag{30}$$

Acknowledgements The authors would like to thank the anonymous referee whose comments have considerably improved the writing of the paper. Marcone C. Pereira was partially supported by FAPESP #2013/22275-1, CAPES DGU #127/07, CNPq #302847/2011-1, and #471210/2013-7, Brazil. Ricardo P. Silva was partially supported by FAPESP #2012/06753-8 and #2014/16165-1, CNPq #440371/2014-7, Brazil.

References

1. Arrieta, J.M., Pereira, M.C.: Homogenization in a thin domain with an oscillatory boundary. J. Math. Pures Appl. **96**, 29–57 (2011)
2. Arrieta, J.M., Pereira, M.C.: Thin domain with extremely high oscillatory boundaries. J. Math. Anal. Appl. **404**(1), 86–104 (2013)
3. Arrieta, J.M., Carvalho, A.N., Pereira, M.C., Silva, R.P., Semilinear parabolic problems in thin domains with a highly oscillatory boundary. Nonlinear Anal. Theory Methods Appl. **74**(15), 5111–5132 (2011)
4. Boukrouche, M., Ciuperca, I.: Asymptotic behaviour of solutions of lubrication problem in a thin domain with a rough boundary and Tresca fluid-solid interface law. Quart. Appl. Math. **64** 561–591 (2006)
5. Capelato, E., Schiabel-Silva, K., Silva, R.P.: Perturbation of a nonautonomous problem in $\mathbb{R}^n$. Math. Methods Appl. Sci. **36**(12), 1625–1630 (2013)
6. Carbone, V.L., Gentile, C.B., Schiabel-Silva, K.: Asymptotic properties in parabolic problems dominated by a p-Laplacian operator with localized large diffusion. Nonlinear Anal. **74**, 4002–4011 (2011)
7. Carvalho, A.N., Cholewa, J.W., Dlotko, T.: Global attractors for problems with monotone operators. Boll. Unione Mat. Ital. Sez. B Artic. Ricerche Mat. **2**(8), 693–706 (1999)
8. Cioranescu, D., Jean-Paulin, J.S.: Homogenization of Reticulated Structures. Springer, New York (1980)
9. Cringanu, J.: Variational and topological methods for Neumann problems with p-Laplacian. Commun. Appl. Nonlinear Anal. **11**(1), 1–38 (2004)
10. Díaz, J.I.: Nonlinear Partial Differential Equations and Free Boundaries. Elliptic Equations, vol. 1. Pitman, London (1985)
11. de Figueiredo, D.G., Gossez, J-P., Ubilla, U.: Local superlinearity and sublinearity for the p-Laplacian. J. Funct. Anal. **257**, 721–752 (2009)
12. Elsken, T.: Continuity of attractors for net-shaped thin domain. Topological Meth. Nonlinear Anal. **26**, 315–354 (2005)
13. Geredeli, P.G.: On the existence of regular global attractor for p-Laplacian evolution equation. Appl. Math. Optim. doi: 10.1007/s00245-014-9268-y
14. Hale, J.K., Raugel, G.: Reaction-diffusion equations on thin domains. J. Math. Pures Appl. **9**(71), 33–95 (1992)
15. Lewinsky, T., Telega, J.: Plates, Laminates and Shells. Asymptotic Analysis and Homogenization. World Scientific, Singapore (2000)
16. Lions, J.L.: Quelques méthodes de résolution des problèmes aux limites non lineáires. Dunod, Paris (1969)
17. Mocenni, C., Sparacino, E., Zubelli, J.P.: Effective rough boundary parametrization for reaction-diffusion systems. Applicable Anal. Discrete Math. **8**, 33–59 (2014)
18. Pazanin, I., Suárez-Grau, F.J.: Effects of rough boundary on the heat transfer in a thin-film flow. Comptes Rendus Mécanique **341**(8), 646–652 (2013)
19. Pereira, M.C.: Parabolic problems in highly oscillating thin domains. Annali di Matematica Pura ed Applicata. doi: 10.1007/s10231-014-0421-7
20. Pereira, M.C., Silva, R.P.: Rates of convergence for a homogenization problem in highly oscillating thin domains. Dyn. Syst. Appl. **6**, 337–340 (2012)

21. Pereira, M.C., Silva, R.P., Error estimatives for a Neumann problem in highly oscillating thin domain. Discrete Contin. Dyn. Syst.-Series A **33**(2), 803–817 (2013)
22. Prizzi, M., Rybakowski, K.P.: The effect of domain squeezing upon the dynamics of reaction-diffusion equations, J. Differ. Equ. **173**(2), 271–320 (2001)
23. Raugel, G.: Dynamics of partial differential equations on thin domains. Lecture Notes in Mathematics, vol. 1609. Springer, New York (1995)
24. Showalter, R.E.: Monotone Operators in Banach Space and Nonlinear Partial Differential Equations. Mathematical Surveys and Monographs, vol. 49. American Mathematical Society, Providence (1997)
25. Silva, R.P.: Behavior of the p-Laplacian on thin domains. Int. J. Differ. Equ. 2013 (2013)
26. Silva, R.P.: A note on resolvent convergence on a thin domain. Bull. Aust. Math. Soc. **89**(01), 141–148 (2014)
27. Sun, C., Yang, M., Zhong, C.: Global attractors for p-Laplacian equation. J. Math. Anal. Appl. **327**(2), 1130–1142 (2007)

Spatial trajectories and convergence to traveling fronts for bistable reaction-diffusion equations

Peter Poláčik

Dedicated to Professor Djairo Guedes de Figueiredo on the occasion of his 80th birthday

1 Introduction

Consider the Cauchy problem

$$u_t = u_{xx} + f(u), \qquad x \in \mathbb{R},\ t > 0, \tag{1}$$

$$u(x,0) = u_0(x), \qquad x \in \mathbb{R}, \tag{2}$$

where $f \in C^1(\mathbb{R})$ and $u_0 \in C(\mathbb{R}) \cap L^\infty(\mathbb{R})$. We assume that f is of a bistable type and $u_0(x)$ takes values between the two stable zeros of f and has a "front-like" shape (see below for precise hypotheses). Classical results then tell us that, under additional conditions on f and u_0, the solution of (1), (2) approaches the orbit of a traveling front. The main purpose of this note is to give a new proof of this result and relax its hypotheses somewhat.

To discuss the large-time behavior of solutions in more specific terms, we introduce two kinds of limit sets. Assuming that the solution of (1), (2) is bounded we set

$$\omega(u) := \{\varphi : u(\cdot, t_n) \to \varphi \text{ for some sequence } t_n \to \infty\}, \tag{3}$$

$$\Omega(u) := \{\varphi : u(\cdot + x_n, t_n) \to \varphi \text{ for some sequences } t_n \to \infty \text{ and } x_n \in \mathbb{R}\}, \tag{4}$$

where the convergence is in $L^\infty_{loc}(\mathbb{R})$ (the locally uniform convergence) in both cases. Since the solution u is determined uniquely by its initial value, we sometimes use the symbols $\omega(u_0)$, $\Omega(u_0)$ for $\omega(u)$, $\Omega(u)$.

P. Poláčik (✉)
School of Mathematics, University of Minnesota, Minneapolis, MN 55455, USA
e-mail: polacik@math.umn.edu

A.N. Carvalho et al. (eds.), *Contributions to Nonlinear Elliptic Equations and Systems*, Progress in Nonlinear Differential Equations and Their Applications 86,
DOI 10.1007/978-3-319-19902-3_24

By standard parabolic regularity estimates, the set $\{u(x+\cdot,t) : t \geq 1,\ x \in \mathbb{R}\}$ is relatively compact in $L^{\infty}_{loc}(\mathbb{R})$. This implies that both $\omega(u)$ and $\Omega(u)$ are nonempty, compact, and connected in $L^{\infty}_{loc}(\mathbb{R})$. Clearly, $\omega(u_0) \subset \Omega(u_0)$, but the opposite inclusion is not true in general. Both these limit sets give a useful information on the solution u: while $\Omega(u)$ gives a picture of the global shape of $u(\cdot,t)$ for large times and is also useful for investigating the behavior of $u(\cdot,t)$ in moving coordinate frames; $\omega(u)$ captures its large-time behavior in local regions.

To formulate our results, we first make precise our hypotheses. We assume the following conditions on f:

(Hf) $f \in C^1(\mathbb{R}), f(0) = f(1) = 0$ and there is $\alpha \in (0,1)$ such that

$$f < 0 \text{ in } (0,\alpha); \quad f > 0 \text{ in } (\alpha, 1). \tag{5}$$

Since we only investigate solutions satisfying $0 \leq u \leq 1$, the values of $f(s)$ for $s \notin [0,1]$ are irrelevant. For convenience, we shall assume that

$$f > 0 \text{ in } (-\infty, 0); \quad f < 0 \text{ in } (1,\infty); \quad f' \text{ is bounded.} \tag{6}$$

Thus 0, α, 1 are all the equilibria of the ordinary differential equation (ODE) $\dot{\xi} = f(\xi)$; 0, 1 are stable, whereas α is unstable, both from above and below (thus the name "bistable nonlinearity"). Obviously, the specific choice of the interval $[0,1]$ does not restrict generality; other bistable nonlinearities are brought to this form by a suitable scaling and translation. We often view 0, α, and 1 as constant functions and then they become steady states of (1).

Hypothesis (Hf) implies (see [2, 6, 16], for example) that there is a traveling front of (1) joining 0 and 1, that is, a solution U of the form

$$U(x,t) = \phi(x - \hat{c}t), \text{ where } \hat{c} \in \mathbb{R}, \phi \in C^2(\mathbb{R}), \text{ and } \phi' > 0. \tag{7}$$

Moreover, both the increasing "profile" function ϕ and the "speed" $\hat{c}$ are uniquely determined, up to translations of ϕ, and $\operatorname{sign} \hat{c} = -\operatorname{sign} F(1)$, where

$$F(u) = \int_0^u f(s)\, ds. \tag{8}$$

For definiteness we shall assume that

$$F(1) = \int_0^1 f(s)\, ds \geq 0. \tag{9}$$

This means that the front "travels to the left" ($\hat{c} < 0$) or is a standing wave ($\hat{c} = 0$). Again, assumption (9) is at no cost to generality; the other case is completely analogous (or, one simply interchange the roles of the two stable equilibria). Note that $\tilde{U}(x,t) = U(-x,t)$ is also a traveling front; it has a decreasing profile function (namely, $\tilde{\phi} = \phi(-x)$) and the opposite speed.

With α as in (Hf), we assume the following conditions on u_0:

(Hu) $u_0 \in C(\mathbb{R})$, $0 \le u_0 \le 1$, and

$$\limsup_{x\to-\infty} u_0(x) < \alpha < \liminf_{x\to\infty} u_0(x). \tag{10}$$

In this sense, u_0 has a "front-like" shape.

Theorem 1.1. *Assume that* (Hf), (Hu) *hold and* $F(1) > 0$. *Let* u *be the solution of* (1), (2). *Then*

$$\Omega(u) = \{0,1\} \cup \{\phi(\cdot - \xi) : \xi \in \mathbb{R}\}, \tag{11}$$

where ϕ *is as in* (7).

Below we also give a theorem for $F(1) = 0$ under an additional assumption on u_0.

Similar results on the approach of solutions to traveling fronts for bistable nonlinearities can be found in [6, 7], among many other publications. Let us discuss the relation of Theorem 1.1 to these classical results in more detail. It is not difficult to show (see Sect. 3) that (11) implies the following

Corollary 1.2. *Assume that the hypotheses of Theorem 1.1 hold. Then there is a* C^1*-function* $\gamma(t)$ *such that* $\gamma'(t) \to 0$ *as* $t \to \infty$, *and*

$$u(\cdot,t) - \phi(\cdot - \hat{c}t - \gamma(t)) \to 0 \text{ as } t \to \infty, \tag{12}$$

where the convergence is in $L^\infty(\mathbb{R})$.

This conclusion was proved in [7] under the extra assumption that u_0 is monotone. Note that Corollary 1.2 only says that the translation group orbit $\{\phi(\cdot - \xi) : \xi \in \mathbb{R}\}$ of ϕ attracts the solution; it does not say that the solution approaches a particular traveling front, or, in other words, that $\gamma(t)$ has a limit as $t \to \infty$. The latter was proved in [6] under the nondegeneracy condition

$$f'(0) < 0, \quad f'(1) < 0. \tag{13}$$

In this case, the monotonicity of u_0 is not assumed and one even gets the exponential rate of convergence in (12). There are many extensions of this convergence results, see, for example, [2, 3, 9, 11–14, 16] and the references therein (for more bibliographical notes and a discussion of classical results for bistable and other types of nonlinearities, see [16, Sect. 1.6]). Usually, the convergence is proved by first showing that the solution gets close to a particular traveling front at some time (this property follows from (12); in the nondegenerate case (13), Fife and McLeod [6] proved it by way of a Lyapunov functional) and then employing an asymptotic stability property of the front. Conditions (13), or similar nondegeneracy conditions, are typically needed to establish the linearized stability of the front. The convergence with the exponential rate is then obtained from the principle of linearized stability for parabolic equations [8, 15].

Here, we only treat the more general case with the weaker conclusion, as in Theorem 1.1, Corollary 1.2, without assuming the nondegeneracy conditions. Our objective is to give a relatively simple geometric proof of the result. The main technical tools of our method are intersection comparison (or zero number) arguments and analysis of spatial trajectories of solutions of (1). If u is a solution, then its *spatial trajectory at time* t is the set $\{(u(x,t), u_x(x,t)) : x \in \mathbb{R}\} \subset \mathbb{R}^2$. Note that if u is a steady state, then its spatial trajectory is independent of t and it is a trajectory, in the usual sense, of the first-order system corresponding to the ODE $u_{xx} + f(u) = 0$. Likewise, if u is a traveling wave, then its spatial trajectory is independent of t and is a trajectory of the first order system corresponding to the equation $u_{xx} + cu_x + f(u) = 0$, where c is the speed of the wave. Our proof depends on a good understanding of how spatial trajectories of the solution of (1), (2) can intersect spatial trajectories of steady states and traveling waves.

We remark that spatial trajectories also appear, though not under this name, in [7]. In that paper, given a solution u with $u_x > 0$, the authors consider a function $p(u,t) = u_x(\zeta(u,t),t)$, where $\zeta(\cdot,t)$ is the inverse function to $u(\cdot,t)$. They show that p satisfies a degenerate parabolic equation and prove the attraction to traveling fronts by delicate comparison arguments for this equation. Observe that for any t, the graph of $p(\cdot,t)$ is the spatial trajectory $\tau(u(\cdot,t))$ of u. Obviously, for the spatial trajectory to be such a graph, the monotonicity in x is necessary. In contrast, we work with the spatial trajectories as curves in $\mathbb{R}^2$, thus we do not need the monotonicity assumption. Also, we do not use any transformed partial differential equations similar to the equation for p. All the essential information from which we can rather easily prove Theorem 1.1 is contained in phase diagrams of the ODEs $u_{xx} + cu_x + f(u) = 0$, for various c.

Theorem 1.1 can probably be proved in several different ways. For example, in [16] it is suggested that, under the additional assumption that

$$\lim_{x\to-\infty} u_0(x) = 0, \quad \lim_{x\to\infty} u_0(x) = 1, \tag{14}$$

the following approach should work. First one proves that there is a function $\epsilon(t) > 0$ such that $\epsilon(t) \to 0$ as $t \to \infty$ and u is increasing in x in the set $\{(x,t) : \epsilon(t) < u(x,t) < 1 - \epsilon(t)\}$. Once this is established, one can modify the arguments in the monotone case, to get the conclusion in this more general situation. The method we use in the present paper is more direct and seems to be simpler than this suggested approach (and (14) is not needed).

Our method applies, with minor modifications, to more general situations where the existence and uniqueness of the traveling front can be established (see [6]), but for simplicity we just consider (5). The method is also useful in other problems; for example, in [10] we will use similar techniques in the proof of quasiconvergence of solutions with localized initial data. On the other hand, the scope of the method seems to be limited to the one-dimensional spatially homogeneous equations.

In the case $F(1) = 0$, we prove the same results as in Theorem 1.1 and Corollary 1.2, but we need a stronger assumption on u_0. For example, the following will do.

(Ha) Either $u_0 - \alpha$ has a unique zero, or the limits $u_0(\pm\infty)$ exist and one has

$$u_0(-\infty) \leq u_0(x) \leq u_0(\infty) \quad (x \in \mathbb{R}). \tag{15}$$

Note that (15) is trivially satisfied if (Hu) and (14) hold.

Theorem 1.3. *Assume that the hypotheses* (Hf), (Hu), (Ha) *hold and* $F(1) = 0$ *(so also* $\hat{c} = 0$*). Let u be the solution of* (1), (2). *Then the conclusions of Theorem 1.1 and Corollary 1.2 hold.*

Note that (11) in particular gives $\omega(u) \subset \{0, 1\} \cup \{\phi(\cdot - \xi) : \xi \in \mathbb{R}\}$. One has $\{0, 1\} \cap \omega(u) \neq \emptyset$ if and only if the function γ in (12) is unbounded. If $\gamma(t)$ has a finite limit ξ when $t \to \infty$, then $\omega(u)$ consists of the single equilibrium $\phi(\cdot - \xi)$ (and one can even take the uniform convergence in the definition of $\omega(u)$). This is the case if the stable zeros are nondegenerate: $f'(0), f'(1) < 0$ [6], but without this assumption the situation is not so clear. As far as we know, examples of solutions satisfying the present hypotheses for which $\omega(u)$ is not a singleton are not available.

The paper is organized as follows. In the next section, we recall several useful results concerning the zero number, Ω-limit sets, and solutions of the ODEs $u_{xx} + cu_x + f(u) = 0$, $c \leq 0$. The proofs of the main results are given in Section 3.

2 Preliminaries

2.1 *Phase space and traveling fronts*

In this section we examine the solutions of the ODE

$$v_{xx} + cv_x + f(v) = 0, \quad x \in \mathbb{R}. \tag{16}$$

This is the equation satisfied by steady states of (1) (if $c = 0$) and by the profile functions of traveling fronts. Throughout the section we assume that the hypotheses (Hf), (6), (9) are satisfied.

The first-order system associated with (16) is

$$v_x = w, \quad w_x = -cw - f(v). \tag{17}$$

Its solutions are all global, by the Lipschitz continuity of f (see (6)). For $c = 0$, we obtain a Hamiltonian system,

$$v_x = w, \quad w_x = -f(v), \tag{18}$$

with the Hamiltonian energy

$$H(v, w) := w^2/2 + F(v).$$

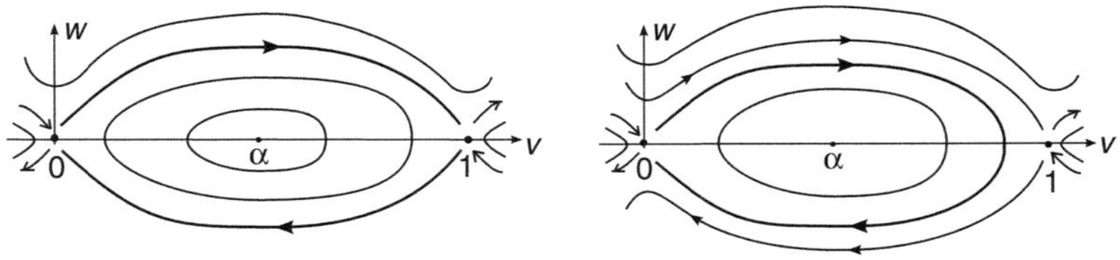

Fig. 1 The phase diagram of system (18); the balanced case ($F(1) = 0$) is on the left, the unbalanced case ($F(1) > 0$) on the right

In this case, the trajectories of (18) are contained in the level sets of H. Note in particular that the level sets are symmetric about the v axis. We now summarize a few basic properties of trajectories of system (18) (see Fig. 1); they are all proved easily by an elementary phase plane analysis using the Hamiltonian and the standing hypotheses (Hf), (6), (9). System (18) has only four types of bounded orbits: equilibria (stationary solutions)—all of them on the v axis, nonconstant periodic orbits, homoclinic orbits—which exist only in the case $F(1) > 0$, and heteroclinic orbits—only in the case $F(1) = 0$. All bounded orbits, other than the equilibria $(0,0)$ and $(1,0)$ are contained in the open strip

$$S := \{(v,w) : 0 < v < 1\}.$$

This strip is covered by the level sets

$$L_\gamma := \{(v,w) : H(v,w) = \gamma\}, \ \gamma \in [F(\alpha), \infty).$$

For each $\gamma \in (F(\alpha), \infty)$, the level set L_γ intersects the vertical line $\{(\alpha, w) : w \in \mathbb{R}\}$, at exactly two points $(\alpha, \pm\sqrt{2(\gamma - F(\alpha))})$; for $\gamma = F(\alpha)$ there is just one intersection, the equilibrium $(\alpha, 0)$. For $\gamma > F(1)$, L_γ consists of two curves not intersecting the v-axis. The solutions v whose trajectories $\tau(v)$ are given by these curves are strictly monotone with infinite limits at $\pm\infty$. If $F(1) > 0$ and $\gamma \in (0, F(1))$, then the part of L_γ intersecting S coincides with a trajectory of a solution v with limits $v(\pm\infty) = -\infty$. If $\gamma = F(1) > 0$, then L_γ consists of $(1,0)$ and the trajectories of solutions which converge to 1 as $x \to \infty$ or $x \to -\infty$. For $\gamma \in (F(\alpha), 0)$, the set $L_\gamma \cap S$ coincides with a nonstationary periodic orbit (or, closed orbit) of (18). For $\gamma = 0$, $L_\gamma \cap S$ is a homoclinic orbit to $(0,0)$ (this is the case if $F(1) > 0$) or the union of two heteroclinic connections between the equilibria $(0,0)$, $(1,0)$ (if $F(1) > 0$). If (v, v_x) is a nonstationary periodic solution of (18), then $v - \alpha$ has infinitely many zeros. Of course, all these zeros are simple by the uniqueness for the Cauchy problem.

Let us now consider system (17) with $c < 0$. In this case, H is increasing along the solutions:

$$\frac{dH(v(x), w(x))}{dx} = -cw^2. \tag{19}$$

In particular, any bounded nonstationary solution of (17) with $c < 0$, is a heteroclinic solution between two different equilibria. For $c = \hat{c}$, and for this value only, (17) has a heteroclinic solution from $(0,0)$ to $(1,0)$, given by the profile function of the traveling front: $(v,w) \equiv (\phi, \phi_x)$ [2, 6, 16]. Obviously, for any solution (v,w) of (17), v is increasing (resp. decreasing) when $w > 0$ (resp. $w < 0$). One also shows easily that the sets

$$Q_1 := \{(v,w) : v \geq 1, w \geq 0\} \setminus \{(1,0)\},$$
$$Q_3 := \{(v,w) : v \leq 0, w \leq 0\} \setminus \{(0,0)\}$$

are positively invariant in the sense that if a solution satisfies $(v(0), w(0)) \in Q_i$, for $i = 1$ or $i = 3$, then for all $x > 0$ one has $(v(x), w(x)) \in \operatorname{Int} Q_i$ (the interior of Q_i). Similarly, the sets

$$Q_2 := \{(v,w) : v \leq 0, w \geq 0\} \setminus \{(0,0)\},$$
$$Q_4 := \{(v,w) : v \geq 1, w \leq 0\} \setminus \{(1,0)\}$$

are negatively invariant.

Let now $c = \hat{c}$, so that $\tau(\phi)$ is a heteroclinic orbit from $(0,0)$ to $(1,0)$. It is well known that, regardless of whether $f'(0), f'(1)$ vanish or are negative, $\tau(\phi)$ contains all initial data $(\xi, \eta) \in S$ such that the solution ψ of (16) with

$$\psi(0) = \xi, \quad \psi_x(0) = \eta \tag{20}$$

satisfies either $(\psi(x), \psi'(x)) \to (1,0)$ as $x \to \infty$, or $(\psi(x), \psi'(x)) \to (0,0)$ as $x \to -\infty$ (see, for example, [2, Sect. 4]). Likewise, there is a solution $\tilde{\phi}$, such that $S \cap \tau(\tilde{\phi})$ is precisely the set of initial data $(\xi, \eta) \in S$ such that the solution ψ of (16), (20) satisfies $(\psi(x), \psi'(x)) \to (1,0)$ as $x \to -\infty$. If $\hat{c} = 0$, then $\tilde{\phi}$ is given simply by $\tilde{\phi}(x) = \phi(-x)$ and $\tau(\tilde{\phi})$ is a heteroclinic orbit from $(1,0)$ to $(0,0)$. If $\hat{c} < 0$, then $\tau(\tilde{\phi})$ intersects the halfline $\{(0,w) : w < 0\}$, hence by the positive invariance of Q_3 one has $(\tilde{\phi}(x), \tilde{\phi}_x(x)) \in Q_3$ for all large enough x (see Fig. 2). Since different trajectories of the autonomous system (17) cannot intersect, using the above properties of ϕ, $\tilde{\phi}$ and the invariance properties of the sets $Q_1 - Q_4$, we obtain the following characterization of the solutions of (16), (20) with

$$(\xi, \eta) \in S \setminus (\tau(\phi) \cup \tau(\tilde{\phi})). \tag{21}$$

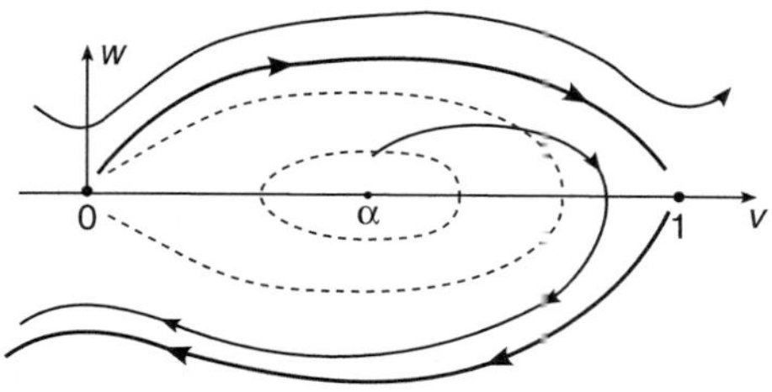

Fig. 2 The phase diagram of system (17) with $c = \hat{c} < 0$; the dashed curves represent orbits of (18).

Lemma 2.1. *Let* $c = \hat{c}$ *and let* ψ *be the solution* (16), (20), *where* (ξ, η) *is as in* (21). *Consider the statements (ai)–(aiii) below. If* $\hat{c} = 0$, *then one of the statements (ai), (aii) holds; if* $\hat{c} < 0$, *then either* $\psi \equiv \alpha$ *or one of the statements (aii), (aiii) holds.*

(*ai*) ψ *is a periodic solution with* $0 < \psi < 1$ *(that is, either* $\psi \equiv \alpha$ *or it is a nonconstant periodic solution).*

(*aii*) *There are numbers* $x_1 < x_2$ *such that*

$$\begin{aligned} &\psi(x) \in (0,1) \quad (x \in (x_1, x_2)),\\ &\psi(x) \notin (0,1) \quad (x \in \mathbb{R} \setminus (x_1, x_2)),\\ &\psi(x_1) \neq \psi(x_2). \end{aligned} \tag{22}$$

(*aiii*) $(\psi(x), \psi'(x)) \to (\alpha, 0)$ *as* $x \to -\infty$, *and there is* $x_0 \in \mathbb{R}$ *such that*

$$0 < \psi(x) < 1 \ (x \in (-\infty, x_0)); \quad \psi(x) < 0 \ (x \in (x_0, \infty)). \tag{23}$$

Note that in (aii) and (aiii), (ψ, ψ') is not an equilibrium, hence $\psi'(x) \neq 0$ whenever $\psi(x) = 0$ or $\psi(x) = 1$. This and (22) imply that in (aii) we have

$$\psi(x_1), \psi(x_2) \in \{0,1\}; \ \ \psi'(x_1) \neq 0, \ \ \psi'(x_2) \neq 0, \tag{24}$$

and in (aiii)

$$\psi(x_0) = 0, \ \ \psi'(x_0) < 0. \tag{25}$$

Corollary 2.2. *Assume that* $\hat{c} < 0$ *and fix* $(\xi, \eta) \in S \setminus \tau(\phi)$. *If* $c \in (\hat{c}, 0)$ *is sufficiently close to* $\hat{c}$ *and* ψ *is the solution* (16), (20), *then either* $\psi \equiv \alpha$ *or one of the statements (aii), (aiii) holds.*

Proof. Denote the solution of (16), (20) by ψ^c; ψ being the solution for $c = \hat{c}$ as in Lemma 2.1. If $(\xi, \eta) = (\alpha, 0)$, then of course $\psi^c \equiv \alpha$.

Assume that $(\xi, \eta) \neq (\alpha, 0)$. For now assume also that (ξ, η) is as in (21), so that (aii) or (aiii) holds for $c = \hat{c}$. We claim that these are robust properties, so, due to the continuous dependence of the solution ψ^c on c, they remain valid—possibly with slightly perturbed x_1, x_2, or x_0—if ψ is replaced with ψ^c and $c > \hat{c}$, $c \approx \hat{c}$. This is obviously the case with (aii) because of (24) and the invariance properties of the sets $Q_1 - Q_4$. If (aiii) holds for $c = \hat{c}$, there is $x_3 < x_0$ such that $(\psi(x_3), \psi_x(x_3))$ is contained inside a periodic orbit of the Hamiltonian system (18). The same is then true if ψ is replaced with ψ^c and $c \approx \hat{c}$. Then $(\psi^c(x), \psi_x(x))$ is "trapped" inside this periodic orbit, by the monotonicity of H, and $(\psi^c(x), \psi^c_x(x))$ has to converge to the equilibrium $(\alpha, 0)$ as $x \to -\infty$. The rest of the properties in (aiii) are clearly robust, due to (25) and the positive invariance of Q_3.

It remains to consider the case $(\xi, \eta) \in \tau(\tilde{\phi})$. Shifting $\tilde{\phi}$, we may assume that $(\xi, \eta) = (\tilde{\phi}(0), \tilde{\phi}_x(0))$, so that

$$(\tilde{\phi}(0), \tilde{\phi}_x(0)) = (\psi^c(0), \psi^c_x(0)). \tag{26}$$

It is clear that if $c \approx \hat{c}$, then, going forward (that is, increasing x), the trajectory of (ψ^c, ψ^c_x) leaves S and then stays in Q_3, just as the trajectory of $(\tilde{\phi}, \tilde{\phi}_x)$ does. Going backward, one can use comparison of solutions of (17) with different values of c (as carried out in [2, Sect. 4], for example), to conclude from (26) that if $c \in (\hat{c}, 0)$ then the trajectory of (ψ^c, ψ^c_x) leaves S through the halfline $\{(1, w) : w < 0\}$. Then it stays in Q_4 by the negative invariance. Hence (aii) holds with ψ replaced by ψ^c if $c \in (\hat{c}, 0)$ and $c \approx \hat{c}$. □

The following lemma will be used in a comparison argument below

Lemma 2.3. *If $\hat{c} < 0$, then there exists numbers $c_n \in (\hat{c}, 0)$, $n = 1, 2, \ldots$, and functions $\psi_n \in C(\mathbb{R})$, $n = 1, 2, \ldots$, such that for each n, ψ_n is a solution of* (16) *with $c = c_n$, $\psi_n < 1$, $\limsup_{|x|\to\infty} \psi_n(x) < 0$, and, as $n \to \infty$, one has $c_n \to \hat{c}$ and* $\max_{x\in\mathbb{R}} \psi_n \to 1$.

Proof. Take any $(\xi, \eta) \in \tau(\phi)$ with $\xi > \alpha$ and let ψ^c be the solution of (16) with $\psi^c(0) = \xi$, $\psi^c_x(0) = \eta$ (at this point, $c \in (\hat{c}, 0)$ is arbitrary). A comparison of solutions of (17) with different values of c [2, Sect. 4] shows that there are $x_1 < 0 < x_2$ (depending on c) such that

$$\psi^c(x_1) = 0,\ \psi^c_x(x_1) > 0, \quad \psi^c(x_2) \in (\alpha, 1),\ \psi^c_x(x_2) = 0. \tag{27}$$

By the negative invariance of Q_2, $(\psi^c(x), \psi^c_x(x)) \in Q_2$ for $x < x_1$. Now, if c is close to $\hat{c}$, then, by the continuity with respect to initial data, x_2 is large and $\psi^c(x_2)$ is close to 1. Using this and the structure of the level set of H for $F(1) = -\hat{c} > 0$ (cf. Fig. 1), one shows easily that for $x > x_2$, $(\psi^c(x), \psi^c_x(x))$ stays in the lower half plane and eventually enters the positively invariant quadrant Q_3. Then $\psi^c(x_2) \approx 1$ is the maximum of ψ^c, hence $\psi^c < 1$. Taking a sequence $\{c_n\}$ in $(\hat{c}, 0)$ with $c_n \to \hat{c}$ and setting $\psi_n := \psi^{c_n}$, we obtain sequences with the stated properties. □

2.2 Properties of $\Omega(u)$

In this section we consider bounded solutions of the problem

$$u_t = u_{xx} + cu_x + f(u), \qquad x \in \mathbb{R},\ t > 0, \tag{28}$$

$$u(x, 0) = u_0(x), \qquad x \in \mathbb{R}, \tag{29}$$

where $f \in C^1(\mathbb{R})$, $u_0 \in C(\mathbb{R}) \cap L^\infty(\mathbb{R})$, and $c \in \mathbb{R}$ (other assumptions are not needed in this section). We define the Ω-limit set of a bounded solution u as in (4). We shall denote this set by $\Omega(u)$ or $\Omega(u_0)$, but it is useful to clarify the following. If u is a bounded solution of (1), then the function $\tilde{u}(x, t) := u(x + ct, t)$ is a bounded solution of (28). Obviously, $\Omega(u) = \Omega(\tilde{u})$ and $\tilde{u}(\cdot, 0) \equiv u(\cdot, 0)$. In other words, if u_0 is given, then $\Omega(u_0)$ is independent of the choice of c in the problem (28), (29).

Assume that the solution u of (28), (29) is bounded. Then, the usual parabolic regularity estimates imply that the derivatives u_t, u_x, u_{xx} are bounded on $\mathbb{R} \times [1, \infty)$ and they are globally α-Hölder on this set for each $\alpha \in (0, 1)$. The following results are standard consequences of this regularity property: $\Omega(u_0)$ is a nonempty, compact, connected subset of $L^\infty_{loc}(\mathbb{R})$. Moreover, in (4) one can take the convergence in $C^1_{loc}(\mathbb{R})$, and $\Omega(u_0)$ is compact and connected in that space as well. The latter implies that the set

$$\{(\varphi(x), \varphi_x(x)) : \varphi \in \Omega(u_0),\ x \in \mathbb{R}\}$$

is compact and connected in $\mathbb{R}^2$.

We now recall the invariance property of $\Omega(u_0)$. Let $\varphi \in \Omega(u)$, so that $u(x_n + \cdot, t_n) \to \varphi$ for some sequence $\{(x_n, t_n)\}$ with $t_n \to \infty$. Then, passing to a subsequence if necessary, one can show that the sequence $u(x_n + \cdot, t_n + \cdot)$ converges in $C^1_{loc}(\mathbb{R}^2)$ to a function U which is an entire solution of (28) (that is, a solution of (28) on $\mathbb{R}^2$). Obviously, $U(\cdot, 0) = \varphi$.

Finally, we note that $\Omega(u_0)$ is also translation-invariant: with each $\varphi \in \Omega(u_0)$, $\Omega(u_0)$ contains the whole translation group orbit of φ, $\{\varphi(\cdot + \xi) : \xi \in \mathbb{R}\}$. This follows directly from the definition of $\Omega(u_0)$.

2.3 *Zero number*

Here we consider solutions of the linear equation

$$v_t = v_{xx} + cv_x + a(x, t)v, \quad x \in \mathbb{R},\ t \in (s, T), \tag{30}$$

where $-\infty < s < T \le \infty$, a is a bounded continuous function on $\mathbb{R} \times [s, T)$, and c is a constant. In the next section we use the following fact, often without notice. If u, $\tilde{u}$ are bounded solutions of the nonlinear equation (28), then their difference $v = u - \tilde{u}$ satisfies a linear equation (30).

We denote by $z(v(\cdot, t))$ the number, possibly infinite, of the zero points $x \in \mathbb{R}$ of the function $x \to v(x, t)$.

The following intersection-comparison principle holds (see [1, 4]).

Lemma 2.4. *Let $v \in C(\mathbb{R} \times [s, T))$ be a nontrivial solution of* (30) *on* $\mathbb{R} \times (s, T)$. *Then the following statements hold true:*

(i) *For each $t \in (s, T)$, all zeros of $v(\cdot, t)$ are isolated.*
(ii) *$t \mapsto z(v(\cdot, t))$ is a monotone nonincreasing function on $[s, T)$ with values in* $\mathbb{N} \cup \{0\} \cup \{\infty\}$.
(iii) *If for some $t_0 \in (s, T)$, the function $v(\cdot, t_0)$ has a multiple zero and $z(v(\cdot, t_0)) < \infty$, then for any $t_1, t_2 \in (s, T)$ with $t_1 < t_0 < t_2$ one has*

$$z(v(\cdot, t_1)) > z(v(\cdot, t_2)). \tag{31}$$

If (31) holds, we say that $z_I(v(\cdot,t))$ drops in the interval (t_1,t_2).

Remark 2.5. It is clear that if $z(v(\cdot,s_0)) < \infty$ for some $s_0 \in (s,T)$, then $z(v(\cdot,t))$ can drop at most finitely many times in (s_0,T) and if it is constant on (s_0,T), then $v(\cdot,t)$ has only simple zeros for each $t \in (s_0,T)$.

Corollary 2.6. *Assume that v is a solution of (30) such that for some $s_0 \in (s,T)$ one has*

$$\liminf_{|x|\to\infty} |v(x,s_0)| > 0. \tag{32}$$

Then there is $t_0 > 0$ such that for $t \geq t_0$ the function $v(\cdot,t)$ has only finitely many zeros and all of them are simple.

Proof. Since the zeros of $v(\cdot,s_0)$ are isolated, (32) implies that there is only a finite number of them. The conclusion now follows directly from Lemma 2.4 and Remark 2.5. □

The next lemma shows that the property for a solution to have multiple zeros is robust.

Lemma 2.7. *Assume that v is a nontrivial solution of (30) such that for some $s_0 \in (s,T)$ the function $v(\cdot,s_0)$ has a multiple zero at some x_0: $v(x_0,s_0) = v_x(x_0,s_0) = 0$. Assume further that for some $\delta,\epsilon > 0$, v_n is a sequence in $C^1([x_0-\delta,x_0+\delta]\times[s_0-\epsilon,s_0+\epsilon])$ which converges in this space to the function v. Then for all sufficiently large n the function $v_n(\cdot,t)$ has a multiple zero in $(x_0-\delta,x_0+\delta)$ for some $t \in (s_0-\epsilon,s_0+\epsilon)$.*

This can be proved using a version of Lemma 2.4 on a small interval around x_0 and the implicit function theorem, see [5, Lemma 2.6] for details. Note that the v_n are not required to be solutions of any equation.

3 Proofs of Theorems 1.1, 1.3, and Corollary 1.2

Throughout this section we assume hypotheses (Hf), (Hu), (6), and (9) to be satisfied, and let u be the solution of (1), (2). Recall that $\hat{c} \leq 0$ is the speed of the traveling front and ϕ is its profile function. Here, we choose the specific translation of the profile function such that $\phi(0) = \alpha$.

If ψ is a nonconstant periodic steady state of (1), we denote by $\mathrm{Int}(\tau(\psi))$ the interior of $\tau(\psi)$ (viewing $\tau(\psi)$ as a Jordan curve).

We start with the following estimates.

Lemma 3.1. *One has*

$$\lim_{t\to\infty}\left(\liminf_{x\to\infty} u(x,t)\right) = 1, \qquad \lim_{t\to\infty}\left(\limsup_{x\to\infty} |u_x(x,t)|\right) = 0. \tag{33}$$

$$\lim_{t\to\infty}(\limsup_{x\to-\infty} u(x,t)) = 0, \qquad \lim_{t\to\infty}(\limsup_{x\to-\infty} |u_x(x,t)|) = 0, \tag{34}$$

Moreover, if $\hat{c} < 0$ *and* $c \in (\hat{c}, 0]$, *then for any* $x_0 \in \mathbb{R}$ *one has*

$$\inf_{x\ge x_0} u(x+ct,t) \to 1 \text{ as } t \to \infty. \tag{35}$$

Proof. We prove (33) and omit the proof of (34), which is completely analogous. It is sufficient to prove the first relation in (33), the second one then follows by standard parabolic regularity estimates for the function $1-u$ (which solves a linear equation (30)). We can always replace u_0 by a nondecreasing function $\tilde{u}_0$, which still satisfies the assumptions of Theorem 1.1 and is such that $\tilde{u}_0 \le u_0$. By the comparison principle, if we prove the first relation in (33) for $\tilde{u}_0$, then it also holds for the original function u_0. We thus proceed assuming that u_0 itself is nondecreasing. Then $u(x,t)$ is nondecreasing in x for each $t \ge 0$. Therefore the limit $\rho(t) := \lim_{y\to\infty} u(y,t)$ exists for each $t \ge 0$. The function ρ is continuous on $[0,\infty)$ and it solves the ODE $\dot{\rho} = f(\rho)$ on $(0,\infty)$ (see, for example, [16, Theorem 5.5.2]). Since $\rho(0) \in (\alpha, 1]$ by assumption, we have $\rho(t) \to 1$, as $t \to \infty$. This completes the proof of (33).

We now prove (35). Again, without loss of generality, we may assume that u_0 is nondecreasing. Let c_n and ψ_n be as in Lemma 2.3. Given any $\epsilon > 0$, we can choose n such that $c_n \in (\hat{c}, c)$ and $\max \psi_n \in (1-\epsilon, 1)$. Shifting ψ_n, we may assume that $\psi_n(0) = \max \psi_n$. By (33) and Lemma 2.3, we can further choose positive constants t_0 and y_0 such that

$$u(x + c_n t_0, t_0) > \psi_n(x - y_0) \quad (x \in \mathbb{R}).$$

Since the functions $u(x + c_n t, t)$ and $\psi_n(x - y_0)$ satisfy the same equation, equation (28) with $c = c_n$, the comparison principle gives

$$u(x + c_n t, t) > \psi_n(x - y_0) \quad (x \in \mathbb{R},\ t \ge t_0).$$

Using the monotonicity of $u(\cdot, t)$, we in particular obtain

$$u(x + ct, t) \ge u(y_0 + c_n t) > \psi_n(0) > 1 - \epsilon \quad (x \ge y_0 + (c_n - c)t,\ t \ge t_0).$$

Since $c_n < c$ and ϵ can be taken arbitrarily small, it is clear that (35) holds for any x_0. □

Relations (33), (34), and the definition of $\Omega(u)$ immediately give the following.

Corollary 3.2. *The constant steady states* 0 *and* 1 *are elements of* $\Omega(u)$.

The next lemma comprises the crux of the proof of Theorem 1.1.

Lemma 3.3. *Let $c \in [\hat{c}, 0]$ and let ψ be a solution of* (16). *Assume that either one of the statements (ai), (aii) in Lemma 2.1 holds, or $\hat{c} < c < 0$ and statement (aiii) holds. Then there is T such that*

$$\tau(u(\cdot,t)) \cap \tau(\psi) = \emptyset \quad (t \geq T). \tag{36}$$

Proof. The proof is by contradiction. We assume that

$$\tau(u(\cdot,t_n)) \cap \tau(\psi) \neq \emptyset \text{ for some sequence } t_n \to \infty. \tag{37}$$

First we show that this leads to a contradiction if (ai) holds. If $\psi \equiv \alpha$, then $\tau(\psi) = (\alpha, 0)$ and (37) means that $u(\cdot,t_n) - \alpha$ has a multiple zero for $n = 1, 2, \dots$. We know that this is not possible due to Corollary 2.6 and Lemma 3.1. Thus, we can proceed assuming that ψ is a nonconstant periodic solution (and $c = 0$). Let $\rho > 0$ be the minimal period of ψ. According to (37), for each n there is $y_n \in [0, \rho)$ such that the function $u(\cdot,t_n) - \psi(\cdot - y_n)$ has a multiple zero, say z_n. Consequently, $x = 0$ is a multiple zero of the function $u(\cdot + z_n, t_n) - \psi(\cdot + z_n - y_n)$. Write $z_n = k_n\rho + \zeta_n$, where $k_n \in \mathbb{Z}$ and $\zeta_n \in [0, \rho)$. We may assume, passing to a subsequence if necessary, that $\zeta_n \to \zeta_0 \in [0, \rho]$ and $y_n \to y_0 \in [0, \rho]$, hence

$$\psi(\cdot + z_n - y_n) = \psi(\cdot + \zeta_n - y_n) \to \psi(\cdot + \zeta_0 - y_0) \text{ in } C^1_b(\mathbb{R}).$$

We may also assume that $u(\cdot + z_n, t_n) \to \varphi$ for some $\varphi \in \Omega(u)$, and $u(\cdot + z_n, \cdot + t_n) \to U$ in $C^1_{loc}(\mathbb{R}^2)$, where U is an entire solution of (1) with $U(\cdot, 0) = \varphi$ (see Sect. 2.2). Clearly, the function $U(\cdot,0) - \psi(\cdot + \zeta_0 - y_0) = \varphi - \psi(\cdot + \zeta_0 - y_0)$ has a multiple zero at $x = 0$ and $u(\cdot + z_n, \cdot + t_n) - \psi(\cdot + z_n - y_0) \to U - \psi(\cdot - \zeta_0 - y_0)$ in $C^1_{loc}(\mathbb{R}^2)$. Now, $V := U - \psi(\cdot - \zeta_0 - y_0)$ is an entire solution of a linear equation (30) (with $c = 0$) and we verify in a moment that $V(\cdot,0) = \varphi - \psi(\cdot - \zeta_0 - y_0) \not\equiv 0$. Therefore, Lemma 2.7 implies that for each sufficiently large n, the function $u(\cdot + z_n, s + t_n) - \psi(\cdot + z_n - y_0)$ has a multiple zero (near $x = 0$) for some small s. However, by Corollary 2.6 and Lemma 3.1, $u(\cdot,t) - \psi(\cdot - y_0)$ has only simple zeros for all sufficiently large t. Since $t_n + s \to \infty$, we have a desired contradiction.

To verify that $\varphi - \psi(\cdot - \zeta_0 - y_0) \not\equiv 0$, we note that for $t > 0$, the function $u(\cdot,t) - \alpha$ has a finite number of zeros and this number is independent of t if t is large enough (see Corollary (2.6)). On the other hand, as $\psi(\cdot - \zeta_0 - y_0) - \alpha$ has infinitely many simple zeros (see Sect. 2.1), the relations $\psi(\cdot - \zeta_0 - y_0) \equiv \varphi = \lim u(\cdot, t_n)$ would give a contradictory conclusion that $\lim z(u(\cdot,t_n) - \alpha) \to \infty$. This shows that $\psi(\cdot - \zeta_0 - y_0) \equiv \varphi$ cannot hold. The proof under condition (ai) is now complete.

Now assume that (aii) holds. Let $\tilde{u}(x,t) = u(x + ct, t)$, so that $\tilde{u}$ and ψ satisfy the same equation (28). Obviously, $\tau(\tilde{u}(\cdot,t)) = \tau(u(\cdot,t))$ for any t, thus (37) means that there is $y_n \in \mathbb{R}$ such that

$$\tilde{u}(\cdot,t_n) - \psi(\cdot - y_n) \text{ has a multiple zero } z_n. \tag{38}$$

In particular, $\psi(z_n - y_n) = \tilde{u}(z_n, t_n) \in (0, 1)$, which implies that $z_n - y_n \in (x_1, x_2)$ (cp. (aii)). We distinguish the following two possibilities regarding the sequence $\{y_n\}$:

(a) $\{y_n\}$ is bounded (b) $\{y_n\}$ is not bounded.

If (a) holds, then $\{z_n\}$ is bounded as well. We now use similar arguments as above for (ai). Passing to subsequences we may assume that for some $y_0, z_0 \in \mathbb{R}$ and $\varphi \in \Omega(u)$, one has $y_n \to y_0$, $z_n \to z_0$, $\tilde{u}(\cdot, t_n) \to \varphi$ in $C^1_{loc}(\mathbb{R})$. Also, we may assume that $\tilde{u}(\cdot, \cdot + t_n) \to \tilde{U}$ in $C^1_{loc}(\mathbb{R}^2)$, where $\tilde{U}$ is an entire solution of (28) with $\tilde{U}(\cdot, 0) = \varphi$ (see Sect. 2.2). Clearly, z_0 is a multiple zero of $\tilde{U}(\cdot, 0) - \psi(\cdot - y_0)$ and one has $\tilde{u}(\cdot, \cdot + t_n) - \psi(\cdot + y_0) \to U - \psi(\cdot - y_0)$. The function $V := \tilde{U} - \psi(\cdot - \zeta_0 - y_0)$ is a entire solution of a linear equation (30) and $V \not\equiv 0$ by (aii) and the fact that $0 \leq \tilde{U} \leq 1$. Lemma 2.7 implies that for each sufficiently large n, the function $\tilde{u}(\cdot, s + t_n) - \psi(\cdot + y_0)$ has a multiple zero for some $s \approx 0$. However, by Corollary (2.6) and (aii), $\tilde{u}(\cdot, t) - \psi(\cdot + y_0)$ has only simple zeros for all sufficiently large t, and we have a contradiction.

Next we consider the possibility (b). For definiteness we assume that, after passing to a subsequence, one has $y_n \to -\infty$; the case $y_n \to \infty$ can be treated in an analogous way. By (22), (24), there is $\epsilon > 0$ such that $|\psi'(x)| > \epsilon$, whenever $x \in [x_1, x_2]$ and $\psi(x) < \epsilon$. By Lemma 3.1, there are positive constants r and t_0 such that $u(x, t_0) + |u_x(x, t_0)| < \epsilon$ if $x < r$. For $\tilde{u}$ this means that $\tilde{u}(x, t_0) + |\tilde{u}_x(x, t_0)| < \epsilon$ if $x < \tilde{r} := r - ct_0$. Consequently, if n is so large that $x_2 + y_n < \tilde{r}$, then $\tilde{u}(\cdot, t_0) - \psi(\cdot - y_n)$ has a unique zero in the interval $[x_1 + y_n, x_2 + y_n]$. Of course, by (22), $\tilde{u}(\cdot, t_0) - \psi(\cdot - y_n)$ has no zero outside this interval, hence $z(\tilde{u}(\cdot, t_0) - \psi(\cdot - y_n)) = 1$. Clearly, by (22), $z(\tilde{u}(\cdot, t) - \psi(\cdot - y_n)) \geq 1$ for all t, hence the equality must hold here by the monotonicity of the zero number (see Lemma 2.4). The unique zero of $\tilde{u}(\cdot, t) - \psi(\cdot - y_n)$ has to be simple for all $t > t_0$ (see Remark 2.5). This holds for all sufficiently large n, in particular, we can choose n so that also $t_n > t_0$. We thus have a contradiction to (38).

Finally, we assume that $\hat{c} < c < 0$ and (aiii) holds. As above, (38) holds with $\tilde{u}(x, t) := u(x + ct, t)$. The possibilities that $\{y_n\}$ is bounded, or $\{y_n\}$ has a subsequence converging to $-\infty$, can be treated similarly as in the case (aii); the only possibility that requires a different consideration is that y_n (replaced by a subsequence) converges to ∞. Assuming that $y_n \to \infty$, choose $\epsilon > 0$ such that $1 - \epsilon > \psi$ everywhere. By (35), there is t_0 such that

$$\tilde{u}(x, t) > 1 - \epsilon \quad (x \geq 0,\ t \geq t_0).$$

This implies that if n is large enough, then all zeros of $\tilde{u}(\cdot, t_n) - \psi(\cdot - y_n)$ are located in $(-\infty, 0]$; in particular, $z_n \leq 0$, where z_n is the multiple zero in (38). Hence, by (aiii) and the assumption that $y_n \to \infty$,

$$(\tilde{u}(z_n, t_n), \tilde{u}_x(z_n, t_n)) = (\psi(z_n - y_n), \psi_x(z_n - y_n)) \to (\alpha, 0). \tag{39}$$

We now take a periodic steady state $\tilde{\psi}$ of (1) such that $0 < \tilde{\psi} < 1$ and $(\alpha, 0) \in \mathrm{Int}(\tau(\tilde{\psi}))$ (see Sect. 2.1 and cp. Fig. 1). Then (39) implies that for large n the spatial trajectory $\tau(\tilde{u}(\cdot, t_n)) = \tau(u(\cdot, t_n))$ has to intersect $\tau(\tilde{\psi})$ (it cannot be contained entirely in $\mathrm{Int}(\tau(\tilde{\psi}))$ because of Lemma 3.1). Thus we have a contradiction to the result proved above in the case (ai). □

Corollary 3.4. *Let c and ψ be as in Lemma 3.3. Then for any $\varphi \in \Omega(u)$ one has $\tau(\varphi) \cap \tau(\psi) = \emptyset$.*

Proof. Assume this is not true. Then for some y_0 the function $\varphi - \psi(\cdot - y_0)$ has a multiple zero. There is an entire solution U of (28) (with the same c as in the statement of the lemma) such that $U(\cdot, 0) = \varphi$ and $u(\cdot + x_n + ct_n, \cdot + t_n) \to U$ in $C^1_{loc}(\mathbb{R}^2)$ for some sequences $x_n \in \mathbb{R}$, $t_n \to \infty$ (see Sect. 2.2). Then $V := U - \psi(\cdot - y_0)$ is a solution of a linear equation (30) and $V \not\equiv 0$, as noted in the proof of Lemma 3.3 (see case (ai) in the proof; if (aii) or (aiii) holds, then $V \not\equiv 0$ is trivial). Thus, using Lemma 2.7 as in the previous proof, we find sequences $\tilde{t}_n \approx t_n, \tilde{x}_n \in \mathbb{R}$, $n = 1, 2, \ldots$, such that $\tilde{t}_n \to \infty$ and $u(\cdot + \tilde{x}_n, \tilde{t}_n) - \psi(\cdot + y_0)$ has a multiple zero for $n = 1, 2, \ldots$. This contradicts (36). □

We next consider the set

$$K_\Omega := \cup_{\varphi \in \Omega(u)} \tau(\varphi) = \{(\varphi(x), \varphi_x(x)) : \varphi \in \Omega(u),\ x \in \mathbb{R}\}. \tag{40}$$

This is a compact, connected subset of $\mathbb{R}^2$ (cp. Sect. 2.2).

Lemma 3.5. *One has $K_\Omega \subset \Sigma$, where*

$$\Sigma := \begin{cases} \{(0,0), (1,0)\} \cup \tau(\phi) & \text{if } \hat{c} < 0, \\ \{(0,0), (1,0)\} \cup \tau(\phi) \cup \tau(\tilde{\phi}) & \text{if } \hat{c} = 0, \end{cases} \tag{41}$$

and $\tilde{\phi}$ is defined by $\tilde{\phi}(x) = \phi(-x)$ (as in Sect. 2.1).

Proof. Assume that $K_\Omega \not\subset \Sigma$. Then there are $(\xi, \eta) \in \mathbb{R}^2 \setminus \Sigma$ and $\varphi \in \Omega(u)$ such that $(\varphi(x_0), \varphi_x(x_0)) = (\xi, \eta)$ for some x_0. Obviously, $0 \le \varphi \le 1$ and the existence of an entire solution through φ (see Sect. 2.2) and the comparison principle show that either $\varphi \equiv 0$, or $\varphi \equiv 1$, or else $0 < \varphi < 1$. Since $(\xi, \eta) \notin \{(0,0), (1,0)\}$, the relations $0 < \varphi < 1$ must hold and, in particular, $0 < \xi < 1$. By Lemma 2.1 and Corollary 2.2, there are $c \in [\hat{c}, 0]$ and a solution ψ, such that $(\psi(0), \psi'(0)) = (\xi, \eta)$ and the assumptions of Lemma 3.3 are satisfied. For this ψ, we have $\tau(\varphi) \cap \tau(\psi) \neq \emptyset$, in contradiction to Corollary 3.4. □

Proof (Completion of the proof of Theorem 1.1). Let $\hat{c} < 0$ Corollary 3.2 implies that K_Ω contains the points $(0, 0)$, $(0, 1)$. Therefore, Lemma 3.5 and the connectedness of K_Ω imply that

$$K_\Omega = \{(0,0), (1,0)\} \cup \tau(\phi). \tag{42}$$

Take now any $\varphi \in \Omega(u)$. As noted in the proof of Lemma 3.5, if φ is not one of the constant steady states 0, 1, then $0 < \varphi < 1$. In this case, (42) implies that $\tau(\varphi) \subset \tau(\phi)$. Since $\phi' > 0$, this means that for each $x \in \mathbb{R}$ there is a unique $\zeta(x)$, such that

$$\varphi(x) = \phi(\zeta(x)), \quad \varphi'(x) = \phi'(\zeta(x)). \tag{43}$$

Moreover, $\zeta \in C^1$ by the implicit function theorem. Differentiating the first identity and comparing to the second one, we obtain that $\zeta' \equiv 1$. Thus there is $\xi \in \mathbb{R}$ such that $\varphi \equiv \psi(\cdot - \xi)$. This proves (11). □

Proof (Proof of Theorem 1.3, Part 1). Assume that $\hat{c} = -F(1) = 0$. Also assume the additional hypothesis on u_0, (Ha), to be satisfied. In this part of the proof we show that (11) holds.

The arguments from the previous proof apply here, the only difference is that the specific statement of Lemma 3.5 for $\hat{c} = 0$ has to be used. Thus, in place of (42), we can only say that one of the following possibilities occurs:

(oi) $\Omega(u) = \{0, 1\} \cup \{\phi(\cdot - \xi) : \xi \in \mathbb{R}\}$ (as stated in Theorem 1.3),
(oii) $\tilde{\phi} \in \Omega(u)$.

We just need to rule out (oii); (11) then follows from (oi), as in the proof of Theorem 1.1. Assume that $\tilde{\phi} \in \Omega(u)$: there are $x_n \in \mathbb{R}$, $t_n > 0$, $n = 1, 2, \ldots$ such that $t_n \to \infty$ and $u(\cdot + x_n, t_n) \to \tilde{\phi}$. From this and Lemma 3.1 it follows that for all large enough n, the function $u(\cdot, t_n) - \alpha$ has at least three zeros whose mutual distances go to infinity as $n \to \infty$. To contradict this conclusion, we employ hypothesis (Ha).

First we note that the monotonicity of the zero number (see Lemma 2.4) implies that $z(u(\cdot, t) - \alpha) \geq 3$. Thus, if $z(u_0 - \alpha) = 1$, we have a contradiction already and we are done. We proceed assuming that the other condition of (Ha) holds. We claim that this condition is preserved at positive times: For each $t > 0$ the limits $u(\pm\infty, t)$ exist and one has

$$u(-\infty, t) < u(x, t) < u(\infty, t) \quad (x \in \mathbb{R},\ t > 0). \tag{44}$$

Indeed, the existence of the limits $u_0(\pm)$ implies that the limits $\rho^{\pm}(t) := u(\pm\infty, t)$ exist for all $t \geq 0$ and they satisfy the ODE $\dot{\rho} = f(\rho)$ with the initial conditions $\rho^{\pm}(0) := u_0(\pm\infty)$ (see [16, Theorem 5.5.2]). Relations (15) give $\rho^{-}(0) \leq u_0 \leq \rho^{-}(0)$. Of course, none of these inequalities is an identity by (Ha). Relations (44) now follow from the strong comparison principle.

By Corollary 2.6, we can choose $t_0 > 0$ such that for $t \geq t_0$, the zeros of $u(\cdot, t) - \alpha$ are all simple, and their number, say k, is finite and independent of t. Let $\zeta_1(t) < \cdots < \zeta_k(t)$ denote the zeros of $u(\cdot, t) - \alpha$ for $t \geq t_0$. Since they are simple, the functions $\zeta_1, \ldots \zeta_k$ are C^1 on $[t_0, \infty)$.

Using (44), one shows easily that there is a smooth increasing function $\tilde{u}_0$ such that

$$
\begin{aligned}
u(-\infty, t_0) < \tilde{u}_0(-\infty) < \min_{\zeta_1(t_0) \le x \le \zeta_k(t_0)} u_0(x), \\
u(\infty, t_0) > \tilde{u}_0(\infty) > \max_{\zeta_1(t_0) \le x \le \zeta_k(t_0)} u_0(x).
\end{aligned} \tag{45}
$$

Clearly, for such $\tilde{u}_0$, if η is large enough, then

$$u_0(x) < \tilde{u}_0(x + \eta) \quad (x \le \zeta_k(t_0)), \tag{46}$$

$$u_0(x) > \tilde{u}_0(x - \eta) \quad (x \ge \zeta_1(t_0)). \tag{47}$$

Let $\tilde{u}$ be the solution of (1) on (t_0, ∞) with the initial condition $\tilde{u}(\cdot, t_0) = \tilde{u}_0$. Then $\tilde{u}(x, t)$ is continuous on $\mathbb{R} \times [t_0, \infty)$ and increasing in x. By (45), the relations $\tilde{u}(-\infty, t) < \alpha < \tilde{u}(-\infty, t)$ hold for $t = t_0$, hence they continue to hold for all for all $t \ge t_0$ (see Lemma 3.1). Therefore, for each $t \ge t_0$ the function $\tilde{u}(x\ t) - \alpha$ has a unique zero $\xi(t)$ and $t \mapsto \xi(t)$ is continuous on $[t_0, \infty)$.

Consider now the relations

$$\xi(t) - \eta < \zeta_1(t), \quad \zeta_k(t) < \xi(t) + \eta. \tag{48}$$

They are both satisfied for $t = t_0$ (use the monotonicity of $u(\cdot, t)$ and the relations (46), (47), with $x = \zeta_1(t_0)$, $x = \zeta_k(t_0)$, respectively). By continuity, they are also satisfied it $t > t_0$ is sufficiently close to t_0. On the other hand, (48) cannot be satisfied for all $t > t_0$ by the properties of the sequence $\{t_n\}$ stated above: $\zeta_k(t_n) - \zeta_1(t_n) \to \infty$. Thus there is $t_1 > t_0$ such that relations (48) hold for all $t \in [t_0, t_1)$ and either $\xi(t_1) - \eta = \zeta_1(t_1)$ or $\zeta_k(t_1) = \xi(t_1) + \eta$. Assume that the former holds (the latter can be dealt with in an analogous way). Then

$$\tilde{u}(\zeta_1(t_1) + \eta, t_1) = \tilde{u}(\xi(t_1), t_1) = \alpha = u(\zeta_1(t_1), t_1). \tag{49}$$

Since $\xi(t) - \eta$ is the unique zero of the function $\tilde{u}(\cdot + \eta, t) - \alpha$ and $\zeta_k(t) > \zeta_1(t)$, the first relation in (48) yields

$$\tilde{u}(\zeta_k(t) + \eta, t) > \alpha = u(\zeta_k(t), t) \quad (t_0 \le t \le t_1). \tag{50}$$

Using this, (46), and the strong comparison principle, we obtain

$$\tilde{u}(x + \eta, t) > u(x, t) \quad (x < \zeta_k(t),\ t_0 \le t \le t_1), \tag{51}$$

contradicting (49).

With this contradiction, the proof of (11) is complete. □

Proof (Proof of Corollary 1.2 and Proof of Theorem 1.3, Part 2). Assume that $\hat{c} = -F(1) > 0$ or that $\hat{c} = F(1) = 0$ and the additional assumption (Ha) is satisfied. Under these assumptions, we have already proved that (11) holds. Since $\phi' > 0$, this implies in particular that if t is large enough, then $u_x(x,t) > 0$ whenever $u(x,t) = \alpha$. Consequently, for large t there is a unique $\gamma(t)$ such that $u(\gamma(t) + \hat{c}t, t) = \alpha$. Moreover, $\gamma \in C^1$, by the implicit function theorem. Denote $\tilde{u}(x,t) := u(x + \hat{c}t, x)$, so $\tilde{u}$ and ϕ solve the same equation (28), with $c = \hat{c}$. Any sequence $t_n \to \infty$ can be replaced by a subsequence such that $\tilde{u}(\cdot + \gamma(t_n), t_n) \to \varphi$ in $L^\infty_{loc}(\mathbb{R})$ for some $\varphi \in \Omega(u)$. Necessarily, $\varphi(0) = \alpha$. Therefore, by (11) and our choice $\phi(0) = \alpha$, we have $\varphi = \phi$. Since this limit is always the same, we have

$$\tilde{u}(\cdot + \gamma(t), t) \to \phi \text{ as } t \to \infty, \tag{52}$$

with the convergence in $L^\infty_{loc}(\mathbb{R})$.

To complete the proof, we need to prove that the convergence takes place in $L^\infty(\mathbb{R})$ and $\gamma'(t) \to 0$ as $t \to \infty$. We start with the latter. Recall that any sequence $t_n \to \infty$ can be replaced by a subsequence such that $\tilde{u}(\cdot + \gamma(t_n), \cdot + t_n)$ converges in $C^1_{loc}(\mathbb{R}^2)$ to an entire solution U of equation (28) with $U(\cdot, 0) = \phi$. Since ϕ is a steady state of (28), we have $U \equiv \phi$, by uniqueness and backward uniqueness for (28). Thus the convergence in $C^1_{loc}(\mathbb{R}^2)$ yields

$$(\tilde{u}(\cdot + \gamma(t_n), \cdot + t_n), \tilde{u}_x(\cdot + \gamma(t_n), \cdot + t_n), \tilde{u}_t(\cdot + \gamma(t_n), \cdot + t_n) \to (\phi, \phi_x, 0).$$

Since this is true for any sequence $t_n \to \infty$, we have, in particular,

$$(\tilde{u}(\gamma(t), t), \tilde{u}_x(\gamma(t), t), \tilde{u}_t(\gamma(t), t) \to (\alpha, \phi_x(0), 0), \tag{53}$$

as $t \to \infty$. Now, differentiating the relation $\tilde{u}(\gamma(t), t) = \alpha$, we obtain $\tilde{u}_x(\gamma(t), t)\gamma'(t) + u_t(\gamma(t), t) = 0$. Since $\phi_x(0) \neq 0$, from (53) we conclude that $\gamma'(t) \to 0$ as $t \to \infty$.

It remains to prove that the convergence in (52) is uniform on $\mathbb{R}$. Assume it is not. Then there exist $\delta > 0$ and sequences $\{x_n\}$, $\{t_n\}$ such that $|x_n| \to \infty$, $t_n \to \infty$, and

$$|\tilde{u}(x_n + \gamma(t_n), t_n) - \phi(x_n)| > 2\delta. \tag{54}$$

Assume for definiteness that $\{x_n\}$ can be replaced by a subsequence so that $x_n \to -\infty$ (the case when $x_n \to \infty$ can be treated similarly). Since $\phi(-\infty) = 0$, (54) in particular implies that for all large enough n one has $\tilde{u}(x_n + \gamma(t_n), t_n) > \delta$. On the other hand, using $\phi(-\infty) = 0$ and (52), we find x_0 such that $\tilde{u}(x_0 + \gamma(t), t) < \delta$ for all sufficiently large t. These relations imply that if n is sufficiently large, then there is y_n between x_n and x_0, such that

$$\tilde{u}(y_n + \gamma(t_n), t_n) = \delta, \quad \tilde{u}_x(y_n + \gamma(t_n), t_n) \leq 0. \tag{55}$$

Take now a subsequence of $\tilde{u}(\cdot + y_n + \gamma(t_n), t_n)$, which converges in $C^1_{loc}(\mathbb{R})$ to some $\varphi \in \Omega(u)$. By (55), $\varphi(0) = \delta$, $\varphi'(0) \leq 0$. However, by (11), $\varphi = \phi(\cdot - \xi)$ for some ξ, hence $\varphi' > 0$. This contradiction completes the proof. □

Acknowledgements This work was supported in part by the NSF Grant DMS–1161923.

References

1. Angenent, S.: The zeroset of a solution of a parabolic equation. J. Reine Angew. Math. **390**, 79–96 (1988)
2. Aronson, D.G., Weinberger, H.: Multidimensional nonlinear diffusion arising in population genetics. Adv. Math. **30**, 33–76 (1978)
3. Chen, X.: Existence, uniqueness, and asymptotic stability of traveling waves in nonlocal evolution equations. Adv. Differ. Equ. **2**(1), 125–160 (1997)
4. Chen, X.-Y.: A strong unique continuation theorem for parabolic equations. Math. Ann. **311**, 603–630 (1998)
5. Du, Y., Matano, H.: Convergence and sharp thresholds for propagation in nonlinear diffusion problems. J. Eur. Math. Soc. **12**(2), 279–312 (2010)
6. Fife, P.C., McLeod, J.B.: The approach of solutions of nonlinear diffusion equations to travelling front solutions. Arch. Ration. Mech. Anal. **65**, 335–361 (1977)
7. Fife, P.C., McLeod, J.B.: A phase plane discussion of convergence to travelling fronts for nonlinear diffusion. Arch. Rational Mech. Anal. **75**, 281–314 (1980/1981)
8. Henry, D.: Geometric Theory of Semilinear Parabolic Equations. Springer, New York (1981)
9. Matano, H., Ogiwara, T.: Monotonicity and convergence in order-preserving systems. Discret. Contin. Dyn. Syst. **5**, 1–34 (1999)
10. Matano, H., Poláčik, P.: Dynamics of nonnegative solutions of one-dimensional reaction-diffusion equations with localized initial data. Part i: a general quasiconvergence theorem and its consequences (in preparation)
11. Muratov, C.B., Novaga, M.: Global exponential convergence to variational traveling waves in cylinders. SIAM J. Math. Anal. **44**, 293–315 (2012)
12. Risler, E.: Global convergence toward traveling fronts in nonlinear parabolic systems with a gradient structure. Ann. Inst. H. Poincaré Anal. Non Linéaire **25**, 381–424 (2008)
13. Roquejoffre, J.-M.: Convergence to traveling waves for solutions of a class of semilinear parabolic equations. J. Differ. Equ. **108**, 262–295 (1994)
14. Roquejoffre, J.-M.: Eventual monotonicity and convergence to traveling fronts for the solutions of parabolic equations in cylinders. Ann. Inst. H. Poincaré Anal. Non Linéaire **14**, 499–552 (1997)
15. Sattinger, D.H.: On the stability of waves of nonlinear parabolic systems. Adv. Math. **22**, 312–355 (1976)
16. Volpert, A.I., Volpert, V.A., Volpert, V.A.: Traveling Wave Solutions of Parabolic Systems. Translations of Mathematical Monographs, vol. 140. American Mathematical Society, Providence, RI (1994)

Regularity estimates for fully non linear elliptic equations which are asymptotically convex

Luis Silvestre and Eduardo V. Teixeira

Mathematics Subject Classification (2010): 35B65, 35J70

1 Introduction

Regularity estimates for viscosity solution to a given fully nonlinear uniformly elliptic equation

$$F(D^2u) = 0 \text{ in some domain } \Omega \subset \mathbb{R}^n \tag{1}$$

have been a primary important line of research since the work of Krylov and Safonov [10, 11] unlocked the theory. By formal linearization, both u and its first derivative, u_ν, satisfy linear elliptic equations in non-divergence form, thus Krylov-Safonov Harnack inequality implies that solutions are *a priori* $C^{1,\beta}$ for some universal, but unknown $\beta > 0$. The language of viscosity solutions allows the same conclusion without linearizing the equation, see [4]. The question whether a viscosity solution is twice differentiable, i.e. classical, turned out to be truly challenging. The first major result in this direct was obtained independently by Evans [6] and Krylov [8, 9], see also [4, Chapter 6]. This is the content of the

L. Silvestre
Department of Mathematics, University of Chicago, Chicago, IL 60637, USA
e-mail: luis@math.uchicago.edu

E.V. Teixeira (✉)
Universidade Federal do Ceará, CEP 60455-760, Fortaleza, Ceará, Brazil
e-mail: teixeira@mat.ufc.br

A.N. Carvalho et al. (eds.), *Contributions to Nonlinear Elliptic Equations and Systems*, Progress in Nonlinear Differential Equations and Their Applications 86,
DOI 10.1007/978-3-319-19902-3_25

Evans-Krylov $C^{2,\alpha}$ regularity theorem that assures that under concavity or convexity assumption on F, viscosity solutions to $F(D^2u) = 0$ are of class $C^{2,\alpha}$ for some $0 < \alpha < 1$. After Evans-Krylov Theorem, many important works attempted to establish a $C^{2,\alpha}$ regularity theory for solutions to special classes of uniform elliptic equations of the form (1), see, for instance, [2] and [19].

Recently Nadirashvili and Vladut [12, 13] showed that viscosity solutions to fully nonlinear equations may fail to be of class C^2. They have also exhibited solutions to uniform elliptic equations whose Hessian blow-up, i.e., that are not $C^{1,1}$. The regularity theory for fully nonlinear equations would turn out to be even more complex: Nadirashvili and Vladut quite recently showed that given any $0 < \tau < 1$ it is possible to build up a uniformly elliptic operator F, whose solutions are not $C^{1,\tau}$, see [14, Theorem 1.1]. These examples are made in high dimensions. In [16] and [15], they showed an example of a non C^2 solution in five dimensions. This is the lower dimension for which such result is available. In two dimensions, however, it is well known that solutions are always C^2. For dimensions $n = 3$ and $n = 4$, the regularity of viscosity solutions to uniformly elliptic equations without further structural assumptions remains an outstanding open problem.

After these stunning examples, it becomes relevant to investigate possible special hidden structures on a given elliptic operator F which might yield further regularity estimates for solutions to (1). In this paper we turn attention to an asymptotic property on F, called the *recession* function. For any symmetric matrix $M \in \mathbb{R}^{n\times n}$, we define

$$F^\star(M) := \lim_{\mu\to 0} \mu F(\mu^{-1}M). \tag{2}$$

The above limit may not exist as $\mu \to 0$. In that case, we say that a recession function $F^\star$ is any one of the subsequential limits.

Heuristically, $F^\star$ accounts the behavior of F at infinity. Recently recession functions appeared in the study of free boundary problems governed by fully nonlinear operators, [1, 17]. The main result we prove in this paper states that the regularity theory for the recession function $F^\star(M)$ grants smoothness of viscosity solutions to the original equation $F(D^2u) = 0$, up to $C^{1,1^-}$.

Theorem 1.1. *Let F be a uniformly elliptic operator. Assume any recession function*

$$F^\star(M) := \lim_{\mu\to 0} \mu F(\mu^{-1}M)$$

has C^{1,α_0} estimates for solutions to the homogeneous equation $F^\star(D^2v) = 0$. Then, any viscosity solution to

$$F(D^2u) = 0,$$

is of class $C^{1,\min\{1,\alpha_0\}^-}_{loc}$. *That is,* $u \in C^{1,\alpha}_{\text{loc}}$ *for any* $\alpha < \min\{1, \alpha_0\}$. *In addition, there holds*

$$\|u\|_{C^{1,\alpha}(B_{1/2})} \leq C\|u\|_{L^\infty(B_1)}, \tag{3}$$

for a constant $C > 0$ *that depends only on* n, α, *and* F.

An immediate Corollary of Theorem 1.1 is the following:

Corollary 1.2. *Let* $F: \mathscr{S}(n) \to \mathbb{R}$ *be a uniform elliptic operator and* u *a viscosity solution to* $F(D^2u) = 0$ *in* B_1. *Assume any recession function* $F^\star(M) := \lim_{\mu\to 0} \mu F(\mu^{-1}M)$ *is concave. Then* $u \in C^{1,\alpha}_{loc}(B_1)$ *for every* $\alpha < 1$.

Clearly the corresponding regularity theory for heterogeneous, non-constant coefficient equations $F(X, D^2u) = f(X)$ is, in general, considerably more delicate. Nevertheless, in this setting, L. Caffarelli in [3] established $C^{1,\alpha}$, $C^{2,\alpha}$ and $W^{2,p}$, *a priori* estimates for solutions to

$$F(X, D^2u) = f(X) \in L^p,$$

for $p > n$, under appropriate continuity assumption on the coefficients. Caffarelli's results are, nevertheless, based upon the regularity theory available for the homogeneous, constant-coefficient equation $F(X_0, D^2u) = 0$. Therefore, it is essential to know the best possible regularity estimates available for equations of the form (1). Of course, combining Caffarelli's regularity theory and Theorem 1.1 it is possible to establish the sharp regularity estimates for heterogeneous non-constant coefficient equations.

An application of Corollary 1.2 concerns local regularity estimates for singular fully nonlinear PDEs:

$$F(D^2u) \sim u^{-\gamma}, \quad 0 < \gamma < 1. \tag{4}$$

In [1] it has been proven that nonnegative minimal solutions are locally uniformly continuous and grow precisely as $\text{dist}^{\frac{2}{1+\gamma}}$ away from the free boundary $\partial\{u > 0\}$. Notice that such an estimate implies that u behaves along the free boundary as a $C^{1,\frac{1-\gamma}{1+\gamma}}$ function. In particular, if γ is small, such an estimate competes with the (unknown) C^{1,α_F} *a priori* estimate. By knowing the recession function, which governs free boundary condition of the problem, it is possible to show that u is locally of the class $C^{1,\frac{1-\gamma}{1+\gamma}}$ and such an estimate does not deteriorate near the free boundary.

Corollary 1.3. *Let* u *be a minimal solution to* $F(D^2u) \sim u^{-\gamma}$, *in* $\Omega \subset \mathbb{R}^n$, *with* $0 < \gamma < 1$. *Assume the recession function* F^* *is unique and has a priori* C^{2,α_*} *estimates. Then* u *is locally of class* $C^{1,\frac{1-\gamma}{1+\gamma}}$ *in* Ω.

The proof of Corollary 1.3 will be delivered in Section 4. Finally, we would like to point out that Theorem 1.1 provides eventual gain of smoothness beyond universal estimates only up to $C^{1,1^-}$. Nevertheless, such a constraint does not come from limitations of the methods employed here. In fact, Nadirashvili and Vladut built up an example of a fully nonlinear operator $\mathfrak{F}$ that admits a viscosity solution $\phi \in C^{1,1} \setminus C^2$. Thus, we could deform $\mathfrak{F}$ outside $B_{\|\phi\|_{C^{1,1}}} \subset \mathcal{S}(n)$ as to assure that $\mathfrak{F}^\star$ is linear, say $\mathfrak{F}^\star = \Delta$. Nevertheless, ϕ would still be a $C^{1,1} \setminus C^2$ solution to an elliptic equation whose recession function is linear. The final result we prove gives $C^{1,\text{Log-Lip}}$ estimates under the uniform limits and under the assumption that F^* has *a priori* $C^{2,\alpha_\star}$ interior estimates. More precisely we have

Theorem 1.4. *Let $F\colon \mathcal{S}(n) \to \mathbb{R}$ be a uniform elliptic operator and u a viscosity solution to $F(D^2u) = f \in \mathrm{BMO}$ in B_1. Assume the recession function $F^\star(M) := \lim_{\mu\to 0} \mu F(\mu^{-1}M)$ exists and has a priori $C^{2,\alpha_\star}$ interior estimates. Assume further that the limit $\lim_{\mu\to 0} \mu F(\mu^{-1}M)$ is uniform in M. Then $u \in C^{1,Log\text{-}Lip}_{loc}(B_1)$, i.e.,*

$$|u(X) - [u(X_0) + \nabla u(X_0)\cdot(X - X_0)]| \le -C|X - X_0|^2 \log|X - X_0|.$$

2 Preliminaries

In this section make few comments about the notion of *recession* function. Throughout this paper, B_r denotes the ball of radius $r > 0$ in the Euclidean space $\mathbb{R}^n$ and $\mathcal{S}(n)$ denotes the space of all real, $n \times n$ symmetric matrices. A function $F\colon \mathcal{S}(n) \to \mathbb{R}$ will always be a uniformly elliptic operator, as in [4]. That is, we assume that there exist two positive constants $0 < \lambda \le \Lambda$ such that, for any $M \in \mathcal{S}(n)$ there holds

$$\lambda\|P\| \le F(M + P) - F(M) \le \Lambda\|P\|, \quad \forall P \ge 0. \tag{5}$$

We will further assume, with no loss of generality, that $F(0) = 0$.

A key information we shall use in the proof of Theorem 1.1 is the fact that solutions to (1) are locally $C^{1,\epsilon}$ for some universal $\epsilon > 0$. Furthermore

$$\|u\|_{C^{1,\epsilon}(B_{1/2})} \le C\|u\|_{L^\infty(B_1)},$$

for a universal constant $C > 0$. As mentioned in the introduction, Nadirashvili and Vladut have proven that $C^{1,\epsilon}$ is the best regularity theory available for general fully nonlinear elliptic equations. The objective of this paper is to show that further smoothness could be assured if we have information on the recession function of F, defined in (2).

Let us discuss a bit about recession functions for fully nonlinear elliptic equations. Initially, it is straightforward to verify that for each μ, the elliptic operator

$$F_\mu(M) := \mu F(\mu^{-1}M)$$

is uniformly elliptic, with the same ellipticity constants as F. Thus, up to a subsequence, F_μ does converge to a limiting elliptic operator $F^\star$ as $\mu \to 0$. Any limiting point $F^\star$ will be called a *recession* function of F. This terminology comes from the theory of Haminton-Jacobi equations, see, for instance, [7].

Initially, let us point out that recession functions may not be unique, as simple 1-d examples show. Nevertheless, if the recession function is unique, it is clearly homogeneous of degree one, that is, for any scalar t, we have

$$F^\star(tM) = tF^\star(M).$$

Also, if F is homogeneous of degree one, then $F = F^\star$. In some applications, it is possible to verify that

$$\lim_{\|M\|\to\infty} D_{i,j}F(M) =: F_{ij}. \tag{6}$$

That is, F has a linear behavior at the ends. Under such condition, it is simple to check that $F^\star$ is a linear elliptic operator, and, in fact,

$$F^\star(M) = \mathrm{tr}\left(F_{ij}M_{ij}\right).$$

A particularly interesting example is the class of Hessian operators of the form

$$F_\iota(M) = f_\iota(\lambda_1, \lambda_2, \cdots \lambda_n) := \sum_{j=1}^{n} (1 + \lambda_j^\iota)^{1/\iota},$$

where ι is an odd natural number. For this family of operators, we have

$$F_\iota^\star = \Delta.$$

A priori F_μ converges pointwisely to $F^\star$. However, the following is a more precise description of how the limit takes place.

Lemma 2.1. *If F is any uniformly elliptic operator and $F^\star(M) = \lim_{\mu\to 0} \mu F(\mu^{-1}M)$ for every symmetric matrix M, then for every $\varepsilon >$, there exists a $\delta > 0$ so that*

$$||\mu F(\mu^{-1}M) - F^*(M)|| \leq \varepsilon(1 + ||M||), \tag{7}$$

for all $\mu < \delta$.

Proof. Since the function F is uniformly elliptic, we have that $F(X + Y) - F(X) \leq \Lambda||Y||$ for some constant Λ and F is Lipschitz. This Lipschitz norm is conserved by the scaling $\mu F(\mu^{-1}M)$. By the Arzela-Ascoli theorem we have that up to a subsequence $\mu F(\mu^{-1}M)$ converges uniformly in every compact set. Since $\mu F(\mu^{-1}M)$ converges pointwise to F^*, then all its subsequential limits must coincide with $F^\star$ and therefore it converges to $F^\star$ uniformly over every compact set.

That means that for every $\varepsilon > 0$ there exists a $\delta > 0$ so that

$$||\mu F(\mu^{-1}M) - F^\star(M)|| \leq \varepsilon,$$

for all matrices M such that $||M|| \leq 1$ and all $\mu < \delta$. This already shows that (7) holds if $||M|| \leq 1$.

Now let M be a matrix with $||M|| > 1$. For any $\mu < \delta$, we can consider also $\mu_1 = ||M||^{-1}\mu < \mu < \delta$. Therefore

$$\left\| \mu_1 F\left(\mu_1^{-1}\frac{M}{||M||}\right) - F^\star\left(\frac{M}{||M||}\right)\right\| \leq \varepsilon,$$

Observing that $\mu_1^{-1}\frac{M}{||M||} = \mu^{-1}M$, and using that F^* is homogeneous of degree one, we obtain

$$\left\|\mu F\left(\mu^{-1}M\right) - F^\star\left(M\right)\right\| \leq \varepsilon||M||.$$

This proves (7) for $||M|| > 1$. □

A model case though is when F equals $F^\star$ outside a ball $B_R \subset \mathrm{Sym}(n)$, for some $R \gg 1$. In this case, the convergence $F_\mu \to F^\star$ is uniform with respect to M—compare with the hypothesis of Theorem 1.4.

3 C^{1,α_0^-} estimates

In this section we prove Theorem 1.1. We start off the proof by fixing an aimed Hölder continuity exponent for gradient of u between 0 and α_0, more precisely, we fix

$$0 < \alpha < \min\{1, \alpha_0\}. \tag{8}$$

We will show that $u \in C^{1,\alpha}$ at the origin. It is standard to pass from pointwise estimate to interior regularity. Initially, as mentioned in the introduction, it follows from Krylov-Safonov Harnack inequality that $u \in C^{1,\epsilon}$ for some universal $\epsilon > 0$. We may assume, therefore, by normalization and translation, that

$$|u| \leq 1 \text{ in } B_{9/11} \tag{9}$$

$$u(0) = |\nabla u(0)| = 0. \tag{10}$$

Our strategy is based on the following reasoning: proving that $u \in C^{1,\alpha}$ at the origin is equivalent to verifying that either there exists a constant $C > 0$ such that

$$\sup_{B_r} |u(X)| \leq C r^{1+\alpha}, \quad \forall r < 1/5,$$

or else, by iteration, that for some $\ell > 0$ and some $r > 0$, there holds

$$\sup_{B_r} |u(X)| \leq 2^{-(1+\alpha)\ell} \sup_{B_{2^\ell \cdot r}} |u(X)|,$$

see [5], Lemma 3.3 for similar inference. Therefore, if we suppose, for the purpose of contradiction, that the thesis of the Theorem fails, there would exist a sequence of viscosity solutions $F(D^2 u_k) = 0$, satisfying (9) and (10), and a sequence of radii $r_k \to 0$ such that

$$(\sup_{B_{r_k}} |u_k|)^{-1} \cdot r_k^{(1+\alpha)} \longrightarrow 0 \tag{11}$$

$$\sup_{B_{r_k}} |u_k| \geq 2^{-(1+\alpha)\ell} \sup_{2^\ell \cdot r_k} |u_k|. \tag{12}$$

For notation convenience, let us label

$$s_k := \sup_{B_{r_k}} |u_k|.$$

In the sequel, we define the normalized function

$$v_k(X) := \frac{1}{s_k} u_k(r_k X).$$

Immediately, from definition of v_k, we have

$$\sup_{B_1} |v_k| = 1. \tag{13}$$

Also, it follows from (12) that v_k growths at most as $|X|^{1+\alpha}$, i.e.,

$$\sup_{B_{2^\ell}} v_k \leq 2^{(1+\alpha)\ell}. \tag{14}$$

In addition, if we define the uniform elliptic operator

$$F_k(M) := (s_k^{-1} \cdot r_k^2) F\left((s_k \cdot r_k^{-2}) M\right),$$

we find out that v_k solves

$$F_k\left(D^2 v_k\right) = 0, \tag{15}$$

in the viscosity sense. By uniform ellipticity and (11), up to a subsequence, F_k converges locally uniformly to a recession function $F^\star$. Thus, letting $k \to \infty$, by $C^{1,\epsilon}$ universal estimates, $v_k \to v_\infty$ locally in the $C^{1,\epsilon/2}(\mathbb{R}^n)$ topology. Clearly v_∞ is a viscosity solution to

$$F^\star(D^2 v_\infty) = 0 \text{ in } \mathbb{R}^n.$$

Taking into account (10), (13), (14), we further conclude that v_∞ satisfies

$$v_\infty(0) = |\nabla v_\infty(0)| = 0, \tag{16}$$

$$\sup_{B_1} |v_\infty| = 1, \tag{17}$$

$$|v_\infty(Y)| \le |Y|^{1+\alpha}. \tag{18}$$

Hereafter let us label

$$\kappa := \min\{1, \alpha_0\} - \alpha > 0.$$

Recall any recession function $F^\star$ is homogeneous of degree one for positive multipliers. Therefore, fixed a large positive number $\ell \gg 1$, the auxiliary function

$$\mathcal{V}_\infty(Z) := \frac{v_\infty(\ell Z)}{\ell^{1+\alpha}},$$

too satisfies

$$F^\star(D^2 \mathcal{V}_\infty) = 0.$$

From (18) we verify that $\mathcal{V}_\infty$ is bounded in B_1 and, hence, from the regularity theory for the recession function, $F^\star$, there exists a constant $C^\star$, depending on dimension and $F^\star$, such that

$$|\nabla \mathcal{V}_\infty(Z)| \le C^\star |Z|^{\alpha+\kappa}, \quad \forall Z \in B_{1/5}. \tag{19}$$

Finally, estimate (19) gives, after scaling,

$$\begin{aligned} \sup_{B_{\frac{\ell}{5}}} \frac{|\nabla v_\infty(Y)|}{|Y|^{\alpha+\kappa}} &= \ell^{-\kappa} \sup_{B_{\frac{1}{5}}} \frac{|\nabla \mathcal{V}_\infty(Z)|}{|Z|^{\alpha+\kappa}} \\ &= \mathrm{o}(1), \end{aligned} \tag{20}$$

as $\ell \to \infty$. Clearly (20) implies that v_∞ is constant in the whole $\mathbb{R}^n$. However, such a conclusion drives us into a contradiction, since, from (16), $v_\infty \equiv 0$ which is incompatible with (17). The proof of Theorem 1.1 is concluded.

4 Proof of Corollary 1.3

In this section we comment on the proof of Corollary 1.3. Given a point $X \in \{u > 0\}$, with

$$d := \text{dist}(X, \partial\{u > 0\}) < \frac{1}{2}\text{dist}(X, \partial\Omega),$$

we consider $Y \in \partial\{u > 0\}$, such that $d = |X - Y|$. Applying Corollary 1.2 we can estimate

$$[u]_{C^{1,\frac{1-\gamma}{1+\gamma}}(B_{d/4}(X))} \lesssim \frac{1}{d^{\frac{2}{1+\gamma}}} \left(\|u\|_{L^\infty(B_{d/2}(Z))} + d^2 \cdot \|u^{-\gamma}\|_{L^\infty(B_{d/2}(Z))}\right). \tag{21}$$

It then follows by the optimal control

$$u(\xi) \sim \text{dist}(\xi, \partial\{u > 0\})^{\frac{2}{1+\gamma}},$$

see [1, Theorem 9], that we can estimate, in $B_{d/2}(Z)$,

$$\|u\|_{L^\infty(B_{d/2}(Z))} \lesssim d^{\frac{2}{1+\gamma}}, \tag{22}$$

$$\|u^{-\gamma}\|_{L^\infty(B_{d/2}(Z))} \lesssim d^{\frac{-2\gamma}{1+\gamma}}. \tag{23}$$

Plugging (22) and (23) into (21) gives

$$[u]_{C^{1,\frac{1-\gamma}{1+\gamma}}(B_{d/4}(X))} \lesssim 1,$$

and therefore u is locally of class $C^{1,\frac{1-\gamma}{1+\gamma}}$, up to the free boundary. The proof is complete. □

5 Proof of Theorem 1.4

For this section we assume that $\lim_{\mu\to 0} \mu F(\mu^{-1}M)$ exists and equals $F^*(M)$ for every matrix M. It is also part of our assumption that the limit is uniform in M. In particular, given $\varepsilon > 0$, we can find $\delta > 0$, such that

$$\left|F_\mu(M) - F(M)\right| \le \varepsilon, \quad \forall M$$

provided $0 < \mu \le \delta$.

Lemma 5.1. *Assume F and F^* are two fully nonlinear uniformly elliptic operators such that*

$$|F(M) - F^*(M)| \leq \varepsilon, \qquad \textit{for any symmetric matrix } M. \tag{24}$$

Assume moreover that $F^*(0) = 0$ and F^* has C^{2,α_*} estimates in the form that any solution u^* of $F^*(D^2u^*) = 0$ in B_1 satisfies

$$||u^*||_{C^{2,\alpha_*}(B_{1/2})} \leq C_*||u||_{L^\infty(B_1)}. \tag{25}$$

Then there exist two constants ε and r (depending only on the ellipticity constants, dimension, C_* and α_*) so that for any solution u of $F(D^2u) = f$ in B_1 with $||f||_{L^\infty} \leq \varepsilon$ and $||u||_{L^\infty} \leq 1$, there exists a second order polynomial P, such that $||P|| \leq C$ and $||u - P||_{L^\infty(B_r)} \leq r^2$.

Proof. The value of r will be specified below in terms of the C^{2,α_*} estimate (25) only. For that value of r, we prove the lemma by contradiction. If the result was not true, there would exist a sequence F_n, F_n^* f_n, u_n so that

$$\begin{aligned}
|F_n(M) - F_n^*(M)| &\leq \frac{1}{n} \qquad \text{for any symmetric matrix } M, \\
||f_n||_{L^\infty(B_1)} &\leq \frac{1}{n}, \\
||u_n||_{L^\infty(B_1)} &\leq 1, \\
F_n(D^2u_n) &= f_n \qquad \text{in } B_1, \\
F_n \text{ and } F_n^* &\text{ have uniform ellipticity constants } \lambda, \Lambda,
\end{aligned}$$

where F_n^* has C^{2,α_*} estimates as in (25) but such polynomial P cannot be found for any u_n.

Since the F_n^* are uniformly elliptic, in particular they are uniformly Lipschitz. Up to extracting a subsequence, they will converge to some uniformly elliptic function F^* which will also have C^{2,α_*} estimates (25). Thus, we can assume that all F_n^* are the same by replacing them by F^* (and taking a subsequence if necessary).

Since the F_n are uniformly elliptic, the functions u_n are uniformly C^α in the interior of B_1 and there must be a subsequence that converges locally uniformly to some continuous function u_*. We extract this subsequence, and by abuse of notation we still call it u_n. Since $u_n \to u_*$ locally uniformly, $F_n \to F^*$ locally uniformly, and $f_n \to 0$ uniformly, we have that $F^*(D^2u^*) = 0$ holds in the viscosity sense. From the C^{2,α_*} estimates (25), if we choose P to be the second order Taylor's expansion of u^* at the origin we will have

$$||u^* - P||_{L^\infty(B_r)} \leq C_* r^{2+\alpha_*}.$$

We choose r small enough so that $C_* r^{\alpha_*} < 1/2$. Note that this choice depends on C_* and α_* only. We thus have

$$||u^* - P||_{L^\infty(B_r)} \leq \frac{r^2}{2}.$$

However, since $u_n \to u_*$ uniformly in B_r, then for n large enough we also have.

$$||u_n - u^*||_{L^\infty(B_r)} \leq \frac{r^2}{2}.$$

Combining the last two previous inequalities we obtain that

$$||u_n - P||_{L^\infty(B_r)} \leq r^2,$$

and so we arrive to a contradiction since we were assuming that such polynomial P did not exist for any n. □

Proof (Proof of Theorem 1.4). We prove the result for $x_0 = 0$ and assuming $f \in L^\infty$—see [18] for the adjustments requested when $f \in$ BMO. From uniform convergence hypothesis, we can find $\delta > 0$ so that for all $\mu < \delta$ the inequality

$$||\mu F(\mu^{-1}M) - F^\star(M)|| \leq \varepsilon,$$

holds, where $\varepsilon > 0$ is the number from Lemma 5.1. We start off now with a convenient rescaling of the problem. We find an r_0, depending only on $||u||_{L^\infty}$ and δ, and consider the scaling

$$u_0(x) = \varepsilon \max\{1, \|u\|_{L^\infty}, \|f\|_\infty\}^{-1} u(r_0 x).$$

We choose $r_0 \sim \sqrt{\delta}$, where δ is the number above. For this choice we have

$$\|u_0\|_{L^\infty(B_1)} \leq 1;$$
$$\mu F(\mu^{-1} D^2 u_0) = \tilde{f}(x),$$

for a $\mu < \delta$ and $\|\tilde{f}\|_\infty \leq \varepsilon$. Now we proceed to show that u_0 is $C^{1,\text{Log-Lip}}$ at the origin. The strategy is to show the existence of a sequence of quadratic polynomials

$$P_k(X) := a_k + \mathbf{b}_k \cdot X + \frac{1}{2} X^t M_k X,$$

where $P_0 = P_{-1} = 0$, and for all $k \geq 0$,

$$F^\star(M_k) = 0, \tag{26}$$

$$\sup_{Q_{r^k}} |u_0 - P_k| \leq r^{2k}, \tag{27}$$

$$|a_k - a_{k-1}| + r^{k-1}|\mathbf{b}_k - \mathbf{b}_{k-1}| + r^{2(k-1)}|M_k - M_{k-1}| \leq C r^{2(k-1)}. \tag{28}$$

The radius r in (27) and (28) is the one from Lemma 5.1. We shall verify (26)–(28) by induction. The first step $k = 0$ is immediately satisfied. Suppose we have verified the thesis of induction for $k = 0, 1, \cdots, i$. Define the re-scaled function $v\colon B_1 \to \mathbb{R}$ by

$$v(X) := \frac{(u_0 - P_i)(r^i X)}{r^{2i}},$$

It follows by direct computation that v satisfies $|v| \leq 1$ and it solves

$$\mu F\left(\mu^{-1}\left(D^2 v + M_i\right)\right) = \tilde{f}(r^i x).$$

If we define

$$F_i(M) := F(M + M_i) \quad \text{and} \quad F_i^\star(M) := F^\star(M + M_i),$$

it follows from uniform convergence that

$$F_i \text{ is close to } F_i^\star.$$

Furthermore, since $F^\star(M_i) = 0$, the homogeneous equation

$$F_i^\star(D^2\xi) = 0$$

satisfies the same conditions as the original $F^\star$. We now apply Lemma 5.1 to v and find a quadratic polynomial $\tilde{P}$ such that

$$\|v - \tilde{P}\|_{L^\infty(B_r)} \leq r^2. \tag{29}$$

If we define

$$P_{i+1}(X) := P_i(X) + r^{2i}\tilde{P}(r^{-i}X)$$

and rescale (29) back, we conclude the induction thesis. In the sequel, we argue as in [18]. From (28) we conclude that $a_k \to u_0(0)$ and $\mathbf{b}_k \to \nabla u_0(0)$, in addition

$$|u_0(0) - a_k| \leq C\rho^{2k} \tag{30}$$

$$|\nabla u_0(0) - \mathbf{b}_k| \leq C\rho^k. \tag{31}$$

From (28) it is not possible to assure convergence of the sequence of matrices $(M_k)_{k\geq 1}$; nevertheless, we estimate

$$|M_k| \leq Ck. \tag{32}$$

Finally, given any $0 < r < 1/2$, let k be the integer such that

$$\rho^{k+1} < r \leq \rho^k.$$

We estimate, from (30), (31) and (32),

$$\begin{aligned}\sup_{Q_r} |u(X) - [u(0) + \nabla u(0) \cdot X]| &\leq \rho^{2k} + |u(0) - a_k| + \rho|\nabla u(0) - \mathbf{b}_k| \\ &\quad + \rho^{2k}|M_k| \\ &\leq -Cr^2 \log r,\end{aligned}$$

and the Theorem is proven. □

References

1. Araújo, D., Teixeira, E.: Geometric approach to nonvariational singular elliptic equations. Arch. Rational Mech. Anal. **209**(3), 1019–1054 (2013)
2. Cabre, X., Caffarelli, L.A.: Interior $C^{2,\alpha}$ regularity theory for a class of nonconvex fully nonlinear elliptic equations. J. Math. Pures Appl. **82**(9), 573–612 (2003)
3. Caffarelli, L.A.: Interior a priori estimates for solutions of fully nonlinear equations. Ann. Math. (2) **130**(1), 189–213 (1989)
4. Caffarelli, L.A., Cabré, X.: Fully Nonlinear Elliptic Equations. American Mathematical Society Colloquium Publications, vol. 43. American Mathematical Society, Providence, RI (1995)
5. Caffarelli, L., Karp, L., Shahgholian, H.: Regularity of a free boundary with application to the Pompeiu problem. Ann. Math. **151**(1), 269–292 (2000)
6. Evans, L.C.: Classical solutions of fully nonlinear, convex, second-order elliptic equations. Commun. Pure Appl. Math. **35**(3), 333–363 (1982)
7. Giga, Y., Sato, M.-H.: On semicontinuous solutions for general Hamilton-Jacobi equations. Commun. Partial Differ. Equ. **26**, 813–839 (2001)
8. Krylov, N.V.: Boundedly nonhomogeneous elliptic and parabolic equations. Izv. Akad. Nauk. SSSR Ser. Mat. **46**, 487–523 (1982) [English Translation in Math USSR Izv. **20**, 459–492 (1983)]
9. Krylov, N.V.: Boundedly nonhomogeneous elliptic and parabolic equations in a domain. Izv. Akad. Nauk. SSSR Ser. Mat. **47**, 75–108 (1983) [English Translation in Math USSR Izv. **22**, 67–97 (1984)]
10. Krylov, N.V., Safonov, M.V.: An estimate of the probability that a diffusion process hits a set of positive measure. Dokl. Akad. Nauk. SSSR **245**, 235–255 (1979) [English Translation in Soviet Math Dokl. **20**, 235–255 (1979)]
11. Krylov, N.V., Safonov, M.V.: Certain properties of solutions of parabolic equations with measurable coefficients. Izv. Akad Nauk. SSSR **40**, 161–175 (1980)
12. Nadirashvili, N., Vladut, S.: Nonclassical solutions of fully nonlinear elliptic equations. Geom. Funct. Anal. **17**(4), 1283–1296 (2007)
13. Nadirashvili, N., Vladut, S.: Singular viscosity solutions to fully nonlinear elliptic equations. J. Math. Pures Appl. (9) **89**(2), 107–113 (2008)
14. Nadirashvili, N., Vladut, S.: Nonclassical solutions of fully nonlinear elliptic equations II. Hessian equations and octonions. Geom. Funct. Anal. **21**, 483–498 (2011)
15. Nadirashvili, N., Vladut, S.: Singular solutions of Hessian elliptic equations in five dimensions. J. Math. Pures Appl. (9) **100**(6), 769–784 (2013)

16. Nadirashvili, N., Tkachev, V., Vladut, S.: A non-classical solution to a Hessian equation from Cartan isoparametric cubic. Adv. Math. **231**(3–4), 1589–1597 (2012)
17. Ricarte, G., Teixeira, E.: Fully nonlinear singularly perturbed equations and asymptotic free boundaries. J. Funct. Anal. **261**(6), 1624–1673 (2011)
18. Teixeira, E.: Universal moduli of continuity for solutions to fully nonlinear elliptic equations. Arch. Rational Mech. Anal. **211**(3), 911–927 (2014)
19. Yuan, Y.: A priori estimates for solutions of fully nonlinear special Lagrangian equations. Ann. Inst. H. Poincaré Anal. Non Linéare **18**(2), 261–270 (2001)

MIX
Papier aus verantwortungsvollen Quellen
Paper from responsible sources
FSC® C105338

If you have any concerns about our products,
you can contact us on
ProductSafety@springernature.com

In case Publisher is established outside the EU,
the EU authorized representative is:
Springer Nature Customer Service Center GmbH
Europaplatz 3, 69115 Heidelberg, Germany

Printed by Libri Plureos GmbH
in Hamburg, Germany